W0036725

X-RAY AND INNER-SHELL PROCESSES

Previous Proceedings
in the Series of International Conferences on X-Ray and Inner-Shell Processes

	Year	Held in	Publisher	ISBN
18th	1999	Chicago, Illinois, USA	AIP Conference Proceedings 506	1-56396-713-8
17th	1996	Hamburg, Germany	AIP Conference Proceedings 389	1-56396-563-1
16th	1993	Debrecen, Hungary	Nucl. Instr. & Methods **87**, 1994	
15th	1990	Knoxville, Tennessee, USA	AIP Conference Proceedings 215	0-88318-790-6

Other Related Titles from AIP Conference Proceedings

645 Spectral Line Shapes: 16th International Conference on Spectral Line Shapes
Edited by Christina A. Back, December 2002, 0-7354-0100-4

641 X-Ray Lasers 2002: 8th International Conference on X-Ray Lasers
Edited by Jorge J. Rocca, James Dunn, and Szymon Suckewer, November 2002,
0-7354-0096-2

636 Atomic and Molecular Data and Their Applications: 3rd International Conference on
Atomic and Molecular Data and Their Applications - ICAMDATA
Edited by David R. Schultz, Predrag S. Krstić, and Fay Ownby, October 2002, 0-7354-0091-1

635 Atomic Processes in Plasmas: 13th APS Topical Conference on Atomic Processes
in Plasmas
Edited by David R. Schultz, Fred W. Meyer, and Fay Ownby, October 2002, 0-7354-0090-3

604 Correlations, Polarization, and Ionization in Atomic Systems: Proceedings of the
International Symposium on (e,2e), Double Photoionization and Related Topics and the
Eleventh International Symposium on Polarization and Correlation in Electronic and Atomic
Collisions, Edited by Don Madison and Michael Schulz, January 2002, 0-7354-0048-2

584 Resonance Ionization Spectroscopy 2000: Laser Ionization and Applications
Incorporating RIS; 10th International Symposium
Edited by James E. Parks and Jack P. Young, August 2001, 0-7354-0024-5

551 Atomic Physics 17: XVII International Conference on Atomic Physics; ICAP 2000
Edited by Ennio Arimondo, Paolo De Natale, and Massimo Inguscio, February 2001,
1-56396-982-3

507 X-Ray Microscopy: VI International Conference on X-Ray Microscopy
Edited by Werner Meyer-Ilse, Tony Warwick, and David Atwood, March 2000,
1-56396-926-2

500 The Physics of Electronic and Atomic Collisions: XXI International Conference
Edited by Yukikazu Itikawa, Kazuhiko Okuno, Hiroshi Tanaka, Akira Yagishita, and Michio
Matsuzawa, February 2000, 1-56396-777-4

To learn more about these titles, or the AIP Conference Proceedings Series, please visit the
webpage **http://proceedings.aip.org/proceedings**

X-RAY AND INNER-SHELL PROCESSES

19th International Conference on
X-Ray and Inner-Shell Processes

Rome, Italy   24–28 June 2002

EDITORS
Antonio Bianconi
Università di Roma "La Sapienza", Rome, Italy
Augusto Marcelli
INFN-LNF, Rome, Italy
Naurang L. Saini
Università di Roma "La Sapienza", Rome, Italy

SPONSORING ORGANIZATIONS
Università di Roma "La Sapienza", Rome, Italy
INFN-LNF, Laboratori Nazionali di Frascati, Frascati, Italy
ENEA, Ente per le Nuove Tecnologie l'Energia e l'Ambiente, Frascati, Italy
INRM, Istituto Nazionale per la Ricerca Scientifica e Tecnologica
   sulla Montagna, Rome, Italy

Melville, New York, 2003
AIP CONFERENCE PROCEEDINGS ■ VOLUME 652

Editors:

Antonio Bianconi
Dipartimento di Fisica
Università di Roma "La Sapienza"
Piazzale Aldo Moro, 2
00185 Roma
ITALY
E-mail: Antonio.Bianconi@roma1.infn.it

Augusto Marcelli
INFN-LNF
P.O. Box 13
00044 Frascati
ITALY
E-mail: Augusto.Marcelli@lnf.infn.it

Naurang L. Saini
Dipartimento di Fisica
Università di Roma "La Sapienza"
Piazzale Aldo Moro, 2
00185 Roma
ITALY
E-mail: Saini@roma1.infn.it

Cover photo by Archivio Folco Quilici, Rome.

Authorization to photocopy items for internal or personal use, beyond the free copying permitted under the 1978 U.S. Copyright Law (see statement below), is granted by the American Institute of Physics for users registered with the Copyright Clearance Center (CCC) Transactional Reporting Service, provided that the base fee of $20.00 per copy is paid directly to CCC, 222 Rosewood Drive, Danvers, MA 01923. For those organizations that have been granted a photocopy license by CCC, a separate system of payment has been arranged. The fee code for users of the Transactional Reporting Service is: 0-7354-0111-X/03/$20.00.

© 2003 American Institute of Physics

Individual readers of this volume and nonprofit libraries, acting for them, are permitted to make fair use of the material in it, such as copying an article for use in teaching or research. Permission is granted to quote from this volume in scientific work with the customary acknowledgment of the source. To reprint a figure, table, or other excerpt requires the consent of one of the original authors and notification to AIP. Republication or systematic or multiple reproduction of any material in this volume is permitted only under license from AIP. Address inquiries to Office of Rights and Permissions, Suite 1NO1, 2 Huntington Quadrangle, Melville, N.Y. 11747-4502; phone: 516-576-2268; fax: 516-576-2450; e-mail: rights@aip.org.

L.C. Catalog Card No. 2002116277
ISBN 0-7354-0111-X
ISSN 0094-243X
Printed in the United States of America

CONTENTS

V.   ATOMIC AND NUCLEAR X-RAY PROCESSES

VI. X-RAY SCATTERING

VII. X-RAY APPLICATIONS TO SOLIDS AND SURFACES

VIII. BIOLOGICAL APPLICATIONS

PREFACE

This book is a collection of the invited papers presented at the *19th International Conference on X-Ray and Inner-Shell Processes* (X-02). The Conference was held at the University of Rome *La Sapienza*, Italy on June 24-28, 2002. The main scientific sessions were conducted in the *Aula Magna*, decorated with the frescos of Mario Sironi, while the parallel sessions were held in the *Aula Amaldi* of the physics department, a building designed by the Italian architect Pagano in 1935.

The X-02 conference was the latest meeting in the long-standing series of conferences *"X-Ray and Inner-Shell Processes"*, dedicated to x-ray science all around the world. The current series is a combination of the earlier series of *"Conferences on X rays"*, started in 1965 (held both in Ithaca and Leipzig) and continued until 1976 (Washington), and the series on *"Inner-Shell Ionization"* (held 1972 in Atlanta and 1976 in Freiburg). The meetings in the joint series were held in Sendai, Japan (1978), Stirling, Scotland (1980), Leipzig, Germany (1984), Paris, France (1987), Knoxville, USA (1990), Eugene, USA (1992), Debrecen, Hungary (1993), Hamburg Germany (1996) and Chicago USA (1999).

The X-02 conference was jointly hosted by the University of Rome *La Sapienza* and the Frascati National Laboratories of the INFN. This INFN laboratory is the home of a new synchrotron radiation facility at the *DAΦNE* storage ring complex. In the early sixties, the laboratory started the first pioneering research with x-rays using the synchrotron light emitted first by an electro-synchrotron and then by the ADONE storage ring.

The X-02 conference in Rome had a participation of about 300 scientists from 30 different countries with special interest shown by young participants. A total of 280 contributed abstracts were received, of which 150 were presented in four poster sessions. The program including 16 Plenary Lectures and 84 talks, covered recent advances and developments in both experimental and theoretical studies of X-ray and inner-shell processes and x-ray applications.

The invited speakers, chosen among the many suggestions from the members of the International Scientific Committee and the Advisory Committee, reviewed the latest developments in the field. The development over recent years of tunable high-brilliance hard x-ray beams, delivered by the last generation of synchrotron radiation sources, has given added impetus to the x-ray science, the subject at the focus of this conference series. At the X-02, for the first time, a session was dedicated to the next generation of x-ray sources: the free electron laser, the new interdisciplinary project that represents the future and the dream of many scientists.

This volume also contains a few selected papers presented as oral contributions, in addition to the invited papers. All scientific contributions were included in the book of abstracts, distributed to the X-02 participants. Additional copies may be obtained by contacting the organizing institutions or through the web site of the conference at the following URL: http://www.superstripes.com.

The conference organizers gratefully acknowledge the suggestions and advice of the International Scientific and Advisory Committees in preparing the program and all participants and contributors. A sincere acknowledgement is due to A. De Grossi and G. Cibin for the professional and enthusiastic contribution given to the preparation of this volume.

Antonio Bianconi and Augusto Marcelli
Conference Chairmen

I. HISTORICAL REVIEWS

Inner-Shell Photoionization:
Teijo Åberg's Concept of RRRS

G. Bradley Armen

Department of Physics, University of Tennessee
Knoxville, Tennessee 37996, U.S.A.

Abstract. Teijo Åberg died on December 29, 2000 — and with his passing we have lost a gifted and important member of our community. Over the years Teijo has made many major contributions to the science of x-ray and inner-shell physics. Perhaps the most visionary of these was his idea of radiative and radiationless resonant Raman scattering (RRRS) [1]. This arose as an outgrowth of his earlier thoughts on Auger [2] and x-ray [3] processes. It is visionary in that it attempts to unify the various aspects of inner-shell photoexcitation and decay as different aspects of the same basic process, *i.e.* resonant inelastic x-ray scattering. In honor and memory of Teijo, and as a last attempt to promote his ideas, I describe here the theory and some of its consequences. First, RRRS is outlined in a semi-rigorous form, with emphasis placed on the physical nature of the concepts. Then, using the example of atomic K-shell photoionization, we see how features such as fluorescence and photoelectron lines arise in a natural way as manifestations of resonance scattering.

THE GENERAL IDEA

A key feature in the RRRS formalism is the inclusion of the radiation field as part of the scattering problem. In this way inner-shell photoionization effects are viewed as the inelastic scattering of a photon. The Hamiltonian of the system is written as a sum of three terms $H = H_{atom} + H_{rad} + V_{int}$. Here, H_{atom} and H_{rad} are the Hamiltonians of the isolated atom and free-radiation field respectively, and $V_{int} = \sum (\mathbf{p} \cdot \mathbf{A} + \mathbf{A} \cdot \mathbf{A}/2)$ is the (non-relativistic) interaction between the atomic electrons and the radiation field.

One starts [2, 4] with single-channel, generalized wave functions $\chi_{\alpha,E}$ which are a direct product of atomic and free-radiation (photon) states. A given index α $(=1,...N)$ labels a particular scattering channel, characterized by quantum numbers describing the atomic core, any ejected electrons, and the radiation field. The index E indicates the total energy of a particular state of the channel, and can take on both continuous and discrete values (any integrals over E imply a summation over these latter values also).

The single-channel wave functions form a beginning basis from which the scattering calculation proceeds. The atomic part of $\chi_{\alpha,E}$ might correspond to a Hartree-Fock or Dirac-Fock many-electron wave function. Any continuum orbitals satisfy standing-wave boundary conditions, *i.e.* asymptotically, a real-valued combination of radial Coulomb functions. The radiation part of $\chi_{\alpha,E}$ specifies the number, energy, wave vector, and polarization of any photons present.

Before proceeding, a few examples of such states are useful to examine: (A) Our initial scattering state, a direct product of the atomic ground state g (energy E_0) and a

CP652, X-Ray and Inner-Shell Processes: 19th International Conference on X-Ray and Inner-Shell Processes
edited by A. Bianconi, A. Marcelli, and N. L. Saini
© 2003 American Institute of Physics 0-7354-0111-X/03/$20.00

single (incident) photon state of energy ω, so that $E = E_0 + \omega$. (B) A state we use to describe an Auger channel, comprised of the zero-photon state, a doubly ionized atomic core state containing holes in the f and f' shells (notation $[ff']$), and two ejected continuum electrons (ε_1 and ε_2). This state is continuously degenerate: $E = E_0 + I_{[f,f']} + \varepsilon_1 + \varepsilon_2$, where $I_{[f,f']}$ is the ionization energy of the ionic core state. (C) A 'fluorescence' state containing a photon (energy ω'), a singly ionized core $[f]$, and an ejected electron with $E = E_0 + I_{[f]} + \varepsilon_1 + \omega'$.

The Hamiltonian is not diagonal within the standing-wave basis, which of course is necessary if there is to be any scattering between channels. Diagonalization of H yields the multi-channel functions

$$\chi_{\alpha,E}^{\pm} = \chi_{\alpha,E} + \sum_{\beta=1}^{N}\left[PV \int dE' \frac{Y_{\beta,\alpha}^{\pm}(E',E)\chi_{\beta,E'}}{E - E'} \mp i\pi Y_{\beta,\alpha}^{\pm}(E,E)\chi_{\beta,E} \right].$$

These new channel functions are a configuration-interaction mixture of our original basis. Here PV indicates the principal value for the integral. Examination of the asymptotic behavior of these new channel wave functions identifies the unknown quantities Y with the scattering T-matrices ($\mathbf{Y}^+ = \mathbf{T}^+$, $\mathbf{Y}^- = \mathbf{T}^{-\dagger}$), and can be determined (at least formally) by solving the Lippmann-Schwinger equations for $\mathbf{T}$. To lowest order $T_{\beta,\alpha}^{\pm} \approx \langle \chi_{\beta,E'}|H - E|\chi_{\alpha,E'}\rangle$ if $\beta \neq \alpha$. Hence, in the absence of any interaction, the scattering $\alpha \to \beta$ has zero amplitude — and we recover the original standing-wave basis.

The states $\chi_{\alpha,E}^{\pm}$ satisfy the ingoing ($-$) or outgoing ($+$) wave boundary conditions. Their physical nature becomes apparent only when the time evolution of wave packets constructed from them is examined. In this light, the "in" states $\chi_{\alpha,E}^{+}$ describe the physical situation in which the system is in channel α as $t \to -\infty$, and scatters into a mixture of the various channels β as $t \to +\infty$. Similarly, the "out" states $\chi_{\alpha,E}^{-}$ describe the 'entrance' of a coherent mixture of channels, ensuring the system escapes into a single channel α as $t \to +\infty$.

If the set of basis states $\chi_{\alpha,E}$ are complete our scattering problem is solved: Determining the multi-channel functions $\chi_{\alpha,E}^{+}$ requires solving for the transition matrix $T_{\beta,\alpha}^{+}$, the probability amplitude for scattering from channel α to β. However, in certain energy regimes resonance states must also be included in the basis set. Strictly speaking, these are *bound-state* solutions of a restricted Hamiltonian. In solving for the scattering basis $\chi_{\alpha,E}$ we make certain demands on the boundary conditions of the solutions. This, in turn, precludes consideration of the resonance states which vanish at large radii. To correctly describe the scattering (which occurs at small radii) their inclusion into the problem is essential.

When considering inner-shell ionization processes, an important example of such a set (φ_m) is the Rydberg series of states associated with an inner-shell vacancy: $[i]ml$. The energy of each such state is $E_m = E_0 + I_{[i]} - \tau_m$, where τ_m is the (positive) ionization energy of the ml electron. In the RRRRS theory, the idea of a resonance state is (formally) extended to include the continuum states of the series as well: a set φ_τ is defined by a continuous label τ, with each state having energy E_τ. Hence, we

4

also include the ionized states $[i]\tau l$ as resonances: the excited mp orbital is replaced by a continuum orbital of energy τ (and $E_\tau = E_0 + I_{[i]} + \tau$).

The existence of these resonance states modifies the scattering process. To calculate how, the Hamiltonian is re-diagonalized in the extended basis set that includes both $\chi^\pm_{\alpha,E}$ and $\varphi_{v,\tau}$ (the index τ implicitly includes discrete states τ_m, and an additional label v is included for the case of more than a single series of resonances). This procedure produces new scattering states $\psi^\pm_{\alpha,E}$. Application of scattering boundary conditions to these new functions identifies the (resonant) transition matrix $T^{(R)+}_{\beta,\alpha}$ in terms of the old $T^+_{\beta,\alpha}$ (scattering in the absence of resonances) and an additional term:

$$
T^{(R)+}_{\beta,\alpha} \;=\; T^+_{\beta,\alpha} \;+\; \sum_\mu \int d\tau \frac{\left\langle \chi^-_{\beta,E}\middle|H-E\middle|\Phi_{\mu,\tau}\right\rangle\left\langle\Phi_{\mu,\tau}\middle|H-E\middle|\chi^+_{\alpha,E}\right\rangle}{E-U_{\mu,\tau}} \;.
$$

Here, the resonant states $\Phi_{\mu,\tau}$ are linear combinations of the $\varphi_{v,\tau'}$. In the case of discrete resonances they are determined by diagonalizing the level-shift matrix, which describes the second-order interaction between $\varphi_{v,\tau}$ and $\varphi_{\mu,\tau}$ — arising through their mutual interaction with $\chi^\pm_{\alpha,E}$. The complex eigenvalues $U_{\mu,\tau}$ are the result of this diagonalization. In the case of continuous resonances, the discrete definitions are formally extended. From the practical point of view, the determination of $\Phi_{\mu,\tau}$ and $U_{\mu,\tau}$ is generally intractable. Hence, the diagonal values are used as an approximation: *i.e.* $\Phi_{v,\tau} \approx \varphi_{v,\tau}$ and

$$
U_{v,\tau} \approx E_{v,\tau} + \Delta_{v,\tau}(E) - \frac{i}{2}\Gamma_{v,\tau}(E) ,
$$

where $\Delta_{v,\tau}(E)$ is the level shift, which can either be assumed negligible, or absorbed into the definition of $E_{v,\tau}$. In radiative cases Δ can be infinite, and becomes negligible only after renormalization. The level width $\Gamma_{v,\tau}(E)$ is a measure of the interaction strength between the resonance state $\varphi_{v,\tau}$ and *all* scattering channels:

$$
\Gamma_{v,\tau}(E) \;=\; \sum_\beta \Gamma_{v,\tau;\beta}(E) \;=\; \sum_\beta 2\pi\left|\left\langle\varphi_{v,\tau}\middle|H-E\middle|\chi^\pm_{\beta,E}\right\rangle\right|^2 .
$$

Formally, the cross sections are derived by squaring the T-matrix and multiplying by an energy-conserving delta function. Experimental scattering can then be obtained by a suitable averaging over the actual situation at hand. In the present case we have been relaxed for the sake of clarity, but energy conservation is included implicitly by considering only on-shell elements of the scattering matrix, $T^{(R)+}_{\beta,\alpha}(E,E)$.

The interpretation of the scattering process as a two-step event, and the width as a total decay probability (a sum of partial decay probabilities), can be inferred from the general formula by making a number of approximations. Assuming the nonresonant scattering is negligible, and that only a single resonant state Φ couples with both the in and out states, the scattering amplitude becomes a single product of 'excitation' ($\alpha \to \Phi$) and 'decay' terms ($\Phi \to \beta$);

$$T_{\beta,\alpha}^{(R)+} \approx \frac{\left\langle \chi_{\beta,E}^{-} \middle| H - E \middle| \Phi \right\rangle \left\langle \Phi \middle| H - E \middle| \chi_{\alpha,E}^{+} \right\rangle}{E - E_{\Phi} + i\Gamma_{\Phi}/2}.$$

The probability of observing outgoing particles in channel β is then the square of this expression, averaged over the specific experimental conditions. If it is assumed that the average (experimental) incident energy is near resonance, the spread (bandpass) ΔE of energies around this average is large compared to Γ_{Φ}, and that the various matrix elements are constant over this energy range, then the probability becomes

$$P_{\beta} = \int_{\Delta E \gg \Gamma_{\Phi}} dE \left| T_{\beta,\alpha}^{(R)+} \right|^2 \approx \frac{\Gamma_{\Phi;\beta}}{\Gamma_{\Phi}} \sigma_{\Phi,\alpha} ,$$

where $\sigma_{\Phi,\alpha} = \left| \left\langle \Phi \middle| H - E \middle| \chi_{\alpha,E}^{+} \right\rangle \right|^2$ can be considered the excitation probability of the 'intermediate' state Φ. The ratio $\Gamma_{\Phi;\beta}/\Gamma_{\Phi}$ assumes its traditional role as the branching ratio for 'decay' from the excited intermediate state to a final channel β.

How is this excitation-decay viewpoint impaired when more than a single resonance contributes to the scattering amplitude? Consider the extension of the previous ideas to the case of two resonance states Φ_1 and Φ_2. The amplitude then becomes a sum of two terms, and an interference term appears in the probability:

$$P_{\beta} \approx \frac{\Gamma_{1;\beta}}{\Gamma_1} \sigma_{1,\alpha} + \frac{\Gamma_{2;\beta}}{\Gamma_2} \sigma_{2,\alpha} + \frac{4(\Gamma_1 + \Gamma_2)}{4(E_2 - E_1)^2 + (\Gamma_1 + \Gamma_2)^2} \sqrt{\Gamma_{1;\beta}\sigma_{1,\alpha}\Gamma_{2;\beta}\sigma_{2,\alpha}} .$$

In this situation, we loose the *interpretation* of the process as a two-step event. If the resonance spacing is large in comparison with the widths (or the bandpass small enough to exclude the second resonance) we might speak of being in a certain regime for which the 'two-step' approximation is valid. However, this is an exercise in semantics rather than physics. While it is convenient for us to *think* in terms of scattering through virtual intermediate states, the actual process is direct scattering to a final state. It is the amplitudes for each such 'virtual process' which must be added, rather than the probabilities.

EXAMPLE: K-SHELL RESONANCE SCATTERING

As a concrete example of the above ideas, in this section we look at K-shell photoionization and decay. Within the viewpoint of RRRRS outlined above, this process is considered as inelastic scattering of an incident photon whose energy is in the region of, or above, the K-shell threshold.

The initial channel is then $\chi_i = \left| g; \omega \right\rangle$, a direct product of the atomic ground state and a single-photon state of energy ω. The total energy is $E = E_0 + \omega$. Since the vector potential $\mathbf{A}$ contains terms proportional to both photon creation and annihilation operators ($a^{\dagger}$ and a), to lowest order in the interaction operator V_{int} we must consider both zero- and two-photon intermediate states. The atomic part of these states are the K-shell vacancy states, either excited ($[1s]ml$, energy $E_0 + I_{[1s]} - \tau_{ml}$) or ionized ($[1s]\tau$, energy $E_0 + I_{[1s]} + \tau$). The scattering amplitude to some final channel χ_F then becomes

$$
T_{F,i}^{(R)+} \approx \left\langle \chi_F \left| V_{int} \right| g; \omega \right\rangle + \sum_l \int d\tau \frac{\left\langle \chi_F \left| H - E \right| [1s]\tau; 0 \right\rangle \left\langle [1s]\tau; 0 \left| \mathbf{p} \cdot \mathbf{A} \right| g; \omega \right\rangle}{E_{exc} - \tau + i\Gamma/2}
$$

$$
+ \sum_l \sum_{\omega'} \int d\tau \frac{\left\langle \chi_F \left| H - E \right| [1s]\tau; \omega, \omega'' \right\rangle \left\langle [1s]\tau; \omega, \omega'' \left| \mathbf{p} \cdot \mathbf{A} \right| g; \omega \right\rangle}{E_{exc} - \tau - \omega - \omega'' + i\Gamma/2}
$$

The *excess energy* $E_{exc} \equiv \omega - I_{[1s]}$ proves a convenient measure of incident-photon energy relative to the K-shell ionization threshold. Again, implicit in the integral over τ is a summation over bound states m, with $\tau \to -\tau_m$. The width, Γ, is taken as the (total) width of the atomic $[1s]$ hole state, and assumed independent of τ. For simplicity we also make the dipole approximation $\mathbf{p} \cdot \mathbf{A} \approx D \equiv (2\pi/V\omega)^{1/2} \mathbf{p} \cdot \hat{\mathbf{\varepsilon}}$, which then limits our resonances to the $l = p$ series.

Radiative Final States

If we consider final states containing a single photon, and the various matrix elements above to lowest order, we are left with channels of the form $\chi_F = \left| [2p]xp; \omega' \right\rangle$. The question of neutral ($x = n$) or ionized ($x = \varepsilon$) final atomic states is left general. The final state energy is $E_0 + I_{[2p]} + \varepsilon_x + \omega'$, where $\varepsilon_x = \varepsilon$ ($x = \varepsilon$) or $\varepsilon_x = -\varepsilon_n$ ($x = n$). Energy conservation between the initial and final states then requires that $\omega' = \omega_{KL} + (E_{exc} - \varepsilon_x)$ where $\omega_{KL} \equiv I_{[1s]} - I_{[2p]}$ is the nominal K_α fluorescence energy.

The amplitude formula reduces to a generalized Kramers-Heisenberg formula [5]

$$
T_{F,i}^{(R)+} \approx \left\langle [2p]xp; \omega' \left| \tfrac{1}{2} \mathbf{A} \cdot \mathbf{A} \right| g; \omega \right\rangle
$$

$$
+ \int d\tau \frac{\left\langle [2p]xp \left| D \right| [1s]\tau p \right\rangle \left\langle [1s]\tau p \left| D \right| g \right\rangle}{E_{exc} - \tau + i\Gamma/2} + \int d\tau \frac{\left\langle [2p]xp \left| D \right| [1s]\tau p \right\rangle \left\langle [1s]\tau p \left| D \right| g \right\rangle}{(\varepsilon_x - \tau) - (\omega + \omega_{KL}) + i\Gamma/2}
$$

The direct-scattering amplitude, second-order in $\mathbf{A}$, corresponds to Compton scattering, which we will neglect here as weak when considering the resonant amplitudes. To proceed further we must examine the atomic parts of the intermediate-to-final state matrix elements. For most purposes the spectator approximation can be used:

$$
\left\langle [2p]xp \left| D \right| [1s]\tau p \right\rangle \approx \left\langle [2p] \left| D \right| [1s] \right\rangle \left\langle xp \left| \tau p \right\rangle \approx \sqrt{\Gamma_{KL}/2\pi} \; \delta_{x,\tau},
$$

where $\Gamma_{KL} \equiv 2\pi \left| \left\langle [2p] \left| D \right| [1s] \right\rangle \right|^2$. The last step is based on the fact that the intermediate- and final-state Rydberg orbitals both see approximately the same Coulomb ($Z=1$) potential, and are very nearly orthogonal. Hence, $\tau \approx \varepsilon_x$ and we see

7

that the third term is never in resonance. This is a consequence of considering scattering from the ground state [5]. Note that in molecular cases, the factorization above includes vibrational Frank-Condon factors, and the simple results that follow need extension.

We are thus reduced to considering the middle, resonant term. Employing all of the above approximations, we can examine the cases of bound and continuous final atomic states:

Resonant Raman Scattering

Consider the bound final atomic state $\chi_F = |[2p]np; \omega'\rangle$. Energy conservation requires $\omega' = E_{exc} + \omega_{KL} + \varepsilon_n$: the so-called linear dispersion. The probability is

$$\sigma(E_{exc}) \approx \frac{\frac{1}{2\pi}\Gamma_{KL}\sigma_{np}}{(E_{exc} + \tau_n)^2 + \Gamma^2/4},$$

where $\sigma_{np} \equiv |\langle [1s]np|D|g\rangle|^2$. Just below threshold we thus see a number of Raman x-ray lines. Each line exhibits linear dispersion with the incident energy, and reaches a maximum intensity when the resonance condition $E_{exc} = -\tau_n \approx -\varepsilon_n$ is fulfilled.

The natural resonant Raman line shape is a delta function, and the line shows appreciable intensity over only a limited region of excitation energy; *i.e.* within ± a few 'Γ' of resonance.

Resonant Fluorescence

If the final atomic state is ionized, ejecting an electron $\chi_F = |[2p]\epsilon p; \omega'\rangle$, the final state is infinitely degenerate. The electron and photon share the available energy $\omega' + \varepsilon = E_{exc} + \omega_{KL}$. The (differential) probability becomes

$$d\sigma(E_{exc}) \approx \frac{\frac{1}{2\pi}\Gamma_{K\alpha}\sigma_{ep}}{(E_{exc} - \varepsilon)^2 + \Gamma^2/4}d\varepsilon = \frac{\frac{1}{2\pi}\Gamma_{K\alpha}\sigma_{ep}}{(\omega' - \omega_{KL})^2 + \Gamma^2/4}d\omega'.$$

If the electron and photon are not measured in coincidence, we observe a distribution of energies for the ejected particles. Consider first the behavior of these distributions at large incident energy: The photoelectron line shape is Lorentzian with a width Γ, and an average energy E_{exc} which exhibits linear dispersion (the photoelectric effect). The fluorescence line is also a Lorentzian of width Γ, with an average energy fixed at ω_{KL}. Once above threshold, the intensity of these lines is determined by $\Gamma_{KL}\sigma_{ep}$.

When the incident energy is near threshold, the lines begin to distort and narrow. This is purely a consequence of energy conservation. Consider the behavior of the Fluorescence line: Because $\varepsilon \geq 0$, the fluorescence photon can only have energies between $\omega' = 0$ and $\omega_{KL} + E_{exc}$. As E_{exc} becomes low, or negative, parts of the Lorentzian are not accessible, *e.g.* at threshold ($E_{exc} = 0$) the Lorentzian's high-energy

8

half is omitted. Below threshold the line is only a portion of the lower-energy flank of the Lorentzian (see *e.g.* fig. 4, [3]). As this happens, resonant Raman lines begin to appear in the energy-excluded regions of the spectrum.

Radiationless Final States

Consider now final scattering states with at least one ejected electron, but no photons: $\chi_F = |F\varepsilon l; 0\rangle$. In this case, the two-photon intermediate states do not contribute, and the lowest-order amplitude is

$$T_{F,i}^{(R)+} \approx \langle F\varepsilon l|D|g\rangle + \int d\tau \frac{\langle F\varepsilon l|H_{atom} - E|[1s]\tau p\rangle\langle[1s]\tau p|D|g\rangle}{E_{exc} - \tau + i\Gamma/2} \; .$$

The resonance term is now a direct consequence of electronic configuration interaction between the resonance and final states.

Autoionization

In this situation the final atomic states involve a single hole in an outer shell. Here we consider the states $\chi_F = |[2p]\varepsilon d; 0\rangle$ as an example. Energy conservation requires $\varepsilon = \omega - I_{[2p]}$, and the amplitude becomes

$$T_{F,i}^{(R)+} \approx \langle [2p]\varepsilon d|D|g\rangle + \sum_m \frac{\langle[2p]\varepsilon d|H_{atom} - E|[1s]mp\rangle\langle[1s]mp|D|g\rangle}{E_{exc} + \tau_m + i\Gamma/2} \; ,$$

where we ignore the continuous intermediate states. Here the direct scattering amplitude can be strong, even at energies near the *1s* threshold. In general, the resonance amplitude interferes strongly with the (slowly varying) direct term, and the intensity of the ejected electron line (zero natural width) can exhibit a wide variety of behavior as E_{exc} is varied through the resonances (see *e.g.* [6]).

The key distinction of autoionization is the strength of the direct scattering term. The final atomic states coincide with a diagram photoionization process. To connect the intermediate-to-final state matrix element, this implies that the excited resonances must 'decay' by a *participator* Auger process; *e.g.* here $mp \to 1s$ accompanied by $2p \to \varepsilon d$.

Radiationless Resonant Raman Scattering (Resonant Auger)

The next type of final state to consider is a double hole state with an excited Rydberg electron, *i.e.* $\chi_F = |[2p^2]np\varepsilon l; 0\rangle$. Energy conservation requires that $\varepsilon = \varepsilon_{KLL} + E_{exc} + \varepsilon_n$, where $\varepsilon_{KLL} \equiv I_{[1s]} - I_{[2p,2p]}$ is the nominal $K - L_{2,3}L_{2,3}$ Auger energy. Hence, the resonant-Auger-electron line shape is also a delta function, and shows linear dispersion as in the autoionization and radiative Raman cases.

The relaxation of atomic orbitals becomes important in the Auger case; the excited, intermediate-state electrons move in a Coulomb field with $Z=1$ whereas the final-state orbitals see $Z=2$ (and $\varepsilon_n \approx 4\tau_n$). The spectator approximation now takes on the form

$$\left\langle [2p^2]np\,\varepsilon l \,\middle|\, H_{atom} - E \,\middle|\, [1s]xp \right\rangle \approx \sqrt{\Gamma_{KLL}/2\pi} \left\langle np^{++} \middle| xp^{+} \right\rangle,$$

where $\Gamma_{KLL} \equiv 2\pi \left| \left\langle [2p^2]\varepsilon l \,\middle|\, H_{atom} - E \,\middle|\, [1s] \right\rangle \right|^2$ is the $K - L_{2,3}L_{2,3}$ Auger amplitude, which we assume here as energy independent. The amplitude becomes

$$T_{F,i}^{(R)+} \approx \left\langle [2p^2]np\,\varepsilon l \,\middle|\, D \,\middle|\, g \right\rangle$$
$$+ \sqrt{\frac{\Gamma_{KLL}}{2\pi}} \left[\sum_m \frac{\left\langle np^{++} \middle| mp^+ \right\rangle\!\left\langle [1s]mp \middle| D \middle| g \right\rangle}{E_{exc} + \tau_m + i\Gamma/2} + \int d\tau \frac{\left\langle np^{++} \middle| \tau p^+ \right\rangle\!\left\langle [1s]\tau p \middle| D \middle| g \right\rangle}{E_{exc} - \tau + i\Gamma/2} \right]$$

The direct term describes $2p$ excitation accompanying $2p$ ionization, *i.e.* shake-up. Generally this term is regarded as negligible near resonance, though this need not be the case (see. *e.g.* [7]).

The two resonance terms take on different degrees of importance for different final states. The first can be thought of as shake-modified resonant Auger transitions, the second as post-collision-interaction (PCI) 'recapture' from the continuum. For final states of low n, and resonance spacing large compared to Γ, it may be that only a single resonance contributes strongly. In this case the resonant Raman Auger line behaves in a similar fashion as the radiative Raman line outlined above. However, for larger n many values of m contribute to the total amplitude, and coherence between the different virtual paths causes noticeable features in the cross section [8]. Finally, for very large-n final states the continuum recapture amplitudes dominate, giving rise to a broad, asymmetric cross section [9].

The distinction between autoionization and the resonant Auger effect can be vague. For open-shell atoms, an excited resonant state may lie in the valence shell, where the difference between participator and spectator 'decay' is lost. Perhaps the best distinction between the two processes is in the range of similar final states: Raman Auger lines form a series of photopeaks whose limit is the Auger line, while autoionization is a single, diagram photopeak.

Resonant Double Ionization (Auger Decay)

Finally, for the case of double ionization, $\chi_F = \left| [2p^2]\varepsilon_1 p\varepsilon_2 l; 0 \right\rangle$, the final state is again infinitely degenerate — as in the resonant fluorescent case. Energy conservation can now be written as $\varepsilon_1 + \varepsilon_2 = E_{exc} + \varepsilon_{KLL}$. The amplitude is

$$T_{F,i}^{(R)+} \approx \left\langle [2p^2]\varepsilon_1 p\,\varepsilon_2 l \,\middle|\, D \,\middle|\, g \right\rangle + \sqrt{\frac{\Gamma_{KLL}}{2\pi}} \int d\tau \frac{\left\langle \varepsilon_1 p^{++} \middle| \tau p^+ \right\rangle\!\left\langle [1s]\tau p \middle| D \middle| g \right\rangle}{E_{exc} - \tau + i\Gamma/2}.$$

The direct scattering term now refers to double ionization, and we regard it as weak – especially since it is distributed over a continuum of energies. In the resonant term the sum over excited states is excluded, since the overlaps $\langle \varepsilon_1 p^{++} | mp^+ \rangle$ are generally negligible. (For the sake of simplicity, we ignore the exchange expression ($c_1 \langle \rangle \varepsilon_2$) that appears in a more rigorous discussion [10].)

Consider first the situation far above threshold. For resonance, we must have $\tau \approx E_{exc}$. At large values of τ we can approximate $\langle \varepsilon_1 p^{++} | \tau p^+ \rangle \approx \delta(\varepsilon_1 - \tau)$. Hence at large E_{exc} we have the same general behavior as in resonant fluorescence: now a Lorentzian photopeak (width Γ) centered at $\varepsilon \approx E_{exc}$, and an Auger line at $\varepsilon \approx \varepsilon_{KLL}$.

At low values of τ however, relaxation effects become important. The continuum orbitals are far from orthogonal and near threshold coherence between different intermediate-state amplitudes is strong. This is the source of the PCI distortion and shift of the Auger and photoelectron lines [10]. As E_{exc} is further decreased to (and below) threshold the line begins to narrow as, in the fluorescence case discussed above, energy constraints restrict the resonance condition.

Natural Line Widths

The results here were based on the idea that the final-state channels lead to (stable) physical states. In fact this is yet another approximation (*i.e.* in the above example the *2p* shell may decay) and our approximate final-state functions are really resonances embedded in yet higher continua.

The present theory can make allowances for this to a limited degree. This is done by taking a statistical average of the results over the *final-state* core binding energies [8], which are spread by the natural width of the specific final state Γ_F. Defined quantities that depend on the final-state binding energies, such as ω_{KL} and ε_{KLL} assume a statistical spread of values (but not E_{exc}). Averaging yields new natural widths for the various lines: the autoionization and Raman lines now have $FWHM = \Gamma_F$, Auger and fluorescence lines (at higher energy) $FWHM = \Gamma + \Gamma_F$, but the photo-peak remains unaffected ($FWHM = \Gamma$ at large energies).

SUMMARY

We have seen how some of the major 'decay' processes — which we envisage as following the creation of an atomic inner-shell hole by a photon — are all the result of inelastic resonant photon scattering. When a single particle (photon or electron) is ejected, energy conservation demands that the 'line' exhibit linear dispersion (the photoelectric effect) and a natural line-shape of zero width. The resonant nature of these lines is reflected in the finite range of incident energy for which they have appreciable intensity. When several particles are ejected, they share the available energy. However, the resonant nature of the process restricts their individual energies to certain characteristic values, dispersed by a natural width Γ reflecting the strength of interaction between the resonance and all finial states.

The following table attempts to organize and compare the possibilities, hopefully emphasizing their common origins while recognizing their differences.

TABLE 1. $\omega + A \rightarrow X^{q+}$ + particle(s).

Process	X	Out	Resonance line(s)		Resonance condition	Direct process
Autoionization	[f]	e⁻	Photo-peak:	$\varepsilon = \omega - I_{[f]}$	$E_{exc} \approx -\tau_m$ $(m=m_{min},...,\infty)$	Ionization
X-ray Raman	[f]nl	γ	X-ray line:	$\omega' = \omega - I_{[f]nl}$ $= E_{exc} + \varepsilon_n + \omega_{if}$	$E_{exc} \approx -\tau_n$	Compton (excitation)
Auger Raman	[f,f']nl	e⁻	Photo-peak:	$\varepsilon = \omega - I_{[f,f']nl}$ $= E_{exc} + \varepsilon_n + \varepsilon_{iff'}.$	$E_{exc} \approx -\tau_n$	Ionization with excitation
Fluorescence	[f]	e⁻,γ	Photo-peak: X-ray line:	$\varepsilon \approx E_{exc}$ $\omega' \approx \omega_{if}$	$E_{exc} \geq 0$	Compton (ionization)
Auger	[f,f']	e⁻,e⁻	Photo-peak: Auger peak:	$\varepsilon \approx E_{exc}$ $\varepsilon \approx \varepsilon_{iff'}.$	$E_{exc} \geq 0$	Double ionization

where $E_{exc} = \omega - I_{[i]}$, $\omega_{if} = I_{[i]} - I_{[f]}$, and $\varepsilon_{iff'} = I_{[i]} - I_{[f,f']}$.

ACKNOWLEDGMENTS

I would like to thank Tejio for all his friendship and help over the years. My thoughts on this subject also result from interactions with many other people. I would particularly like to thank Helena and Seppo Aksela, Bernd Crasemann, Sami Heinäsmäki, Jon Levin, Jukka Tulkki, Steve Southworth, and Scott Whitfield.

REFERENCES

1. Åberg T., and Crasemann, B., "Radiative and Raditionless resonant Raman Scattering" in *Resonant Anomalous X-ray Scattering: Theory and Applications*, edited by G. Materlik, C.J. Sparks, and K. Fischer, Amsterdam: North-Holland, 1994, p. 431.
2. Åberg, T., and Howat, G., "Theory of the Auger Effect" in *Corpuscles and Radiation in Matter* I, edited by S. Flügge and W. Mehlhorn, Handbuch der Physik **31**, Berlin: Springer, 1982, p. 469.
3. Åberg, T., and Tulkki, J., "Inelastic X-Ray Scattering including Resonance Phenomena" in *Atomic Inner-Shell Physics*, edited by B. Crasemann, New York: Plenum, 1985, p. 419.
4. Åberg, T., *Phys. Scr.* **21**, 495 (1980), and *Phys. Scr.* **T41**, 71 (1992).
5. Åberg, T., *Applied Quantum Mechanics: Collisions and Half-Collisions*, Uppsala: Nordplus Lecture Series, 1994.
6. Sorensen, S. L., Åberg, T., Tulkki, J., Rachlew-Källne, E., Sundström, G., and Kirm, M., *Phys. Rev. A* **50**, 1218 (1994).
7. Marinho, R. R. T., Björneholm, O., Sorensen, S. L., Hjelte, I., Sundin, S., Bässler, M., Svensson, S., and Naves de Brito, A., *Phys. Rev. A* **63**, 32514 (2001).
8. Armen, G. B., Aksela, H., Åberg, T., and Aksela, S., *J. Phys. B* **33**, R49 (2000).
9. Armen, G. B., and Levin, J. C., *Phys. Rev. A* **56**, 3734 (1997).
10. Tullki, J., Armen, G. B., Åberg, T., Crasemann, B., and Chen, M. H., *Z. Phys. D* **5**, 241 (1987).

Ugo Fano and shape resonances

Antonio Bianconi

Dipartimento di Fisica, Università di Roma "La Sapienza",
P.le Aldo Moro 2, 00185 Roma, Italy

Abstract. Ugo Fano has been a leader in theoretical Physics in the XX century giving key contributions to our understanding of quantum phenomena. He passed away on 13 February 2001 after 67 years of research activity. I will focus on his prediction of the quantum interference effects to understand the high-energy photoabsorption cross section giving the "Fano lineshapes". The Fano results led to the theoretical understanding of "shape resonances" (called also "Feshbach resonances") that should be better called "Fano resonances". Finally I will show that today this Fano quantum interference effect is behind several new physical phenomena in different fields.

Ugo Fano was born in Turin, Italy, on 28 July 1912. His father Gino Fano (1871-1952) was professor of mathematics at Turin, Italy, specializing in differential geometry. He has spent his childhood mostly in Verona at "villa Fano" where he developed a love for mountains hiking and rock climbing. He studied Mathematics at University of Turin, but after the "Laurea" he turned his interests on physics following the lectures of Enrico Persico (coming from the Fermi's group in Rome) on theoretical physics and discussions with his cousin Giulio Racah (1909-1965), a theoretical physicist known for the powerful theory of angular momentum. He moved on 1934 to Rome to work with Enrico Fermi where he worked in the period 1934-1936. In these years the Fermi's group shifted from atomic physics to the new experimental nuclear physics. They focused on systematic researches on the absorption and scattering properties of slow neutrons and discovered the artificial radioactivity induced by slow neutrons [1]. The fact that the cross-section is high for small neutron velocities [2] was interpreted as due to the capture of a neutron by a nucleus at a scattering resonance called "risonanza di forma" ("shape resonance").

Fano addressed his interests not on the experimental work, that was the main interest of the Fermi's group, but on the application of quantum theory for interpreting strange looking shapes of spectral absorption lines. Fano investigated the stationary states with configuration mixing under conditions of autoionization introduced by Rice [3] and he pointed out the basic physics of the quantum interference phenomenon between a discrete level and a continuum [4]. This theoretical result in atomic physics is related with the resonant scattering of a slow neutron in a nucleus, the "shape resonances", found by Fermi. In fact it deals with processes inverse to those

CP652, *X-Ray and Inner-Shell Processes: 19th International Conference on X-Ray and Inner-Shell Processes*
edited by A. Bianconi, A. Marcelli, and N. L. Saini
© 2003 American Institute of Physics 0-7354-0111-X/03/$20.00

considered in the Fano's theory of autoionization. This is a scattering process, in which the system is formed by combining an incident particle b with the "rest" and then the system breaks up releasing alternatively either the same particle b or another particle c and ... . In this process the interference of resonance and potential scattering amplitudes gives a large reaction cross section. The theory for the nuclear scattering cross-section was developed by Breit and Wigner in 1936 [5].

In 1936-1937 Fano worked in Leipzig with Werner Heisenberg and he visited Arnold Sommerfeld, Niels Bohr, Edward Teller and George Gamow. From Germany Fano moved to Paris with the Joliot - Curie group. Fano returned to the University of Rome as a lecturer in 1938.

In these years Fano started to address his interests to radiation biology and genetics that will become central topics for his later research. It is noteworthy that, after a seminar in Rome by P. Jordan on x-ray effects on genetic material, Fermi had suggested to Fano that the biological action of radiation would be an important and suitable topic for study. He was in close contact with his school friend Salvatore Luria who moved from Turin to Rome where from 1935-1940 was in charge of Medical Physics and Radiology under the direction of Enrico Fermi and Edorado Amaldi. In 1939 Fano married Camilla ("Lilla") Lattes, who collaborated with him in science and worked as a teacher for many years. In the same year the couple immigrated to the United States in 1939 to escape the racial laws.

He was at the University of Michigan summer school at Ann Harbor in 1939 when Werner Heisenberg and Edoardo Amaldi visited Fermi in the states. Once he told me that during a party in August he became aware that for everybody there it was clear that Fermi and Heisenberg would become the leaders of the USA and German projects for the nuclear bomb. At this point he decided not to work in nuclear physics being interested on other science issues. He focused on the interaction of Radiation with matter and in particular on effects of radiation on living organisms (genetic resistance to radiation effects). In 1940-1944 worked with the help of his wife in what was later to be called radiation biology at the Department of Genetics of the Carnegie Institution at Cold Spring Harbor. Fano's papers in this period concerned chromosomal rearrangement mutations, lethal effects, and genetic effects of X-rays and neutrons on Drosophila melanogaster, as well as theoretical analysis of genetic data. His work also included the discovery of bacteriophage-resistant mutants in Escherichia coli, following up earlier studies by Salvador E. Luria who moved to USA in 1939 and visited the Carnegie Institution.

After the war in the years 1946-1966, he joined the staff of National Bureau of Standards (NBS), initially working in the radiological physics group led by Lauriston S. Taylor and after in the basic physics of atoms, molecules, and condensed matter elucidating fundamental physical processes.

In the fifties the scientific interest for "shape resonances" was coming up again, in fact the availability of monochromatic neutron beams at nuclear reactors allowed the measure of the neutron cross section on selected nuclei as a function of neutron energy. The trapping of the slow neutrons for long time inside the nuclei has been clearly shown to be at the origin of resonances for neutron capture. The theory of "shape resonances" in the frame of potential scattering was developed by J.M. Blatt

FIGURE 1. Ugo Fano

and V. F. Weisskopf [6] and later by Herman Feshbach in 1958 [7] for a general nuclear reaction theory, based on the projection of the nuclear state into direct and compound channels, following the method introduced by Fano [4]

The "shape resonance" occurs when a quantum particle with energy E and wave vector $k=2\pi/\lambda$ is trapped within a potential well with finite barrier of a given size R given by the radius of the nucleus, with the generic condition for the shape resonance: $R = n\ \lambda/2$ where n is an integer. The name "shape resonance" indicates the fact that the shape of the potential barrier determines the energy of the resonance, therefore they have been used in 1949-1954 by Edoardo Amaldi and others to measure the size of the nuclei of several elements with the precision of $\pm 10^{-13}$ cm [8].

In these years the interest of Fano returns to this phenomenon following the extensive investigation of line profiles of high energy levels of excitation in the far-UV absorption spectra in atomic and molecular spectroscopy undertaken by means of far-ultraviolet light of electron bombardment and also of energy transfer in molecular collisions. The Fano prediction and interpretation of the experiments on the excitation of quasi bound states buried in continua was his major outcome in these years [9] and the review paper on the *"The theory of atomic photoionization"* remains as a relevant milestone in the physics of XX century [10]. He has shown that lineshape of the absorption lines (Fig. 2) in the ionization continuum of atomic (and molecular) spectra are represented by the formula

$$\sigma(\varepsilon) = \sigma_a \left[\frac{(q+\varepsilon)^2}{1+\varepsilon^2} \right] + \sigma_b \tag{1}$$

where $\varepsilon = \dfrac{E-E_r}{\Gamma/2}$ indicates the deviation of the incident photon energy E from the idealized resonance energy E_r which pertains to a discrete auto-ionizing level of the atom. This deviation is expressed in a scale whose unit is the half-width $\Gamma/2$ of the line ($\hbar/\Gamma$ is the mean life of the discrete level with respect to autoionization). $\sigma(\varepsilon)$

represents the absorption cross section for photons of energy E whereas σ_a and σ_b are two portions of the cross section corresponding to transitions to states of the continuum that do and do not interact with the discrete auto-ionizing state respectively. Finally q is a numerical index which characterizes the line profile.

Thus, he encouraged Robert P. Madden and co-workers at NBS to use synchrotron radiation for spectroscopic studies. He convinced his old friend Edoardo Amaldi and Mario Ageno to push the Italian scientific community to use the Frascati synchrotron as a synchrotron radiation source for high energy spectroscopy of atoms and solids.

In 1966 he joined the Physics faculty of Chicago where he continued his research on the interaction of radiation with matter. I was still in my twenties when I had the possibility to meet him regularly in the years 1971-1976. He used to come to Rome each summer in June-July, spending his time in Frascati Laboratories with the small group of synchrotron radiation researchers: Adalberto Balzarotti, Emilio Burattini, Mario Piacentini and myself. I remind very nice days discussing the x-ray absorption spectra measured at the new Frascati soft x-ray synchrotron radiation beam line. He informed me on the resonances observed in the scattering of electron on nitrogen molecule that were described by Dehmer and Dill with the same formalism of his "shape resonances". These discussion led me to the interpretation of the x-ray absorption near edge structure (XANES) in complex solids and metalloproteins in term of "shape resonances" of the excited photoelectron within a finite cluster of atoms surrounding the absorbing atom [11,12].

The Fano quantum interference effects have been observed in modern experiments of photo-fragmentation [13]. Here the Fano quantum interference appears when a quantum state is formed at the resonance energy E_r above the fragmentation threshold

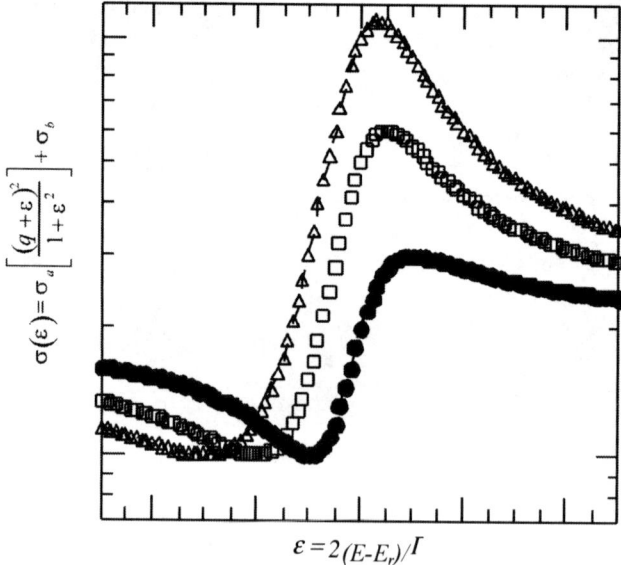

$$\varepsilon = 2(E-E_r)/\Gamma$$

FIGURE 2. The Fano lineshape of the absorption cross near resonance energy E_r of a discrete state buried in the continuum for various values of the q parameter.

of the system. The variation of the cross section in the neighborhood of the energy of fragmentation resonance follows the Fano lineshape. In the field of photo-fragmentation these resonances are usually classified into "Feshbach resonances" if several (n>1) electronic transitions are required to emit one electron or "shape resonances" if only one electronic transition is required to emit one electron. However this classification makes sense only if the lifetimes of the resonances are not larger than the typical vibrational time of the system, otherwise the distinction is not clear. The quantum mechanical phase shift and consequent interference effects encountered on passing through a shape resonance energy have been studied by many authors and they can be seen as a movie on the web [14].

In 1963 it was pointed out by Thompson and Blatt [15] that the basic physics of the quantum interference between a discrete state and a continuum introduced by Fano predicts the amplification of the superconducting critical temperature in a superconducting film of thickness R if $R=\lambda_F/2$ x integer, where λ_F is the wavelength of the electrons at the Fermi level. This prediction did not work since in a single film phase fluctuations suppress the condensate phase coherence. In 1993 we have shown that the "shape resonance" amplification works in a superlattice of superconducting wires [16] where the enhancement is obtained by tuning the Fermi level at a "shape resonance" of the superlattice of period $L=\lambda_F/2$ x integer. The quantum interference effects between the pairs in a narrow band and in a wide band give two superconducting gaps in two different bands in momentum space and in real space, that has been recently confirmed in the new MgB_2 superconductor made by a superlattice of boron layers [17].

In 1995 the Bose Einstein condensation (BEC) of atoms trapped magnetically in a vacuum and cooled to a few billionths of a degree above absolute zero has been achieved [18]. This ultracold atomic gas is dilute system in which the interparticle interactions are weak and easy to be treated theoretically. It has been found that using a magnetic field it is possible to trap the atoms into Feshbach resonances (19) and it is possible to control the fundamental interactions (20). Moreover the inhomogeneous trapping potential leads to spatial separation oh high- and low-energy atoms, giving rise to a Fermi surface that is manifest in the real as well as in the momentum space. Recently the onset of Fermi degeneracy in a ultracold gas of fermions has been reported [21]. Here the pairing interaction between fermions could be controlled by tuning the system to a "Feshbach resonance" and the condensation of pairs (as in He_3) could become possible, similar to Cooper pair formation in superconductivity. This process will be similar to the T_c amplification by "shape resonance" in high T_c superconductors [16].

The Fano lineshapes have been seen also in the zero bias conductance as a function of the gate voltage in a single electron transistor [22]. Here an artificial atom is created by making a layer of GaAs on top of which is a layer of AlGaAs doped with Si. The electrons from the dopants fall into the GaAs, and the resulting positive charge on the Si atoms creates a potential that holds the electrons at the GaAs/AlGaAs interface, creating a two dimensional electron gas. The quantization of energy and charge makes the confined droplet of electrons closely analogous to an atom. The resonance component in the Fano interference appears to come from the single electron charging of the artificial atom interacting with a continuous component.

In conclusion I have shortly focused only on a particular aspect [4,9,10] of the wide scientific activity of Ugo Fano since he had a relevant influence on my personal scientific activity and today its 1935 paper [4] is stimulating new physics.

REFERENCES

1. Fermi, E., Amaldi, E., D'Agostino, O., Rasetti,F. and Segrè, E., *Proc. Roy. Soc.* **A146**, 483 (1935).
2. Amaldi, E., d'Agostino, O., Fermi, E., Pontecorvo, B., Rasetti, E., and Segrè, E. *Proc. Roy. Soc.* **A149**, 522 (1935).
3. Amaldi, E. and Fermi E., *Phys. Rev.* **50**, 899 (1936).
4. Fano, U., *Nuovo Cimento* **12**, 156 (1935).
5. Breit, G. and Wigner, E., *Phys. Rev.* **49**, 519 (1936).
6. Blatt, J.M. and Weisskopf, V. F., *Theoretical Nuclear Physics,* John Wiley & Sons Inc. New York (1952), pp.379 ff. in particular p. 401.
7. Feshbach, H., *Ann. Phys. (N.Y.)* **5**, 357 (1958)..
8. de Shalit, A. and Feshbach, H., *Theoretical Nuclear: Volume on Nuclear structure,* J. Wiley and Sons Inc., New York (1974) p. 87
9. Fano, U., *Phys. Rev.* **124**, 1866 (1961); Fano, U. and Cooper, J. W., *Phys. Rev.* **137**, A1364 (1965).
10. Fano, U. and Cooper, J. W., *Rev. Mod. Physics* **40**, 441 (1968).
11. Bianconi, A., *Appl. of Surf. Science* **6**, 392 (1980).
12. Bianconi, A., "XANES spectroscopy" in *X-Ray Absorption: principles, Applications, Techniques of EXAFS, SEXAFS and XANES* edited by R. Prinz and D. Koningsberger (J. Wiley Interscience, New York, 1988).
13. Reid, S. A. and Reisler, H., *Annu. Rev. Phys. Chem.* **43**, 591 (1996); Shapiro, M., *J. Phys. Chem.* **102**, 9570 (1998); Fyodorov, V. V. and Sommers, H.-J., *J. Math. Phys.* **38**, 1918 (1997); Moller, K. B., Henriksen, N. E. and Zewail, A. H., *J. Chem. Phys.* **113**,10477 (2000); Brems, V., Desouter-Lecomte, M., *Journal of Chemical Physics* **116**, 6318 (2002)
14. Powis, I., Chem. Phys., J., **106**, 5013 (1997); and on the web site http://www.chem.nott.ac.uk/movies/cf3clm.gif
15. Thompson, C. J. and Blatt, J. M., *Phys. Lett.* **5**, 6 (1963).
16. Bianconi, A., *Sol. State Commun.* **89**. 933 (1994); Bianconi, A., Valletta, A., Perali, A., Saini, N.L., *Physica C* **296**, 269 (1998).
17. Bianconi, A., Di Castro, D., Agrestini, S., Campi, G., Saini, N. L., Saccone, A., De Negri, S. and Giovannini, M., *Phys.: Condens. Matter* **13**, 7383 (2001).
18. Cornell, E. A. Ensher, J. R. Wieman, C. E. online abstract available at http://xxx.lanl. gov/abs/cond-mat/9903109; Ketterle, W. Durfee, D. S. Stamper-Kurn, D. M. online abstract available at http://xxx.lanl.gov/abs/cond-mat/9904034
19. Inouye, S., *et al., Nature* **392**, 151 (1998); Courteille, Ph., Freeland, R. S., Heinzen, D. J., van Abeelen, F. A., Verhaar, B. J., *Phys. Rev. Lett.* **81**, 69 (1998); Roberts J. L. *et al., ibid.*, p. 5109; Vuletic, V., Kerman, A. J., Chin, C., Chu, S. *ibid.* **82**, 1406 (1999).
20. Donley, E. A., Claussen, N. R., Cornish, S. L., Roberts, J. L., Cornell, E. A. and Wieman, C. E., *Nature* **412**, 295 (2001); Holland, M. J. et al., *Phys. Rev. Lett.* **87**, 120406 (2001); Chiofalo, M.L. et al. *Phys. Rev. Lett.* **88**, 090402 (2002).
21. De Marco, B., and Jin, D. S., *Science* **285**, 1703 (2002).
22. Kastner, M. A., *Rev. Mod. Phys.* **64**, 849 (1992).

Richard Day Deslattes, 21 Sept 1931 - 16 May 2001: Calibration of light, matter and fundamental constants

C. T. Chantler

School of Physics, University of Melbourne, Vic. 3010, Australia

Abstract.
Richard Deslattes passed away on 16 May 2001 after a life dedicated to fundamental metrology. Although the themes of calibrating light, matter and fundamental constants can give three guiding principles through his career, the wide-ranging nature of his areas of interest are encompassed by over 165 refereed publications with several cited over 100 times. He has left an enduring legacy to science.

INTRODUCTION AND EARLY WORK

Richard (Dick) Deslattes was a great scientist, a good colleague and a genius in developing applications to solve intractable problems. His life was his work, so much so that he continued to fight cancer until the very end, remaining active in the laboratory at NIST until his retirement just three weeks before his death. He was a tough boss, who epitomised the dedication he expected from his staff in the pursuit of perfection, and which lead to his labs and his Quantum Metrology Division achieving an abundance of major scientific results compared to much larger divisions at NIST. As a guest researcher and Fellow of the English-Speaking Union of the Commonwealth, this author learned from him and his group the meaning of dedication to the methodical pursuit of precision science, and the author is very grateful for his patience with a peculiar Australian theoretician (at the time) from arguably a very different culture. He was a sincere father of five children. His wife Mary has also been dedicated to his causes, and together they demonstrated the care of their visitors with numerous delightful parties at their home, always beautifully catered and always engaging.

Dick (Fig. 1) was born on 21 September 1931 and grew up in New Orleans. He received a B.Sc. degree from Loyola University in New Orleans in 1952 and a Ph.D. from the Johns Hopkins University in 1959, strongly influenced by J.A. Bearden. Following a postdoctoral position at Cornell University, associated with Lyman Parratt, Deslattes joined the staff of the National Bureau of Standards (now the National Institute of Standards and Technology) in 1962. Dick's early years at NIST were devoted to the spectroscopy of molecules, the development of high-resolution x-ray spectrometers and powerful x-ray sources, and the characterization of solution-grown single crystals.

Dick carried out the first definitive study of the photon excitation functions of the principal satellites of the valence emission band of chlorine in KCl closing a chapter on

CP652, *X-Ray and Inner-Shell Processes: 19th International Conference on X-Ray and Inner-Shell Processes*
edited by A. Bianconi, A. Marcelli, and N. L. Saini
© 2003 American Institute of Physics 0-7354-0111-X/03/$20.00

FIGURE 1. Dick as we will remember him.

these controversial spectra and showing their similarity to the corresponding spectra of atomic argon, whose gas phase fluorescence spectrum was recorded for the first time. He continued these studies with colleagues and staff in later years. [1] [2]

He showed the applicability of the, then novel, technique of double crystal diffraction topography to the characterization of solution grown single crystals and demonstrated that this technology yielded dislocation free, dynamically perfect specimens. [3] Dick developed high resolution x-ray spectrometers, powerful x-ray sources, and new approaches to the operation of gaseous detectors in the x-ray region.[4] He pioneered the application of high resolution x-ray spectroscopy to the study of the electronic structure of molecules and discovered extraordinary resonance structures in absorption spectra and the systematic interpretation of the emission spectra from families of gas phase molecules with diverse structures.[5] This was the beginning of an almost 39 year career that was extremely varied and productive.

His subsequent research continued these interests progressively and centered on recurring themes of high precision metrology, x-ray spectroscopy, and development of novel experimental technology and devices. From 1980-1981, he was the director of the Division of Physics at the National Science Foundation. He was the Chief of the successful Quantum Metrology Division at NIST from 1987-1996, with numerous similar positions and titles in different divisions and sections of NIST, and was a Senior Fellow Emeritus on his retirement.

He was unstinting in his activity and support of national scientific bodies, and was a Fellow of the American Physical Society, Fellow of the American Association for the Advancement of Science, Fellow of the Washington Academy of Sciences, and a Member of SPIE. He became Chair, 1986-1987, of the American Physical Society, Division of Electron and Atomic Physics after extensive service; was on six different committees or panels of the National Academy of Sciences/National Research Council; Member, International Advisory Board, Journal of Physics B, 1992-2001; Member, Editorial Board for Physical Review A, 1999-2001; Member, Editorial Board of Review of Scientific Instruments, 2000-2003; Member at Large of the Section on Physics (B) for the American Association for the Advancement of Science, 1999-2001; Member, IUCr Commission on

Crystallographic Apparatus of the International Union of Crystallography, 1994-2002; and Fellow and Chartered Physicist, Institute of Physics.

He chaired or sat on many organising committees of many international and national conferences, and many workshops and related groups throughout his career, especially including seven International Conferences on the Physics of X-ray Spectra; CPEM; SPIE; International Nuclear Physics Conference, Wiesbaden, Germany, July 26 - August 1, 1992; International Committee, Nobel Symposium on "Heavy Ion Spectroscopy and QED Effects in Atomic Systems," Stockholm, Sweden, June 29 - July 3, 1992; Ninth International Conference on Vacuum Ultraviolet Radiation Physics (VUV9) 1989; and CODATA Task Group on Fundamental Constants.

He received many honours including the Department of Commerce Meritorious Service Award (Silver Medal), 1967; the Arthur S. Fleming Award, 1969; the NBS Samuel W. Stratton Award, 1974; the Department of Commerce Gold Medal, 1979; the Alexander von Humboldt Foundation Senior Scientist Award, 1983; the Presidential Rank Award, 1988; the SUN-AMCO Medal, 1990; and was awarded two patents.

He provided generic and friendly support of the whole field of precision measurement and fundamental constants, and of young international researchers in the fields. Many comments from colleagues at X2002 support and confirm this from many fields and sub-fields of physics and crystallography.

X-RAY AND OPTICAL INTERFEROMETER

The motivation for XROI came from the first X-ray interferometer by Bonse & Hart, then at Cornell, in 1965 [6]. Dick was excited by the opportunity here for a real calibration of X-ray spectra, and developed his first X-ray and Optical Interferometer (XROI) from 1968-1969 [7]. Dick's early metrology efforts at NIST were therefore directed in two lines of research toward the development of x-ray interferometry and the iodine-stabilized laser. [9] Dick produced the first combined x-ray and optical interferometer that was able to demonstrate the feasibility of accurate measurements of the lattice periods of silicon single crystals tied to the SI definition of the meter. A schematic of XROI is given in Fig. 2.

The key problem in the calibration of any research using X-rays is the link between the X-ray wavelengths and the optical definition of the metre. This is compounded by the broad natural widths of atomic lines. Consequently the interferometric methods, even when successful, cannot be used as a high-precision calibration of an X-ray standard wavelength. Instead Dick pursued the calibration of the lattice spacing in silicon, which is robust and transferable, via interferometry. The complexity here is indicated by figure 2 where the optical oscillation must be divided by 1600 to be at all sensitive to the X-ray wavelength. Conversely, the X-ray fringe counting must be perfectly linear along the same axis as the optical interferometer, and there can be no fringes lost or missed in 10000 if any accuracy in the final result is to be obtained.

Any who haven't seen XROI will find it hard to comprehend the sensitivity both for any reasonable measurement across such a large range of length scale, and also the sensitivity to systematics of every kind. XROI and NIST led the way in precision

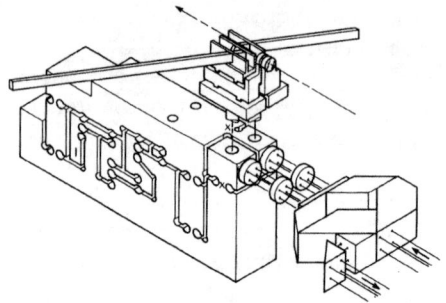

FIGURE 2. Schematic of the first X-ray and Optical Interferometer by Deslattes, Henins et al. Note the optical and X-ray paths must be perfectly aligned, and the motion of the second face is controlled and amplified by a complex but beautifully machined series of weak links.

Optical vs X-ray Fringes

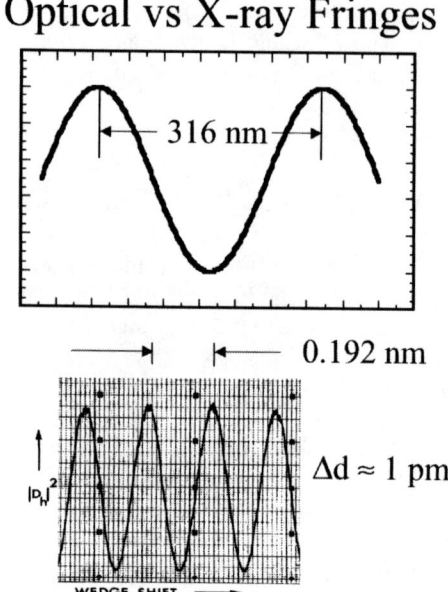

FIGURE 3. The precision needed for any useful link between the X-ray and optical wavelengths is demonstrated schematically in the above comparison between a visible interference wave and a corresponding X-ray period.

measurement of the wavelength transfer standard with the work from PTB, and with the resolution of the discrepancy between the two results which required the development of XROI2 to address. [8]

The room holding XROI is isolated from the basement of the physics building, and vibrations due to the laminar flow of air into the room are shielded by a series of physical covers and enclosures. Thermal vibrations are measured to high accuracy. The author had the pleasure of working on XROI2 for three years at the end of the 1980's and it is a little known fact that the NIST precision at that time had already exceeded that of the original XROI by over an order of magnitude. It is a mark of Dick's determination to achieve the best possible standard that this was not published at the time.

Silicon lattice spacings have sensitivities of 10^{-6} per degree or per atmosphere, and similar sensitivities to impurity concentrations. Hence for a precision transfer standard these variables must be carefully controlled and measured against a comparator to transfer the standards to other laboratories. The lattice comparator was developed by Kessler, Henins, and Deslattes over several years and was crucial in relating standards measurements of different laboratories and identifying artifacts due to the crystal samples in those measurements. [10] This was reported relatively recently, but had been developed and used with enormous success for many years prior to the dates of publication, in support of standards work and researchers around the world.

SYNCHROTRON DEVELOPMENTS AND PHYSICS

Dick was a pioneer in the use of intense synchrotron radiation for atomic physics studies. He and his colleagues showed the existence and schematic interpretation of the richly detailed supra-threshold absorption spectrum of atomic argon during parasitic operation at the Stanford Synchrotron Radiation Project. He subsequently pursued the design and optimization of primary monochromators and secondary spectrometers for such threshold studies using synchrotron radiation. [4] This early experience with synchrotron radiation led to the establishment of an innovative beamline, X24A, at the Brookhaven National Synchrotron Light Source. This facility has been an active source of new discoveries such as x-ray selection of oriented molecules and polarization spectroscopy. [1]

AVOGADRO'S CONSTANT, AND FUNDAMENTAL CONSTANTS

Early work led to a new methodology for density measurements based on solid objects, [11] and absolute isotopic abundance measurements to determine, for the first time, a value for the Avogadro constant with a defensible error budget near the 1 ppm level. [12]

Uncertainties for the (silicon) unit cell dimension a_0 fell from the previous standard of a ruled grating of 5×10^{-6} to 0.15×10^{-6} in the first round of XROI. Uncertainties in density ρ fell from the previous water standard of 5×10^{-6} to 0.7×10^{-6} in the first round of solid object density standards. Uncertainties of the mean molar mass M fell from the previous geochemical average (not sample specific) of 10×10^{-6} to 0.7×10^{-6} in the first round of calibration with mixtures of separated isotopes. All this was achieved in the 1970's at NIST, leading to an overall major improvement in the X-ray Crystal

Density (XRCD) approach to the determination of Avogardo's constant

$$N_A = \frac{nM}{\rho\,(a_0)^3}$$

which in turn is helping towards a possible redefinition of the kilogram as (for example) the mass of approximately 5.018×10^{25} free ^{12}C atoms at rest and in their ground state. The current status of this research remains quite complex and quite exciting, and Dick has continued to lead and develop the research towards these fundamental standards goals. The NIST work culminating in 1976 was developed over the years by other groups including PTB, IMGC and MNIJ but the error budget was not improved upon until 1995 in a large scale collaboration across the world's standard laboratories including CSIRO's National Measurement Laboratory with the silicon sphere's, which also involved major efforts and direction from Dick.

The iodine-stabilized laser research led by Deslattes included characterization of this potential new length standard and comparisons with infra-red radiation and the krypton length standard. This effort, along with contributions from other national measurement laboratories, led to the effective replacement of the krypton standard and the ultimate elimination of a separate wavelength standard through a redefinition of the meter in 1983 that fixed the speed of light.

GAMMA-RAY SPECTROSCOPY

Deslattes regarded the accurately-measured, nearly-perfect crystals as laboratory artifacts that were needed to measure x-ray and gamma-ray wavelengths, the real invarients of nature. This led to major efforts in precision angle measurements and to more accurate x-ray and gamma-ray wavelength standards consistent with the SI definition of the meter.

With Ernie Kessler and others, Dick extended optically-based wavelength measuring technology first to low energy gamma-rays at NBS, and later to higher energy gamma-rays at the Institut Laue-Langevin in Grenoble.[13] Already in its earliest phase this work showed that the then current gamma-ray standard was in error and that discrepancies between theory and experiment for high-Z muonic atoms were artifacts of this erroneous scale.

In 1983-1984, collaboration at the ILL developed into a world-class gamma-ray spectroscopy facility that is still active today with emphasis on nuclear and solid state physics as well as precision measurements. At the ILL Deslattes and his colleagues made an accurate measurement of the deuteron binding energy that led to an improved value for the neutron mass and extended the SI based gamma-ray wavelength measurements into the 6 MeV region.

Related work in this area led to new values for the mass of the neutron. [14] The improved standards also resolved discrepancies between theory and experiment for the mass of the negative kaon. [15]

The ILL team in Grenoble exploited the high resolution capabilities of the NIST-developed gamma-ray spectrometer with NIST to obtain sub-picosecond lifetime measurements for nuclear excited states and used the sharp lines from long-lived states to obtain accurate values for elemental scattering factors at gamma-ray energies. [16]

TESTS OF QUANTUM ELECTRO-DYNAMICS

Early work on spectrometers has already been mentioned, but particular value was found in the development of two-dimensional backgammon detectors for synchrotron and precision physics investigations. [17]

With this development but using a variety of techniques, Dick pursued high-resolution, high-precision spectroscopic X-ray tests of quantum electro-dynamics (QED). In collaborations with the University of Heidelburg and GSI in the early 1980's, Deslattes applied crystal-diffraction spectroscopy to spectra of highly-stripped ions produced by large nuclear accelerators and developed a scheme to reduce the large Doppler corrections associated with these spectra.

Initial investigations pursued recoil ions, [18] but it was observed that satellite contamination in these systems cast doubt on the profile fitting of complex spectra, and limited the final results.

The accel-decel method pioneered by himself and Mokler is being developed and pursued actively today for highly stripped ions produced by large accelerators, and has borne fruit in several related fields. It decelerated bare elemental nuclei prior to capture of electrons into excited few-electron states, leading to a clean single-interaction capture process and largely satellite-free spectra. Arguably the best results for X-ray tests of QED, as measured by the quoted error bars (12 ppm or 1.2×10^{-5}), were those first measurements with Dick, providing stringent tests of the theory of quantum electrodynamics for Argon and hydrogenic nickel. [18, 19, 20, 21]

The quality of these results attracted both John Schweppe from California and the author from Oxford to work with Dick at NIST. It was clear that the two most promising techniques in medium-Z QED tests at the time were those of the Lyman α - Balmer β intercomparison technique in Silver's group in Oxford, and the absolute calibration technique of Dick at NIST, both of which were in separate collaborations with GSI.

These investigations and these collaborations continued to produce exciting results and develop techniques at accelerators [22, 23], and have also led to major contributions to exotic atom spectroscopy, particularly aniprotonic hydrogen. [24]

It also led to a major series of efforts in X-ray spectroscopy and QED tests using the novel sources of electron-beam ion traps (EBITs). In this area the author's collaboration with Dick has been very fruitful, and has developed with John Gillaspy, Larry Hudson and other members of Dick's group over the past years. [25] These investigations continue to be pursued, providing additional new results on EBIT sources recently.

In the helium-like ions in particular, there remains a perplexing anomalous discrepancy from the traditional theory of Drake for these ions, [26] which invites further investigation.

PHOTOIONIZATION, EXAFS AND SCATTERING

Dick's early work on xenon and selenium on photoionisation [27] laid the groundwork for further developments linking up with synchrotron studies [4] and X-ray spectrometry [5] already mentioned.

However, the problem of the interaction of X-rays (light) with neutral matter continued to motivate his research and led to studies of extended X-ray anomalous fine structure. [28] This research simply expressed his command of other fields and his ability to make incisive contributions with ease.

The area of scattering investigations [29] was taken up and pursued further by group members such as Cowan, Levin and Southworth in references already cited and in their own work subsequently. This has been a rich field of endeavour and has led to complex and beautiful experiments.

NEUTRAL ATOMIC PHYSICS: X-RAY DIFFRACTION THEORY, ATOMIC FORM FACTORS AND CHARACTERISTIC ENERGIES

Such extensive X-ray spectrometry could not develop without continued and progressive developments in the understanding of theory in general and of X-ray diffraction theory in particular.

Dick contributed directly to some of the earlier developments and ideas, and always had keen intuition and insight in his encouragement and collaborations with later progress.[30]

Dick also enthusiastically supported the author's development of new form factor theory addressing some major problems in the X-ray regime. This is the subject of current development and has become a NIST database. [31]

In the early 1980s, Deslattes and his colleagues initiated a long-term study of the systematics of neutral-atom x-ray spectra that included comparison with progressively refined theoretical calculations. The progressively enhanced reference theoretical structures led finally to the possibility of a new, all-Z, x-ray wavelength database.[32] This now serves as a key reference for future calibration explorations.

APPLICATION OF X-RAY PHYSICS AND THEORY TO MAMMOGRAPHY AND MAJOR FACILITIES

Dick and his staff developed a number of novel experimental devices, techniques, and applications. One of these applications was directed toward a precision calibration device to permit radiologists to record better quality mammograms. We rapidly and effectively addressed the problem of critical high voltage measurement for quality control in mammographic x-ray radiology by application of diffraction spectrometry (US Patents 5295176, 5381458). [33] This was since extended to chest X-ray applications.

Hudson, Henins, Deslattes and others also provided major X-ray diagnostic equipment for NASA, and the mamographic developments led to the provision of major X-ray diagnostic equipment for one of the largest laser facilities (OMEGA and DOE). [34]

POWDER DIFFRACTOMETRY, GRAZING INCIDENCE REFLECTOMETRY, X-RAY MULTILAYER FABRICATION

With Jean-Louis Staudenmann and Larry Hudson, he initiated a complete reappraisal of powder diffraction standards in light of major anomalies in the characterisation and use of these standards by a wide crystallographic and mineral science community. Major new results have come out of this which address some key issues for this field. [35] Some key issues remain unresolved and invite further investigation.

Dick introduced new technology for the characterization of multi-layer optics in the x-ray region and with Joe Pedulla established a new, advanced technology facility for the production of such structures with a level of perfection not achieved elsewhere. [36]

CONCLUSION

Some select scientists in their career have published more than 165 papers in refereed journals, and a few have been cited as many as 281 times on selected publications (with several publications cited over 100 times); but few in the modern era have had the wide-ranging and fundamental impact across so many fields, in part due to his dedication to issues of fundamental and applied significance, and in part due to his encouragement of different scientists and different areas of study.

His contributions to physics will be cited for an extremely long time but his enormous energy, his ability to lead, and his remarkable creativity will be greatly missed.

ACKNOWLEDGMENTS

In preparing this summary I must explicitly acknowledge his obituary in Physics Today, 55, Jan. 2002, 71, with special thanks to E. G. Kessler, Jr and L.T. Hudson of NIST and M. Sanchez del Rio of the ESRF for supplying crucial materials and comments.

REFERENCES

1. Lindle, D. W., Cowan, P. L., La Villa, R. E., Jach, T., Deslattes, R. D., Karlin, B., Sheehy, J. A., Gil, T. J., Langhoff, P. W., Phys. Rev. Lett. (1988), 60(11), 1010-13; Lindle, D. W., Cowan, P. L., LaVilla, R. E., Jach, T., Deslattes, R. D., Perera, R. C. C., Karlin, B., J. Phys., Colloq. (1987), (C9, Vol. 1), C9-761/C9-763; Lindle, D. W., Cowan, P. L., Jach, T., LaVilla, R. E., Deslattes, R. D., Nucl. Instrum. Methods Phys. Res., Sect. B (1989), B40-B41(1), 257-61; Lindle, D. W., Cowan, P. L., Jach, T., La Villa, R. E., Deslattes, R. D., Perera, R. C. C., Phys. Rev. A (1991), 43(5), 2353-66; Perera, R. C. C., Cowan, P. L., Lindle, D. W., LaVilla, R. E., Jach, T., Deslattes, R. D., Phys. Rev. A (1991), 43(7), 3609-19
2. Deslattes, R. D., La Villa, R. E., Cowan, P. L., Henins, A., Phys. Rev. A (1983), 27(2), 923-33

3. Deslattes, Richard D., Torgesen, John L., Paretzkin, Boris; Horton, Avery T. Advan. X-Ray Anal. (1965), 8 315-24; Deslattes, Richard D., Torgesen, John L., Paretzkin, Boris; Horton, Avery T. J. Appl. Phys. (1966), 37(2), 541-8

4. Deslattes, Richard D., Simson, Bert G., Rev. Sci. Instr. (1966), 37(6), 753-5; Deslattes, Richard D., Simson, Bert G., LaVilla, Robert E. Rev. Sci. Instr. (1966), 37(5), 596-9; Deslattes, Richard D., Rev. Sci. Instrum. (1967), 38(5), 616-20; Deslattes, Richard D., Rev. Sci. Instrum. (1967), 38(6), 815-20; X-ray monochromators and resonators from single crystals. Deslattes, Richard D., Appl. Phys. Lett. (1968), 12(4), 133-5; Small set of reference crystals for double-crystal topography. Deslattes, Richard D., Paretzkin, B., J. Appl. Crystallogr. (1969), 2(Pt. 2), 81-2; Performance of a tunable secondary x-ray spectrometer. Brennan, S., Cowan, P. L., Deslattes, R. D., Henins, A., Lindle, D. W., Karlin, B. A., Rev. Sci. Instrum. (1989), 60(7, Pt. 2B), 2243-6

5. La Villa, Robert E., Deslattes, Richard D., J. Chem. Phys. (1966), 44(12), 4399-400; Deslattes, Richard D., La Villa, Robert E., Appl. Opt. (1967), 6(1), 39-42; Lavilla, R. E., Deslattes, R. D., J. Phys. (Paris), Colloq. (1971), (4), 160-4; Lavilla, R. E., Deslattes, R. D., Colloq. Int. Cent. Nat. Rech. Sci. (1971), No. 196 160-4.

6. Bonse, Hart, Applied Physics Letters 6, 1965, 155-156; Bonse, Hart, Zeitschrift f§r Physik 188, 1965, 154-164.

7. Deslattes, Richard D., Appl. Phys. Lett. 15(11), (1969), 386-8

8. Deslattes, Richard D., Henins, Albert, Phys. Rev. Lett. (1973), 31(16), 972-5; Deslattes, Richard D., Tanaka, Mitsuru; Greene, Geoffrey L., Henins, Albert; Kessler, Ernest G., Jr, IEEE Trans. Instrum. Meas. (1987), IM-36(2), 166-9

9. Schweitzer, W. G., Jr., Kessler, E. G., Jr., Deslattes, R. D., Layer, H. P., Whetstone, J. R., Appl. Opt. (1973), 12(12), 2927-38; Layer, H. P., Deslattes, R. D., Schweitzer, W. G., Jr., Appl. Opt. (1976), 15(3), 734-43

10. Kessler, E. G., Henins, A., Deslattes, R. D., Nielsen, L., Arif, M., J. Res. Natl. Inst. Stand. Technol. (1994), 99(1), 1-18; Kessler, Ernest G., Jr., Schweppe, John Edward; Deslattes, Richard D., IEEE Transactions on Instrumentation and Measurement (1997), 46(2), 551-555; Kessler, Ernest G., Jr., Owens, Scott M., Henins, Albert; Deslattes, Richard D., IEEE Transactions on Instrumentation and Measurement (1999), 48(2), 221-224

11. Deslattes, Richard D., Peiser, H. Steffen, Bearden, Joyce A., Thomsen, Metrologia (1966), 2(3), 104-11

12. Deslattes, R. D., Henins, A., Bowman, H. A., Schoonover, R. M., Carroll, C. L., Barnes, I. L., Machlan, L. A., Moore, L. J., Shields, W. R., Phys. Rev. Lett. (1974), 33(8), 463-6; Deslattes, R. D., Henins, A., Schoonover, R. M., Carroll, C. L., Bowman, H. A., Phys. Rev. Lett. (1976), 36(15), 898-900; Kessler, Ernest G., Jr., Schweppe, John Edward; Deslattes, Richard D., IEEE Transactions on Instrumentation and Measurement (1997), 46(2), 551-555

13. Kessler, E. G., Jr., Deslattes, R. D., Henins, A., Sauder, W. C., Phys. Rev. Lett. (1978), 40(3), 171-4; Kessler, E. G., Jr., Jacobs, L., Schwitz, W., Deslattes, R. D., Nucl. Instrum. Methods (1979), 160(3), 435-7; Deslattes, R. D., Kessler, E. G., Sauder, W. C., Henins, A., Ann. Phys. (N. Y.) (1980), 129(2), 378-434; Kessler, E. G., Greene, G. L., Dewey, M. S., Deslattes, R. D., Borner, H., Hoyler, F., J. Phys. G: Nucl. Phys. (1988), 14(Suppl.), S167-S174; Borner, H. G., Jolie, J., Hoyler, F., Robinson, S., Dewey, M. S., Greene, G., Kessler, E., Deslattes, R. D., Phys. Lett. B (1988), 215(1), 45-9

14. Greene, G. L., Kessler, E. G., Jr., Deslattes, R. D., Borner, H. Phys. Rev. Lett. (1986), 56(8), 819-22; Kessler, E. G. , Jr., Dewey, M. S., Deslattes, R. D., Henins, A., Borner, H. G., Jentschel, M., Doll, C., Lehmann, H., Physics Letters A (1999), 255(4-6), 221-229

15. Lum, G. K., Wiegand, C. E., Kessler, E. G., Jr., Deslattes, R. D., Jacobs, L., Schwitz, W., Seki, R., Phys. Rev. D (1981), 23(11), 2522-32.

16. Dewey, M. S., Kessler, E. G., Jr., Greene, G. L., Deslattes, R. D., Sacchetti, F., Petrillo, C., Freund, A., Borner, H. G., Robinson, S., Schillebeecks, P., Phys. Rev. B: Condens. Matter (1994), 50(5), 2800-8

17. Duval, B. P., Barth, J., Deslattes, R. D., Henins, A., Luther, G. G., Nucl. Instrum. Methods Phys. Res., Sect. A (1984), 222(1-2), 274-8

18. Deslattes, R. D., Beyer, H. F., Folkmann, F., J. Phys. B (1984), 17(21), L689-L694

19. Beyer, H. F., Mokler, P. H., Deslattes, R. D., Folkmann, F., Schartner, K. H., Z. Phys. A (1984), 318(2), 249-50; Beyer, H. F., Deslattes, R. D., Folkmann, F., LaVilla, R. E., J. Phys. B (1985), 18(2), 207-15

20. Richard, Patrick, Stockli, Martin, Deslattes, R. D., Cowan, P., LaVilla, R. E., Johnson, B., Jones, K., Meron, M., Mann, Rido, Schartner, K., Phys. Rev. A (1984), 29(5), 2939-42; Deslattes, R. D., Schuch, R., Justiniano, E., Phys. Rev. A (1985), 32(3), 1911-13

21. Beyer, H. F., Indelicato, P., Finlayson, K. D., Liesen, D., Deslattes, R. D., Phys. Rev. A (1991), 43(1), 223-7 [also GSI-Rep. (1990), (GSI 90-1), 143]

22. Beyer, H. F., Finlayson, K. D., Liesen, D., Indelicato, P., Chantler, C. T., Deslattes, R. D., Schweppe, J., Bosch, F., Jung, M., et al., J. Phys. B: At., Mol. Opt. Phys. (1993), 26(9), 1557-67

23. Suleiman, J., Berry, H. G., Dunford, R. W., Deslattes, R. D., Indelicato, P., Phys. Rev. A (1994), 49(1), 156-60

24. Borchert, G. L., Gotta, D., Schult, O. W. B., Simons, L. M., Elsener, K., Rashid, K., Reidy, J. J., Deslattes, R. D., Kessler, E. G., Mooney, T., Ettore Majorana Int. Sci. Ser.: Phys. Sci. (1990), 52(Electromagn. Cascade Chem. Exot. At.), 295-300

25. Paterson, D., Chantler, C. T., Tran, C. Q., Hudson, L. T., Serpa, F. G., Deslattes, R. D. Physica Scripta, T (1997), T73 400-402; Chantler, C. T., Paterson, D., Hudson, L. T., Serpa, F. G., Gillaspy, J. D., Deslattes, R. D. Physica Scripta, T (1997), T73, 87-89; Takacs, E., Meyer, E. S., Gillaspy, J. D., Roberts, J. R., Chantler, C. T., Hudson, L. T., Deslattes, R. D., Brown, C. M., Laming, J. M., et al. Phys. Rev. A: At., Mol., Opt. Phys. (1996), 54(2), 1342-1350; Gillaspy, J. D., Aglitskiy, Y., Bell, E. W., Brown, C. M., Chantler, C. T., Deslattes, R. D., Feldman, U., Hudson, L. T., Laming, J. M., et al., Phys. Scr., T59 (1995), 392-5.

26. Drake, G. W., Can. J. Phys. 66, (1988) 586

27. Photoionization of the M shell of xenon. Deslattes, Richard D., Phys. Rev. Lett. (1968), 20(10), 483 5; K absorption edge of selenium. Deslattes, Richard D., DeBen, Hillery S., Phys. Rev. (1959), 115 71-4; K-absorption fine structures of sulfur in gaseous SF6. La Villa, Robert E., Deslattes, Richard D., J. Chem. Phys. (1966), 44(12), 4399-400; Estimates of x-ray attenuation coefficients for the elements and their compounds. Deslattes, Richard D., Acta Cryst. A25 (1969), 89-93

28. Application of a high intensity laboratory x-ray source to EXAFS spectroscopy. Cohen, Gabrielle G., Deslattes, Richard D., Nucl. Instrum. Methods Phys. Res. (1982), 193(1-2), 33-9; Extended fine structure in x-ray absorption spectra of certain perovskites. Perel, Joseph; Deslattes, Richard D., Phys. Rev. B (1970), [3] 2(5), 1317-23

29. MacDonald, M. A., Southworth, S. H., Levin, J. C., Henins, A., Deslattes, R. D., LeBrun, T., Azuma, Y., Cowan, P. L., Karlin, B. A., Phys. Rev. A: At., Mol., Opt. Phys. (1995), 51(5), 3598-603

30. Crystal reflectivity for bent crystal spectrometers. Kaerts, E., Van Assche, P. H. M., Greene, G. L., Deslattes, R. D. Nucl. Instrum. Methods Phys. Res., Sect. A (1987), A256(2), 323-8; Chantler, C. T., Deslattes, R. D., Rev. Sci. Instrum. (1995), 66(11), 5123-47; Owens, S. M., Deslattes, R. D., Pedulla, J., Advances in X-Ray Analysis (2000), Volume Date 1999, 43 249-253; Owens, S. M., Deslattes, R. D., Pedulla, J., Advances in X-Ray Analysis (2000), Volume Date 1999, 43 254-259

31. C. T. Chantler, J. Phys. Chem. Ref. Data 24, 71 (1995); C. T. Chantler, J. Phys. Chem. Ref. Data, 29, 597 (2000); see also http://physics.nist.gov/PhysRefData/FFast/Text/cover.html

32. Systematics of x-ray transition energies for high-Z atoms. Deslattes, R. D., Kessler, E. G., Kim, Y. K., Indelicato, P., J. Phys., Colloq. (1987), (C9, Vol. 1), C9-591/C9-595; Mooney, T., Lindroth, E., Indelicato, P., Kessler, E. G., Jr., Deslattes, R. D., Phys. Rev. A (1992), 45(3), 1531-43; Schweppe, J., Deslattes, R. D., Mooney, T., Powell, C. J., J. Electron Spectrosc. Relat. Phenom. (1994), 67(3), 463-78

33. Deslattes RD, Levin JC, Walker MD, Henins A, Medical Physics 21 (1) (1994) 123-126; Chantler CT, Deslattes RD, Henins A, Hudson LT, British Journal of Radiology 69 (823) (1996) 636-649; Hudson LT, Deslattes RD, Henins A, Chantler CT, Kessler EG, Schweppe JE, Medical Physics 23 (10) (1996) 1659-1670

34. Hudson, L. T., Henins, A., Deslattes, R. D., Seely, J. F., Holland, G. E., Atkin, R., Marlin, L., Meyerhofer, D. D., Stoeckl, C., Review of Scientific Instruments (2002), 73(6), 2270-2275;

35. Deslattes, R. D., Staudenmann, J.-L., Hudson, L. T., Henins, A., Cline, J. P., Adv. X-Ray Anal. (1998), 40 221-231

36. Egelhoff, W. F., Jr., Chen, P. J., McMichael, R. D., Powell, C. J., Deslattes, R. D., Serpa, F. G., Gomez, R. D., Journal of Applied Physics (2001), 89(9), 5209-5214; Prudnikov, I. R., Matyi, R. J., Deslattes, R. D., Journal of Applied Physics (2001), 90(7), 3338-3346

Yvette Cauchois and her contribution to X-ray and inner-shell ionization processes

François J. Wuilleumier

Laboratoire d'Interaction du Rayonnement Avec la Matière (LIXAM, ex-LSAI), Paris-Sud University, B. 350, 91405-Orsay, France

Abstract. At the end of 1999, Mademoiselle Yvette Cauchois passed away. For over 50 years, she has contributed in a profound way to our understanding of x-ray physics and chemistry. The main aspects of her accomplishments will be briefly outlined.

INTRODUCTION

Born on December 1908, Yvette Cauchois received her bachelor's degree in physics in 1928. Then she was admitted in the Physical Chemistry Laboratory of the Unversity of Paris "La Sorbonne", whose director was Jean Perrin, who had win the Nobel Prize in 1926. Her thesis was dedicated to the invention, the design and the use of a new type of x-ray spectrograh, i. e., the bent crystal transmission spectrograph. After having defended successfully her PhD in 1933, she became research associate at the Centre National de la Recherche Scientifique, a newly organization fonded by Jean Perrin. She dedicated the next ten years of her life to a thorough investigation of these x-ray emission lines whose energies did not fit into the energy diagram of x-ray levels, the so-called satellites or non-diagram lines. In 1947, together with H. Hulubei, she published an exhaustive table summarizing the status of the experimental data available on the wavelengths of X-ray emission lines and absorption edges. She became full professor at La Sorbonne in 1951, and was apppointed as Director of the Physical Chemistry Laboratory in 1953. At this position, she deeply influenced generations of young students and researchers. As the "grand patron", she was able to develop in many directions new program of research not only in x-ray physics but also in various fields of physical chemistry. She continued to be personnaly involded in research, especially in the studies of the influence of chemical bonds on photoabsorption and x-ray emission processes. Early in Europe, she was the first to understand that the newly discovered synchrotron radiation would become a major tool of investigation for fundamental and applied research. Under her vigorous leadership, several of her collaborators, such as Christiane Bonnelle and Pierre Jaegle, were among the first to begin to use, in 1963, the synchrotron light emitted by the Frascati electron synchrotron

CP652, *X-Ray and Inner-Shell Processes: 19th International Conference on X-Ray and Inner-Shell Processes*
edited by A. Bianconi, A. Marcelli, and N. L. Saini
© 2003 American Institute of Physics 0-7354-0111-X/03/$20.00

in Italy. She founded a new laboratory dedicated to physical chemistry in Orsay, twenty miles south of Paris, in

a place that would become later the University Paris-Sud (Paris XI). She was involved in the organization of many scientific activities, such as X-70 in Paris in 1970. She retired in 1978 and was appointed as emeritus professor. She died in Romania in 1999. Before trying to illustrate the various aspects of her scientific life, I would like to present a nice portrait, shown in Figure 1. It was taken in 1987, when she received the gold metal of University Paris VI. I think that many of her collaborators and students remember her exactly as she is looking in this picture.

FIGURE 1. A photograph of Yvette Cauchois taken when she received the Gold Medal of University Paris VI in 1987 ; on the wall, a painting of the Cardinal de Richelieu by Philippe de Champaigne (1602-1674).

THE FIRST YEARS: A NEW X-RAY SPECTROGRAPH

During the second decade of the twentieth century, x-ray spectroscopy was one of the major developing technique helping to better understand the atomic structure and the interaction processes occuring between radiation and atoms. Born in 1912 from the discoveries of Laue [1] and Bragg [2], the use of this new spectroscopic method grew rapidly in many laboratories, and was continuously improved by people such as Manne Siegbahn and L. G. Parratt. The first extensive x-ray data base was established by Siegbahn [3] in compiling all available measurements of absorption edges and X-ray emission lines. Owing to Bohr [4] atomic theory and further developments by Moseley [5], it was rapidly understood that characteristic X-ray emission lines originate from the radiative decay of electronically excited states in atomic inner-shells. The electronic states involved as initial and final states of a process, either absorption or emission process, were called X-ray levels. The X-ray lines emitted after the removal of a single electron from the electronic configuration of the neutral ground state were called X-ray diagram lines, while the term of satellite lines or non-diagram lines was reserved for any line that did not fit into the X-ray energy level. A quasi-universally adopted notation was proposed by Siegbahn. As an example, the diagram lines emitted in a transition of an atomic electron belonging to either the L_{III}- or M_V-subshells into a vacancy in the K- or L_{III} subshell were called $K\alpha_1$ and $L\alpha_1$ lines, respectively.

What was the status of the experimental techniques at the beginning of 1930? X-rays are produced by the interaction of charged particles with an electromagnetic field. At the time where our story begins, the almost exclusive way to produce X-rays on earth was to bombard a solid target by fast electrons, thus producing a continuous X-ray spectrum called bremsstrahlung, and a number of discrete emission lines characteristic of the target material. The decay of radioactive isotopes was, sometimes, used for the calibration of X-rays detectors. But X-rays from synchrotron radiation sources of from highly-charged laboratory plasmas were only part of a far and unknown future. The photon energy range accessible for spectrographic studies was the hard x-ray region, i.e., with x-ray wavelengths below 1500 xu or 1.5 angström. This wavelength border limits roughly the part of the spectrum which is not absorbed by the air. Above 2000 xu, the radiation is absorbed by the air and the study of these so-called soft x-rays necessitates the whole spectrograph, including the detector, to be placed in good vacuum.

The most simple spectrograph built for X-ray spectroscopy in this early stage was equipped with a plane crystal. Based on the well known relation nl = 2d sinθ established by Bragg [2], it makes use of a flat crystal to diffract the x-rays and of a photographic plate to record the monochromatized spectrum. The selective reflection of X-rays against the atomic planes of the crystal is comparable to the reflection of a beam of visible light at the surface of a plane mirror. It is possible to use this selective reflection without the help of any slit, but the resolution is very small in this case. Thus, it was soon realized that it was necessary to limit the size of the incident x-ray beam. In the first X-ray spectrographs [6], there was a fixed entrance slit. The crystal was rotated continuously about an axis parallel to the slit. During the rotation, the various monochromatic diffracted beams cross a definite point of a circle whose center coincides with the center of rotation of the crystal and which crosses the slit. Later instruments were derived from this system, making used of one or two slits to limit the X-ray beam. In fact, in a plane crystal spectrograph, only one slit is necessary to get a good as resolution as is possible. There are mainly two possible positions for this slit. It can either be placed in front, or behind the crystal. In case where an edge is placed in the middle of the crystal itself, the edge and its image in the crystal constitute the slit.

With a slit placed in front of the crystal, like in the case of the de Broglie spectrometer [7], an X-ray spectrum is registered all around the circle previously described. For this purpose, the crystal has to be rotated with its axis of rotation perpendicular to the plane of the circle and through its centre. Rays entering through the slit in a divergent beam strike the rotating crystal at the glancing angle corresponding to their wavelength. The use of the rotation was favorable also from the point of view that the crystals were not always of the highest quality. Spectrographs with a slit in front of the crystal have been used through the whole X-ray region [8]. When hard X-rays are concerned, the slit has to be made of a very absorbent material, such as gold or lead. One example of spectrograph often used over the world was designed by Siegbahn [9]. Photographic registration was used at the beginning rather than ionization chamber or counter detectors. Several methods were developed to measuring very accurately the Bragg angle, by reducing the adjustment errors. Errors due to the penetration of the radiation into a not perfect crystal, however, can never be completely eliminated, but at very long wavelengths. When the slit is placed behind the crystal, a diaphragm in front of the crystal limits the width of the beam from the X-ray tube, and the slit itself is placed on the other side of the crystal at the same distance from the crystal center. The diffracted spectrum is registered on a photographic plate far from the slit. Lead plates must be used to prevent the direct beam from penetrating the slit. A good example of this tube spectrometer was built by Siegbahn and Larsson [10] in 1925. The need to limit the incident x-ray beam by a slit reduces considerably the intensity available on the photographic plate. The experimentalists have always to find the best compromise between resolving power and intensity. Several new systems were proposed around 1930 to increase both factors. Higher resolution was needed to better resolve x-ray lines very close in energy, higher intensities were strongly required to study weak and unexplained lines appearing frequently on the high energy side of the diagram lines.

The principle of the two-crystal spectrometer was to use two consecutive crystals to diffract and measure the X-ray spectra. With such a device, photographic recording was excluded. In the early time, ionization chambers served to measure the intensity of the X-ray spectra, allowing in fact a better comparison of the relative intensities of weak and strong x-ray lines than the photographic plate. Counters were introduced later ad provided more reliable intensity data. First proposed by Ehrenberg and Mark [11] in 1927, the idea to have the X-ray beam reflected by two crystals raised may hopes. In the so-called antiparallel position, the dispersion and the resolving power was expected to be twice that of a single crystal instrument. The resolution does not depend anymore on the width of the slit, only the height of the slits plays a significant role. Even though this expected goal was never be reached because of the limiting factor introduced by the diffraction pattern of each crystal, the improved resolving power of this instrument helped greatly to determine, with a better accuracy, the energies of many x-ray lines, diagram or non-diagram lines, and to observe weak lines immerged in the tail of a neighborhood intense X-ray line. The price to pay for this improved resolution was the lost of intensity in the recorded diffracted spectrum and the difficulty to measure very weak lines. Early examples of two-crystal spectrometers were built by Du Mond [12], Allisson [13], and Parratt [14].

A fruitful idea to increase considerably the luminosity of x-ray spectrographs was to concentrate the X-ray beam in the direction perpendicular to that of the dispersion. Following an early suggestion by Gouy [15], Johann [16] was the first to build an instrument with a bent crystal to produce by reflection a monochromatic x-ray beams from widely divergent incident x-ray radiation. A scheme of his spectrograph is shown in the left part of Figure 2. The crystal is bent to a radius twice that of the focussing circle. A divergent X-ray beam is reflected by the system of atomic planes parallel to the curved surface. The X-rays originating from a focus spot placed on the circle are reflected into a fairly sharp line on the same circle. No slit is needed to reduce the divergence of the emitted X-ray beam. The focussing of the reflected beam is not perfect, however, since the image of the source formed by reflection on the surface of the crystal has some extension along the focussing circle. The focussing defect increases with the distance from the point of reflection to the center of the crystal. Thus, the line shape is asymetric and depends on the aperture of the crystal. It decreases with increasing values of the Bragg angle. For a given reticular distance of atomic planes, the highest is the Bragg angle, the highest is the wavelength of the reflected x-rays. The instrument is the most efficient for longer wavelengths, namely in the soft x-ray range. The luminosity is significantly enhanced because of the convergence of the reflected rays. Later, Johansson [17] improved the focussing of the instrument by grounding the crystal before curvature, making possible, at least theoretically, to get an exact focusing

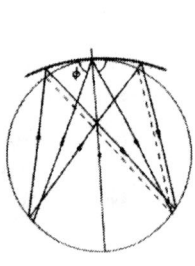 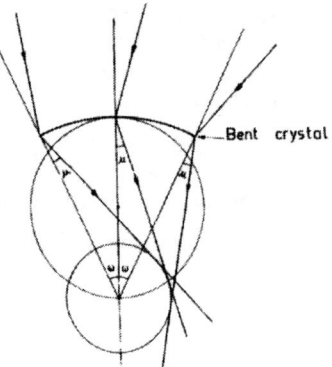

FIGURE 2. Schemes of the bent crystal spectrographs working by reflection (left part) and by transmission (right part) (from Ref. 20).

34

To work in the hard x-ray range, typically above 8 keV photon energy, Y. Cauchois [18, 19] proposed an alternative way to increase the luminosity of the spectrographs. She suggested to use the transmission of a widely divergent ray beam through a bent crystal. The scheme of this new instrument is shown in the right part of Figure 2. The rays are incident on the convex side of the crystal and are focussed, on the concave side. The selective reflection occurs on reticular planes whose orientation is perpendicular or oblique relative to the surface of the bent crystal. In the instrument built by Y. Cauchois, the diffracted x-ray spectrum was registered on a photographic plate placed along the focusing circle. In her thesis [19], she described in details the theory of the diffraction of the x-ray beams in this spectrograph in which all lines transmitted by the crystal are located on a circle which is tangent to the crystal circle and whose diameter is equal to the radius of the crystal. More exact formulae were derived later [20]

In her first spectrograph, Y. Cauchois chose R = 20 cm, with gypsum and mica crystals. Thus, the dispersion in first order was in the order of 25 xu/mm and the resolving power was better than 1 xu. The instrument could be used over the photon energy range extending above 6 keV. In practice, the resolving power of a bent crystal should be slightly inferior to that of a one crystal spectrometer, because of the mosaic effect [21], but the luminosity is considerably enhanced by a factor of about 100 as compared to a Bragg spectrometer [19]. Thus, the use of a bent crystal spectrometer is highly recommended to measure faint x-ray lines of very weak intensity.

Y. Cauchois made use of her instrument to study the x-ray emission lines for atoms in the gas phase, while most of the existing measurements had been made on solid target for evident reasons of intensity. She already had the idea that it should be possible to follow the influence of the chemical bond through a large number of various compounds of a given Z-element. To start with, she decided to measure the K-emission spectrum of krypton. To achieve her goal, she had a high-power fast-electron tube built in her laboratory. Combining its use with the high luminosity of her spectrograph, she succeeded to measure the diagram K-emission lines of krypton [19, 22], as shown in Figure 3.

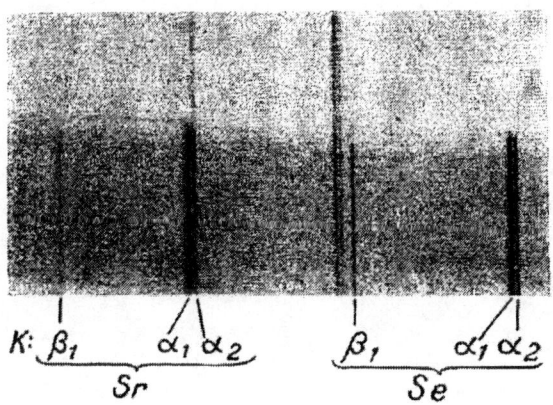

FIGURE 3. K-X-ray emission spectrum of krypton (from Ref. 22)

The geometrical width of the spectral line on the photographic plate was between 0.03 and 0.05 mm, providing an accuracy of ± 0. 25 xu. The $K\beta_1$ and $K\alpha_1\alpha_2$ lines of Sr and Se served as reference lines. In Table I are shown the results of her measurements. The values measured for Kr fit rather well within the Moseley diagram showing the square root of the energy linear as a function of the atomic number Z. In the last column of Table 1, I like to show also the results of the most recent and most accurate experiment carried out about 60 years later [23]. Since these new measurements were referred directly to the angström scale, I had to convert the values with the conversion factor taken equal to 1.00209(1) kxu/A according to the values quoted by R. D. Deslattes et al. [24]. In the latest experiment, only the $K\beta_2$ line was not re-measured. The accuracy is evidently greatly improved, by more than one order of magnitude, but the difference between the values measured by Y. Cauchois in 1932 and 1933 and the latest results are well within the error bars she quoted, demonstrating, if needed, the high quality of her early measurements.

Table I. Wavelengths of the diagram K-emission lines in krypton.

Transition	Cauchois kxu (1933)	1999 value [Ref.24] kxu
$K\alpha_2$	982.10 (25)	982.306 (14)
$K\alpha_1$	978.10 (25)	978.223 (14)
$K\beta_1$	876.70 (25)	876.690 (14)
$K\beta_3$	not resolved	877.178 (16)
$K\beta_2$	864.30 (25)	864.30 (16)

A PERIOD OF INTENSE RESEARCH ACTIVITY : THE SATELLITE STORY

During the 10-12 years following her PD, until 1945 when she became associate professor at La Sorbonne, Y. Cauchois was a full time research scientist, free of any teaching duty, thanks to her position at the Centre National de la Recherche Scientifique. She could then be fully dedicated to research. Having in hands the beautiful tool she has created, she could explore various aspects of x-ray physics and the application of x-ray spectroscopy in chemistry. A field which required as high as possible a luminosity was the study of non-diagram lines, also called satellite lines. Another goal she has in mind was the compilation of all data available for the x-ray emission lines and absorption edges, with the special care of defining as best as possible the chemical and physical states of the measured elements, as well as the accurate identification of the observed transitions.

What was, known about x-ray satellite lines in 1935 ? They had been discovered by Siegbahn and Stenstrom [25] in 1916 when they observed, on the high-energy side of the K $\alpha_{1,2}$ doublet of Zn (Z = 30), a weak line whose energy did not fit into the X-ray energy level diagram. This faint x-ray emission line was becoming more prominent for elements of lower Z, until Na (Z = 11). It was perhaps fortunate, as mentioned by Richtmeyer [26] years after, that the spectrographs available in this early time did not have high sensitivity and resolving power. The diagram x-ray lines were the most intense and the most easily resolved lines in the x-ray spectra. These lines were well represented by the energy level diagram of singly ionized atoms and constituted a firm basis to go further into the interpretation of additional emission lines. Thanks to the improvements in instrumental techniques, many more faint lines were subsequently discovered, and did not fit into this diagram. Thus, they were called non-diagram or, preferably, satellite lines.

Within a few years, it became evident that the more experiments were performed, the more was the number of discovered lines which were not diagram lines. Wentzel [27-29] proposed a theory of the origin of satellite lines, suggesting that they were originating from transitions in multiply-ionized atoms. As an example, the K α_4 satellite line was supposed to arise from a KK $\rightarrow$ KL transition where KK refers to a doubly ionized atomic state in which the two K-electrons are missing. The probability of having two electrons ejected from an atom by successive impacts of the electrons in the target of an x-ray tube was much too low to account for the observed satellite intensities [30]. The double ionization process had to occur by a single impact, i. e., by simultaneous ionization of two electrons. The original Wentzel theory was questioned by the determinations of the excitation voltages of satellites [31, 32]. The measured values for the K-satellite lines gave values considerably less than twice that of the parent diagram lines. With a wealth of new data becoming available, Druyvesteyn [33-35] modified the Wenzel's theory by proposing that the initial state of the atom giving rise to satellite lines is one in which the double ionization never refers to the same atomic shell. In the now so-called Wentzel-Druyvesteyn theory, the main satellite lines of the K α_1 diagram line are produced by radiative decay in a KL $\rightarrow$ LL transition. The excitation potential should be approximately equivalent to the total energy required to remove the K-electron of the atom Z plus the energy required to remove an L-electron of atom (Z + 1). The x-ray transition producing the emission of the satellites is a single-jump transition. (We know now that, in the double ionization process, two electrons can be ejected simultaneously from the same shell, according the shake off theory which was developed 30 years after the time where the case of satellite lines was a hot topic. Even double K-electron ejection is possible and gives rise to the hyper-satellite lines, a subject well documented by recent theoretical and experimental works [36]). Further improved studies [37, 38] of the excitation potential needed to produce satellite lines of either the parent K- or L- diagram lines were in good agreement with the Wentzel-Druyvesteyn theory.

Keeping the idea that x-ray satellites have their origin in x-ray transitions within multiply-ionized atoms, Richtmeyer [39] proposed that the satellite emission were resulting from a two-electron transition. Assuming that the initial doubly ionized state is a type in which one of the initial ionizations occurs in an inner shell, and the other not far from the outermost filled shell, he suggested that the two electrons jumping simultaneously into the two vacancies produce the emission of one satellite whose frequency should be equal to the sum of the frequencies of the two transitions taken separately. This two-jump theory explained the low excitation potentials of the satellite lines, while the variation of the relative intensities of the most intense K-satellites were found in agreement with the one jump theory. Both theories were able to explain the measured energies of the satellite lines. The key point, however, was the comparison of the intensities of the satellites relative to the parent lines and their variation with the atomic number Z.

At the end of 1934, the number of discovered satellites was still increasing rapidly, to the point that the number of known satellites was exceeding the total number of diagram lines. Most of the data available to date had been obtained using the Siegbahn vacuum type spectrometer and photographic detection. The use of the high luminosity Cauchois spectrometer and of the two-crystal spectrometer with ionization chambers or counters was barely starting and many more data were still to come. With the low luminosity Siegbahn spectrograph and the photographic detection, only rough estimates of the intensities of satellites had been made. To bring out the satellites on a photographic plate required that the parent lines were usually much overexposed, so that the direct comparison of intensities was difficult. In addition, many of the satellites were lying in the shadow of the parent line, the shape of which had to be estimated in order to make an even semi-quantitative determination of the relative intensities. In summary, the relative intensities of the most intense Ka satellites had been measured for Z between 13 and 29. Calculations of the K-L ionization probability, based on the Born approximation of collision theory, and using self-consistent-field wave functions for sodium [40] and potassium [41] atoms predicted variation of the relative intensities for Ka satellites in good agreement with the measurements. Extension of these calculations from Z = 17 to Z = 29 gave relative intensities whose Z-dependence was in excellent agreement with the measured data. These results supported the Wentzel-Druyvesteyn interpretation of the Ka satellites. For the Lα and Lβ satellites the answer was raising puzzling questions. In the low and intermediate Z-range (37 to 52), the ratio of the intensities of the satellites to that of the parent line was shown [42, 43) to rise from a few percent for Z = 40 to a large maximum (nearly 50 percent) in the neighbourhood of Z = 45 (for Lα) and 47 (for Lβ), and to fall rapidly to zero for Z = 52. For Z = 53 and above, no satellites could be observed until they reappear for Z = 73 to 90, with a significant intensity of 5 to 6 percent [44]. The L-satellites observed for the heavy elements seemed also to be different from the ones measured below Z = 52. These results were in full disagreement with the one jump theory which predicted a continuous decrease in satellite intensities with increasing Z atomic number. It seemed impossible to correlate these rapid changes of the relative intensities with any know change in the atomic electron configuration .

Seen from far away in the future, it is surprizing that it took so much time to understand the behavior of the L-satellite lines. Ten years before, Auger [45, 46] had discovered the non-radiative decay of core-hole ionized atomic states by observing

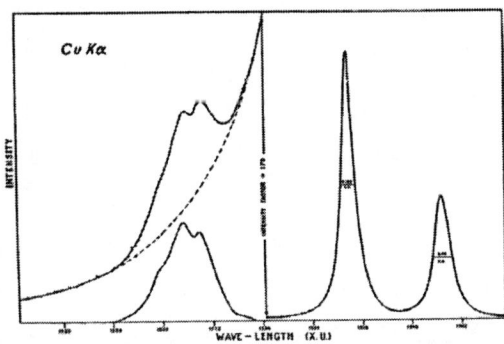

FIGURE 4. The Kα doublet of Cu (right part) with accompanying satellites (left part, intensity scale increased 170 times) as measured with a two-crystal spectrometer (from Ref. 50).

short tracks in a Wilson cloud chamber ionized by X-rays. Coster and Kronig [47] established that some doubly ionized atomic states were arising from a so-called internal photoelectric effect, redistributing the vacancies within the L-subshells. Assuming that an L$_I$ electron is missing, a L$_{III}$-electron can drop into the vacancy, freeing an energy equal to the differences between both binding energies. This energy is then used to ionize an outer electron provided that the energy released is higher than the binding energy of this electron. Between Z = 52 and Z = 73, the energy released by Coster Kronig transitions was not sufficient to expel M$_{IV}$ or M$_V$ electrons. This interpretation was in complete agreement with all experimental results and brought a strong support to the Wentzel-Druyvesteyn theory. The puzzling problem risen by the widths of K-, L-, and M-x-ray lines was also illuminated by consideration of the Coster-Kronig transitions. From 1935, the way was then paved for an accurate interpretation of more and more experimental data.

To make the results on x-ray satellites fully quantitative, more accurate data were needed. Some people, such as Paratt [48, 49], chose to use systematically the two-crystal spectrometer to make profit of the high-resolving power of the instrument and of the linearity in the dynamical response of the detection system. Soon new satellites were measured. It was noted [26] that and atom, which is capable of being singly ionized in 16 different ways (K to N shells), can be doubly ionized in 120 different ways. As an example of the high quality of the data obtained with the two-crystal spectrometer, I show in Figures 4 and 5 the K$\alpha_{3,4}$ satellites of copper and the Lα satellites of silver, respectively. In the Kα spectrum, there appear to be at least four components in the satellite group. In the La spectrum recorded with a resolving power of 11 000, the group of satellites near the Lα_1 diagram line is made of at least five components (21 satellite lines were measured in total). The difficulty in tracing accurately the background on the tail of the parent line, still maintains some significant uncertainties on the relative intensity of the satellites, even with such a high resolution instrument.

Cauchois decided to use her high luminosity apparatus in order to explore known and unknown K- and L-satellites. During the 10 following years, her activity was mainly devoted to measurements and identifications of L-satellite spectra in heavy

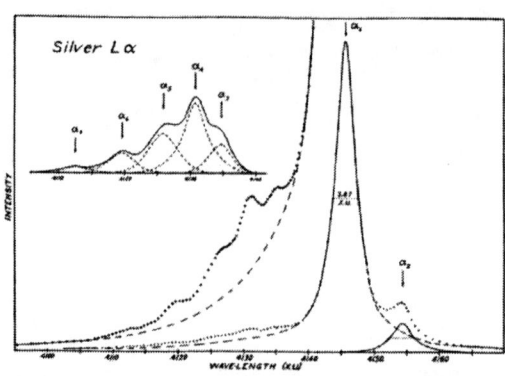

FIGURE 5. Intensity curve of the Ag Lα region. Five component satellites lines are sketched in the left upper part of the figure with the intensity scale increased 6 times (from Ref. 51).

elements [52], from W (Z = 74) to thorium (Z = 90), including some radiactive elements : As [53], Po [54] and Np [55]. She also studied in details the K-absorption spectrum of lighter elements [56], below Se (Z = 34). Over a period of ten years, she published not less than forty papers. To illustrate the results she obtained, I show in Figure 6 the L-x-ray emission spectra of gold (Z = 79) and W. (Z = 74), registered with a mica (201) crystal bent under a radius of 40 cm . The plates were overexposed to bring out the faintest L-satellites. She found many new lines in the Lα and Lβ spectra, with some of them being possibly attributed to a double electron jump (such as $L\alpha_a$ and $L\alpha_s$). In particular, she demonstrated the existence of a new series of satellites correlated with the $L\beta_5$ diagram lines ($L_{III} \rightarrow O_{IV, V}$), made at least of two components and relatively intense in lead and gold. A comprehensive summary of her results was published in 1944 [52]. From the data she accumulated, she was able to propose the first table of the energy levels of doubly ionized atoms.

One of her goal was to compile and publish the results of all wave lengths measurements for diagram and satellite x-ray lines in a widely available table. She wanted also to include the values of the x-ray absorption edges in order to have a consistent set of data for the binding energies of all atomic electrons. To select the data, she took a great care of the physical and chemical state in which was the measured elements. The completion of this additional part rose additional problems. The main question was, at a time where there did not exist accurate data from low-energy electron spectrometry, what does one actually measure when studying an X-ray absorption edge? The answer is quite different, depending on the status of the element under investigation. For an atomic gas (rare gases, metallic vapors), the answer is relatively simple. It is well illustrated by the X-ray absorption spectrum of argon measured by Parratt at the K-absorption edge [57] which is shown in Figure 7. Recorded with a two-crystal spectrometer, the measured absorption curve was analyzed in terms of the resonant $1s \rightarrow np$ (n > 3) absorption lines. The data were corrected for the finite resolving power of he spectrometer. The resonant structures observed below

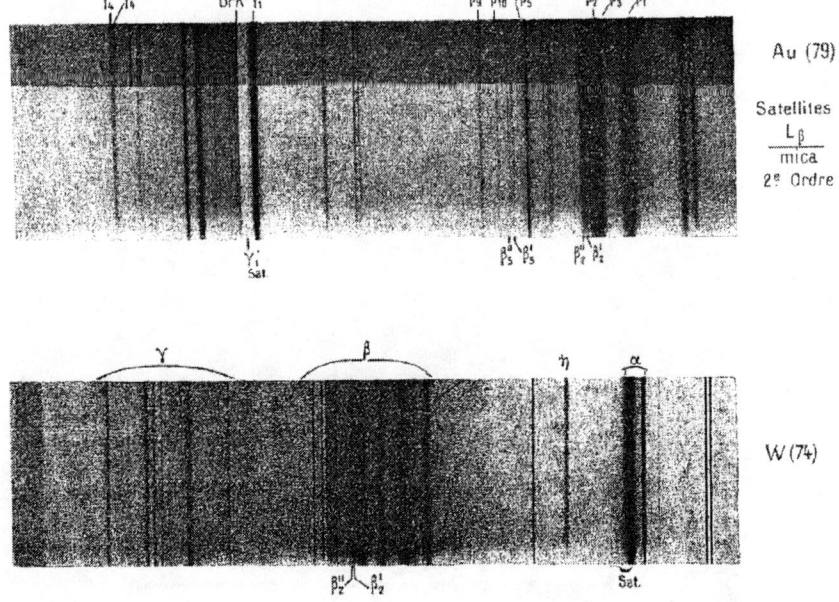

FIGURE 6. L spectrum of gold (upper panel) and tungsten (lower panel) showing the overexposed diagram lines and many of the weaker satellites (from Ref. 52]

the K-edge (located at 3866 kxu) were supposed to be of equal width (0.58 eV, equal to the natural width of the K-hole state) as given by the width of the fully resolved first resonant line. The data were analyzed assuming that the splitting between the np levels is equal to the separation of the optical terms of potassium. The theoretical shape of the main edge is represented by an arctangent curve. The analysis places the main edge (transfer of the 1s-electron to infinity with zero energy) at 3.5 to 4 xu below the wavelength of the maximum absorption. Nothing is remarkable in the spectrum at this wavelength of the true edge value which it would not be possible to determine without being able to decompose the structure of the absorption edge. The situation would then be more difficult to analyze if the resolving power would be worse.

In the case of a metal, shown in Figure 8, the lowest energy transition in the photoabsorption process corresponds to the transfer of the K-electron to the bottom of the empty part of the conduction band. Thus, there is no narrow absorption line, but a broad absorption band at the K-edge. How to compare the energy of the electron excited in the conduction band with the binding energy as defined above? If the elements under investigation are available only in different chemical states, the actually measured binding energy of the electrons can be spread over several electron volts.

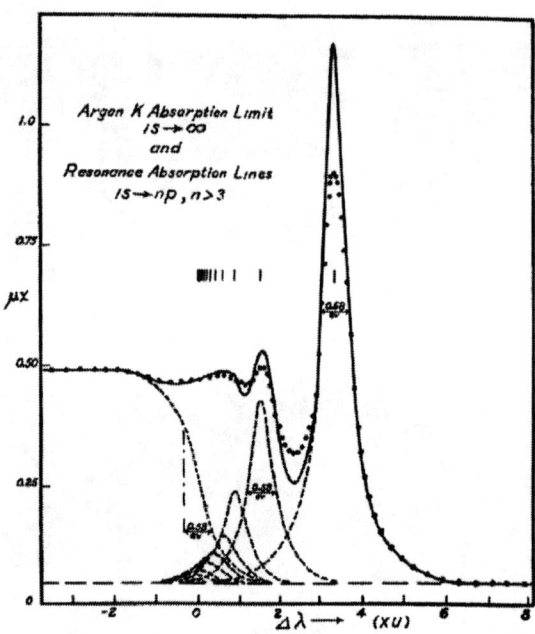

FIGURE 7. The K-absorption edge of argon. The broken curves represent the main edge and resonance absorption lines (from Ref. 57).

Finally, Cauchois published, together with H. Hulubei, their Table des Longueurs d'Onde des Emissions X et des Discontinuités d'Absorption X [58]. When available, the selected values were chosen from the works carried out in her laboratory. This table was the first one available. Let's mention that a new edition was published in 1978 with the help of Christiane Senemaud.

Cauchois established later reliable data for the binding energies of most elements. She published two separate Tables of these energy levels [59, 60] covering a large part of the Mendeliev Table.

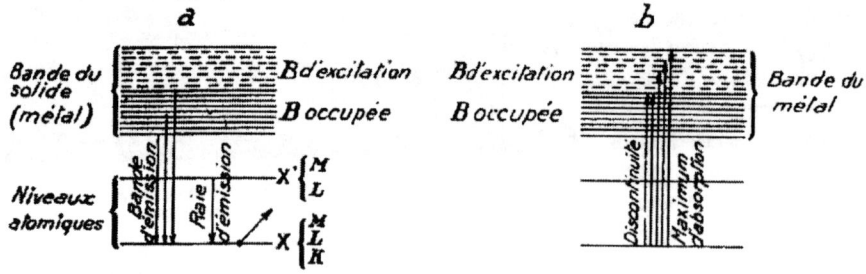

FIGURE 8. Scheme of absorption and emission processes occurring in a metal.

THE PROFESSOR AND THE DIRECTOR OF A LABORATORY

Y. Cauchois was appointed at the University of Paris "La Sorbonne" as an associate professor in 1945, and as a full professor in 1953. She became the director of the Physical Chemistry Laboratory of the University of Paris in 1953. She was succeeding Edmond Bauer who was the director of this laboratory after the death of his founder Jean Përrin in 1926. In such a position, she had to considerably diversify her activities as it is the duty of the director of any laboratory of significant size, and as a worldwide recognized leader in X-ray spectroscopy. As a professor she was teaching undergraduate and graduate students. Her graduate school was well attended by many students who prepared their PhD in her laboratory or in an associate institution. To help the students, she wrote various university books dedicated to X-ray scattering, X-ray spectroscopy, X-rays in Chemistry, X-ray absorption elementary particles, all of them being published by the Centre de Documentation Universitaire de Paris [61]. She wrote several books for researchers and students such as "Les spectres de rayons X et la structure électronique de la matière [62], Atomes, Spectres, Matière [63], Cheminement des particules chargées [64] ".

As the leader of a scientific school, she continued to push the development of x-ray studies into new directions. One important new field was the exploration of the soft and ultra-soft photon energy ranges. Emission and absorption processes in atomic gases as well as in many chemical compounds were among her favored studies. She had built in 1945 a so-called universal spectrometer [65] which could be used either in transmission or in reflection. Several other instruments were developed and built in her laboratory by collaborators such as C. Bonnelle and P. Jaeglé, to continuously cover the range between the hard X-rays and the VUV region. She was personally involved in several areas: the chemical shift in X-ray absorption and emission processes, the transuranium elements, the extraterrestrial X-ray radiation, together with many people who became her collaborators after having been his students. I would like to recall here the names of C. Bonnelle and Y. Héno, who were her main collaborators (C. Bonnelle succeeded her as the director of the Physical Chemistry Laboratory when she retired in 1978), P. Jaeglé who became the first director of the Laboratoire de Spectroscopie Atomique et Ionique in Orsay at the University Paris XI, the X-UV part of the laboratory founded by Y. Cauchois in Orsay, C. Senemaud, a bright scientist, director of research in her laboratory who died prematurely in 1997, A. Maquet and C. Hague two very active scientists who are presently director and deputy director of the Physical Chemistry Laboratory after C. Bonnelle retired, R. Barchewitz who was professor at Paris VI University, and some well known theoreticians such as F. Combet Farnoux and A. Sureau, who were developing sophisticated multi-configuration codes to treat photo-absorption processes in atoms and plasmas.

Y. Cauchois was also deeply involved in many national and international organizations. In particular, she was active in creating the series of International Conferences dedicated to X-ray processes. It might be interesting for the jung newcomers in the field and for the participants to X-ray-2002 in Roma to know that she

FIGURE 9. Photograph of Y. Cauchois and H. Curien at the opening ceremony of X-70 at the CNRS in Paris

was the chaiman of X-1970 in Paris, called "Processus électroniques simples et multiples des domaines X et X-UV" (I helped her to organize this 4th edition of the series) after Gatlinburg (1962), Leipzig (1965) and Kiev (1968). I like to show in Figure 9 a photograph of the chairs at the opening ceremony of X-70. Hubert Curien, seated on the right side, was the General Director of the CNRS before becoming later Minister of Research in the Government.

The people who would like to know more about the life of Y. Cauchois, will read with interest the bibliography published by C. Bonnelle in Physics Today [66].

FIRST EXPERIMENTS USING SYNCHROTRON RADIATION

Y. Cauchois was interested quite early in the perspective open by the possible use of synchrotron radiation. In one of her book, she wrote a long chapter on the characteristics of synchrotron radiation [64]. Since the discovery of the existence of this radiation in 1947, and the first characterization of this new source in 1956 by Tomboulian [67], the National Bureau of Standards in Washington, USA, was the first center to develop some programs to use the radiation emitted by the 180 MeV-SURF synchrotron in the VUV range. The first major physics result using synchrotron radiation was obtained by Madden and Codling, when they discovered the famous doubly-excited states in helium [68]. Predicted by Fano, the existence and properties of the newly observed autoionizing states were soon interpreted. I remember that we had

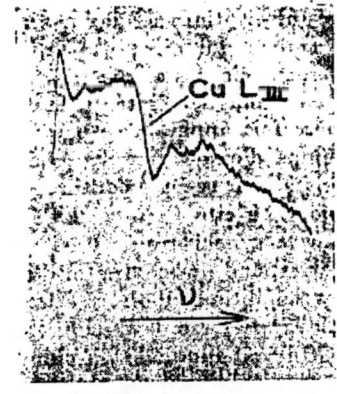
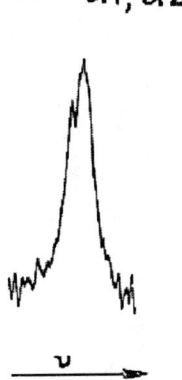

Al Kα_1, α_2

FIGURE 10. The L$_{III}$ absorption edge of Cu (left part, from Ref. 69) and the fluorescent Ka emission line of Al (right part, from Ref. 69) measured with the synchrotron radiation emitted by the 1.1 GeV electron synchrotron in Frascati.

the visit of Cooper and Fano in her laboratory in 1962 or 1963. We were all excited by the lively discussion she organized with them, and I was particularly impressed when she invited a graduate student like me to participate to the working lunch.

Together with C. Bonnelle and P. Jaeglé, she established a long term collaboration with the Istituto di Sanitaria di Roma to prepare a program using the radiation emitted by the Frascati electron synchrotron. In 1963 was published [69] the first observation of x-ray processes due to interaction of this radiation with various targets. Figure 10 shows examples of the spectra obtained with a soft x-ray spectrometer at the L$_{III}$ absorption edge of Cu, and the K$\alpha_{1,2}$ emission lines of Al produced by fluorescence of an irradiated target. Even though the experimental conditions were not yet fully optimized during these first experiments, the comparison of the exposure time to record these spectra with the time needed to obtain equivalent spectra with the bremsstrahlung emitted by an x-ray tube shows that the gain in effective intensity was already several thousands at the Al K-edge, suggesting very exciting developments in the future.

A little later, P. Jaeglé and his collaborators installed a grazing incidence spectrograph in Frascati to work in the vacuum ultraviolet range. Soon they discovered a new effect in photoabsorption [70] as shown in Figure 11. The photoabsorption spectra of some heavy elements (from Ta to Bi) show an unexpected behavior over the 200 eV photon energy range. When the energy of incident photons increased, the absorption coefficient show large variations well away from any ionization threshold, passing through a deep minimum before reaching a weaker maximum, and then followed by a slower decrease towards high photon energies. Photoionization cross sections performed in the central field approximation explained these results by the combined effects of the 4f- and 5d subshells.

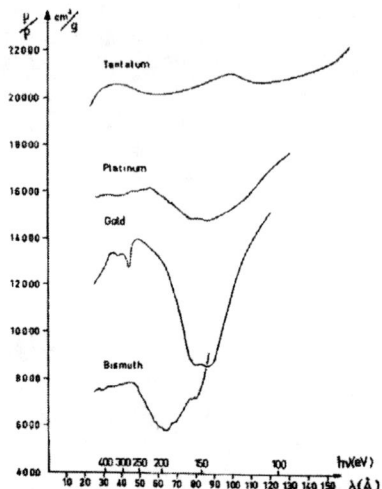

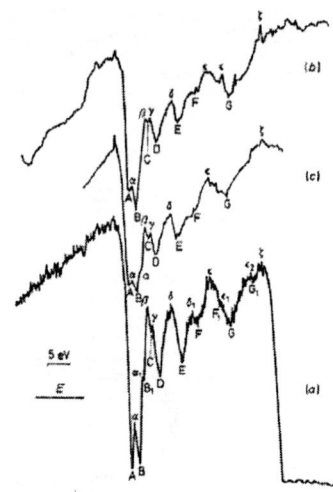

FIGURE 11. Mass absorption coefficients measurements for tantalum, platinum, gold, and bismuth recorded with the synchrotron radiation available for the Frascati synchrotron (from Ref. 70).

FIGURE 12. Comparison between the specular reflection at 8 mrad (a), and the photoabsorption spectra obtained by Barchewitz (b) and Rule (c) for Na K-absorption (from Ref. 72, with permission from IOP Publishing).

Finally, I would like to show as a third example of studies achieved at Frascati, some results obtained by R. Barchewitz and co-workers [71, 72]. They developed a new method to determine photoabsorption spectra of a bulk sample from the measurements of the specular reflection of continuous radiation on its surface. The apparatus used in Frascati consisted of two plane parallel mirrors placed in the path of the beam so that the axis of rotation of the first mirror was in the orbital plane and perpendicular to the beam direction. The second mirror was located so that its edge lies on a line passing through the center of the first mirror. The reflected radiation was analyzed by means of a bent crystal spectrograph using a mica crystal. Sodium was evaporated under vacuum onto the mirrors to give a film of nearly uniform thickness. Figure 12 presents a comparison between the specular reflection spectrum measured for a glancing angle of 8 mrad (curve a), the photoabsorption spectrum measured also in this work (curve b) and the Na K absorption spectrum measured previously by Rule (curve c, Ref. 73). The dispersion between the values deduced from reflection and absorption curves is of the order of the experimental errors, showing that reflection measurements are a good alternative to absorption measurements when a solid of convenient thickness cannot be prepared. Moreover, under equivalent resolution, some features are better seen in the reflected than in the transmitted spectrum or are observable only in the reflected spectrum (they are noted by the suffixed letters).

In 1968, Y. Cauchois and myself wrote a report on the use of synchrotron radiation to be included in the "Rapport de Prospective du CNRS". It was suggested to use the newly built storage ring in Orsay ACO for synchrotron radiation experiments. At this time the proposal was turned down by the particle physicists. A few years later,

46

however, another attempt made by a group of atomic and molecular scientists, several of them from her laboratory, was successful, opening the way to the creation of the LURE laboratory [74]. People working since years at the french synchrotron radiation center owe some credit to Yvette Cauchois, a great pioneer in the field.

CONCLUSION

The end of the life of this outstanding scientist occurs in Romania where she had kept many friends, when she died of pneumonia during a short visit in 1999. I show in Figure 13 the place where she was buried in the country side, not far from the monastery of Barsana. The stone on her tomb was engraved with the following dedication :

OUTSTANDING SCIENTIST
FAITHFUL AND JOYFUL FRIEND
LOVING ALL GREAT AND BEAUTIFUL THINGS
REST IN THE PEACE AND LOVE OF GOD

FIGURE 13. Photograph of the grave of Yvette Cauchois in Romania (courtesy of Viorica Florescu)

ACKNOWLEDGMENTS

The author thanks C. Bonnelle, and R. Barchewitz for communication of some documents, and V. Florescu for the photographs taken in Romania, and B. Chauveau for the photographs of Y. Cauchois. He would like also to express his warm thanks to Chantal Jucha for preparing the camera-ready version of this manuscript.

REFERENCES

1. Friedrich, W., Knipping, W., and Laue, M, *Proc. Bavarian Acad. Sci.*, 1912, p. 303.
2. Bragg, W. H., and Bragg, W. L., *Proc. Roy. Soc. London*, **A 88**, 428 (1913).
3. Siegbahn, M., *Spektroskopie der Röntgenstrahlen*, 2nd edition, Julius Springer, Berlin, 1931.
4. Bohr, N., *Phil. Mag.* **26**, 476 (1913).
5. Moseley, H. G. J., *Phil. Mag.* **26**, 1024 (1913); ibid. **27**, 703 (1914).
6. Compton, A. H., *Phil. Mag.* **45**, 1121 (1923).
7. De Broglie, M., and Lindeman, *C. R. Ac. Sc. Paris* **158**, 944-946 (1914).
8. Seeman, H. *Ann. Physique* **49**, 479 (1916).
9. Thoraeus, S., and Siegbahn, M., *Rev. Sci. Instr.* **13**, 235 (1926).
10. Siegbahn, M., and Larsson, A., *Arkiv f. Math. Astr. och Fysik* **18**, 11 (1926).
11. Ehrenberg, W., and Mark, H., *Zeits. f. Physik* **42**, 807 (1927).
12. Du Mond, J. W. H., and Hoyt, A., *Phys. Rev.* **36**, 1702 (1930).
13. Allison, K., *Phys. Rev.* **38**, 203 (1931).
14. Parratt, L. G., *Rev. Sci. Instr.* **5**, 395 (1934).
15. Gouy, *Ann. d. Phys.* **5**, 241 (1916).
16. Johann, H., *Zeits. f. Physik* **69**, 185 (1931).
17. Johansson, T., *Zeits. f. Physik* **82**, 507 (1933).
18. Cauchois, Y., *C. R. Ac. Sci. Paris*, **194**, 362-364 (1932).
19. Cauchois, Y., *Extension de la spectrographie des rayons X*, Thèse de Doctorat, Masson, Paris, 1933.
20. Cauchois, Y., and Bonnelle, C., in *Atomic Inner-Shell Physics*, edited by B. Crasemann, Academic Press, New York, 1950, pp. 83-121.
21. Lind, D. A., West, W. J., and Du Mond, J. W. M., *Phys. Rev.* **77**, 475 (1950).
22. Cauchois, Y., and Hulubei, H., *C. R. Ac. Sc. Paris* **197**, 681-683 (1933).
23. Indelicato, P., and Lindroth, E., *Phys. Rev. A* **46**, 2426-2436 (1992).
24. Deslattes, R. D., Kessler, E. G., Jr, Indelicato, P., and Lindroth, E., in *International Tables for Crystallography*, 2nd edition, edited by A. J. C. Wilson and E. Prince, Kluwer Academic, Dordrecht, 1999, pp. 200-213.
25. Siegbahn, M., and Stenstrom, W., *Phys. Zeits.* **17**, 48 and 318 (1916).
26. Richtmeyer, R. D., *Phys. Rev.* **49**, 1-8 (1936).
27. Wentzel, G., *Ann. d. Physik* **66**, 437 (1921).
28. Wentzel, G., *Ann. d. Physik* **73**, 647 (1924).
29. Wentzel, G., *Zeits. f. Physik* **31**, 445 (1925).
30. Coster, D., *Phil. Mag.* **45**, 65 (1923).
31. Du Mond, J. W. H., and Hoyt, A., *Phys. Rev.* **36**, 799 (1930).
32. Coster, D., and Thijssen, W. J., *Zeits. f. Physik* **84**, 686 (1933).
33. Druyvesteyn, M. J., *Zeits. f. Physik* **43**, 707 (1927).

34. Coster, D., and Druyvesteyn, M. J., *Zeits. f. Physik* **40**, 765 (1927).
35. Druyvesteyn, M. J., *Het Röntgenspectrum van de tweede Soort*, Dissertation, Groningen, 1928.
36. Aberg, T., Jamson, K. A., and Richard, P., *Phys. Rev. Lett.* **37**, 63-67 (1976).
37. Parratt, L. G., *Phys. Rev.* **49**, 132-139 (1936).
38. Parratt, L. G., *Phys. Rev.* **49**, 502-507 (1936).
39. Richtmeyer, F. K., *J. Franklin Inst.* **208**, 325 (1929).
40. Kennard, E. H., and Ramberg, E., *Phys. Rev.* **46**, 1040 (1934).
41. Wolfe, H. C., *Phys. Rev.* **43**, 221 (1933).
42. Pearsall, A. W., *Phys. Rev.* **46**, 694-697 (1994).
43. Richtmeyer, F. K. *Rev. Mod. Phys.* **9**, 391-402 (1937).
44. Richtmeyer, F. K., and Kaufman, S., *Phys. Rev.* **44**, 605-609 (1933).
45. Auger, P., *C. R. Ac. Sci. Paris* **180**, 65-67 (1925).
46. Auger, P., *C. R. Ac. Sci. Paris* **182**, 773-775 (1925).
47. Coster, D., and Kronig, R. L., *Physica* **2**, 13-26 (1935).
48. Parratt, L. G., *Phys. Rev.* **41**, 553-563 (1932).
49. Parratt, L. G., *Rev. Sci. Instr.* **6**, 387-395 (1935).
50. Parratt, L. G., *Phys. Rev.* **50**, 1-15 (1936).
51. Parratt, L. G., *Phys. Rev.* **50**, 598-602 (1936).
52. Cauchois, Y., J. *Physique Rad. S-VIII*, **5**, 1-11 (1944).
53. Hulubei, H., and Cauchois, Y., *C. R. Ac. Sc. Paris*, **210**, 696-698 (1940).
54. Hulubei, H., and Cauchois, Y., *C. R. Ac. Sc. Paris*, **210**, 761-763 (1940).
55. Hulubei, H., and Cauchois, Y., *C. R. Ac. Sc. Paris*, **209**, 476-478 (1939).
56. Hulubei, H., and Cauchois, Y., *C. R. Ac. Sc. Paris*, **211**, 316-318 (1940).
57. Parratt, L. G., *Phys. Rev.* **56**, 295-297 (1939).
58. Cauchois, Y., and Hulubei, H., *Longueurs d'Onde des Emissions X et des Discontinuités d'Absorption X*, Hermann, Paris, 1947.
59. Cauchois, Y., *J. Physique Rad.* **8**, 113-121 (1952).
60. Cauchois, Y., *J. Physique Rad.* **16**, 1253-262 (1955).
61. Cauchois, Y., *Les Cours de Sorbonne*, Centre de Documentation Universitaire, Paris, 1956.
62. Cauchois, Y., *La structure électronique de la matière*, Gauthiers-Villars, Paris, 1948.
63. Cauchois, Y., *Atomes, Spectres, Matière*, Albin Michel, Paris, 1952.
64. Cauchois, Y., and Héno, Y., *Cheminement des particules chargées*, Gauthiers-Villars, Paris, 1964.
65. Cauchois, Y., *J. Physique Rad. Ser VIII*, **6**, 59-96 (1945).
66. Bonnelle, C., *Phys. Today* **54**, 88-89 (2001).
67. Tomboulian, D. H., and Hartman, P. L., *Phys. Rev.* **102**, 1423-1436 (1956).
68. Codling, K., and Madden, R. P., *Phys. Rev. Lett.* **10**, 516-519 (1963).
69. Cauchois, Y., Bonnelle, C. and Missoni, G., *C. R. Ac. Sc. Paris*, **257**, 409-11 and 1242-44 (1963).
70. Jaeglé, P., Combet Farnoux, F., Dhez, P., Cremonese, M., and Onori, G., *Phys. Rev.* **188**, 30-35 (1969).
71. Barchewitz, R., Bonnelle, C., Cremonese, M., and Onori, G., *C. R. Ac. Sc. Paris* **268**, 151-154 (1969).
72. Barchewitz, R., Cremonese-Visicato, M., and Onori, G., J. *Phys. C* **11**, 4439-4445 (1978).
73. Rule, K. C., *Phys. Rev.* **166**, 199-204 (1944).
74. Dagneaux, P., Depautex, C., Dhez, P., Durup, J., Farge, Y., Fourme, R., Guyon, P.-M., Jaeglé, P., Leach, S., Lopez-Delgado, R., Morel, G., Pinchaux, R., Thiry, P., Vermeil, C., and Wuilleumier, F. J., *Ann. d. Physique* **9**, 9-61 (1975).

49

II. NEW X-RAY SOURCES
AND TECHNIQUES

Conceptual Design of a Soft X-ray SASE-FEL Source

D. Alesini, S. Bertolucci, M.E. Biagini, C. Biscari, R. Boni, M. Boscolo,
M. Castellano, A. Clozza, G. Di Pirro, A. Drago, A. Esposito, M. Ferrario,
V. Fusco[+], A. Gallo, A. Ghigo, S. Guiducci, M. Incurvati, P. Laurelli,
C. Ligi, F. Marcellini, M. Migliorati[+], C. Milardi, L. Palumbo[+],
L. Pellegrino, M. Preger, P. Raimondi, R. Ricci, C. Sanelli, F. Sgamma,
B. Spataro, M. Serio, A. Stecchi, A. Stella, F. Tazzioli, C. Vaccarezza,
M. Vescovi, C. Vicario, M. Zobov[#],
E. Acerbi, F. Alessandria, D. Barni, G. Bellomo, C. Birattari, M. Bonardi,
I. Boscolo, A. Bosotti, F. Broggi, S. Cialdi, C. De Martinis, D. Giove,
C. Maroli, P. Michelato, L. Monaco, C. Pagani, V. Petrillo, P. Pierini,
L. Serafini, D. Sertore, G. Volpini[##],
E. Chiadroni, G. Felici, D. Levi, M. Mastrucci, M. Mattioli, G. Medici,
G.S. Petrarca[###],
L. Catani, A. Cianchi, A. D'Angelo, R. Di Salvo, A. Fantini, D.
Moricciani, C. Schaerf[°],
R. Bartolini, F. Ciocci, G. Dattoli, A. Doria, F. Flora, G.P. Gallerano,
L. Giannessi, E. Giovenale, G. Messina, L. Mezi, P.L.Ottaviani,
L. Picardi, M. Quattromini, A. Renieri, C. Ronsivalle[°°]
L. Avaldi, C. Carbone, A. Cricenti, A. Pifferi, P. Perfetti, T. Prosperi,
V. Rossi Albertini, C. Quaresima and N. Zema.[°°°]

[#]*INFN -Laboratori Nazionali di Frascati, Via E. Fermi 40, I-00044 Frascati (Roma), Italy*
[##]*INFN, Sezione di Milano, Via Celoria 16, I-20133 Milano, Italy*
[###]*INFN -Sezione di Roma1, University of Roma "La Sapienza", P.le A. Moro 5, - I-00185 Roma,Italy*
[°]*INFN - Sezione di Roma2 and Università degli Studi "Tor Vergata",*
Via della Ricerca Scientifica 1, I-00133 Roma, Italy
[°°]*ENEA,CR Frascati/FIS, Via E Fermi 45, I-00044 Frascati (Roma), Italy*
[°°°]*Istituto di Struttura della Materia, CNR, Area della Ricerca di Roma-Tor Vergata - Roma, Italy*
[+]*University of Roma "La Sapienza", Dip. Energetica,.Via A. Scarpa 14, I-00161 Roma, Italy*

Abstract. FELs based on SASE are believed to be powerful tools to explore the frontiers of
basic sciences, from physics to chemistry to biology. Intense R&D programs have started in the
USA and Europe in order to understand the SASE physics and to prove the feasibility of these
sources. The allocation of considerable resources in the Italian National Research Plan (PNR)
brought about the formation of a CNR-ENEA-INFN-University of Roma "Tor Vergata" study
group. A conceptual design study has been developed and possible schemes for linac sources
have been investigated, bringing to the SPARX proposal. We report in this paper the results of a
preliminary start to end simulation concerning one option we are considering based on an S-band
normal conducting linac with high brightness photoinjector integrated in a RF compressor.

CP652, *X-Ray and Inner-Shell Processes: 19th International Conference on X-Ray and Inner-Shell Processes*
edited by A. Bianconi, A. Marcelli, and N. L. Saini
© 2003 American Institute of Physics 0-7354-0111-X/03/$20.00

THE SPARX PROPOSAL

Driven by the large interest that X-ray SASE FEL's light sources have raised world-wide in the synchrotron light scientific community, as well as in the particle accelerator community and following solicitations arising from several Italian national research institutions, the Italian Government launched in 2001 a long-term initiative devoted to the realization in Italy of a large scale ultra-brilliant and coherent X-ray source. The allocation of considerable resources in the Italian National Research Plan (PNR) brought about the formation of a CNR-ENEA-INFN-University of Roma "Tor Vergata" study group. A conceptual design study has been developed and possible schemes for linac sources have been investigated bringing to the SPARX proposal.

Two spectral complementary regions around 13.5 nm and 1.5 nm, are considered for the radiation source. In order to generate the SASE-FEL at these wavelengths, it is necessary to produce a high brightness beam to inject inside two long undulators. A preliminary analysis of the beam parameters required for such a source leads to values reported in Tab. 1.

We report in the next sections the results of a preliminary start to end simulation concerning one option we are considering based on an S-band normal conducting linac.

The basic scheme is shown in Fig. 1 and consists of an advanced high brightness photoinjector followed by a first linac that drives the beam up to 1 GeV with the correlated energy spread required to compress the beam in a subsequent magnetic chicane. The second linac drive the beam up to 2.5 GeV while damping the correlated energy spread tacking profit of the effective contribution of the longitudinal wake fields provided by the S-band accelerating structures. A peculiarity of this linac design is the choice to integrate a high brightness photoinjector in a rectilinear RF compressor, as recently proposed [1], thus producing a 300-500 A beam in the early stage of the acceleration. The potentially dangerous choice to compress the beam at low energy (<150 MeV) when it is still in the space charge dominated regime, results to be not a concern provided that a proper emittance compensation technique is adopted [2], a possibility that is not viable in a magnetic chicane. In addition the propagation of a shorter bunch in the first linac reduces the potential emittance degradation caused by transverse wake fields and longitudinal wake fields results to be under control by a proper phasing of the linac.

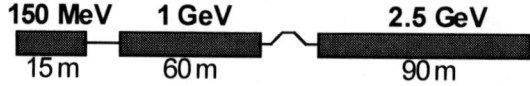

FIGURE 1. Linac scheme of SPARX project.

TABLE 1. Electron Beam parameters

Beam Energy	2.5	GeV
Peak current	2.5	kA
Emittance (average)	2	mm-mrad
Emittance (slice)	1	mm-mrad
Energy spread (correlated)	0.1	%

TABLE 2. RF compressor parameters

TW Section	I	II	II
Gradient [MV/m]	15	25	25
Phase [Deg]	-88.5	-67	0 (on crest)
Solenoid field [G]	1120	1400	0

HIGH BRIGHTNESS PHOTOINJECTOR WITH RF COMPRESSOR

The injector preliminary design considers a 1.6 nC bunch 10 ps long (flat top) with 1.2 mm radius, generated inside a 1.6-cell S-band RF gun of the same type of the BNL-SLAC-UCLA one [3] operating at 140 MV/m peak field equipped with an emittance compensating solenoid. Three standard SLAC 3-m TW structures each one embedded in a solenoid boost the beam up to 150 MeV. With a proper setting of accelerating sections phase and solenoids strength it is possible, applying the compression method described in [2], to increase the peak current preserving the beam transverse emittance. In the present case we have got with PARMELA simulation a bunch average current of 440 A with a normalized rms emittance below 1 mm mrad. The low compression ratio (a factor 3) has been chosen to keep the longitudinal emittance as low as possible in order to simplify the second compression stage. We used the first two TW sections as compressor stages in order to achieve a gradual and

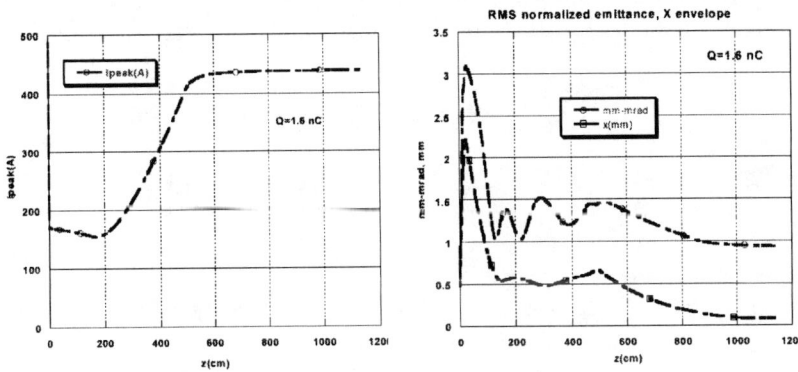

FIGURE 2. Rms current (left), rms norm. emittance and rms beam envelope (right) along the injector, up to 150 MeV.

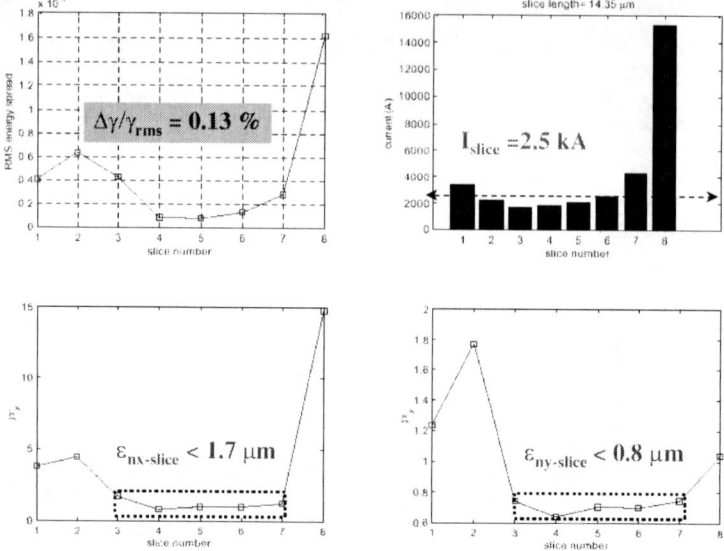

FIGURE 3. Energy spread, peak current and transverse emittances along the bunch.

controlled bunching, the current has to grow about at the same rate of the energy, and we increased the focusing magnetic field during the compression process. An optimized RF compressor parameters set is reported in Tab. 2.

Fig. 2 (left) shows the current growth during bunch compression until 150 MeV, envelope and emittance evolution are also reported (right), showing the emittance compensation process driven by the solenoids around the accelerating section that keep the bunch envelope close to an equilibrium size during compression [2].

A dedicated R&D program (SPARC project [5]) is envisaged at LNF-INFN in collaboration with CNR-ENEA-INFM-ST-Tor Vergata University. Its aim is the generation of electron beams with ultra-high peak brightness to drive a SASE-FEL experiment at 520 nm, performed with a 12m undulator following the linac.

THE LINAC

The accelerator dedicated to the FEL-SASE source has the task of accelerating high brightness electron bunches up to the energy of 2.5 GeV including a second compression stage. Linac1 consists of 15 S-band TW structures, operating at 20 MV/m and the beam is propagated 20 degrees off crest. In the Linac2 additional 24 accelerating structures are foreseen with the same gradient and the beam is propagating on crest. The beam optics consists in a FODO lattice. The nominal values for the proposed source have been reported in Tab. 1.

The 10k macro-particles beam generated by PARMELA has been propagated through Linac1, Magnetic Compressor and Linac2 with the code ELEGANT. The

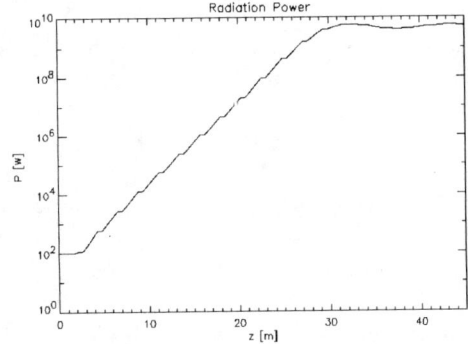

FIGURE 4. FEL signal evolution (λ=1.5 nm) along the undulator 2. The typical "steps" in the exponential rise are due to beam focusing regions where there are no undulators.

correlated energy spread induced by Linac1 is 0.6% in order to compress the beam by a factor 6 in the 15 m long magnetic chicane with an R_{56} = 48 mm. At the exit of the Linac2 the required parameters for FEL operation have been achieved over more than 50% of the bunch length, as shown in Fig. 3.

A further improvement is expected by fully optimizing the compression scheme and by using a 4th harmonic cavity [4] for the linearization of the longitudinal phase space distribution.

THE FEL SASE SOURCE

We envisage using the same beam to feed two undulators whose characteristics are

TABLE 3. Undulators characteristics

	Undulator 1 @1.5 nm	Undulator 2 @13.5 nm
Type	Halbach	Halbach
Period	3 cm	5 cm
K	1.67	4.88
Gap	12.67 mm	12.16 mm
ResidualField	1.25 T	1.25 T

TABLE 4. FEL-SASE expected performances

Wavelength (λ)	1.5 nm	1.5 nm
Saturation length	24.5 m	24.5 m
Peak Power	10^{10} W	10^{10} W
Peak Power 3° harm.	$2\ 10^{8}$ W	$2\ 10^{8}$ W
Peak Power 5° harm.	$3\ 10^{7}$ W	$3\ 10^{7}$ W
Brilliance *	$1.8\ 10^{31}$	$1.8\ 10^{31}$
Brilliance * 3° harm.	10^{29}	10^{29}
Brilliance * 5° harm.	$9\ 10^{28}$	$9\ 10^{28}$

* The brilliance is given in photons/sec/0.1%bw/(mm mrad)

reported in Tab. 3. The characteristics of the FEL-SASE radiation up to the 5th harmonics, have been investigated by means of several codes: GINGER, GENESIS, MEDUSA, PROMETEO, PERSEO, and the results are shown in Tab. 4 and Fig. 4.

With the two undulators it is possible to cover a bandwidth from 1.2nm to 13.5nm, with the first harmonic, and a bandwidth from about 0.4nm to 4nm, using the 3rd harmonic, which exhibits still a considerable peak power, as reported in Tab. 4.

Time dependent FEL simulations, performed using the particle distributions produced by the start-to-end simulations presented in previous section, are in progress, showing saturation for 50% of bunch slices after 30 m of active undulator length. These first preliminary results are encouraging and will be the starting point for further optimizations.

CONCLUSIONS

A preliminary start to end simulation of the SPARX proposal has been presented. The possibility to integrate an RF compressor into a linac for FEL application has been investigated for the first time. The 1 nC case with 120 MV/m peak field on the gun is also under investigation.

ACKNOWLEDGMENTS

We like to thanks for the many helpful discussion C. Pellegrini, J.B. Rosenzweig (UCLA), M. Cornacchia, P. Emma and D.T. Palmer (SLAC). Concerning the FEL source proposal, we are in debt with W.B. Fawley (LBNL), H.P. Freund (NRL), S.G. Biedron, S.V. Milton (ANL) and the EXOTICA international working group.

REFERENCES

1. L. Serafini and M. Ferrario, *Velocity Bunching in PhotoInjectors*, *AIP CP* **581**, 87 (2001)
2. M. Boscolo et al., Beam Dynamics Study of RF Bunch Compressors for High Brightness Beam Injectors, *Proc. of EPAC* 2002, Paris.
3. D.T. Palmer, *The next generation photoinjector* , PhD. Thesis, Stanford University
4. M. Ferrario, K. Flottmann, T. Limberg, Ph. Piot, B. Grygorian, TESLA-FEL 2001-03.
5. D. Alesini *et al.*, The SPARC Project, *to be published in Proc. of the 24rd FEL conference*, *Chicago,USA, September 9-13,2002.*

VUV and Soft X-Ray Projects

Giuseppe Dattoli and Alberto Renieri

ENEA –UTS FIS, Centro Ricerche di Frascati, C.P. 65 – 00044 Frascati, Rome (Italy)

ABSTRACT

We review the state of the art of the VUV-X ray FEL sources and analyze the limit of the relevant technologies and the future perspectives. Possible strategies aimed at developing the design of a road map toward very short wavelengths, very high brilliance and very short pulses are discussed. Within such a framework we report and comment on the proposed devices.

1. INTRODUCTION

Synchrotron Radiation (S.R.) facilities had a tremendous impact on our understanding of different fields of science, going from biocrystallography to spectromicroscopy [1]. Their status is still prosperous and the importance and the amplitude of their contributions to the development of Technology and Science is far from being considered close to any saturation point [2].

It has become quite a commonplace to underline that S.R. developed from the stage of parasitic devices to the so called 3rd generation sources, which have now been operated for about eight years, and are capable of providing photon beams with peak brilliance of at least 10^{18} c.u. (*photons/sec/0.1%bw/(mm mrad)^2*). Such an evolution in terms of brilliance has been made possible by the tremendous effort in the conception, design and technology advancement of the accelerators providing the electron beam (e.b.) and of the insertion devices where the radiation is generated.

In a typical 3rd generation light source, the horizontal (vertical) r.m.s. emittance ε_x (ε_y) is about 10^{-9} (10^{-10}) m*rad, thus ensuring good transverse coherence in the UV region and partial coherence down to tens of Å or shorter. Notwithstanding the development and the operation of 3rd generation sources did not determine automatically the death of the 2nd generation counterparts [3]. Third generation sources are indeed coexisting with 2nd generation devices, which have recently provided very important and crucial results in condensed matter Physics [4]. This is a clear indication that the number of figures of merit characterizing a specific application is so wide that 3rd generation sources cannot be considered indisputably better for any experiment involving synchrotron light. Such a lesson from the past is not of secondary importance and is a key point to be kept in mind when discussing of any further improvement beyond 3rd generation.

Since 1976 Free Electron Lasers (FEL) produce highly coherent radiation beams with brightness many orders of magnitude larger than that of undulator radiation. To

CP652, X-Ray and Inner-Shell Processes: 19th International Conference on X-Ray and Inner-Shell Processes
edited by A. Bianconi, A. Marcelli, and N. L. Saini
© 2003 American Institute of Physics 0-7354-0111-X/03/$20.00

date the shortest wavelength record of a FEL oscillator (1st harmonic) is 190 nm from the Storage Ring based FEL oscillator in Trieste [5]. The present wavelength limit of FEL oscillators is mainly due to mirror availability and the relevant technological improvements will certainly allow, in the next future, the operation at shorter wavelengths. By keeping as figure of merit the e.b. characteristics of third generation sources we can safely conclude that storage ring FEL oscillators are technologically mature devices, capable of providing highly coherent radiation with peak brilliance exceeding 10^{28} c. u. in the region around 150 nm. We believe that storage ring based FELs have provided us with an important tool of research not fully explored and recognized. We want to mention (apart from the possibilities, not yet exploited as they should be, offered by FEL light combined with those from the ordinary S.R. sources in pump probe synchronous experiments) the important contribution to the understanding of microwave instability evolution in storage rings, made possible by the study of e.b.-FEL radiation combined dynamics [6]. These results could provide a deeper insight into the phenomenology of coherent synchrotron radiation [7] and on its cures.

2. WHAT WE LEARNT FROM THE PAST

In the previous section we have given just an idea of what is the present scenario of coherent, or partially coherent sources in the VUV-X range. Scientific community is however pushing for new solutions going far beyond the present status by designing devices capable of delivering X-ray beams with much larger brilliance and shorter pulses. Just to give some references numbers, we note that X-ray flashes in the Å region with a time duration of 100 fs and with brilliance 10-11 orders of magnitude larger than the presently available sources, could provide a real breakthrough in biomolecular imaging, because such a beam would prevent the problems associated with the radiation damage [8].

Let us now list what we do expect from the next generation sources
   a) Coherence (transverse, longitudinal)
   b) Pulses with short time duration
   c) Tunability towards the Å region
   d) Brilliance (peak and average), values significantly exceeding the present levels
   e) Access to a wide number of users
   f) Long operation time.

The possible options as next generation S.R. sources are
   i)   Storage Ring Upgrade
   j)   Linac Based Sase FELs
   k) Energy Recovery linacs.

Regarding the first solution, going in the direction of developing S.R. sources without any dramatic change of technology, we must emphasize that so far storage rings provide e.b. with a brightness which is a result of the damping and quantum excitation processes occurring in the ring bending magnets. The minimization of the emittance requires the design of a lattice which compensates the dilution effects induced by the quantum excitation. Unfortunately, the demands for high brilliance coherent radiation at shorter wavelengths is conflicting with the request of small emittance, which increases quadratically with energy. Its minimization requires the design of a lattice which compensates the dilution effect induced by the quantum excitations. The price to be paid to have a low emittance is to have a high energy storage ring with a large circumference to allow the insertion of the optical elements necessary for the quantum diffusion compensation. To give a more quantitative idea we note that the number of cells (N_c) necessary to fulfill the condition (K is the usual pinch parameter and γ is the electron energy in rest mass unity)

$$\varepsilon = \lambda/(4\pi), \quad \lambda = \lambda_u(1+K^2/2)/(2\gamma^2) \tag{1}$$

is [9]

$$N_c = (2\pi \cdot 7.7 \cdot 10^{-4}(1+K^2/2) \ \ (\lambda_u/\lambda^2))^{1/3} \tag{2}$$

With λ_u (length of undulator period) and λ (operating wavelength) given in mm. It is clear that such a condition becomes more and more difficult to satisfy at shorter wavelengths. E.g., a design of an ultimate Hard X-ray Source (UHXRS) [9] foresees a circumference of 2200 m necessary to allocate 160 bending magnets and 720 quadrupoles. Such a magnet array would provide at 7 GeV, with respect to ESRF at 6 GeV, e.b. with horizontal (vertical) emittance 20 (2) times smaller with a bunch length about 1/3 shorter at 12 keV of photon with average and peak brilliance more than 2 order of magnitudes larger (i.e. $3.5*10^{22}$ c.u. and $1.0*10^{25}$ c.u. respectively) [10].There is a second tribute which should be considered when dealing with solutions of this type. The large number of insertion devices will inevitably increase the machine longitudinal impedance, thus determining current intensity limitations associated with microwave type instabilities, and the increase of Touscheck intra-beam scattering as a consequence of the smallness of the e.b. emittance, will cause the reduction of the beam life-time to few hours. This fact will require re-injection of a small current every few tens of minutes. In any case the beam transport lattice prevents the achievement of pulse length shorter than few tens of ps.

The use of Energy Recovery Linacs (ERL) is an interesting candidate which prevents all the unpleasing features of storage ring sources. There are indeed no problems associated with the impedance of the device, the emittance depends essentially on the cathode emittance and the pulse, being limited by longitudinal emittance only, can be very short. Namely, after the successful operation of the high power IR FEL at the Jefferson Lab., in which a quite efficient e.b. recovery scheme has been exploited, interesting proposals have been issued by many laboratories, aimed at the enhancement of the brilliance, together with the possibility to generate very short radiation pulses (hundred of fs or even less) in the X – ray spectral region. Problems arise in the handling of high power e-beams and possible degradation during the transport.

3 SCENARIO OF THE PROPOSED VUV AND SOFT X-RAY FEL PROJECTS

In Tables. 3.I and 3.II have been reported the scenarios of the under development and proposed VUV and soft X – ray FEL projects respectively.

All these projects, with the exception of SOLEIL FEL, are based on a SASE FEL scheme driven by r.f. linac, which appears to be the only suitable for operation in the soft X ray region. Storage ring based FEL oscillator are limited to the VUV region, due to the lack of suitable mirrors (which, in addition, would have to support a quite high power radiation flux too) and to the extreme difficulty to produce very short (sub-ps) radiation pulses due to the impossibility in a storage ring to shape the e.b. in a suitable way (no memory system). This fact precludes also the SASE operation (long undulators in by-pass sections) in storage rings. Anyway storage ring FELs, as stressed in the previous section, maintain a quite large interest, being complementary to the SASE devices in the VUV spectral region. In the above list we reported the SOLEIL FEL, which could be the more advanced device of this type.

In this list there is no mention of ERL devices, in which *only* normal (i.e. spontaneous) undulator radiation is generated with a peak brilliance expected to be up to 5 order of magnitude higher than the present 3rd generation synchrotron radiation sources. As to the devices devoted to FEL operation, only the 4GLS proposal is based on an ERL scheme in a kind of ring configuration, in which XUV (SASE FEL), VUV (FEL oscillator), IR (FEL oscillator) and spontaneous undulator radiation sources are operated and, eventually, utilized together for multiple photon applications.

TABLE 3.I.   Under development VUV and soft X-ray projects scenario

project	location	type	e-beam energy	λ	notes	Ref.
TTF2	DESY, Hamburg (D)	SASE	1 GeV	6 nm	*start of operation: 2004*	[11]
SCSS	SPring-8 (J)	SASE	230 MeV (phase I)	40 nm	"compact": - short period in vacuum undulator - high gradient C-Band accelerator - low emittance beam injector *start of operation at 40 nm: 2005*	[12]
			1 GeV (phase II)	3.6 nm		
DUV-FEL	NSLS, BNL (USA)	SASE HGHG (laser seed)	200 MeV	200-50 nm	achieved 400-nm light by SASE on Febr. 2002	[13]
LEUTL	APS, Argonne (USA)	SASE HGHG	220 MeV	660-130 nm	achieved 130 nm light by HGHG	[14]

In the case of a LINAC based SASE FEL the electron beam parameters are, to a large extent, determined by the gun. The present technology could provide 1 nC electron bunches with invariant emittance of the order of 1 mm*mrad in a pulse length of the order of 10 ps. The other factors enabling LINAC based SASE are:

1)   the e.b. qualities preservation through compression and acceleration processes
2)   development of long undulators and beam transport handling with adequate precision.

At the moment the most delicate point is associated with 1), especially in connection with the degradation due to coherent synchrotron radiation (CSR) emission. This aspect of the problem, which may be an intrinsic limitation for the performances of SASE FEL devices at short wavelengths, deserves a careful understanding of the mechanisms underlying its origin and its possible cures. What we have learned on the saw-tooth instability and its interplay with FEL in storage ring based devices may provide a useful tool to explore the CSR dynamics.

TABLE 3.II. Proposed VUV and soft X-ray projects scenario

project	location	type	e-beam energy	λ	notes	Ref.
TESLA X-FEL	DESY, Hamburg (D)	SASE	25 GeV	0.1 nm	considered to be worthy of support by German Science Council (Pressemitteilung 20/2002) *start of operation: 2011*	[15]
Soft X-ray FEL	BESSY, Berlin (D)	SASE	2.25 GeV	1.2 nm		[16]
LCLS	SLAC, Stanford (USA)	SASE	15 GeV	0.15 nm	The project has been receiving funding of $1.5 Million per year from the Department of Energy, Basic Energy Sciences for the four year period FY1999-FY2002. Funding is expected to increase in FY2003 to $6.0 Million as listed in the Presidents Budget for FY2003 *start of operation: 2008*	[17]
SPARC	ENEA,INFN,INFM, CNR,Sync. Trieste, Un. Roma II (I)	SASE	150 MeV	VIS-VUV	(funded) *start of operation: 2005*	[18]
SPARX	ENEA,INFN,CNR, Un. Roma II (I)	SASE	2.5 GeV	1.5 nm	It is intended as a second step after SPARC (awaiting for Government decision) *start of operation: 2008*	[18]
VXFEL	INFM, Sync. Trieste, Pirelli S.p.A. (I)	SASE FEL oscillator + harmonic generation	1 GeV	6.4 nm		[19]
FERMI	INFM, Sync. Trieste (I)	SASE	3 GeV	1.2 nm	(awaiting for Government decision) *start of operation: 2008*	[19]
4GLS	Daresbury (UK)	SASE (XUV) FEL oscillator (VUV)	600 MeV	VUV-XUV (IR FEL osc.) ERL scheme	Secretary of State for Trade and Industry: on 16th April 2002 it was agreed that 4GLS should proceed to the next stage of the Office of Government Commerce Gateway review process	[20]
SOLEIL FEL	SOLEIL (F)	FEL oscillator	1.5 GeV	150 nm up to 30 nm (5th harmonic)	Storage ring FEL	[21]

4   ROAD MAP TOWARD VUV-X SPECTRAL REGION

The challenge associated with the realization of very high brilliance short pulse VUV – X sources is considerable. The cost of this kind of R&D activity is close to that of the so-called "Big Science". This means that it does not appear quite convenient that just a single laboratory (whatever big it is) would invest in this field, still in strong evolution, the required large amount of money and man power. For this reason the need of a more or less coordinated effort was clear from the very beginning. Namely, e.g., both TTF1 and TTF2 in Hamburg are based on large collaborations among European and American laboratories. From the analysis of what has been done up to now it is possible to recognise *a posteriori* a kind of road map in which we can single out the directions along which the past R&D work has been developed. In this moment many laboratories are studying new strategies in which a reliable road map toward VUV – X spectral region has to be designed. Milestones of this road map must be those of all the laboratories involved in, in order to share at the best experience and results. We believe that, in a quite natural way, due to the formal and informal collaborations and mutual interactions between all the people of the FEL community, this common road map appears now quite well outlined, as we shall try to show in this Section.

We shall limit ourselves, in this Section to the case of single passage SASE FEL devices. Alternative options do not appear suitable, at the present status of the art, for operation up to the Angstrom spectral region.

The main key-issues relevant to a SASE FEL device can be summarized as it follows:

- *Physics of the SASE FEL process*
- *Generation and shaping of suitable electron beams*
- *Electron beam acceleration and transport*
- *Undulator magnets*
- *Electron beam and X-ray diagnostics*

It is out of the scope of this paper to enter into the technical details of these topics. What we shall do is to try, for each of them, to single out some relevant milestones measuring the path along the way in the direction toward X –ray spectral region.

4.1 Physics of the SASE FEL Process

Physics of SASE FEL has been investigated many years ago in series of papers [22] in which the main features of the process have been clarified.

In a FEL the active medium consists of a beam of free electrons, propagating at relativistic velocities in a magnetic undulator with spatial period λ_u and on axis magnetic field B_0. In the undulator the electrons emit radiation at frequency

$$\omega = \frac{2\gamma^2 \omega_u}{1+K^2/2}, \quad \omega_u = 2\pi e/\lambda_u$$

$$K = \frac{eB_0\lambda_u}{2\pi m_0 c^2}$$

(3)

and if the e-beam current is sufficiently large, the emitted radiation grows, in SASE devices according to the relation

$$I(z) = \frac{1}{9}I_0 e^{\frac{z}{L_G}}$$

(4)

where z is the longitudinal coordinate, L_G is the so-called gain length, namely

$$L_G = \frac{\lambda_u}{4\pi\sqrt{3}\rho}$$

(5)

and ρ is the Pierce parameter linked to the FEL small signal gain coefficient g_o by

$$\rho = \frac{1}{4\pi}\left(\frac{\pi g_0}{N^3}\right)^{1/3} \qquad (6)$$

with N being the total number of undulator periods.

The parameter ρ is of the order of 10^{-3} and the saturation length and relative bandwidth are linked to ρ by

$$L_s \cong \frac{\lambda_u}{\rho}$$
$$\left(\frac{\Delta\omega}{\omega}\right) \cong \rho \qquad (7)$$

while the saturated power is linked to the e-beam power by

$$P_{SASE} \cong \rho P_E \qquad (8)$$

The main experimental milestones which provided the benchmark for the previous scaling relations can be singled out as follows:
- **1984** LLNL (Livermore, USA) [23]: 34.6 GHz radiation propagating along a wave-guide, driven by induction accelerator (very high efficiency: saturation with tapered undulator)
- **1989** MIT (USA) [24]: 240 – 470 GHz radiation propagating along a wave-guide (FEL (weakly) collective Raman regime)
- **1998** UCLA (Los Angeles, USA) [25]: λ =12 μm, free space propagation (not saturated)
  With this last experimental result there was a quantitative and qualitative improvement of our knowledge on SASE process. Namely the operating wavelength was about three orders of magnitude shorter than LLNL 1984 device and a factor 50 shorter than MIT 1989, and, in addition, the optical radiation propagates in free space and not along a wave-guide.
- **2001** TTF1 (Hamburg, D) [26]: λ =80-120 nm (saturation): **good agreement with the theory**.
  Lasing at 100 nm was a quite big step toward VUV – X ray spectral region (more than two orders of magnitude with respect to first UCLA 1998 result).
  The state of the art of SASE FEL theory is now mature, i.e. we can be confident that the process is well understood and, if the requested experimental conditions will be correctly satisfied, it will be possible to operate with the desired spectral and brilliance characteristics up to the soft X-ray region. A large part of the device reported in Tab 3.I is foreseen to operate in the region between 10 and 0.1 nm (i.e. only between one and three orders of magnitude shorter wavelength with respect to TTF I).

In Fig. 1 the operating wavelengths of existing and proposed SASE FEL sources are reported. It is worth noticing that between the 1984 LLNL result and the future soft X - ray sources there are about eight order of magnitude in wavelength.

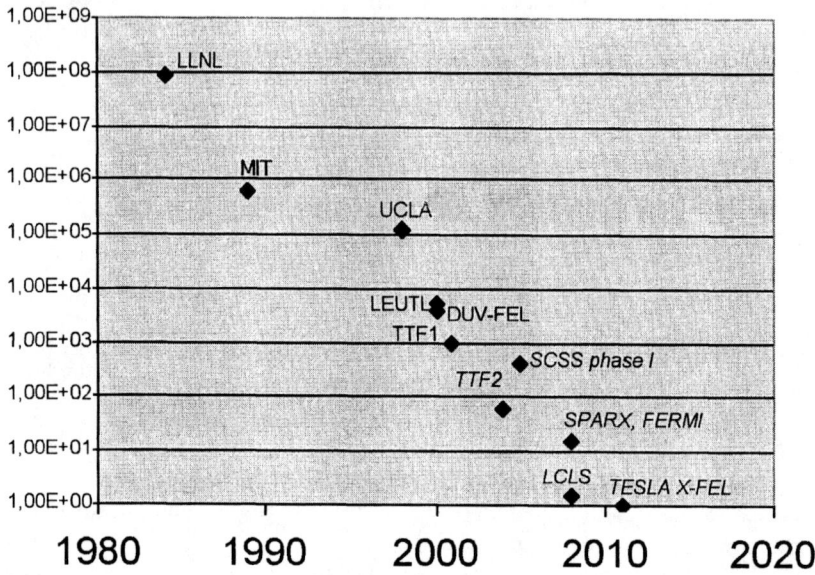

FIGURE 1. SASE FEL devices operating wavelength (Ångstrom) vs (foreseen) year of start of the operation

On the other side, a lot of additional work is still required for the investigation and analysis of new schemes, in particular aimed at improving spectral and/or coherence properties of the generated VUV – X radiation. Specific experimental activity is foreseen in the presently under development devices (Table 3.I) and in the new proposal (Table 3.II). Relevant future milestones can be outlined as it follows:

- HGHG (High Gain Harmonic Generation) theoretical and experimental R&D (see also Sect. 4.4)
- Spectral purification and coherence enhancement (seeding, filtering, low energy electron beam feedback , ...)

4.2 Generation and shaping of suitable electron beams

The important advances in the field of generation and shaping of very high brightness e.b. are mainly related to the development of photo-injectors [27]. FEL community shared this technology and pushed a strong R&D activity in order to reach goals of specific interest for Visible and UV FEL oscillator, at the beginning, and, now, for SASE FEL devices.

For the past we can single out the following milestones:

- Photo-cathodes and driver lasers development: first devices (at that time called "Lasertron") were developed at SLAC (USA), KEK (J), LAL (Orsay, F) and LANL

(USA) in the middle of 80's. In particular it must be underlined the quite big effort made in LANL, where high efficiency photocathodes have been developed just for FEL applications [28].

- Magnetic beam compression: it is interesting to note that this technique have been utilized in a FEL device many years ago (in the middle of 80's) in order to have just the opposite effect, i.e. to reduce the energy spread at the expense of bunch length and peak current (FELIX (Glasgow, UK) [29]).

For the future, a first milestone certainly concerns further important improvements of performance and reliability of the photoinjectors and R&D on alternative cathode schemes (e.g. single crystal thermionic gun, plasma ore "needle" cathodes). Furthermore we must underline a milestone related to the realization of a superconducting (s.c.) r.f. photo-injector. Namely the possibility to exploit at the best the high duty cycle of a s.c. linac is related to the availability of a full s.c. device, gun included. Finally, as to the e.b. shaping, due to the fact that the CSR emitted in the curved electron path can generate severe damage to the e.b. emittance and energy spread, a further milestone related to a non-magnetic beam compression scheme [30] must be foreseen.

- High brilliance cathode development
  - o SCSS: CeB6 single crystal thermionic gun [12]
  - o TESLA X FEL –first measurements at the PITZ (photo-injector test facility at DESY Zeuthen) [31]
  - o S.c. r.f. photo-injector: first results of collaboration FZR-BINP-DESY-ACCEL-MBI [32,33]
- Non magnetic compression schemes
  - o UCLA [34]: Velocity bunching observed
  - o DUVFEL[35]: RF compression (factor 5 of current enhancement)
  - o SPARC – SPARX: longitudinal bunch compressor R&D [18]

4.3   Electron beam acceleration and transport

The acceleration and transport of a quite bright, short and monochromatic e.b. is a challenging task. Namely it is mandatory to preserve the high quality characteristics of the beam in all the stages of injection, acceleration and steering into the undulator magnet. All the possible sources of perturbation must be well understood and minimized. The problems related to the magnetic beam compression have been mentioned in the previous subsection. But the same problems hold for the other curved parts of the electron trajectory, in particular at the exit of the accelerator for the steering of the e.b. into the undulator. Another quite dangerous source of troubles is the interaction of the e.b. with the beam pipe surface (resistive and corrugation effects). Finally it must be assured a quite good shot to shot stability, in order to provide a radiation beam with the required spectral quality, stable in power and alignment. A quite big theoretical and numerical effort has been done in this subject, but there is the need of clear experimental results, in order to correctly evaluate these effects. The relevant milestones to are:

- Compact Linac R&D

- o SCSS: C- band linac (40 MV/m) [12]
- Understanding of CSR effects (micro-bunches instability)
  - o UCLA: numerical and theoretical effort [36]
- Characterizing the interaction between electrons and pipe surface (resistive wall, corrugation)
  - o TTF (DESY): experimental activity and comparison with predictions on wake fields generated by ps electron bunches on artificially corrugated narrow beam pipes [37]
  - o ATF (BNL): experimental activity and comparison with predictions on wake fields generated on periodically and randomly corrugated beam pipes [38]. Results show that the energy loss is greatly reduced in the random case.

4.4　Undulator magnets

The technology of undulator magnets made quite big advances in the last years. In spite of this fact the realization of undulators for SASE FEL devices operating in the VUV – X region is still a challenging task. Namely, just to mention some of the most important issues, we have to consider that tight tolerances required for SASE FEL operation have to be maintained along quit long devices (many tens of meters), the undulator design has to allow to insert a large variety of diagnostics systems and, finally, in the undulator too the e.b. high brightness has to be preserved, this means that beam pipe surface quality and dimensions have to be adequate. In addition, due to the interest to operate at very short wavelengths at lower e.b. energy, new undulator schemes have been proposed [39]. A quite promising scheme is the so called bi-harmonic (or dual-harmonic) undulator [40,41] for which theoretical and experimental activity is planned in the frame of the projects LEUTL, SPARX [42,18].

4.5　Electron beam and VUV – X ray diagnostics

Electron beams with normalized emittance of the order of 1 mm x mrad and energy spread less than 0.1 % are not only not easy to generate but they are quite difficult to measure too. The same applies to the X–ray beam. A quite big effort is under way in all the laboratories involved in SASE FEL activity (see, e.g., for TTF ref. [43,44]). Namely reliable on line measuring and diagnostic systems are essential not only for all the research activities related to the SASE devices, but also for the correct commissioning and operation of the final systems.

5 CONCLUSIONS

Theoretical and experimental status of the art in the field of VUV – Soft X ray FELs appears quite adequate for a further step toward the Å spectral region. Namely the first results coming from the devices under development give us a pretty good confidence on the reliability of the theoretical scaling laws governing the SASE FEL process. In the mean time the recent advances in the e.b. generation and manipulation make the technology level quite close to that required for that ambitious goal. It must be again strongly underlined that the 4th generation S.R. source will not replace the previous ones but it can fruitfully complement them. The same applies to the UV-VUV FEL oscillators (storage ring and linac based devices), whose potentialities appear far from being completely exploited.

REFERENCES

1. See, e.g., "Synchrotron Radiation: Perspectives and New Technologies" Atti dei Convegni Lincei 179, Roma 2002.
2. Neil, G.R., *Nucl. Instrum. Methods* **A483**, 14 (2002).
3. Altarelli, M., in ref. [1].
4. see, e.g., W. Lynch, D., Olson, C.G. "Photoemission Studies of High Temperature Superconductors" Cambridge University Press , Cambridge 1998 and references therein.
5. Trovò, M., et al. *Nucl. Instrum. Methods* **A483**, 157 (2002).
6. Bartolini, R., et al., *Phys. Rev. Lett.* **87**, 134801 (2001).
7. Huang, Z. and Je Kim K. *Phys. Rev. Special Topics – Accel. Beams*, **5**, 074401 (2002).
8. Haidn, J., in ref. [1].
9. Ben Zvi, I., *PERL Photoinjector Workshop*, Brookhaven National Laboratory, January 22-23 2001.
10. Ellaume, P., in ref. [1].
11 The TESLA Test Facility FEL Team: "SASE FEL at the TESLA Facility, Phase 2", DESY Report TESLA-FEL 2002-01, June 2002.
12. Shintake, T., et al., "Status of SCSS: SPring-8 Compact SASE Source Project", EPAC2002.
13. Johnson, E., "DUV-FEL Current Status and Plans", TJNAF LPC Workshop, Jan. 2001; see for further up to date details: http://nslsweb.nsls.bnl.gov/nsls/org/SDL.
14. Doyuran, A., et al., *Phys. Rev. Lett.* **86**, 5902 (2001); Biedron S.G. et al. "Exotic Harmonic Generation Schemes in High-Gain Free-Electron Lasers," LASE 2002, Proceedings of SPIE Vol. 4632; see for further up to date details: http://www.aps.anl.gov/aod/mcrops/leutl/.
15. Materlik, G., Th. Tschentscher (eds.), *TESLA Technical Design Report, Part V*, DESY Report TESLA-FEL, **2001-05**, Hamburg (2001).
16. Abo-Bakr, M., et al., *Nucl. Instrum. Methods* **A483**, 470 (2002), and "The BESSY Soft X-ray Single-Pass FEL Design", Proceedings FEL Conference, Argonne (USA) Sept. 2002, to be publ. Nucl. Instr. & Meth..
17. LCLS Design Study Report, Report SLAC-R-521 (1998); Conceptual Design Report for the LCLS project of the U.S. Department of Energy, April 2002 (http://www-ssrl.slac.stanford.edu/lcls/).
18. Renieri, A., "SPARC and SPARX FEL Projects", Proceedings FEL Conference, Argonne (USA) Sept. 2002, to be publ. *Nucl. Instr. & Meth.*.
19. Bocchetta, C.J., et al. "Overview of FERMI@ELETTRA. A proposed Ultra Bright Coherent X-ray Source in Italy", Proceedings FEL Conference, Argonne (USA) Sept. 2002, to be publ. *Nucl. Instr. & Meth.*.
20. Poole, M.W., McNeal, B.W.J. "FEL Options for the Proposed UK Fourth Generation Light Source (4GLS)", Proceedings FEL Conference, Argonne (USA) Sept. 2002, to be publ. *Nucl. Instr. & Meth.*, see for further up to date details: http://www.4gls.ac.uk/home.htm.
21. see: http://www.soleil.u-psud.fr/workshops/LEL/Notice-LEL_2001.pdf.
22. see, e.g., Sprangle, P., et al., *Phys Rev.* **A21**, 302 (1980); Dattoli, G., et al., *IEEE J. Quantum Electron.*, **QE-17**, 1371 (1981); Bonifacio, R., et al., *Opt. Commun.* **50**, 373 (1984); Kwang-Je Kim, *Nucl. Instrum. Methods* **A250**, 396 (1986).
23. Orzechowski, T.J., et al., *Phys Rev Lett.* **54**, 889 (1985).
24. Kirkpatrick, D.A., et al., *Phys. Fluids* **B1**, 1511 (1989).
25. Hogan, M.J., et al., *Phys. Rev. Lett.* **81**, 4867 (1998); see for further up to date details: http://pbpl.physics.ucla.edu/src/neptune.xml.
26. Ayvazyan, V., et al., *Phys. Rev. Lett.* **88**, 104802 (2002); J.Rossbach,, "Demonstration of Gain Saturation and Controlled Variation of Pulse Length at the TESLA Test Facility FEL", Proceedings FEL Conference, Argonne (USA) Sept. 2002, to be publ. *Nucl. Instr. & Meth.*.

27. Russel, S.J "Overview of High-Brightness, High-Average-Current Photoinjectors for FELs", Proceedings FEL Conference, Argonne (USA) Sept. 2002, to be publ. *Nucl. Instr. & Meth.*.
28. Tallerico, P.J., *Nucl. Instrum. Methods* **A272**, 218 (1988) and ref. therein.
29. Poole W. M., et al., **A237**, 207 (1985).
30. Serafini, L. and Ferrario, M., "Velocity bunching in photoinjectors", *AIP CP* **581**, 2001, pag.87.
31. Baher, W.J., "First Measurements at the Photo Injector TEST Facility at DESY Zeuthen", Proceedings FEL Conference, Argonne (USA) Sept. 2002, to be publ. *Nucl. Instr. & Meth.*.
32. Barhels, E., et al., *Nucl. Instrum. Methods* **A445**, 408 (2000).
33. Büttig, H. et al., "First Operation of a Superconducting RF Electron Gun", Proceedings FEL Conference, Argonne (USA) Sept. 2002, to be publ. *Nucl. Instr. & Meth.*.
34. Musumeci, P., and Rosenzweig, J.B., "Velocity Bunching: experiment at Neptune Photo-Injector", to be published in the Proceedings of ICFA Workshop on "The Physics and Applications of High Brightness Electron Beams", Chia Laguna, Italy, July 1-5, 2002.
35. Piot ,P., et al.,"Subpicosecond compression by velocity bunching in the DUV-FEL photo-injector at BNL", to be published in the Proceedings of ICFA Workshop on "The Physics and Applications of High Brightness Electron Beams", Chia Laguna, Italy, July 1-5, 2002.
36. Reiche, S., and Rosenzweig, J.B., "A Fast Method to Estimate the Gain of the Microbunch Instability in a Bunch Compressor ", Proceedings FEL Conference, Argonne (USA) Sept. 2002, to be publ. *Nucl. Instr. & Meth.*; Jan Menzel et al., "Experimental Investigations on Coherent Diffraction, Synchrotron, and Transition Radiation", Proceedings FEL Conference, Argonne (USA) Sept. 2002, to be publ. *Nucl. Instr. & Meth.*.
37. Huening, M., *Phys. Rev. Lett.* **88**, 074802 (2002) and "Measurements and Simulation of Surface Roughness Wake Fields", Proceedings FEL Conference, Argonne (USA) Sept. 2002, to be publ. *Nucl. Instr. & Meth.*.
38. Zhou, F., et al. "Experimental Characterization of Surface Roughness Wakefield at the ATF", Proceedings FEL Conference, Argonne (USA) Sept. 2002, to be publ. *Nucl. Instr. & Meth.*.
39. Dattoli, G., et al. "Bunching and exotic Undulator Configurations in SASE FELs", Proceedings FEL Conference, Argonne (USA) Sept. 2002, to be publ. *Nucl. Instr. & Meth.*.
40. Yang,, Y. Ding, W., *Nucl. Instrum. Methods* **A407**, 60 (1998).
41. Dattoli, G. et al., "Bunching and Exotic Undulator Configurations in SASE FELs", Proceedings FEL Conference, Argonne (USA) Sept. 2002, to be publ. *Nucl. Instr. & Meth.*.
42. Biedron, S.G. et al., "Spectral Properties of Two Harmonic Undulator Radiation and their Influence on FEL Performance", Proceedings FEL Conference, Argonne (USA) Sept. 2002, to be publ. *Nucl. Instr. & Meth.*.
43. Gerth, Ch. et al. "Photon Beam Diagnostics of Intense, Ultra-Short VUV Radiation of a SASE FEL", Proceedings FEL Conference, Argonne (USA) Sept. 2002, to be publ. *Nucl. Instr. & Meth.*.
44. Nölle, D., "Electron Beam Diagnostics for TTF II", Proceedings FEL Conference, Argonne (USA) Sept. 2002, to be publ. *Nucl. Instr. & Meth.*.

Doppler-free Auger resonant Raman spectroscopy on photochemistry beamline at SPring-8

Y. Tamenori[1], M. Kitajima[2], A. De Fanis[3], H. Shindo[2], T. Furuta[2], M. Machida[4], M. Nagoshi[4], K. Ikejiri[5], H. Yoshida[6], H. Ohashi[1], I. Koyano[4], H. Tanaka[2], P. Baltzer[6], and K. Ueda[3]

[1]SPring-8/JASRI, Sayo-gun, Hyogo 679-5198, Japan
[2]Department of Physics, Sophia University, Tokyo Japan
[3]IMRAM (TAGEN), Tohoku University, Sendai 980-8577, Japan
[4]Department of Material Science, Himeji Institute of Technology, Kamigori, Hyogo 678-1297, Japan
[5]Department of Chemistry, Hiroshima University, Higashi-Hiroshima 739-8526, Japan
[6]Department of Physical Sciences, Hiroshima University, Higashi-Hiroshima 739-8526, Japan
[6]HiSOR, Hiroshima University, Higashi-Hiroshima 739-8526, Japan

Abstract. A Doppler-free electron spectroscopy apparatus that consists of a high-resolution electron spectrometer and a molecular beam source has been installed on beamline 27SU at SPring-8 in Japan. The apparatus is described and the Auger resonant Raman spectra of neon atoms recorded at the $1s \rightarrow 3p$ excitation and those of water molecules recorded at the O $1s \rightarrow 2b_2$ excitation are presented to demonstrate the performance.

INTRODUCTION

The resolution of modern soft X-ray monochromators is sufficiently high to allow to perform resonant photoemission study with the excitation photon bandwidth smaller than the natural width of the inner-shell excited states. In this situation, the width of the observed Auger bands is not determined by the natural width of the excited states but only by the experimental conditions [1]. In general, the instrumental width is determined by the convolution of the photon bandwidth and bandwidth of the electron energy analyzer [2,3]. In case of gaseous samples at room temperature, with the decrease of the monochromator and analyzer bandwidths, the experimental line width is eventually limited by the inhomogeneous Doppler broadening due to thermal motion of the sample gas. In some of the highest resolution measurements for resonant Auger spectra of light molecules [4,5], the experimental widths were indeed dominated by the Doppler width.

CP652, *X-Ray and Inner-Shell Processes: 19th International Conference on X-Ray and Inner-Shell Processes*
edited by A. Bianconi, A. Marcelli, and N. L. Saini
© 2003 American Institute of Physics 0-7354-0111-X/03/$20.00

Recently, we installed, on the soft X-ray photochemistry beamline 27SU at SPring-8 in Japan, a Doppler-free electron spectroscopy apparatus that consists of a high-resolution electron spectrometer (Gammadata-Scienta SES2002) and a molecular beam source (MB Scientific MBS JD-01). The molecular beam source reduces the kinetic energy dependent resolution deterioration due to the Doppler effect in gas phase electron spectroscopic experiment. The original vibrational and rotational distributions in the gas are conserved as no expansion cooling occurs. We first describe the apparatus and then, to demonstrate its performances, present the Auger resonant Raman spectra of neon atoms recorded at the 1s -> 3p excitation (867.12 eV) and those of water molecules recorded at the O 1s -> $2b_2$ excitation (535.77 eV).

EXPERIMENTAL

The beam device has an orifice of 1 mm x 8 mm, covered by a 1 mm thick micro channel plate. The channel diameter is 10 μm, and the open area ratio is approximately 70%. The surface facing the excitation region is covered with a conductive layer of graphite with suitable work function. An electrode structure surrounds the orifice, which allows precision modeling of the electrostatic potential in the target volume. The electrode voltages, as well as the voltage applied to the device body, can be externally adjusted to optimize the energy resolution. A multi-channel array permits a much higher target pressure than a single tube device. The inlet pressure of an effusive beam source is limited roughly to 1 Torr, i.e., the level where the mean free path regarding intermolecular collisions equals the channel length. The achievable target pressure scales as the square root of the number of channels when all other parameters are kept constant. The channel length to diameter ratio (i.e., 100 in our beam device) determines the beam divergence angle. The Doppler width is proportional to the width of the velocity distribution of the molecule parallel to the electron emission direction. A reduction of the Doppler width scales linearly with the ratio between the effective exciting beam diameter (typically 1 mm), and the distance to the beam (typically 5 mm). The reduction of more than 10 times can be easily achieved with minimal intensity loss. The molecular beam is directed in the downward vertical direction and enters a turbo molecular pump of 2000 l/sec, mounted off axis as the most efficient pumping action occurs close to the rim of the rotor, where the blade speed is highest.

The beam device described above is combined with the high-resolution electron spectrometer. The lens axis of the analyzer is in the horizontal direction so that the analyzer can see the molecular beam in the perpendicular direction. Figure 1 shows a typical example of the photoelectron spectra excited by Helium I radiation, recorded as a performance test of the apparatus. Here rotational levels of water molecules, having a Doppler contribution of 7 meV at room temperature, can easily be observed with the total line width less than 3 meV. The Doppler contribution to the observed FWHM is approximated as FWHM(Doppler)=0.722 $(E_{kin}T/M)^{1/2}$, where Ekin is the average kinetic energy of the observed electrons, T the temperature of the gas sample

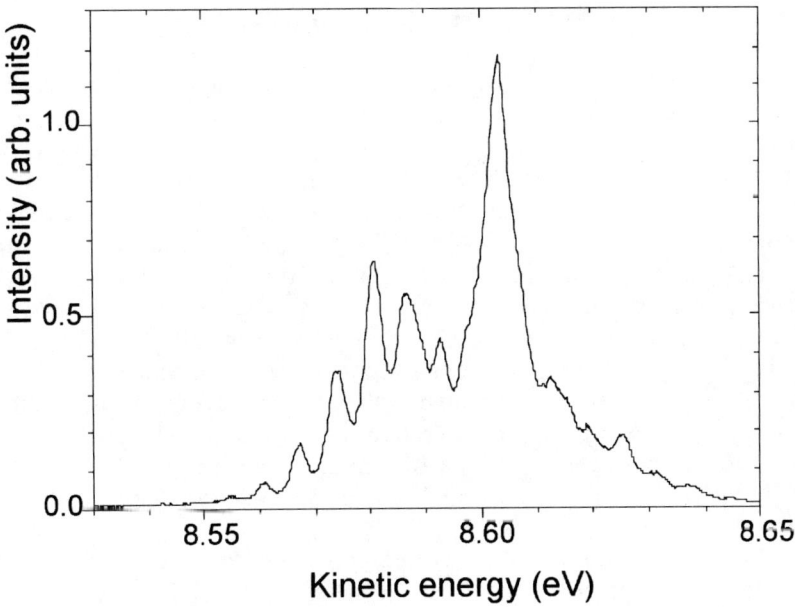

FIGURE 1. Rotational structure of the H_2O^+ X (000) band in the photoelectron spectrum of H_2O molecules excited by the He I radiation.

and M its mass, if E_{kin}, T and M are expressed in eV, K and atomic units, FWHM results in eV.

We installed the Doppler-free electron spectroscopy apparatus described above onto the beamline 27SU, nicknamed soft X-ray photochemistry beamline [6], at SPring-8, a synchrotron radiation facility with an 8-GeV storage ring in Japan. The monochromator installed in the beamline is of Hettrick-type. It has three exchangeable varied line-space plane gratings and covers the photon energy region between 150 eV and 2.5 keV [7]. Typical monochromator resolutions employed in the gas-phase photoemission studies are between 10,000 and 14,000. The Doppler-free electron spectroscopy apparatus sits on an XYZ stage so that the focusing point of the analyzer can be adjusted easily to the fixed photon beam position. The beamline 27SU has a figure-8 undulator as a radiation source and provides linearly polarized light. The direction of polarization vector E is horizontal for the 1st order harmonic and vertical for the so-called 0.5th order harmonic [8]. Thus one can switch the direction of the E vector by changing the gap of the undulator and perform the photoemission angular distribution measurement without rotating the analyzer.

RESULTS AND DISCUSSION

Figure 2 (a) shows a portion of the electron spectra of the Auger resonant Raman transitions to the Ne$^+$ 2p^{-2}(^{1}D$_2$)3p ^{2}P, ^{2}D and ^{2}F states via the Ne 1s^{-1}3p core-excited state. The spectra are recorded at the photon energy of 867.12 eV for horizontal (upper) and vertical (lower) polarizations. The analyzer is operated at the pass energy of 50 eV resulting in the analyzer bandwidth of ~33 meV. The spectra are well fitted to the three Gaussian profiles of ~66 meV FWHM as shown in Fig. 2. Gaussian profiles are chosen because in the electron spectra the broadening caused by instrumental effects is much larger than the natural lifetime width of the Auger final states. The experimental width of 63 meV corresponds to a convolution of the analyzer bandwidth ~33 meV and the monochromator bandwidth ~54 meV. This experimental width should be compared with the experimental width of previous observation at room temperature [4], where the observed widths 100~105 meV correspond to the convolution of the instrumental widths 60~63 meV and the inhomogeneous Doppler broadening of ~79 meV FWHM due to thermal motion of the neon atoms.

Figure 2 (b) shows a portion of the electron spectra of the Auger resonant Raman transitions to the Ne$^+$ 2p^{-2}(^{1}D$_2$)4p ^{2}P, ^{2}D and ^{2}F states via the Ne 1s^{-1}3p state. Here 2p^{-2}(^{1}D$_2$)4p ^{2}D and ^{2}P are completely overlapped even with the present highest resolution ever achieved, but 2p^{-2}(^{1}D$_2$)4p ^{2}F can be seen separately from ^{2}D and ^{2}P, more clearly than in the previous measurement collected with a gas cell [4].

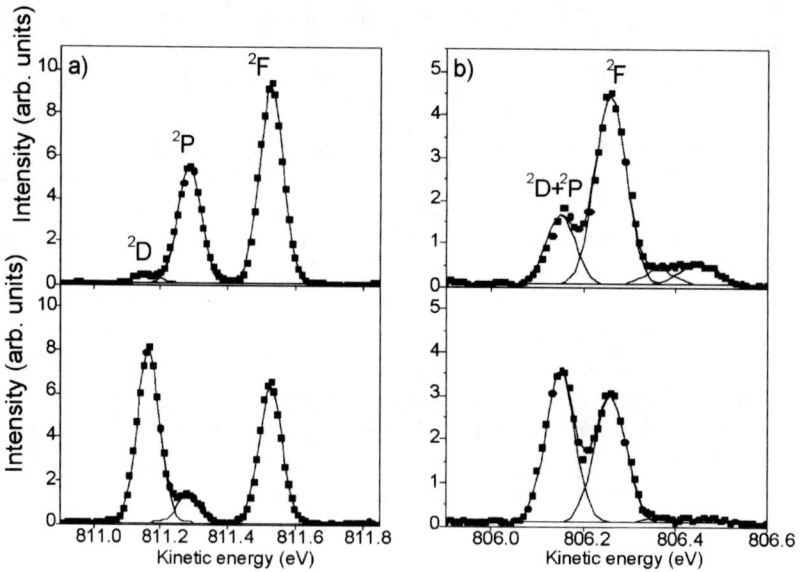

FIGURE 2. A portion of the electron spectra of the Auger resonant Raman transitions to (a) the Ne$^+$ 2p^{-2}(^{1}D$_2$)3p ^{2}P, ^{2}D and ^{2}F states and (b) the Ne$^+$ 2p^{-2}(^{1}D$_2$)4p ^{2}P, ^{2}D and ^{2}F states via the Ne 1s^{-1}3p core-excited state

The branching ratio for the $2p^{-2}(^1D_2)3p$, $4p$ ^2P, ^2D and ^2F final states as well as the anisotropy parameters β can be extracted from the spectra in Fig. 2. The results agree well with those measured previously, the prediction by the MCDF calculations, and the prediction based on the spectator model within the LSJ-coupling scheme [4]. The branching ratios for the $2s^{-1}2p^{-1}np$ final states as well as the anisotropy parameters β will be presented elsewhere.

Figure 3 presents a portion of the electron spectra of the Auger resonant Raman transitions to the H_2O^+ $1b_1$ $^2B_1$ X, $3a_1$ $^2A_1$ A, and $1b_2$ $^2B_2$ B states via the O $1s^{-1}$ $2b_2$ (000) excitation. Both spectra are recorded at room temperature, the upper spectrum is recorded using the gas cell whereas the lower spectrum is recorded with the gas jet device under the Doppler free condition. The monochromator bandwidth is ~40 meV and the bandwidth of the electron analyzer is ~33 meV. Unresolved rotational structure may add additional contribution to the line broadening. The drift of the photon energy during the measurements may also add additional line broadening. The Doppler width due to thermal motion of the sample molecule at room temperature is ~70 meV. One can clearly see the bending vibration (0,1,0) in the X band only in the Doppler free spectrum. The vibrational structure of the B band is also clearer in the Doppler free spectrum. The most significant difference is however seen in the A band, where a long progression of the bending vibrations is clearly seen only in the Doppler free spectrum.

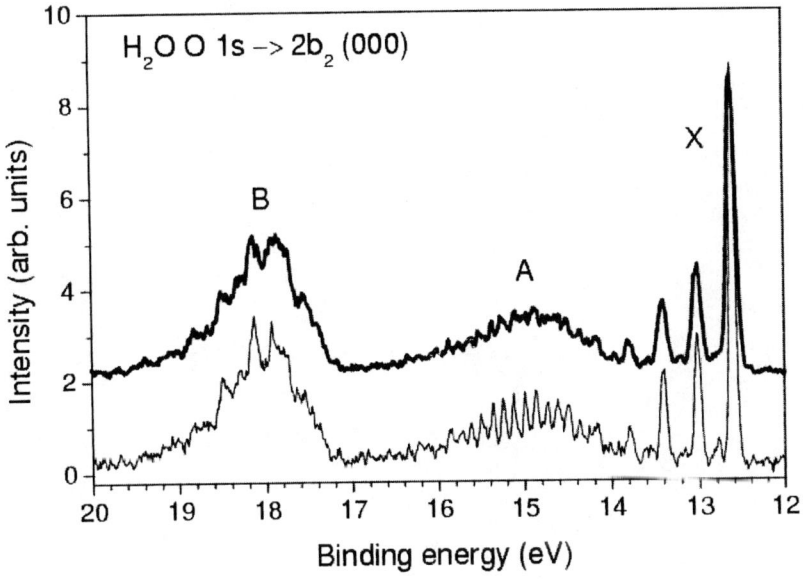

FIGURE 3. A portion of the electron spectra of the Auger resonant Raman transitions to the H_2O^+ $1b_1$ $^2B_1$ X, $3a_1$ $^2A_1$ A, and $1b_2$ $^2B_2$ B states via the O $1s^{-1}$ $2b_2$ state. The upper one, recorded with a gas-cell; the lower one, recorded under the Doppler free condition by using the gas-jet device.

ACKNOWLEDGMENTS

The experiment was carried out with the approval of the SPring-8 program advisory committee and supported in part by Grants-in-Aid for Scientific Research from the Japan Society of the Promotion of Science. The authors are grateful to Mitsuse Matsuki of MBS and Dr. Y. Takata of JASRI/SPring-8 for great assistance during the construction of apparatus and the performance test.

REFERENCES

1. T. Aberg and B. Crasemann, Resonant anomalous X-Ray Scattering, edited by G. Materlik, C. J. Sparks, and K. Fischer (North-Holland, 1994).
2. A. Kivim\"aki , A. Naves de Brito, H. Aksela, S. Aksela, O.-P. Sairanen, A. Ausmees, S. J. Osborne, L. B. Dantas, and S. Svensson , Phys. Rev. Lett. **71**, 4307 (1993).
3. G.B. Armen, H. Aksela, T. Aberg, and S. Aksela, J. Phys. B : At. Mol. Opt. Phys. **33**, R49 (2000).
4. Y. Shimizu, H. Yoshida, K. Okada, Y. Muramatsu, N. Saito, H. Ohashi, Y. Tamenori, S. Fritzsche, N.M. Kabachnik, H. Tanaka, and K. Ueda, J. Phys. B: At. Mol. Opt. Phys. **33**, L685 (2000).
5. A. De Fanis, K. Nobusada, I. Hjelte, N. Saito, M. Kitajima, H. Tanaka, H. Yoshida, A. Hiraya, I. Koyano, M.N. Piancastelli, and K. Ueda, J. Phys. B: At. Mol. Opt. Phys. **35** L23 (2002).
6. H. Ohashi, E. Ishiguro, Y. Tamenori, H. Kishimoto, M. Tanaka, M. Irie, and T. Ishikawa, Nucl. Instrum. Methods A **467**, 529 (2001).
7. H. Ohashi, E. Ishiguro, Y. Tamenori H. Okumura, A. Hiraya, H. Yoshida, Y. Senba, K. Okada, N. Saito, I.H. Suzuki, K. Ueda, T. Ibuki, S. Nagaoka, I. Koyano, and T. Ishikawa., Nucl. Instrum. Methods A **467**, 533 (2001).
8. T. Tanaka and H. Kitamura, J. Synchrotron Radiation **3**, 47 (1996).

New Perspectives for Advanced Science at the Brazilian Synchrotron Light Laboratory

Hélio C. N. Tolentino

LNLS – Laboratório Nacional de Luz Síncrotron, CP 6192, 13084-971, Campinas, Brazil

Abstract. The LNLS (Laboratório Nacional de Luz Síncrotron) is a national laboratory in Brazil that operates a 1.37 GeV storage ring for synchrotron light users since July 1997. Eleven bending magnet beamlines are open to a wide range of possibilities for research in ultra-violet and X-ray spectroscopy, single crystal and powder diffraction, magnetic and anomalous scattering, protein crystallography, X-ray fluorescence, X-ray lithography and small angle X-ray scattering. The recent conclusion of the booster injector opened the way for insertion devices to be accommodated in the four straight sections available. A multipolar wiggler, for protein crystallography using the MAD technique, is the first planned to be installed during 2003. The construction of the first LNLS undulator, for the vaccum ultra-violet and soft X-ray domain, has already started and will expand the possibilities in atomic, molecular and surface physics, as well as in catalysis and magnetism. LNLS has expanded its infra-structure as an open multidisciplinary research laboratory into complementary areas, such as electron and scanning probe microscopy, nanostructure synthesis and molecular biology. Many technological and scientific achievements have been attained in these last five years. Some of them will be highlighted here, with emphasis in the area of nanostructured and magnetic materials.

INTRODUCTION

The decision to set up a national laboratory and build a synchrotron light source in Brazil was taken in 1987 [1], after many years of debate, by the Brazilian National Council for Scientific and Technological Development (CNPq). The important question of why building such a huge facility in a developing country and, more importantly, how the historical, economic and political situation influenced the decision was, and still is, a subject of a warm debate [2,3]. The idea behind the building up of the synchrotron light source was conceptually defined by Cylon Gonçalves da Silva, former Director of LNLS, in a three-fold way: developing accelerator technology and instrumentation in Brazil; performing outstanding science using synchrotron light; introducing in our country the concept of a National Laboratory [3]. The strategy adopted by Cylon and Ricardo Rodrigues, the physicist who lead the whole technical project, was to start from scratch, form young people and interact as much as possible with local industry. The LNLS as a technological project has been a success and the conceptual organization as an open national facility has been accepted and integrated into the scientific community. The establishment of the local community was headed by Aldo Craievich, responsible for the scientific

CP652, *X-Ray and Inner-Shell Processes: 19th International Conference on X-Ray and Inner-Shell Processes*
edited by A. Bianconi, A. Marcelli, and N. L. Saini

© 2003 American Institute of Physics 0-7354-0111-X/03/$20.00

program, by advertising the potential of light sources as research tools through workshops and schools [4]. Since the opening to the scientific community in July 1997 [5], doing good science has been the crucial challenge faced by LNLS. Taking advantage of the success of the concept of national laboratory, it has expanded its capabilities into multi-disciplinary areas, such as electron and scanning probe microscopy, nanostructure synthesis and molecular biology, willing to be a strong research center. Nowadays, the Brazilian Synchrotron Light Laboratory is a reality and represents an important milestone for science and technology in our country.

In this review, I will present the main characteristics of the accelerators and outline the available beamlines. In addition, I will briefly describe other important existing facilities at LNLS. As far as science is concerned, I will focus only on a few examples related to nanostructure and magnetic materials using essentially synchrotron light. I will conclude by presenting the perspectives of implementing new facilities.

The Synchrotron Light Source

The LNLS synchrotron light source is composed of a 1.37 GeV electron storage ring, with a 120 MeV linear accelerator (LINAC) and a 500 MeV booster synchrotron for injection into the ring [6,7]. The flux produced by the bending magnet source, with a current of 160 mA stored, is compared with other similar storage rings (Fig.1).

The commissioning of the synchrotron light source started in May 1996. The full electron energy was attained, with a few mA, in October 1996. The electron current at full energy increased steadily, reaching the design value of 100 mA in July 1997. The

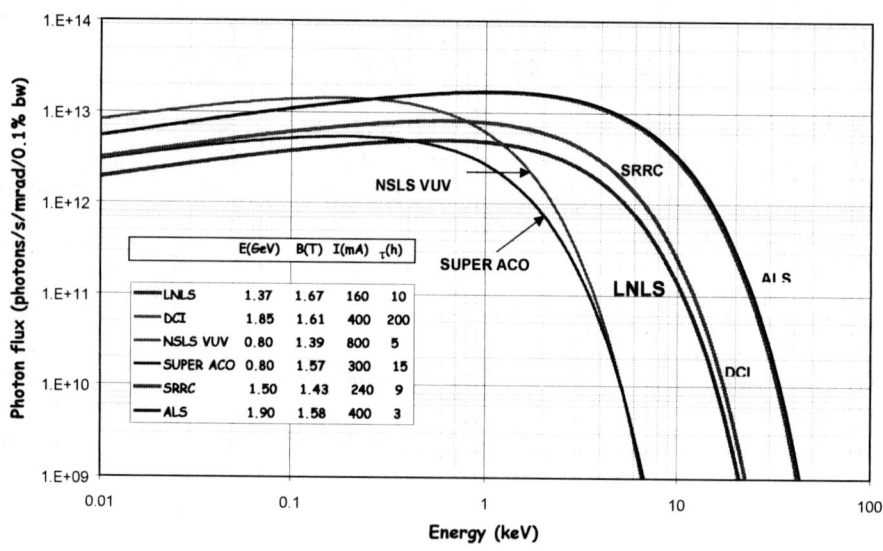

FIGURE 1. The photon flux delivered by the LNLS bending magnets compared to bending magnets sources in other facilities, using typical operating conditions. Courtesy from Liu Lin, LNLS

78

TABLE 1. Light source parameters for LNLS.

Parameter	Value	Unit
Operation energy	1.37	GeV
Injection energy from booster	500	MeV
Injection energy from LINAC	120	MeV
Critical photon energy	2.08	keV
Natural emmitance	100	nm rad
Number of dipoles	12	
Available straight sections	4	
Straight section free-length	2.95	m

significant challenge of injecting at very low energy, 120 MeV, was surpassed [6]. The storage ring operated from July 1997 until May 2001 using this injection scheme, delivering an average current of about 160 mA, with a mean lifetime of about 16 hours. Even though all design parameters of the synchrotron light source had already met their specifications, a booster synchrotron was built and installed between the linear accelerator and the storage ring [7,8]. Two important reasons guided this choice: 1/ improvement of the injection efficiency and 2/ installation of small gap insertion devices. Both the injection efficiency and vertical beam dimensions were limited by the low injection energy. The commissioning of the new injection system, including the 120 MeV LINAC and the 500 MeV booster, was accomplished in April 2001. The beam current in the storage ring achieved a record current of 540 mA at the injector energy of 500 MeV and up to 300 mA could be ramped to 1.37 GeV, being limited by the available power in the storage ring RF system and by the heat load in the dipole vacuum chambers. The upgrade of the RF system and the replacement of all dipole vacuum chambers are under way in order to enable storing higher currents in the main ring [8]. A list of the main parameters of the storage ring is shown in Table 1.

Overview of the Beamlines

The development and manufacturing of complex scientific instruments have been one of the main concerns of LNLS over the years. A total of eleven beamlines are currently in operation at LNLS, eight of them are for X-ray applications and three are for Vacuum Ultra-Violet (VUV) and Soft X-ray applications. All beamlines were built at LNLS, many of them with local original solutions. The list the operating beamlines are summarized in Table 2. Those under construction and projected are in Table 3.

Complementary Facilities

As an open multidisciplinary research laboratory, LNLS has expanded its infra-structure into complementary areas such as electron and scanning probe microscopy, nanostructure synthesis and molecular biology. According to their own characteristics, these facilities are open for users under a submission of proposal and selection. The main equipment for electron microscopy are a Low-Vacuum SEM (JSM-5900LV), a Field Emission Gun SEM (JSM-6330F), a 300 kV High Resolution TEM (JEM 3010) along with a complete sample preparation laboratory. STM and AFM Microscopes are also available as well as a chemistry laboratory for nanostructure synthesis. Deep X-ray lithography for micro-mechanical engineering is available, with a dedicated beamline and the infra-structure of the Microfabrication Laboratory. An important program on biology has been conducted with the creation of the Center for Structural Molecular Biology which has well-equipped laboratories for molecular biology, two NMR spectrometers and, in connection with the synchrotron light source, a bending magnet beamline for protein crystallography and a multipolar wiggler beamline for MAD which is under construction.

TABLE 2. LNLS beamlines in operation in 2002.

Beamlines	Monochromator	Applications
D03B: Protein Crystallography	One single bent crystal 6 – 12 keV	Structural molecular biology
D04A: Soft X-ray Spectroscopy	Double-crystal 800 – 4000 eV	Photoabsorption and Photoemission Spectroscopy
D04B: X-ray Absorption Spectroscopy	Channel-cut 3 – 24 keV	Material science; thin films and diluted systems
D05A: VUV Spectroscopy	Toroidal Grating (TGM) 12 – 300 eV	Surface, atomic and molecular physics; time-of-flight
D06A: Dispersive X-ray Absorption Spectroscopy	One single bent crystal 4 – 12 keV	Material science; in-situ studies; magnetic dichroism
D06B: X-ray lithography	Filtered White beam 5 – 20 keV	Deep X-ray lithography ; LIGA process
D08A: Soft X-ray and VUV spectroscopy	Spherical Grating (SGM) 300 – 1200 eV	Surface and interfaces; atomic and molecular physics
D09A: X-ray Fluorescence	White beam or Double-crystal 4 – 24 keV	Environment and geochemistry; biophysics and agriculture
D10A: X-ray Magnetic Scattering	Focusing Double-crystal 4 – 12 keV	Magnetic scattering; grazing incidence; nanostructures
D11A: Small Angle X-ray Scattering	One single bent crystal 6 – 12 keV	Glasses and nanocrystals, polymers, molecular biology
D12A: Multi-purpose X-ray Diffraction	Focusing Double-crystal 4 – 12 keV	Single crystal diffraction; multiple beam diffraction

TABLE 3. LNLS beamlines under construction and projected.

Beamlines	Monochromator	Applications
W01A: Multipolar wiggler for MAD Protein Crystallography	Focusing Double-crystal 6 – 15 keV	Structural molecular biology using MAD technique
D02A: Small Angle X-ray Scattering	One single bent crystal 6 – 12 keV	Glasses and nanocrystals, polymers, molecular biology
D02B: Powder X-ray Diffraction	Focusing Double-crystal 4 – 15 keV	Powder diffraction studies in material science
D05B: UV Fluorescence for Biology	Gratings	Molecular biology
D08B: X-ray Absorption Spectroscopy II	Focusing Double-crystal 4 – 15 keV	Material science; thin films and diluted systems
U11: Sasaki Undulator High resolution VUV spectroscopy	Plane Grating (PGM) 100 – 1200 eV	Surface and interfaces; atomic and molecular physics; X-ray magnetic dichroism

SCIENCE AT LNLS USING SYNCHROTRON LIGHT

The LNLS facilities have been used by the scientific community, with about 15% coming from abroad, since July 1997. The number of users, which can be measured by the number of performed proposals, is increasing every year, as well as the number of papers in international scientific publications. Many outcomes in science and technology have been documented over the last five years and can be found at www.lnls.br. Here, I will focus on a few highlights using synchrotron light, with emphasis in nanostructure and magnetic materials.

The Nature of Metal-Polymer Interaction in Cu and Co aggregates immersed in Polypyrrole

The mechanism of metal-loading into polymers is a complex subject of intense investigation, especially the very first stages, which starts with the formation of a complex between the metal and the polymer matrix [9]. The performance of these materials in applications, such as microelectronics and catalysis, is largely dependent on the chemical interaction of the metal and polymer, as well as on the crystalline structure of metal aggregates.

The structural and electronic properties of Cu and Co aggregates embedded in composite polypyrrole films were studied by XAS. Results at the Cu K-edge during *in situ* reduction [10] suggest that the reaction starts with the formation of the complex [-$[(C_4H_2N)_3CH_3(CH_2)_{11}OSO_3^-]_yCu^{+2}]_n$ (y=4), in which the copper is bonded to oxygen atoms. The reduction of this complex leads to the synthesis of Cu metal aggregates in the film (Fig.2). The results at the N K-edge evidence that the incorporation of the metal into the polymer network does not disturb the electronic structure nor the

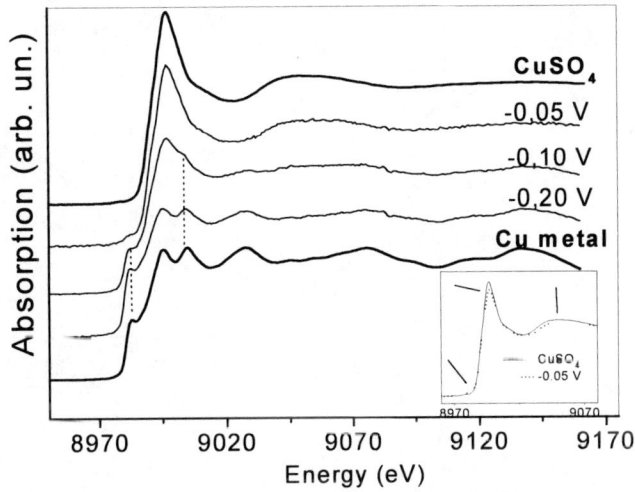

FIGURE 2. XAS results at the Cu K-edge of copper aggregates in polypyrrole. The evolution is shown for a series of applied potentials. In the inset, it is shown the reduction of the main white-line and the appearance of the pre-edge feature. Courtesy from Noêmia Watanabe, LNLS.

environment of the N from the pyrrole unit. However, the O K-edge indicates that the metal/polymer interaction happens via hybridization of O2p and Cu3d orbitals, resulting in a bonding having a quasi-covalent character [11]. On the other hand, Co seems to bond to N and not disturb the O environment. These results are compatible with quite different morphologies, shown by Scanning Electron Microscopy, of Cu and Co aggregates [12].

Structural, magnetic and transport behavior of Co clusters in Cu

Among the most remarkable questions concerning granular magnetic materials are the conditions leading to the phenomenon of giant magnetoresistance (GMR) [13]. The effect is understood in terms of the spin-dependent scattering of the electrons at interfaces, impurities and, to a lesser extent, within the magnetic elements [14].

Cobalt and copper form a granular magnetic system with nanoscale agglomerates of Co atoms embedded in the nonmagnetic Cu matrix. Apart from the magnetic properties, questions related to the shape and size of the particles, to the structural parameters and to the mechanism of crystal nucleation and growth are crucial to tune their properties.

EXAFS has been applied to study the evolution of Co particles in samples containing 3, 10, 12 and 25 at.% Co, annealed at different conditions. The coming out conclusions were that Co atoms segregates into small particles keeping the FCC packing arrangement, with a systematic contraction in the average nearest neighbor distance around Co as the thermal annealing is increased. These results are correlated to the evolution of the superparamagnetic behavior [15]. A simple model, based on spherical particles, permits to obtain the average particle size and to predict a correlation between the nearest neighbor distance contraction and the overall expected

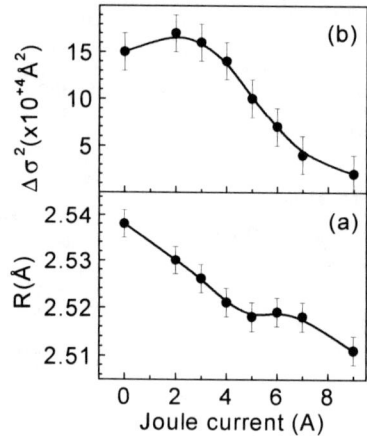

FIGURE 3. Average coordination distance (R) and relative Debye-Waller factor ($\Delta\sigma^2$) from EXAFS around Co atoms for a 10 at.% Co within the Cu matrix. As the annealing current increases, R contracts (a), showing that the Co atoms are segregating. The reduction in the Debye-Waller is related to the hardening of Co bonds to its neighbors. The figure is a courtesy from Julio C. Cezar.

disorder, given by the Debye-Waller term. The analysis leads to the finding that the maximum GMR occurs when the particle diameter is about 4 nm, which corresponds to about 30% of the Co atoms at the interface. This value optimizes GMR because the particles are sufficiently large to change orientation under an applied magnetic field and have a large amount of surface to scatter the conducting electrons responsible for the transport properties. Moreover, the disorder is significantly reduced at that point, due to a hardening of Co bonds, favoring the GMR [16].

Shape Transition of Self-Assembled Ge on Si(001) Using Anomalous X-ray Scattering

Epitaxial Ge on top of Si(001) has been heavily studied due to its simplicity to model self-assembled nanocrystal growth. One of the many morphologies that can be produced is the coherent island formation. The shape transition from pyramid to dome [17] depends on the amount of elastic energy stored in each morphology. In order to accurately evaluate the elastic energy of these self-assembled nanocrystals, detailed information on the atomic positions and chemical composition over the whole nanocrystal is necessary. This has been done by the combination of Atomic Force Microscopy (AFM), grazing incidence and anomalous X-ray scattering. AFM statistics provided a relationship between the island diameter and height. From grazing incidence X-ray scattering the strain relaxation as function of the vertical position in the island was obtained. To complement the study, anomalous x-ray scattering was used as a unique probe of the lattice parameter and composition. From all these data combined, the elastic energy of self-assembled Ge:Si pyramids and domes has been evaluated.

In the case of Ge:Si, a considerable decrease in energy from pyramids to domes was observed. This is induced both by the increase of lattice parameter as well as by and enhancement in intermixing [18]. In addition to the accurate evaluation of the

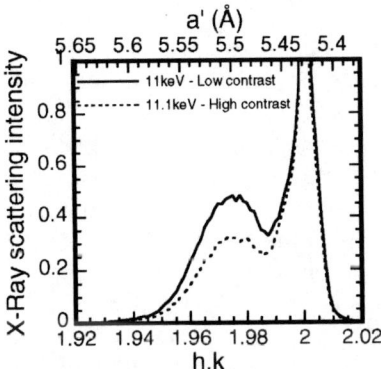

FIGURE 4. Anomalous x-ray scattering on domes, where the difference in scattering in the two curves comes from the presence of Ge atoms inside the islands. Therefore, for lattice parameters closer to the island base, a higher percentage of Si can be inferred, as it can be readily seen on the asymptotic behavior of the two curves towards the Si lattice parameter. Courtesy from Rogério Paniago. UFMG/LNLS.

elastic energy by a direct measurement, the authors have shown that both the use of X-ray anomalous scattering as well as the combination of lattice parameter and composition measurements are essential for the correct evaluation of the elastic energy and that the method can be generally applied to any other heteroepitaxial system.

Local Distortion and Strain in Magnetic Perovskites

Transition metal oxides with perovskite structure have attracted considerable attention due to their interesting physical properties, such as high Tc superconductivity in cuprates, colossal magnetoresistance in manganites and metal-to-insulator transitions in nickelates [19]. The ground state properties of these strongly correlated electron systems reflect the complexity among several correlated processes. Orbital and charge ordering, magnetic phases, electron-phonon interactions and lattice defects seem to be essential components for the understanding of the conducting mechanism.

A lot of work has been done to gather information about the ground state and lattice distortions on those perovskites. In particular, we have used XAS spectroscopy to study the role of distortion and electronic structure in manganese and nickel perovskites. XAS probes the local environment of the 3d-metal atom and has the advantage of being independent on the long-range order of the material.

I recall here the study of the Jahn-Teller (JT) distortion on $LaMnO_3$ polycrystalline sample. At room temperature, this compound is an antiferromagnetic insulator where a cooperative JT effect in the octahedra MnO_6 induces an orbital ordering and a large separation of the lattice parameters with, at the local scale, three different Mn-O distances [20] (1.97 Å, 1.91 Å, 2.18 Å). Heating up $LaMnO_3$ above $T_{JT} \approx 705$ K, the structure measured by neutron diffraction becomes quasi-cubic, with a Mn-O average distance of 2.01Å, suggesting a much more regular MnO_6 octahedron [4]. The local environment of Mn^{3+} ions was studied by EXAFS for several temperatures below and above T_{JT} and the result was that the modifications are essentially related to the MnO_6 polyhedral arrangement without any significant symmetrisation of the coordination shell around Mn atoms [21]. We concluded that the transition at T_{JT} corresponds to a change from a static and cooperative distortion to a non-cooperative one. This example illustrates the fundamental difference between a local and long-range probes.

In this proceeding, another study related to the strain on the structure of $La_{0.7}Sr_{0.3}MnO_3$ thin films grown epitaxially on different substrates is presented. The strain on this film strongly affects its magnetic and transport properties [22]. The results show that the tensile substrate induces either a deformation on the octahedra and an increase on the Mn-Mn distances [23], in such a way that one can hardly conclude about any change in Mn-O-Mn angle.

Rare earth nickelates display a very sharp metal-to-insulator transition as temperature decreases [24]. The EXAFS at Ni K edge of the $PrNiO_3$ perovskite indicate a transition from a structure with two different Ni sites at the insulating phase to a unique distorted Ni site at the metallic phase. These two sites have not been seen by neutron diffraction. The Ni L edge spectra show a remarkable difference between the spectra measured at the insulating and metallic phases. This result comes due to a different degree of hybridization between Ni3d and O2p bands in the metallic and in the insulating phase [25].

FUTURE DEVELOPMENTS AT LNLS

The LNLS users' scientific community has grown since the opening of the first beamlines and the importance of the synchrotron light source for the advancement of science in our country is demonstrated by the increasing number of proposed experiments and publications in international journals (www.lnls.br).

Following the internal and external user demands for beam-time, which is in some cases twice the offer, new beamlines are under construction and/or projected for the near future (Table 3). Research using X-ray Absorption Spectroscopy is one of the most demanding. In addition to the actual XAS beamline, a Dispersive XAS (DXAS) beamline is currently being commissioned and a high throughput XAS beamline for diluted systems is projected for 2003. A dedicated X-ray Powder Diffraction beamline is under construction and will be finished by the end of 2002. Connected to the molecular biology program, an Ultra-Violet Fluorescence (UVF) beamline has been partly funded by the FAPESP, funding agency of the state of São Paulo, and is under construction.

The recent operation with the new injector system, using a 500-MeV booster synchrotron, opened the way for small gap insertion devices to be installed in the ring. A 2 Tesla multipolar wiggler, for a protein crystallography beamline using the MAD (Multiple wavelength Anomalous Diffraction) technique, will be the first insertion device in the storage ring, with installation scheduled by Oct/Nov 2003. The detailed design of an undulator for VUV and Soft X-ray Spectroscopy has already started [26] and the completion of the beamline, scheduled for 2004, will open many possibilities in atomic, molecular, surface and condensed matter physics.

The users' community of synchrotron light in Brazil was almost non-existing a few years ago but it is a reality now. Over the last five years, the challenge of doing good science and becoming an international competitive research center has been faced by LNLS. Taking advantage of the success of the concept of national laboratory, LNLS has expanded its capabilities into multi-disciplinary areas, such as electron and scanning probe microscopy, nanostructure synthesis and molecular biology, willing to be an important actor in the development of science and technology in Brazil and Latin America. The enormous research potential existing at LNLS is still to be fully explored.

ACKNOWLEDGMENTS

I am grateful to the LNLS staff and to many colleagues who have contributed to the research described in this paper. Especial thanks to Liu Lin, Noêmia Watanabe, Julio C. Cezar, Rogério Magalhães-Paniago, Aline Y. Ramos and Cinthia Piamonteze, who have been involved in the research that are presented here and provided figures. The work presented here has been partially supported by LNLS. I would like also to thank funding agencies FAPESP and CNPq. Finally, I thank to the organizers of X02 for inviting me to present the Brazilian Synchrotron Light Laboratory in a such an interesting and fruitful occasion.

REFERENCES

1. Gonçalves da Silva, C.E.T., Rodrigues A.R.D., Craievich, A.F., *Synchrotron Radiation News Compendium*, Vols. **1-3**, 52-56 (1988); Gonçalves da Silva, C.E.T., *Synchrotron Radiation News* **4**, 9-11 (1991).
2. Burgos, M.B., *Ciência na Periferia: a luz síncrotron brasileira*, Juiz de Fora (MG), Brazil, Ed. UFJF, 1999. This is a result of a PhD Thesis on Sociology in the Federal University of Juiz de Fora.
3. Gonçalves da Silva, C.E.T., *Beam Line Magazine*, SLAC, Spring-Summer 1996; Gonçalves da Silva, C.E.T., Revista Mexicana de Física **43**, No. 5, 795-804 (1997).
4. Craievich, A.F., *Synchrotron Light: Applications and Related Instrumentation*, World Scientific, 1990; Craievich, A.F., Rodrigues, A.R.D., *Hyperfine Interact.* **113**, 465-475 (1998);
5. Rodrigues, A.R.D., Craievich, A.F., Gonçalves da Silva, C.E.T., *J. Synchrotron Rad.* **5**, 1157-1161 (1998);
6. Rodrigues, A.R.D., et al., "Commissioning and operation of the Brazilian Synchrotron Light Source" in *Proceedings of the 1997 Particle Accelerator Conference*, Vancouver, Canada, ed. M. Comyn, M.K. Craddock, M. Reiser, J. Thomson: Piscataway, NJ, USA, IEEE, 1997, pp. 811
7. Farias, R.H.A. et al., "Commissioning of the LNLS 500-MeV booster synchrotron" in *Proceedings of the 2001 Particle Accelerator Conference*,
8. Tavares, P.F., "The Brazilian Synchrotron Light Source" in *Proceedings of the International Symposium on the Utilization of Accelerators-2001*, São Paulo, Brazil, Intl. Atomic Energy Agency.
9. Cioffi, N. et al, *J. Mat.Chem.* **11**, 1434 (2001).
10. Alves, M.C.M., Watanabe N., Ramos, A.Y., Tolentino, H.C.N., *J. Synchrotron Rad.* **8**, 517 (2001).
11. Watanabe, N., Moraes, J., Alves, M.C.M., submtited (2002)
12. Watanabe, N., PhD Thesis, LNLS and Instituto de Química at UNICAMP, 2002
13. Berkowitz et al., PRL **68**, 3745 (1992); Xiao et al., PRL **68**, 3749 (1992); Chien et al., J.Appl. Phys. **73**, 5309 (1993)
14. S. Zhang and P.M. Levy, *J. Appl. Phys.* **59**, 4768 (1991); S. Zhang, *Appl. Phys. Lett.* **61**, 1855 (1992); S. Zhang and P.M. Levy, *J. Appl. Phys.* **73**, 5315 (1993)
15. Cezar, J.C., Tolentino, H.C.N., Knobel, M., *J. Magn. Magn. Mater.* **233**, 103-107 (2001);
16. Cezar, J.C., Tolentino, H.C.N., Knobel, M., *Phys. Rev. B*, submitted (2002)
17. Medeiros-Ribeiro et al., Science **279**, 353 (1998)
18. Magalhães-Paniago, R., Medeiros-Ribeiro, G., Kycia, S., Kamins, T.I., Stan Willians, R., *Phys. Rev. Lett.*, submitted (2002);
19. Imada, M., Fujimori, A., Tokura, Y., Rev. Mod. Phys. 70, No 4, 1039-1262 (1998).
20. Rodríguez-Carvajal, J., Hennion, M., Moussa, F., Moudden, A.H., Phys. Rev. B **57** 3189 (1998).
21. Araya-Rodriguez, E., Ramos, A.Y., Tolentino, H.C.N., Granado, E., Oseroff, S.B., *J. Magn. Magn. Mater.* **233**, 88-90 (2001).
22. Ranno, L., Llobet, A, Tiron, R. and Favre-Nicolin E, *Appl. Surf. Sc* **188** (2002) 170
23. Ramos, A.Y., Souza Neto, N.M., Giacomelli, C., Tolentino, H.C.N., Ranno, L., Favre-Nicolin, E., these proceedings.
24. Lacorre, P., Torrance, J.B., Pannetier, J., Nazzal, A.I., Wang, P.W., Huang, T.C., J. Solid State Chem. **91**, 225 (1991)
25. Piamonteze, C., Tolentino, H.C.N., Ramos, A.Y. Ramos, Massa, N.E., Alonso, J.A., Martinez-Lope, M.J., Casais, M.T., these proceedings.
26. De Castro, A.R.B., "An Undulator Beamline for LNLS", Memorando Técnico MeT – 01/98, LNLS, 1998

III. ADVANCES IN X-RAY OPTICS

From Surface Down To Bulk X-Ray Channeling

Sultan B. Dabagov

INFN - Laboratori Nazionali di Frascati, PO Box 13, I-00044 Frascati (RM), Italy
RAS - P.N. Lebedev Physical Institute, 119991 Moscow, Russia

Abstract. The basic point to be considered in this report is that X-ray capillary optics relies on the ability of a tapered and/or bent capillary channel to act as an X-ray waveguide. Recently several coherent phenomena associated with propagation of X-rays in capillary optical elements have been observed. In order to describe coherent phenomena of radiation propagation a quantum-wave theory of X-ray channeling was developed. Simple estimations speak on possibility of X-ray channeling in nano-scale capillaries, but with a significant change in character of channeling. In this connection, discovery of carbon nanotubes opens new opportunities to apply capillary waveguide optics. X-ray propagation in capillary micro- and nanostructures within the frame of wave approach will be discussed.

INTRODUCTION

Presently capillary optics represents a well-established X-ray and neutron optical instrument that allows experiments to be effectively performed on a much smaller scale than conventional devices[1]. These optical elements consist of hollow tapered tubes that transmit the radiation by multiple external reflections from the inner channel surface, efficiently deflecting through the angles that are tens and hundreds of times over the Fresnel's angle[2, 3, 4] . X-ray research activities over the last fifteen years demonstrated that capillary optics is a powerful instrument to guide the beams that makes it useful in numerous branches of applied research. The basic point to be considered in this report is that X-ray capillary optics relies also on the ability of a tapered and/or bent capillary channel to act as an X-ray waveguide[5], in other words, an optical element may be considered as a whispering-gallery X-ray device[6]. Recently several coherent phenomena associated with propagation of X-rays in capillary optical elements have been observed. In order to describe these effects a quantum-wave theory of X-ray channeling was developed. Simple estimations speak on possibility of X-ray channeling in nano-scale capillaries, but with a significant change in character of channeling: instead of surface channeling taken place in micro-size channels we deal with bulk channeling effect. In this connection, discovery of carbon nanotubes opens new opportunities to apply capillary waveguide optics. Theoretical analysis on scattering of X-rays in capillary micro- and nanostructures within the frame of wave approach will be discussed in comparison with experimental data.

As well known, a whispering gallery device is a multiple reflection one with a large number of bounces. The total angular aperture of the device is determined by the number of bounces N and the glancing angle θ, and equals $2N\theta$. While the reflectivity of single bounce light from a substance with the complex dielectric function $\varepsilon = 1 - \delta + i\beta$ (δ

CP652, *X-Ray and Inner-Shell Processes: 19th International Conference on X-Ray and Inner-Shell Processes*
edited by A. Bianconi, A. Marcelli, and N. L. Saini
© 2003 American Institute of Physics 0-7354-0111-X/03/$20.00

and β are the polarization and attenuation parameters, respectively) is defined by

$$R_1 {\scriptsize \begin{Bmatrix} s \\ p \end{Bmatrix}} \simeq 1 - 2Re\left(\begin{Bmatrix} 1 \\ \varepsilon \end{Bmatrix} (\varepsilon - 1)^{-1/2} \right), \tag{1}$$

where s, p are the indexes of radiation polarization, and $Re(x)$ is a real part of the complex variable x, the integral reflection may be estimated by taking the $\theta \to 0$ limit -

$$R {\scriptsize \begin{Bmatrix} s \\ p \end{Bmatrix}} \simeq \exp\left(-R_1 {\scriptsize \begin{Bmatrix} s \\ p \end{Bmatrix}}^{N\theta} \right) \tag{2}$$

The latter confirms quasi independence of the multiple reflection coefficient on the polarization of radiation. Evaluations made by these relations have shown that the whispering galleries offer high efficiency with narrow band pass[7].

Application of wave theory to the radiation passage through capillary structures opens up new prospects for the study[8, 9, 10]. As will be shown below, capillary optical systems act in such a way that radiation propagating in channels consists of two portions: one is scattered by the laws of ray optics, the other one is captured in bound modes by a surface potential. Moreover, when the channel size values approach the transverse wavelength of radiation, the bulk channeling of photons occurs similarly to the channeling of charged particles in crystals[11]. At present, the technology of capillary system manufacturing allows structures of the deep submicron level to be produced [12]. For example, as examined samples one may consider the carbon nanostructures (nanocapillary systems) [13], in the fabricating of which a big progress has been achieved in recent years [14].

A graphite nanotube [15, 16, 17] may be considered as a small-size (nano-size) capillary. The wall surface of it is formed by carbon atoms, the distance among which is estimated to be $1 \div 2$ Å. The typical diameter of a single nanotube is about tens of nanometers (the channel diameter / wall thickness ratio may reach two orders of magnitude), and the length of such a structure may be of submillimeter order. All these features of the nanotube structures result in the utilization of wave propagation theory instead of the ray approximation one, and consequently, nanotubes may be considered as X-ray waveguides. This allows the well known channeling theory to be applied for describing X-ray beam propagation inside these structures [18, 19].

Simulations of particle beam channeling in carbon nanotubes have been recently performed, in order to evaluate the possibilities for experimental observation of channeling effects in both straight and bent nanotubes, considering different charged particle species, such as protons of 1.3 and 70 GeV, and positrons of 0.5 GeV [20]. There the capabilities of a nanotube channeling technique for charged particle beam steering have been discussed, based on earlier Monte Carlo simulations [21].

In this work a unified description for X-ray propagation (note that the same theory is valid for neutrons) through capillaries of various diameters is presented.

QUANTUM-WAVE TREATMENT

As it has been shown recently, the propagation of X-ray photons through capillary systems exhibits a rather complex character [22]. Not all features shown in the experiments can be explained within the geometrical (ray) optics approximation [23, 24, 25]. On the contrary, the application of wave optics methods allows us to describe in details the processes of radiation spreading into capillaries.

The propagation of X radiation through capillary systems is mainly defined by its interaction with the inner channel walls. In the ideal case, when the boundary between hollow capillary and walls represents a smooth edge, the beam is split in two components: the mirror-reflected and refracted ones. The latter appears sharply suppressed in the case of total external reflection. The characteristics of scattering inside capillary structures can be evaluated from the solution of a wave propagation equation. In the first order approximation, without taking into account the roughness correction $\Delta\varepsilon(\mathbf{r}) = 0$ ($\Delta\varepsilon$ is the perturbation in dielectric permittivity induced by the presence of roughness), the wave equation in the transverse plane to the propagation direction reads

$$\left(-\nabla^2 + k^2\delta(\mathbf{r}_\perp) - k_\perp^2\right) E(\mathbf{r}_\perp) = 0 , \tag{3}$$

where E is a function of the radiation field, and $\mathbf{k} \equiv (k_\parallel, k_\perp)$ is a wave vector.

Due to the fact that the transverse wave vector may be presented as $k_\perp \approx k\theta$ under the grazing wave incidence ($\theta \ll 1$), an "effective interaction potential" is estimated by the expression

$$V_{eff}(\mathbf{r}_\perp) = k^2\left(\delta(\mathbf{r}_\perp) - \theta^2\right) = \begin{cases} -k^2\theta^2 & , \quad r_\perp < r_1 \\ k^2\left(\delta_0 - \theta^2\right) & , \quad r_\perp \geq r_1 , \end{cases} \tag{4}$$

where r_1 corresponds to the reflecting wall position. From the latter phenomenon of total external reflection at $V_{eff} = 0$ follows, when $\theta \equiv \theta_c \simeq \sqrt{\delta_0}$ - the Fresnel's angle. When we introduce a curvature in the reflecting surface, the effective potential obtains an additional contribution. This term that corresponds to the additional "potential energy", can be seen physically in the following way. Due to the reflecting surface curvature a photon receives an angular momentum $kr_{curv}\varphi$, where r_{curv} is a curvature radius of the photon trajectory. The latter is supplied by the "centrifugal potential energy" $-2k^2r_\perp/(r_{curv})$

$$V_{eff}(\mathbf{r}_\perp) = k^2\left(\delta(\mathbf{r}_\perp) - \theta^2 - 2\frac{r_\perp}{r_{curv}}\right). \tag{5}$$

The situation is explained in the scheme in Fig. 1. Because of the variation in the spatial system parameters, the interaction potential has been changed from the step potential with the potential barrier of $k^2\delta_0$ to the well potential, with the depth and width defined by the channel characteristics.

In the following we briefly discuss a solution of the wave equation in the case of an ideal reflecting surface (i.e. without roughness), when the reflected beam is basically determined by the coherently scattered part of radiation (for details see Refs. [26, 27]). The evaluation of the wave equation with the boundary conditions of a capillary channel

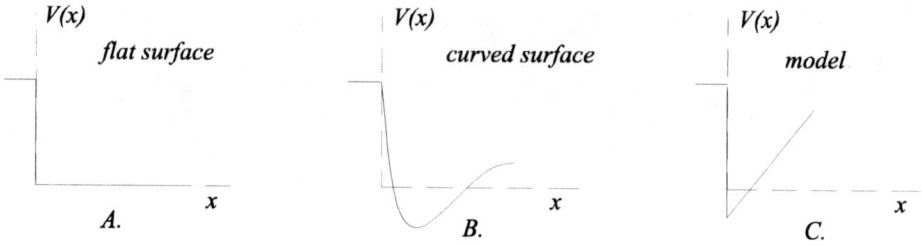

FIGURE 1. The change of the interaction potential between the flat surface (A.) and the curved one (B.). For simplicity, in calculations the real potential (B.) may be replaced by the model potential (C.).

shows that X-radiation may be distributed over the bound state modes defined by the capillary channel potential (see below). It is important to underline here that the channel potential acts as an effective reflecting barrier, and then, the effective transmission of X-radiation by the hollow capillary tubes is observed. While the main portion of radiation undergoes incoherent diffuse scattering, the remaining contribution (usually small) is due to coherent scattering that represents a special phenomenon, extremely interesting to observe and clarify [29].

Let us estimate the upper limit of curvature radius $(r_{curv})_m$ (which is defined by capillary/system of capillaries bending), at which the wave behaviors are displayed under propagation of radiation in channels [8], by considering a photon with the wave vector $\mathbf{k}$ channeling into capillary with curvature radius $(r_{curv})_i$ (i-th layer of capillaries). At small glancing angles, θ, the change of the longitudinal wave vector, $k_{\parallel}$, under reflection from a capillary wall is negligibly small; but one mainly changes the transverse wave vector, $k_{\perp}$,

$$k_{\perp} \simeq k\theta \ (\theta < \theta_c). \tag{6}$$

Correspondingly, from this relation it follows that the transverse wavelength will much exceed the longitudinal wavelength that provides the interference effects to be observ-

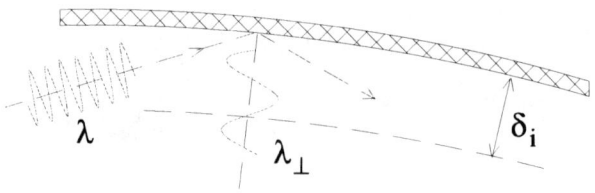

FIGURE 2. Illustration of X-ray reflection from the inner capillary surface. At glancing angles θ, when the cross size of a beam $\delta_i(\theta)$ becomes comparable with the transverse wavelength $\lambda_{\perp}(\theta)$, the radiation is grasped in a mode of surface channeling.

able even for very short wavelengths. Indeed,

$$\lambda_\perp \simeq \lambda/\theta >> \lambda \qquad (7)$$

quantum mechanical principles say that, in order to display the wave properties of a channeling photon, it is necessary that typical sizes of an "effective channel" δ_i, in which waves have been propagating, be commensurable with the transverse wavelength, i.e. $\delta_i(\theta) \simeq \lambda_\perp(\theta)$ (Fig. 2). This condition may be rewritten in the following form:

$$(r_{curv})_i \theta^3 \sim \lambda , \qquad (8)$$

from which we obtain $(r_{curv})_m \sim 10$ cm for a photon of $\lambda \sim 1$ Å wavelength. So, from this simple estimate we can conclude that the relation (8) provides a specific dependence for surface bound state propagation of X-rays - surface channeling - along the curved surfaces (for instance, in capillary systems) (see also [28]).

PROPAGATION EQUATION

Surface channeling

Since the waveguide is a hollow cylindrical tube, if the absorption is considered to be negligible, the interaction potential, in which a wave propagates, is determined by Eq.(5) with the radiation polarizability parameter $\delta_0 \simeq \theta_c^2$. Solving the wave equation we are mainly interested in the surface propagation, which, in fact, defines a wave guiding character inside the channel ($r_\perp \simeq r_1$, $\rho \ll r_1$) [10]

$$E_n(r) \simeq \sum_m C_m u_m(\rho)\, e^{i(kz+n\varphi)} ,$$

$$u_m(\rho) \propto \begin{cases} Ai_m(\rho) & , \quad \rho > 0 \\ \alpha Ai'_m(0) e^{\alpha \rho} & , \quad \rho < 0 \ \ (\alpha > 0) \end{cases} \qquad (9)$$

where $Ai_m(x)$ is the Airy function, and α is the arbitrary unit characterizing the capillary substance. Evidently, these expressions are valid only for the lower-order modes and in the vicinity of a channel surface. The expression (9) characterizes the waves that propagate close to the waveguide wall, or in other words, the equation describes the grazing modal structure of the electromagnetic field inside a capillary (surface bound X-ray channeling states) (Fig. 3). The solution shows also that the wave functions are damped both inside the channel wall and going from the wall towards the center. It should be underlined here that the bound modal propagation takes place without the wave front distortion. The analysis of these expressions allows us also to conclude that almost all radiation power is concentrated in the hollow region and, as a consequence, a small attenuation along the waveguide walls is observed.

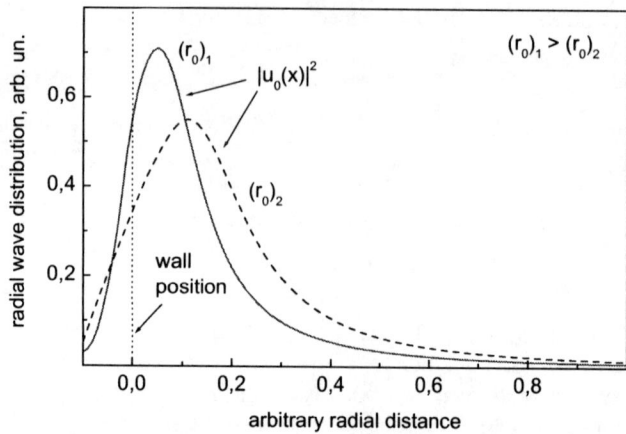

FIGURE 3. The radial distributions of the main bound mode of radiation inside a capillary channel for various channel diameters. The decrease of diameter ($2r_0$) results in a spatial displacement of the distribution away from the channel wall towards the center. The wall surface position is shown by the dotted line.

As for the supported modes of the electromagnetic field, estimating a characteristic radial size of the main grazing mode ($m = 0$) results in

$$\bar{u}_0 \simeq \sqrt[3]{\frac{\lambda^2 r_1}{2\pi^2}} \quad , \tag{10}$$

and we can conclude that the typical radial size $\bar{u}_0$ may overcome the wavelength λ, whereas the curvature radius r_1 in the trajectory plane exceeds the inner channel radius, r_0: $\bar{u}_0 \gg \lambda$ (for example, $\bar{u}_0 \gtrsim 0.1$ μm for a capillary channel with the radius $r_0 = 10$ μm).

Bulk channeling

Above we have considered the transmission of X-ray beams by capillary systems of micron- and submicron-size channels. Obviously, in that case we deal with surface channeling of radiation due to the fact that the channel sizes are larger than the radiation wavelength at least by three orders of magnitude. However, the situation sharply varies in the case when the sizes of channels become comparable with the radiation wavelength. In practice it means, that the angle of diffraction for the given wave, determined as $\theta_d = \lambda/d$ (being d a capillary diameter), becomes comparable with a critical angle of total external reflection. In other words, the transverse wavelength of a photon approaches the diameter of a capillary: $\lambda_\perp/d \sim 1$. In this case channeling of photons (note, not superficial (surface) channeling!) in channels of capillary systems should occur, i.e. we actually deal with a X-ray waveguide optics similar to a light waveguide one. Under the condition of the ordering channels in the system cross section, the capillary nanostructures are similar to crystals in relation to the charged particles, flying by under small angles to the main crystallographic directions.

As the analysis of the wave equation shows, the task cannot be analytically resolved for the real nanotube potential. For the sake of simplicity, let us consider the problem in the radial approximation for the periodic field of a multilayer waveguide with the size d_0 of a central channel and the distance d between the layers composing a waveguide wall. An interaction potential of the radiation in such a waveguide system may be presented as follows:

$$V(r) = \sum_n V_n(r) = k_r^2 \left[1 + \Delta \sum_n \delta\left(|r| - \frac{d_0}{2} - nd\right)\right] , \qquad (11)$$

where $\Delta \equiv \overline{\delta}_0 d$ is the spatially averaged polarizability of the wall substance.

Taking into account the boundary conditions and because of the potential symmetry, one may conclude that for the central channel $|r| \leq d_0/2$ the solution of the propagation equation in the transverse plane will be defined by the simple expression

$$E_0(r) = \begin{cases} a \cos(k_r r) &, \quad even \quad mode \\ a \sin(k_r r) &, \quad odd \quad mode \end{cases} \qquad (12)$$

Then we define the equation solution for the 1st layer $d_0/2 \leq |r| \leq d_0/2 + d$ by superposition of the opposite-directed waves $E(r) = b\, e^{ik_r r} + c\, e^{-ik_r r}$. As it has been done in the previous section, we make the mathematical assumption that all the modes of the total energy operator constitute a complete set of functions in the sense that an arbitrary continuous function can be expanded in terms of them. Then, we have a wave function $E(r)$ at a particular instant in time that obeys the continuous boundary conditions at the walls. In other words, we impose on the solutions the requirements that the wave function E and its transverse derivative E_r' be continuous at the wall-channel boundary

$$\begin{cases} E\left(\frac{d_0}{2} + d - 0\right) = E\left(\frac{d_0}{2} + d + 0\right) , \\ E_r'\left(\frac{d_0}{2} + d\right) = e^{i\kappa d} E_r'\left(\frac{d_0}{2}\right) , \end{cases} \qquad (13)$$

where κ is the quasimomentum determined by the Bloch theorem: $E(r+d) = e^{i\kappa d} E(r)$ - for the periodical potential function $V(r) = V(r+d)$. From these expressions we obtain the dispersion relations for even and odd waves

$$\begin{cases} \tan\frac{k_r d_0}{2} \\ \cot\frac{k_r d_0}{2} \end{cases} = \begin{cases} -\frac{k^2 \Delta}{k_r} + \frac{\cos(k_r d) - e^{i\kappa d}}{\sin(k_r d)} \\ \frac{k^2 \Delta}{k_r} - \frac{\cos(k_r d) - e^{i\kappa d}}{\sin(k_r d)} \end{cases} \qquad (14)$$

that allow the eigenvalue/eigenfunction problem to be solved.

Now it is interesting to write the wave functions of the supported modes for the narrow channel $\{k_r d_0, k_r d\} \ll 1$

$$E_n(r) \propto \begin{cases} \cos(k_r r)\, e^{ikz} & , \quad -\frac{d_0}{2} \le |r| \le \frac{d_0}{2} \\ \cos\frac{k_r d_0}{2} \; \dfrac{e^{i\kappa r}\sin(k_r|\tilde{r}|)-\sin[k_r(|\tilde{r}|-d)]}{\sin(k_r d)}\, e^{i(\kappa nd+kz)} & , \quad \frac{d_0}{2}+nd \le |r| \le \frac{d_0}{2}+(n+1)d \end{cases} ,$$

$$(15)$$

where $|\tilde{r}| \equiv |r| - d_0/2 - nd$. In this case we see that the Eqs.(14) may be solved only for even modes. However, it is more important to underline that the even mode exists for any ratio between the channel size and the layer distance. The spatial distribution of the mode has a maximum at the centre of the channel, and due to the leak through the potential barrier of wall layers we observe the propagation of radiation in substance. The radiation intensity for the successive layer decreases following an exponential law and is characterized by a local maximum far from the layer wall (the radial wave function distribution for the case of quasidistant layer system is shown in Fig. 4).

Because of the small wall thickness of nanotube channels (less than $\lambda_\perp \lesssim 100$ Å) we have to note that part of the radiation, channeling inside a nanotube structure, will undergo "tunneling" through the potential wall barrier. A simple analysis of the radiation propagation in systems both for the case of macroscopic channel and for the case of totally isotropic spatial structure, shows the presence of the main channeling mode (the main bound state) for any structure, whereas the high modes may be suppressed for specific channel sizes. Hence, nanotubes present a special interest as waveguides, which allow the supported modes to be governed. Moreover, there is a special interest in studying the dispersion of radiation in a nanosystem with a multilayered wall. As follows from the analysis of the general equation of radiation propagation considered above, at any correlation between the channel size and the interlayer distance at least one mode (bound state) should be formed in such a structure. In that case the diffraction of waves reflected from various layers of the channel wall should be observed, hence affecting the radiation distribution at the exit of system.

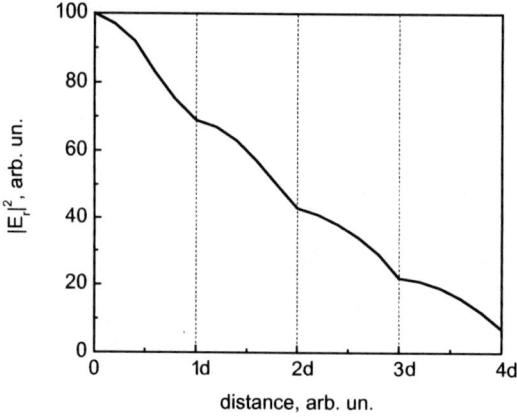

FIGURE 4. The typical radial wave function distribution for a periodic multilayer waveguide, where d is the distance between the layers composing a waveguide wall.

Evidently, the efficiency of these structures for applications have to be analyzed, despite the importance of the nanotube X-ray waveguide phenomenon from the fundamental point of view. The problems associated with X-ray and neutron channeling in capillary nanotubes (single- and multi-wall systems) present a special interest and will be analyzed in a subsequent paper.

CONCLUSION

The reduction in the channel size of capillary structures, as well as the discovery of a new class of natural nanosystems - i.e. carbon nanotubes - puts the problem of passage of X-ray quanta through these systems on a new qualitative level.

In the present work the general unified approach, allowing us to describe processes of radiation passage through capillary systems in a wide range of channel diameters, has been considered. Our analysis shows, that in the case of micron diameters the surface channeling of quanta presents in the mechanism of propagation (the part of radiation is distributed, being trapped by the bent surface of the channel), that can influence essentially the angular radiation distribution behind capillary systems. A decrease in size up to the nano-level results in a transition from surface channeling to bulk channeling. Thus, all radiation is involved in the process of the modal propagation, as opposed to the surface channeling when only part of the radiation is subject to the bound spreading.

Recently, it has been demonstrated for the specific cases that a decrease in the angular divergence of X-radiation behind capillary systems may be observed [30]. Strong differences between the observed and simulated angular distributions were demonstrated. It was shown that the obtained discrepancy in experimental and theoretical results could not be reproduced within the framework of the ray approximation and may be explained using the wave approach method.

The discovery of a new class of ordered systems like nanostructures (fullerenes, nanotubes) opens up interesting opportunities for studying coherent effects of the radiation interacting with nanosystems. The interest is not limited to fundamental research, on the contrary, there is a big potential for using the nanosystem samples, in order to develop new technological ideas (nanotube benders, nanosystem detectors, nanosystems as a source of electromagnetic radiation, nanotube undulators, etc.).

The main purpose of the present work is fundamental - i.e. to study the processes accompanying the X-ray transmission by capillary structures - in spite of the fact that capillary optics applications are covering larger and larger areas in X-ray physics and chemistry, biology and medicine. X-ray and neutron research activities over the last ten years demonstrated that capillary optics is a powerful instrument to guide neutral particle beams. Capillary/polycapillary optics can be applied in numerous branches of X-ray research, e.g. spectroscopy, fluorescence analysis, crystallography, imaging technics, tomography, lithography, etc. [12].

ACKNOWLEDGEMENTS

This work was partly supported by the NANO experiment of the Commissione Nazionale V of the Istituto Nazionale di Fisica Nucleare and by the Russian Federation Federal Program "Integration".

REFERENCES

1. Proc. SPIE **4765** (2002).
2. M.A. Kumakhov, and F.F. Komarov, Phys. Rep. **191.**, 289 (1990).
3. P. Engström, S. Larsson, and A. Rindby, Nucl. Instr. Meth. **A302**, 547 (1991).
4. D.J. Thiel, D.H. Bilderback, A. Lewis, and E.A. Stern, Nucl. Instr. Meth. **A317**, 597 (1992).
5. E. Spiller, and A. Segmüller, Appl. Phys. Lett. **24**, 60 (1974).
6. A. Vinogradov, V. Kovalev, I. Kozhevnikov, and V. Pustovalov, Sov. Phys. - Tech. Phys. **30**, 335 (1985).
7. N.V. Smith, and M.R. Howells, Nucl. Instr. Meth. **A347**, 115 (1994).
8. S.B. Dabagov, Research Report of FIROS for 1992 (Nalchik-Moscow, 1992) (in Russian).
9. S.B. Dabagov, and M.A. Kumakhov, Proc. SPIE **2515**, 124 (1995).
10. Yu.M. Alexandrov, S.B. Dabagov, M.A. Kumakhov, et al., Nucl. Instr. Meth. **B134**, 174 (1998).
11. J. Lindhard, Kgl. Dan. Vid. Selsk. Mat.-Fys. Medd. **34(14)**, 1 (1965).
12. M.A. Kumakhov, Proc. SPIE **4155**, 2 (2000).
13. E. Burattini, and S.B. Dabagov, Nuovo Cimento **B116**, 361 (2001).
14. R. Saito, G. Dresselhaus, M. S. Dresselhaus, "Physical Properties of Carbon Nanotubes" (Imperial College Press, London, 1998).
15. S.Iijima, Nature **354**, 56 (1991).
16. P.M. Ajayan, and S. Iijima, Nature **361**, 333 (1993).
17. S. Iijima, and T. Ichihashi, Nature **363**, 603 (1993).
18. N.K. Zhevago, and V.I. Glebov, Phys. Lett. **A250**, 360 (1998).
19. G.V. Dedkov, Nucl. Instr. Meth. **B143**, 584 (1998).
20. S. Bellucci, V.M. Biryukov, Yu.A. Chesnokov, et al., Nucl. Instr. and Methods (to be published), presented at the COSIRES conference in Dresden, Germany, June 23-27, 2002, physics/0208081.
21. V.M. Biryukov and S. Bellucci, Phys. Lett. **B542**, 111 (2002), physics/0205023.
22. S.B. Dabagov, M.A. Kumakhov, S.V. Nikitina, et al., J. Synchrotron Rad. **2**, 132 (1995).
23. N. Artemiev, A. Artemiev, V. Kohn, and N. Smolyakov, Phys. Scripta **57**, 228 (1998).
24. S.B. Dabagov, and A. Marcelli, Appl. Opt. **38**, 7494 (1999).
25. S.V. Kukhlevsky, F. Flora, A. Marinai, et al., Nucl. Instr. Meth. **B168**, 276 (2000).
26. S.B. Dabagov, V.A. Murashova, N.L. Svyatoslavsky, et al., Proc. SPIE. **3444**, 486 (1998).
27. S.B. Dabagov, A. Marcelli, V.A. Murashova, et al., Appl. Opt. **39**, 3338 (2000).
28. Chien Liu, and J.A. Golovchenko, Phys. Rev. Lett., **79**, 788 (1997).
29. S.B. Dabagov, A. Marcelli, V.A. Murashova, et al., Proc. SPIE **4138**, 79 (2000).
30. G. Cappuccio, S.B. Dabagov, C. Gramaccioni, and A. Pifferi, Appl. Phys. Lett. **78**, 2822 (2001).
31. S. Bellucci, S. Bini, V.M. Biryukov, et al., Phys. Rev. Lett. (submitted), physics/0208028.

Bragg amplification of the Si $K\alpha$ line emitted from a Mo/Si multilayer irradiated by an electron beam

P. Jonnard, J.-M. André, C. Bonnelle* and F. Bridou, B. Pardo[†]

*Laboratoire de Chimie Physique- Matière et Rayonnement, Université Pierre et Marie Curie,
UMR-CNRS 7614, 11 rue Pierre et Marie Curie, F-75231 Paris Cedex 05, France
[†]Laboratoire Charles Fabry de l'Institut d'Optique, UMR-CNRS 8501, Bat. 503, Centre
Scientifique d'Orsay, BP 147, F-91403 Orsay Cedex, France

Abstract. We report on the intensity modulation of the Si $K\alpha$ line emitted from a Mo/Si multilayer as a function of the exit angle of the photons. The observation takes place around the direction corresponding to the Bragg diffraction of Si $K\alpha$ by the multilayer. The sample is irradiated by an electron beam whose the energy is varied between 2 and 6 keV. An important intensity variation is observed within the angular range corresponding to the diffraction pattern of the emitting structure. A 15 % enhancement of the emitted radiation is measured in the Bragg direction of the multilayer, whatever the incident electron energy. This amplification is interpreted on the basis of the reciprocity theorem. A possible application as x-ray resonator is suggested.

INTRODUCTION

The dynamical theory of x-ray diffraction forsees that a characteristic x-ray line emitted from an element inside a periodic heterostructure can be diffracted under the Bragg conditions by the same emitting structure [1]. Interferences due to the diffraction process cause a modulation of the x-ray line intensity as a function of the exit angle in the narrow angular range satisfying the Bragg condition. Such modulations as a function of the exit angle have already been observed for the case of x-rays emitted by multilayer systems irradiated by photons (fluorescence) [2, 3].

We analyze the intensity of a characteristic line emitted by a periodic system under electron excitation as a function of the observation direction, in order to get an insight into the role of the excitation conditions on the x-ray intensity emitted in the Bragg direction. We report our results on the intensity variation of the Si $K\alpha$ line emitted by a Mo/Si multilayer.

EXPERIMENT, THEORY AND RESULTS

The Mo/Si multilayer is made of 40 Mo/Si bi-layers deposited by electron beam evaporation on a Si substrate. The thicknesses of the Mo and Si layers are 1.6 nm and 3.2 nm respectively, leading to a super-lattice period of 4.8 nm. An electron beam is used to ionized the 1s core shell of the silicon atoms (1.84 keV binding energy) present in the

CP652, X-Ray and Inner-Shell Processes: 19th International Conference on X-Ray and Inner-Shell Processes
edited by A. Bianconi, A. Marcelli, and N. L. Saini
© 2003 American Institute of Physics 0-7354-0111-X/03/$20.00

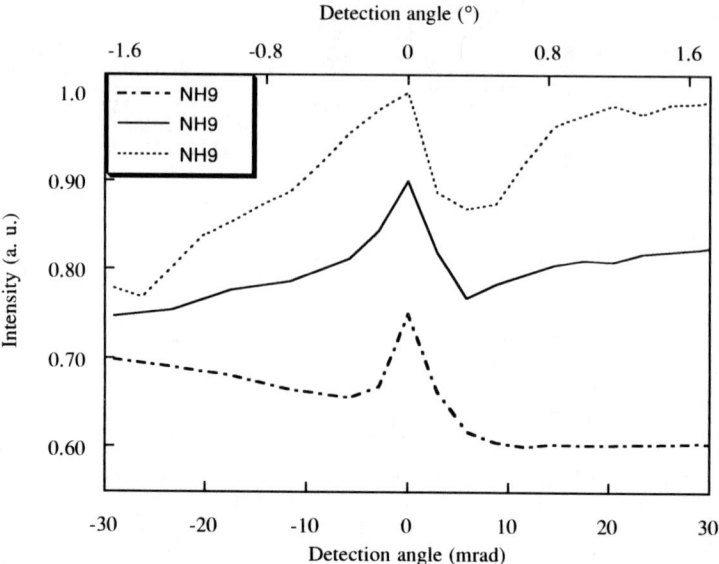

FIGURE 1. Intensity of the Si $K\alpha$ emission as a function of the α exit angle for the Mo/Si multilayer : 2.1 keV (dashed-dotted line); 4 keV (solid line); 6 keV (dotted line).

sample. The experimental setup is such that the directions of the incident electrons and the detected photons are perpendicular. Then the α angle between the sample surface and the direction of the detected photons is equal to the incidence angle of the electrons on the sample. The rotation axis of the sample holder is perpendicular to the plane defined by the electron and photon directions. This device enables to vary the α angle around the Bragg angle of the Mo/Si multilayer for the Si $K\alpha$ emission (76.8 mrad) with a relative precision of $\pm$ 1.5 mrad.

The Si $K\alpha$ emission (0.713 nm) from the Mo/Si multilayer is analyzed with a high-resolution x-ray spectrometer using an InSb (111) crystal at the first reflection order. The acceptance angle by the spectrometer is 0.4 mrad. The spectrometer is positioned at the peak of the Si $K\alpha$ emission and the intensity is measured as a function of the α exit angle. The measurements are performed for 2.1, 4 and 6 keV incident electron energies.

In Figure 1, we plot the distributions of the Si $K\alpha$ radiation between 35 and 105 mrad, emitted by the multilayer under 2.1, 4 and 6 keV electron irradiation. They present a maximum at an exit angle taken as zero in the figure. This angle is near the Bragg angle of the Mo/Si multilayer for the Si $K\alpha$ emission. At 4 and 6 keV, a dip appears towards the larger angles of the maximum. At 2.1 keV, the dip is very shallow. The relative intensity increase at the maximum with respect to the background has been estimated to be around 15 % for the three angular distributions. Similar experiments performed on a silicon single crystal only show a monotonous behavior as a function of the exit angle.

The intensity modulation observed for the multilayer spreads over an angular range of about 30 mrad. By using an optical model [4], the angular width of the multilayer reflection curve has been estimated to be of the same order. Then, the intensity modulation

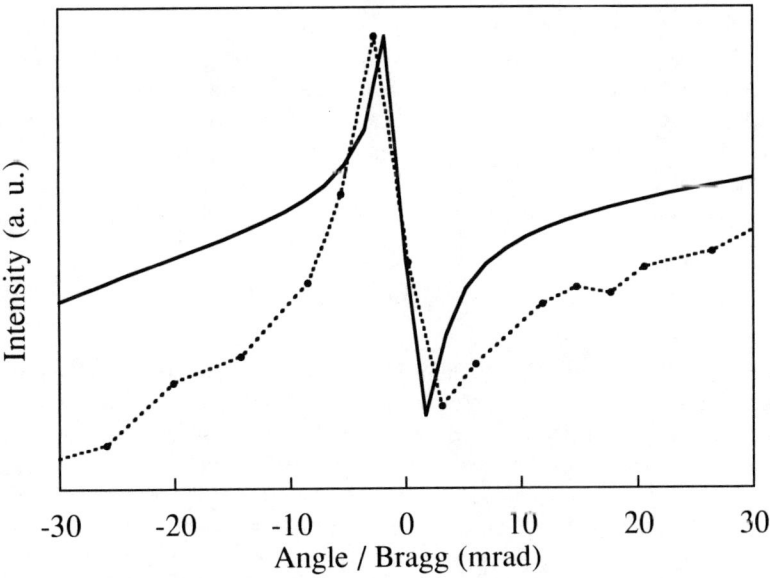

FIGURE 2. Intensity of the Si $K\alpha$ emission as a function of the α exit angle for the Mo/Si multilayer : experiment at 4 keV (dashed-dotted line); simulation (solid line).

extends over the entire Bragg reflection range.

A calculated angular distribution is derived from a model similar to that previously implemented to describe the Bragg diffraction of the fluorescence emission in multilayers [2]. For this calculation, we make the rough assumption that the excitation created by the incident electrons is independent of the depth, i.e. the slowing down of the electrons is neglected. The detected electromagnetic intensity is determined by using the Helmoltz reciprocity theorem [5], which specifies that the electric field on the detector can be deduced from the field produced at the point of the source by a fictitious dipole situated at the detector position. The calculation is performed by using the standard matrix formalism suited for the optics of multilayer medium [4, 6].

The theoretical intensity distribution is plotted in Figure 2, and compared to the experimental one obtained at 4 keV. In this figure, the zero angle is the Bragg angle of the theoretical curve. The experimental and simulated curves are adjusted at the angle corresponding to their intensity jump. The agreement between both curves is satisfactory in the modulation region. >From the lower exit angles, the intensity increases up to a maximum for an α angle very close to the Bragg angle. Beyond, the outgoing intensity decreases rapidly and a dip is observed. The general phenemona is well reproduces by the model. The discreapancy between the experiment and the theory can be attributed to the assumption of the homogeneous distribution of the ionizations inside the sample.

The comparison of the theory with the curves obtained at 2.1 and 6 keV would have been worse, because the shape of these curves is somewhat different from that of the curve obtained at 4 keV. By using a semi-empirical model [7, 8], we estimate that the Si 1s ionizations take place in the first 4 bi-layers at 2.1 keV, in almost all the Mo/Si heterostructure at 4 keV and in all the the multilayer and in a part of the Si substrate at 6 keV. The ionizations are produced within the same thickness in the experiment at 4 keV and in the calculation, which explains the good agreement between the two angular distributions. At 2.1 and 6 keV, because of the different locations of the x-ray generation, the shapes of the experimental angular distribution change and are not in so good agreement with the theory. This shows the importance of using electrons as incident particles to control the generation of the x-rays within the heterostructure.

CONCLUSION

We have shown that it is possible to enhance the x-ray intensity emitted by a multilayer under electron irradiation in the direction of the Bragg diffraction. The observed modulation of the x-ray intensity takes place within the width of the rocking curve of the multilayer. One expects an increase of the x-ray intensity with respect to the background with the narrowing of the multilayer reflection curve. By varying the incident electron energy, it is observed that the shape of the angular distribution changes, but the enhancement of the intensity above the background remains the same, i.e. about 15 % in the case of the studied Mo/si multilayer. This amplification is well interpreted on the basis of the reciprocity theorem.

Following Yariv [9], we think that it is possible to use the multilayer systems simultaneously as emitter and resonator structures in the soft x-ray range. Up to now, experiments developed with the aim of amplifying the soft x-ray emissions have been performed by subjecting the x-ray laser beam to one, or more, reflections on an independent optical system. Then, in all the cases, the emitting and amplifying media were not the same. We propose here to use as an amplifying medium a multilayer mirror and use one of the two bi-layers as the emitting medium. The advantage of such a multilayer based device over the crystal based one proposed by Yariv [10], is that the super-lattice period can be suited to any wavelength range. One, or more, feedback reflections of the radiation can occur in the medium before its outgoing and amplification of the radiation is expected in the Bragg direction.

REFERENCES

1. R. W. James, *The Optical Principles of the Diffraction of X-rays* (G. Bell and Sons, London, 1982) pp.438–456.
2. J.-P. Chauvineau, and F. Bridou, J. Phys. IV (Fr.) **6** , C7-53 (1996).
3. H. P. Urbach, and P. K. de Bokx, Phys. rev. B**63**, 085408 (2001).
4. B. Pardo, T. Megademini, and J.-M. André, Rev. Phys. Appl. **23**, 1579 (1988).
5. H. A. Lorentz, Proc. Amsterdam Acad. **8**, 401 (1905).
6. F. Abelès, Ann. Phys. **12**, 706 (1950).
7. P.-F. Staub, X-Ray Spectrom. **27**, 43 (1998).
8. P.-F. Staub, P. Jonnard, F. Vergand, J. Thirion, and C. Bonnelle, X-Ray Spectrom. **27**, 58 (1998).
9. A. Yariv, and P. Yeh, J. Opt. Soc. Am. **67**, 438 (1977).
10. A. Yariv, Appl. Phys. Lett. **25**, 105 (1974).

Novel methods of synthesis of a-Si(H)/Mo multilayers for Extreme UV applications

V. Rigato[a], A. Patelli[a,b]

[a] INFN, Laboratori Nazionali di Legnaro, Via Romea, 4, 35020 Legnaro (Padova), Italy
[b] Dipartimento di Ingegneria dell'Informazione, Università di Padova, Via Gradenigo 6/A, 35100 Padova (Italy)

Abstract. This report is focused on the control of the plasma sputtering process used for the synthesis of multilayer mirrors for extreme UV and soft X-ray optical devices. The case of a-Si/Mo and of a-Si(H)/Mo multilayers will be discussed, with emphasis on the study of the basic parameters that control the growth of the single Mo and a-Si(H) layers. The deposition apparatus will be described. The effects of the plasma composition and of the bombardment of the growing layers by Ar^+ ions and electrons on the properties of the layers are reported. The hydrogen concentration can be varied up to about 30 at% in the a-Si(H) layers: the hydrogen incorporation leads to a decrease of the material density. The accurate control of the plasma density and of the electron temperature and plasma potential is accomplished through the Langmuir probe method, that permits the accurate measurement of the energy of the Ar^+ ions and of their fluence. The possibility of using the present apparatus for the synthesis of a class of ML designed for smaller wavelength applications (down to the water window spectral region) is discussed.

INTRODUCTION

The aim of the ARCHIMEDE project financed by INFN is to grow multilayer mirrors with high normal-incidence reflectivity for x-ray and EUV wavelengths in the spectral region from 2.4 to 30 nm onto large shaped substrates. In this spectral region the foreseen applications in astrophysics and in projection nanolithography are extremely important. Furthermore the lower wavelength region from about 2.4 to about 4.0 nm is of particular interest for x-ray microscopy of organic structures (viruses, proteins, DNA) in aqueous environment down to 20 nm in size. For any such envisaged application, it is necessary to collect the x-ray radiation emitted from a source and focus it onto an object. This can be accomplished by accurately depositing x-ray multilayer mirrors consisting of alternating high/low refractive index layers with a minimum absorption, onto large, curved substrates.

For maximum reflectivity, abrupt and flat interfaces are essential. For this reason a low substrate temperature and the absence of energetic particle irradiation during growth should be used to minimize bulk diffusion and interface mixing. These

CP652, *X-Ray and Inner-Shell Processes: 19th International Conference on X-Ray and Inner-Shell Processes*
edited by A. Bianconi, A. Marcelli, and N. L. Saini
© 2003 American Institute of Physics 0-7354-0111-X/03/$20.00

growing conditions may lead to a cinematically limited growth with an increased roughness as a consequence. Therefore the ion process needs a careful optimization.

Many research groups are involved in the production of these multilayer structures by different techniques as DC magnetron sputtering [1-3] and evaporation [4]. In this work the multilayers are prepared by RF plasma magnetron sputtering.

In the magnetron sputter deposition process we expect that the energetic particles present in the plasma affect the surface morphology of the growing film. In multilayers' growth these effects will be enhanced due to the large amount of interfaces.

We will summarize our results on the growth of a-Si(H)/Mo multilayer suited for applications at 13 nm and at 30.4 nm wavelength. The emphasis will be mainly on the growth method and on the process variables that permit to control the preparation of these multilayers.

EXPERIMENTAL

The experimental apparatus used at LNL for producing the X-Ray /EUV mirrors consists in a AISI 304/316 sputtering chamber with dimensions 110 x 72 x 65 cm evacuated by a 1000 l/s turbomolecular pump down to about 10^{-7} mbar. Two rectangular 10 x 20 cm^2 titanium magnetrons sputter sources are used for depositing Ti onto part of the chamber walls before multilayer deposition starts. In this way the ultimate pressure of about 10^{-8} mbar can be routinely achieved. The oxygen and water vapor residual pressures are thus kept at a value such that they do not contaminate the growing Mo and Si layers. In the deposition region of the sputtering chamber up to 4 magnetron sputter sources can be installed. These sources are either 2" planar or 2"x6" rectangular. Direct sputter deposition or facing target deposition can be accomplished by internally interchanging the hardware. The samples can be mounted onto different sample holders, capable of different movements as needed.

In the following we will describe the apparatus comprising two planar 2" UHV type II unbalanced magnetron sputter sources (AJA International) and a sample holder with two positions that can be biased positive or negative with respect to ground or it can be left floating. The method is sputtering up-word with the sputter target normal making an angle of 0°, 10°, and 20° with respect to the sample normal. For measuring the temperature of the sample during film deposition a 1.0 mm shielded "K" type thermocouple is fixed onto the sample holder near the sample position. The sample holder motion is controlled by a personal computer that also controls the open/close status of the source shutters. The samples are positioned alternatively over the Mo or Si sources following a time sequence that is pre-determined by the user on the basis of the mirror design and is stored in the computer program. We do not spin the samples around their axis during deposition as it is done in most other laboratories [1-3], instead our deposition is static and the experimental parameters are chosen so that we can guarantee a film uniformity of better that 1% absolute over a region 2x2 cm^2. The distance between target and substrate is set at 16.5 cm.

The plasma gas is an Ar (99.9999%) and H_2 (99.9999%) mixture. The operating total pressure is $5 \cdot 10^{-3}$ mbar as measured by a MKS capacitive gauge, Model 627B, capable of extremely stable readings in the range 10^{-1} to $1 \cdot 10^{-5}$ mbar. This instrument is also used for precise measurement of the H_2 partial pressure. For the deposition of the a-Si(H) layers the H_2 partial pressure has been varied from 1% to 10%. Standard a-Si/Mo multilayers are obtained in pure Ar plasma.

Each source is driven by a RF power supply (RFPP 5S and AE RFX600) at 13.56MHz connected to a low-loss matching box. The two power supplies are regulated in the forward power mode: $W_{Si}=150W$ (V_{B_Si} about -500V), $W_{Mo}=60W$ (V_{B_Mo} about -300V). For molybdenum the net Ar^+ current density during sputtering has been determined to be 5.0 mA/cm^2 by weighting the total amount of deposited Mo onto a domed shaped surface after a given period of time assuming a sputtering Yield $Y_{Mo}(300eV) = 0.45$.

In the experimental setup used in this experiment the samples are biased with the aid of a power amplifier (KEPCO Mod. BOP 100-4M) sourcing currents from –4 A to +4 A at bias voltages (V_B) from –100V to +100V. This amplifier is interfaced to a programmable function generator (NF 1940) that permits to program an arbitrary sample bias profile for the modulation of the ion and electron bombardment during the film growth. Several experiments have been carried out by DC biasing the samples at voltages from –50V (enhance Ar^+ bombardment) to +70V (electron bombardment, sample heating). Some samples have been grown with a pulsed DC waveform at 1.0 kHz and 50Hz (negative –50V, positive +70 V, 25%/75% duty cycle).

Ion Beam Analysis (Rutherford Backscattering Spectrometry –RBS- and Elastic Recoil Detection Analysis – ERD-) is used routinely for determining both deposition rates and contamination of the deposits. The deposition rates of Si is about $2.4 \cdot 10^{14}$ atoms/cm^2/s (0.48 Å/s) and that of Mo is $3.0 \cdot 10^{14}$ atoms/cm^2/s (0.47 Å/s) as measured by RBS.

Four kind of samples are investigated in this paper: 1) 200 nm thick a-Si, a-Si(H) and Mo single layers; 2) bilayers of a-Si (or a-Si(H)) and Mo 200 nm + 200 nm; 3) a-Si(H)/Mo multilayer mirrors designed for $\lambda=30.4$ nm ($\Lambda=d_{Si}+d_{Mo}$ about 16 nm, $d_{Mo} = 3.0$ nm.; 15 periods); 4) a-Si(H)/Mo multilayer mirrors designed for $\lambda=13$ nm ($\Lambda=d_{Si}+d_{Mo}$ about 7 nm, $d_{Mo} = 2.8$nm; 40 periods). All these samples are deposited onto Si single crystals with surface roughness of about 0.1 nm using different bias voltages.

To know the Ar^+ energy and bombardment flux of the growing films as a consequence of the above mentioned bias procedure, we performed the measurement of the plasma density, plasma potential (V_p) and electron temperature (T_e) of the RF glow discharge as a function of the position near the deposition surface at different bias voltages. This has been accomplished by means of a motorized cylindrical Langmuir probe apparatus suited RF plasma diagnostics [5] and characterized by high speed data acquisition (Scientific Systems SmartProbe).

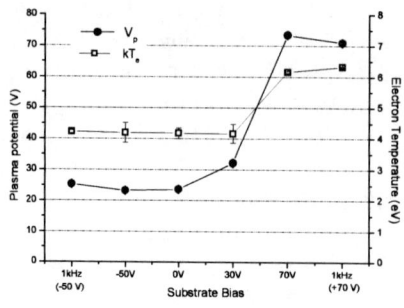 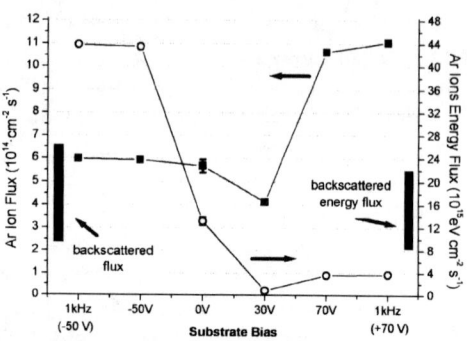

FIGURE 1. Plasma potential and electron temperature as a function of the different bias conditions. The data at 1 kHz refer to a bias profile with –50V(25%)/+70V(75%). The plasma potential and electron temperature are measured both in the negative and positive pulse period and are found to be quite different.

FIGURE 2. Ar$^+$ fluxes bombarding the samples as a function of DC sample bias. The total energy flux to the growing film is shown. The backscattered Ar flux and the corresponding energy flux in the case of Mo deposition is estimated from TRIM.SP calculations and geometrical parameters.

RESULTS

By X-Ray diffraction analysis (not reported in the present paper) and RBS analysis we found that the Si and Si(H) layers are X-ray amorphous and that they incorporate Ar with average concentration varying between 1 to 7 at%. On the other hand Ar incorporation in Mo layers is found to be much lower (of the order of less than 0.5 at% as determined by RBS). Mo layers are nano-crystalline with (110) preferred orientation.

Particle bombardment is generally accomplished by accelerating Ar$^+$ ions into the growing films by applying an appropriate voltage to the substrate. This method permits to control directly the amount of bombardment of the growing film if the plasma potential and ion density are known. In Fig. 1 the plasma potential and electron temperature are shown as a function of the different bias conditions used in this work. As it can be seen the electron temperature is about 4 eV for bias below about +25V, and jumps to more than 6 eV for higher V_B values. The plasma potential V_p is about +23V for bias voltages up to +25V and raises at about V_p=+74V at positive bias of +70V. The Ar$^+$ flux bombarding the Si and Mo samples as a function of sample bias is shown in Fig. 2 together with the total energy flux to the growing films. We notice that the ion flux varies from about 10^{15} ions/cm^2/s (V_B=+70V) to about 4-6 10^{14} ions/cm^2/s at a dc bias of 0V or negative. The energy e(V_p-V_B) of these ions can be very low when the plasma potential and bias potential are very near. From the data reported in Fig. 2 we see that the energy deposited intentionally into the sample surface is maximum when the sample is negatively biased at –50V and reaches its minimum when the sample is positively biased.

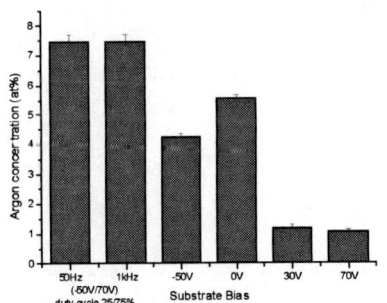

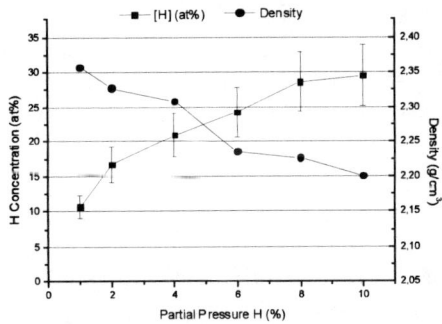

FIGURE 3. Argon atomic concentration in the thick a-Si layers as a function of different bias.

FIGURE 4. Density of a-Si(H) films as determined by X-Ray reflectometry as a function of the H_2 partial pressure in the deposition plasma. The H atomic concentration is determined by ERD.

The measured temperature of the growing surface when positively biased is in the range 100-150 °C while the 0V and –50V biased samples are deposited at a temperature that does not exceed 50°C. The ratio between the deposited atoms and the Ar ion fluxes varies in the range from about 2 to 4.

As a matter of fact the growing film is also bombarded by other energetic particles whose origin is intrinsic to the plasma sputtering process: the sputtered particles themselves and the reflected Ar atoms which are responsible for a complex bombardment process that cannot be easily controlled. The Ar atoms reflected back from the target can be characterized in terms of Particle Reflection Coefficient (R_N) which represents the ratio of the number of back-scattered particles to the number of incident particles and in terms of the Energy Reflection Coefficient (R_E) that quantifies the fraction of the initial input energy to the target that is carried away by the reflected particles.

The process of reflection of Ar atoms from Si is completely different and less efficient than that of Mo. Following the data reported in [6,7] based on TRIM.SP Monte Carlo calculations [8] $R_E^{Si}=5.8 \cdot 10^{-5}$ and $R_N^{Si}=3.8 \ 10^{-3}$ (for energy $E_{Ar}=500eV$, normal incidence) while for Mo we expect $R_E^{Mo}=0.04$ and $R_N^{Mo}=0.24$ (for 300 eV, normal incidence). This means that during the Si deposition the effects due to reflected Ar neutrals are negligible. Furthermore the Si sputtered particles undergo lot of collisions with the Ar gas particles and reach the substrate with energy much lower than the original sputter ejection energy due to the considerable thermalization. On the other hand, when the samples are positioned over the molybdenum target they are subject to the bombardment operated by the high flux of reflected Ar neutrals and by the sputtered Mo atoms less efficiently thermalized than Si atoms because of their higher mass. In order to quantify the flux and energy of the bombardment we performed calculation of the reflected Ar flux at our substrates by integrating the angular end energy distributions that resulted from TRIM.SP [8] calculations using the measured value of Ar^+ impinging current. The results of this calculations show that the flux of reflected neutrals to our substrates can be estimated to vary from about 2 to about 6 10^{14} $Ar/cm^2/s$ (see Fig. 2). The average energy at the substrate of the backscattered Ar is about 35 eV taking into account the effects of thermalization [9].

The estimated contribution of the backscattered particles is comparable to the intentional flux and it is high enough to produce measurable effects on the film microstructure and to the quality of the Si/Mo interfaces. The energy of the backscattered atoms dominates the surface bombardment for bias potentials higher than 0 V.

Ar Incorporation

The incorporation of rare gases at concentration ranging from a small fraction of percent to several atomic percent is often observed in films deposited at low temperature using rare gas plasma sputter deposition. Since rare gases are only weakly physisorbed on solid surfaces, their incorporation can only occur by trapping of incident energetic particles. As said above there are two primary sources of these energetic species: a) ions which are accelerated to the substrate by an induced or applied negative substrate bias, b) ions that are neutralized and reflected from the sputtering target. The measurement of Ar incorporation is important for understanding the effects of the Ar bombardment on the micro-structural properties of the growing films. The argon content in the 200nm thick a-Si films is shown in Fig. 3 as a function of the different bias conditions. As it can be seen when these samples are positively biased they trap much less Ar due to both the increase of the layer temperature caused by the electron bombardment, and to the lower energy of argon ions. The samples obtained at V_B=-50V show a lower Ar incorporation than those obtained at V_B=0V due to the resputtering induced by the energetic Ar flux (Fig. 2). We observe a high Ar incorporation (7.4at%) in thick a-Si layers subject to alternate −50V(25%)/+70V(75%) polarization either at 1 kHz or at 50 Hz. These layers are bombarded by Ar^+ (75eV) for 1/4 of the period (with fluxes equal to those obtained for steady V_B= −50V polarization- see Fig. 2) and by both an intense flux of plasma electrons and by a very high flux of low energy Ar ions, for the remaining fraction of the period. The temperature of the growing layers in these conditions rises to about 100-150°C. The average Ar concentration in the a-Si layers of different multilayers mirrors is reported in Table 1. As it can be seen the average Ar content in the a-Si thin layers decreases with the Si layer thickness. The effective deposition rate of silicon is also dependent on the layer thickness as it can be seen from the data reported in Table 1, the effect being more pronounced for negative bias voltages.

H Incorporation

In Si/Mo mirrors technology silicon is the "spacer" material characterized by low radiation absorption. In mirror design it would be desirable to reduce the density of the spacer material so that less absorption occurs at each layer. The gain in the theoretical reflectance at normal incidence of a a-Si(H)/Mo mirror may be of 1.5% for a reduction of the density of a-Si(H) of about 10%. The incorporation of hydrogen can be accomplished by sputter depositing Si in an Ar+H_2 gas mixture. The most important effect of incorporation of hydrogen in silicon is the reduction of the material density.

TABLE 1. Ar concentration and a-Si effective deposition rates in a-Si layers and a-Si/Mo multilayers deposited at different bias.

Sample	Average Ar Concentration in a-Si (at%)	a-Si Layer Thickness (nm)	Variation of a-Si deposition rate in ML with respect to thick layers
Thick a-Si layer, $V_B = 0V$	5.5	~200	-
Thick a-Si layer, $V_B = -50V$	4.2	~200	-
30.4 nm Mirror , $V_B = 0V$	4.0	13.5	-0.3%
30.4 nm Mirror , $V_B = -50V$	3.2	12.8	-2.2%
13.0 nm Mirror , $V_B = 0V$	3.2	3.9	-2.0%
13.0 nm Mirror , $V_B = -50V$	2.6	2.9	-8.8%

In Fig. 4 the hydrogen incorporation and the density of the thick a-Si(H) layers deposited at different H_2 partial pressures are shown. The progressive incorporation of H is accompanied by a decrease of the film density as expected. The density data have been determined by X-Ray reflectometry [10] on 200 nm thick coating deposited at $V_B=0V$.

Our ERD analysis performed on the Mo thick layers deposited in Ar/H_2 gas mixtures containing 1 to 10% H_2 do not reveal measurable H incorporation into the Mo layers. This is due to the fact that the heat of solution for H in bulk Mo is positive [11].

DISCUSSION

In view of the application of the sputter deposition technology to the preparation of mirrors for wavelength lower that 10 nm, we focussed our attention on the issues concerning nanometer thick layers preparation at low deposition rates: in the present work the deposition rates of Si is about $2.4 \cdot 10^{14}$ atoms/cm^2/s (0.48 Å/s) and that of Mo is about $3.0 \cdot 10^{14}$ atoms/cm^2/s (0.47 Å/s). For the operating pressure of $5 \ 10^{-3}$ mbar the calculations of the backscattered particles flux and of the transport of particles through the plasma gas show that the Si layers at the beginning of the growth of the Mo over-layer, and the Mo layers themselves, are bombarded by Ar reflected particles with energy of about 35 eV at fluxes in the range from 2 to $6 \ 10^{14}$ particles/cm^2/s even when V_B is very near V_P. This displacement energy is high enough for creating bulk and surface defects [12]. The bombarded thickness of Si of about 1 nm as deduced from TRIM.SP calculations. This effect is more important for the a-Si/Mo and a-Si(H)/Mo multilayers designed for $\lambda=13$ nm for which the total thickness of each a-Si (or a-Si(H)) layer is about 3nm. Because the plasma potential is about +23V intermixing also occurs when Si is initially deposited onto Mo with $V_B \leq 0$. Here the bombarding energy $E=e(V_P-V_B)$ is higher than 25 eV and leads to atomic displacements. Furthermore the deposition rate of a-Si is initially much lower than the "bulk" value due to the amplified re-sputtering [13] of Si atoms caused by the presence of the Mo substrate. These facts suggest that a mixed Si_xMo interface is readily formed on both interfaces that may cause a deterioration of the reflectivity properties of the multilayer.

The lower effective deposition rate of a-Si and the lower average incorporation of Ar in a-Si/Mo multilayers with respect to the thick single a-Si layers deposited under the same conditions (see Table 1) may be an experimental evidence of these interfacial effects connected with noble gas particle bombardment. In the measurement of the EUV reflectivity it is not possible, without a direct measurement of the interface roughness, to separate the contributions of atomic intermixing from those induced by interface roughness. As a matter of fact the reflectivity of the multilayers designed for 13 nm radiation shows a marked dependence on the applied bias: the maximum reflectivity of our multilayers (of order of 60-62%) is obtained for positive bias ($V_B > +20V$). This points out a beneficent effect of bombardment characterized by high fluxes of low energy particles. These reflectivity values are well below the maximum ideal value of about 70-72%. The mirrors designed for $\lambda = 30.4$ nm show a reflectivity at normal incidence not higher than 24% that is lower than the maximum ideal value of 26-27%. Despite the excellent periodicity and repeatability of the layer deposition (as deduced by XRR and HR-TEM not reported in this paper) the main concern remains the quality of a-Si/Mo interfaces.

The maximum theoretical values of the normal incidence EUV reflectance of the multilayers are not obtainable without the control of the bombarding energy of Ar ions and of backscattered particles. Experimentally this can be accomplished in several ways. The energy of the bombarding particles could be modulated by applying a bias profile $V_B(t)$ such that the intentional bombardment is optimized (reduced) at the interfaces to avoid intermixing and then increased in order to obtain smoother interfaces [14-15]. As far as the reflected neutrals, TRIM.SP calculations show that R_E and R_N can be reduced by using, as sputtering gas, higher mass noble-gases like Xe and Kr. Furthermore both R_N and R_E at the substrate increase at high incidence angles. Thus by adopting a tilted geometry of deposition a modulation of the backscattered flux can be achieved

CONCLUSIONS

An experimental sputter deposition set-up used for developing new processes and for studying the effects of plasma parameters on the physical properties of X/EUV multilayer mirrors is described. The results presented in this paper concern specifically the preparation of a-Si/Mo and a-Si(H)/Mo multilayers with high normal-incidence reflectivity at 13 and 30.4 nm. The Ar and hydrogen incorporation in a-Si(H) layers are studied. Hydrogen can be successfully incorporated into silicon up to atomic concentrations of about 30%. The incorporation of hydrogen leads to a measurable decrease of the density of the spacer material, that calculations indicate as possible source of reflectance improvement. It is shown that Ar is incorporated only into silicon layers: the Ar content is related to the applied bias. We found that by varying the bias of the substrates we can improve the reflectance of the deposited mirrors of more that 15%, but without reaching the highest possible values due to combined effects of intermixing and interface roughness. The presented results show that the plasma sputter deposition technique when applied to the production of high reflectance X-ray multilayer mirrors needs a very accurate control of intentional and un-intentional energetic particle irradiation of the growing film.

ACKNOWLEDGMENTS

We wish to thank Dr. E. Bontempi of the University of Brescia for the XRR tests for determination of Si(H) density and Dr. M.G. Pelizzo of the University of Padova for the EUV reflectivity.

REFERENCES

1. Feigl, T., Yulin, S., Kuhlmann, T. and Kaiser, N., "Damage resistant and low stress Si-based multilayer mirrors" in *Soft X-Ray and EUV imaging Systems II*, edited by A. Tichenor et al., SPIE Proceedings 4506, 2001, pp. 121-127.
2. Bajt, S., Alameda, J., Barbee Jr., T., Clift, et al., "Improved reflectance and stability of Mo/Si multilayers" in *Soft X-Ray and EUV imaging Systems II*, edited by A. Tichenor et al., SPIE Proceedings 4506, 2001, pp. 65-75.
3. Eriksson, F., Johansson, G. A., Hertz, H. M., and Birch., J., "Enhanced Soft X-Ray Reflectivity of Cr/Sc Multilayers by Ion Assisted Sputter Deposition" in *Soft X-Ray and EUV imaging Systems II*, edited by A. Tichenor et al., SPIE Proceedings 4506, 2001, pp. 84-92.
4. Louis, E., Yakshin, A.E., Görts, P.C.,et al., "Mo/Si multilayer coating technology for EUVL, coating uniformity and time stability" in *Soft X-Ray and EUV imaging Systems*, edited by N. Kaiser et al., SPIE Proceedings 4146, 2000, pp. 60-63.
5. Hopkins, M.B.,. Graham, W.G., *Rev. Sci. Instrum.* **57** (9), 2210-2217 (1986).
6. Coufal, H., Winters, H. F., Bay, H. L., and Eckstein, W., *Phys. Rev. B* **44** (10), 4747-4758 (1991).
7. Eckstein, W., and Biersack, J. P.., *Z. Phys. B* **63**, 471-478 (1986).
8. Eckstein W. Computer Simulation of Ion-Solid Interactions, Springer Verlag, Berlin 1991
9. Somekh, R.E., *J. Vac. Sci. Technol. A* **2** (3), 1285-1291 (1984).
10. Kojima, I., Wei, S., Li, B., Fujimoto, T., and Kojima, I., *J.Surf. Anal.* **4**, 70 (1998)
11. Hjörvarsson, B., Vergnat, M., Birch, J., Sundgren, J.-E., and Rodmacq, B., *Phys. Rev. B* **50**, 11223-11226 (1994).
12. Tsao J.Y et al., *Nucl. Instr. & Meth.* B**39**, 72-80 (1989)
13. Berg S. et al., *J. Vac. Sci. Technol.* A **10**(4), 1592-1596 (1992)
14. Zhou X. W. and Wadley H.N.G., *J. Appl. Phys.* **87**(12), 8487-8496
15. Zhou X. W. and Wadley H.N.G., *J. Appl. Phys.* **88**(10), 5737-5743

A Bragg Magnifier with Sub-μm Resolution Using High Energy Synchrotron Light

Marco Stampanoni[1,2], G. L. Borchert [3*], R. Abela[4], P. Rüegsegger[5]

[1]IBT, ETH Zürich Switzerland and
[2]SLS, Paul-Scherrer-Institut Switzerland
[3]FRM II, Technical University München Germany
[4]SLS, Paul-Scherrer-Institut Switzerland
[5]IBT, ETH Zürich Switzerland

Abstract. X-ray computer microtomography using synchrotron light (SRμCT) has proven to be a highly powerful method in many fields of modern research as medicine, biology and material science. Presently used instruments, however, are limited to about 1μm resolution at a total efficiency of a few percent, due to the properties of the scintillator, the optical light transfer, and the CCD granularity. To overcome these limitations we have realized a novel approach based on extremely asymmetrical Bragg reflection. Our instrument, the "Bragg Magnifier" combines two asymmetrically cut Si crystals, mounted close to each other on two rotation and adjustment units. It is installed at the Materials Science Beamline of the Swiss Light Source (SLS). It operates at favorably high energies between 21 keV and 23 keV. In a first experiment using a human bone trabecula a two-dimensional magnification factor of 100x100 was achieved yielding a spatial resolution of 140nm.

INTRODUCTION

X-ray computer microtomography (**SRμCT**) in conjunction with the availability of high intensity synchrotron light has opened new fields for modern non-destructive research in medicine, biology, and materials science. The most commonly used instrument consists of a sample holder, rotateable around a central axis and a detector unit. The synchrotron light beam passes through the sample and hits onto a scintillator crystal, which converts the X-rays to visible light. By means of a suitable optics it is transferred to a CCD detector which delivers the image information to a computer. When rotating the sample by small angular steps and continuously taking pictures, the stored information can serve to reconstruct a three-dimensional image of the sample [1-4]. The usual tradeoff between resolution and efficiency limits for a state-of-the-art detector of this type the resolution to 0.8 μm at 0.5% efficiency when using 30 keV photons [5]. Alternative solutions as Fresnel plates [6] or refractive lenses [7] reach sub-μm resolution only in the energy range well below 20 keV while many applications request sub-μm resolution at energies above 20 keV maintaining a decent efficiency.

---

[*] corresponding author

CP652, *X-Ray and Inner-Shell Processes: 19th International Conference on X-Ray and Inner-Shell Processes*
edited by A. Bianconi, A. Marcelli, and N. L. Saini
© 2003 American Institute of Physics 0-7354-0111-X/03/$20.00

To overcome these limitations we have realized a novel approach that combines sub-μm resolution with good efficiency at high photon energies: the Bragg Magnifier.

THE INSTRUMENT

The Bragg Magnifier is based on the principle of asymmetrical Bragg reflection applied at high energies. For convenience we briefly summarize the relevant results of dynamical theory.

Using the definitions of fig.1 and by choosing the adequate solutions of Maxwell equations for a medium with triply periodic electricity constant we obtain a rather complicated expression for the crystal response function [8,9,10]. A typical result is shown in fig.2 [11,12]. For the special case of a symmetrical reflection ($\alpha = 0$, $m = 1$) the diffraction width ω_s of an ideal crystal is

$$\omega_s = \frac{2}{\sin(2\theta_B)} \frac{r_0 \lambda^2}{\pi V} C |F_h| e^{-DW}$$

where $r_0 = 2.818 \cdot 10^{-15}$ m is the classical electron radius, λ is the wavelength of the incident radiation, V is the volume of the unit cell, C is the polarization factor, θ_B is the Bragg angle, F_k is the crystal structure factor and e^{-DW} is the Debye-Waller factor.

In this case the widths of the incident and the reflected beams are equal. This is no longer true for asymmetrical reflections ($\alpha \neq 0$, see fig.2). Here the dynamical theory yields for the diffraction width of the incident beam

$$\omega_0 = \sqrt{|m|} \cdot \omega_s$$

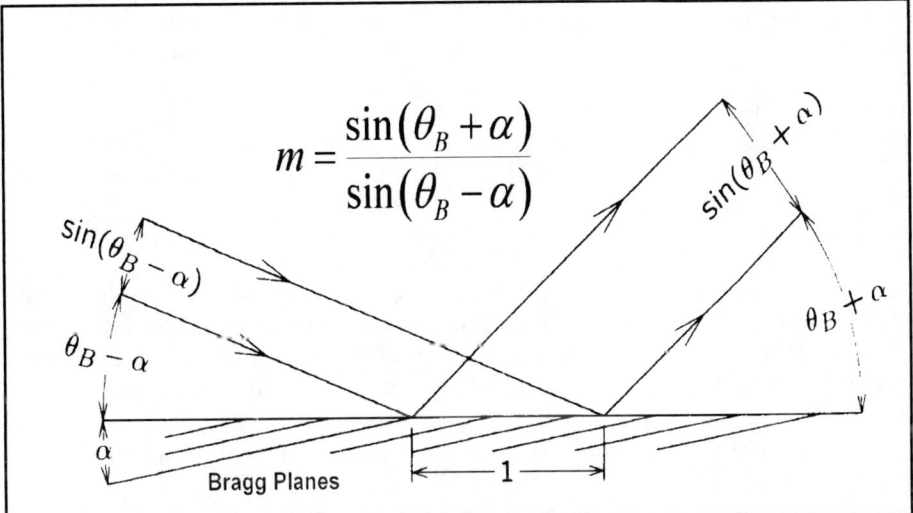

FIGURE 1. The basic definitions of asymmetrical reflection: with the Bragg angle θ_B and the asymmetry angle α between crystal surface and lattice planes the magnification m is determined.

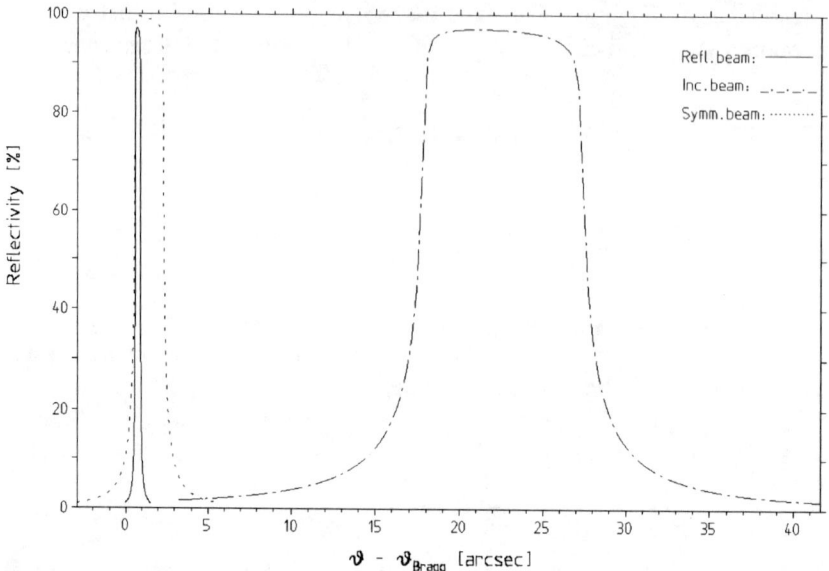

FIGURE 2. The diffraction widths for Si (220) at 21.3 keV. With an asymmetry angle α = 8° the curves for the reflected beam and the incident beam are given according to the above pattern code. For comparison we add the rocking curve for symmetrical reflection (α = 0°). Note that the index-of-refraction shift is different for each curve.

and for the diffraction width of the reflected beam

$$\omega_H = \frac{1}{\sqrt{|m|}} \cdot \omega_s .$$

It is obvious that for m >1 the incident distribution is larger than ω_s which facilitates the matching to the photon beam divergence in order to obtain a high transfer efficiency.

Inspection of fig.2 reveals that for m ≠1 there occurs an offset from Bragg's law which again is different for the incident and the reflected beam. For the cross sections of the incident beam S_0 and the reflected beam S_H Liouville's theorem requires $S_0 \cdot \omega_0 = S_H \cdot \omega_H$ and hence

$$S_H = |m| \cdot S_0 .$$

An X-ray beam from the target area with a one-dimensional width S_0 hitting onto an asymmetrically cut crystal undergoes Bragg reflection with the cross section increased by a factor m in one dimension. The Bragg reflection at a second crystal, which is oriented so that the reflection planes are perpendicular with respect to the first one, produces a beam that in addition is magnified in the other dimension by m′ (see fig.3). Thus the final result is an image with a two-dimensional magnification m·m′. Usually the crystal planes and the asymmetry angles are chosen equal for both crystals so that m = m′. Finally this magnified X-ray image is converted by a scintillator crystal into the corresponding visible light distribution, which is handled by suitable optics and a

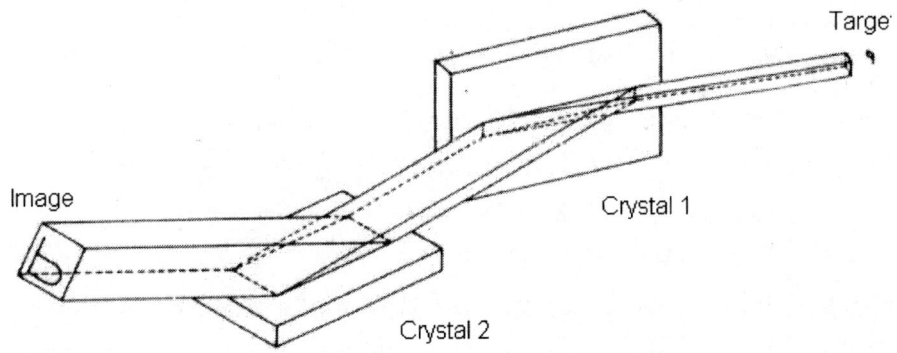

FIGURE 3. The principle of the Bragg Magnifier.

CCD camera in the conventional way. Recently such a device has been operated successfully at 12.3 keV reaching a resolution of about 1.2 µm [13].

Our instrument consists of two identical Si(220) crystals (size: $150 \times 80 \times 15 mm^3$) with an asymmetry angle α of 8° [14]. To avoid stress and lattice plane distortions they were mounted by optical contacting on precision-polished glass plates [15], which have the same thermal expansion coefficient as Si. Those are fixed on top of two adjustment units that consist of two crossed swivels and a high-resolution goniometer [16] to rock the crystals. Its angular resolution of 0.05 arcs has to cope with the

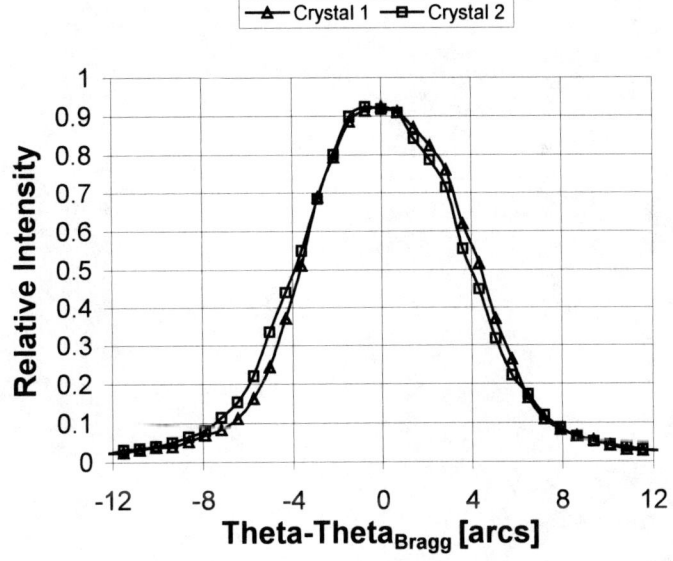

FIGURE 4. The rocking curve of the Si(220) crystals measured at 8 keV (Cu Kα₁)

corresponding width of the rocking curve of 0.3 arcs at 21 keV. As the rocking curves of the two crystals are very similar in width and position (see fig.4) a symmetrical magnification in both dimensions can be expected.

In our instrument the magnified X-ray image is converted to visible light with high efficiency by a 300μm thick CsI(Tl) scintillator (size: 35x35 mm^2). The light is transferred by a 1:1 tandem optics [17] to the CCD chip [18] (2048x2048 pixels with 14 μm pitch).

Our device is designed to operate at the material science beam-line of the SLS [19], which produces a photon beam of $3x10^{13}$ photons/s in an energy range of 21 keV to 23 keV with a divergence of less than 20 μrad [20]. This number coincides very well with the acceptance angles of the Si(220) crystals so that finally 95% of the photon beam is accepted by the Bragg Magnifier. The experimental setup is shown in fig.5.

FIGURE 5. The experimental setup of the Bragg magnifier. The beam coming from the right is reflected by crystal 1 (only glass support and mounting from backside visible). The horizontal crystal 2 acts as a mirror of the adjustment unit above. In the black unit on the left side the scintillator is visible

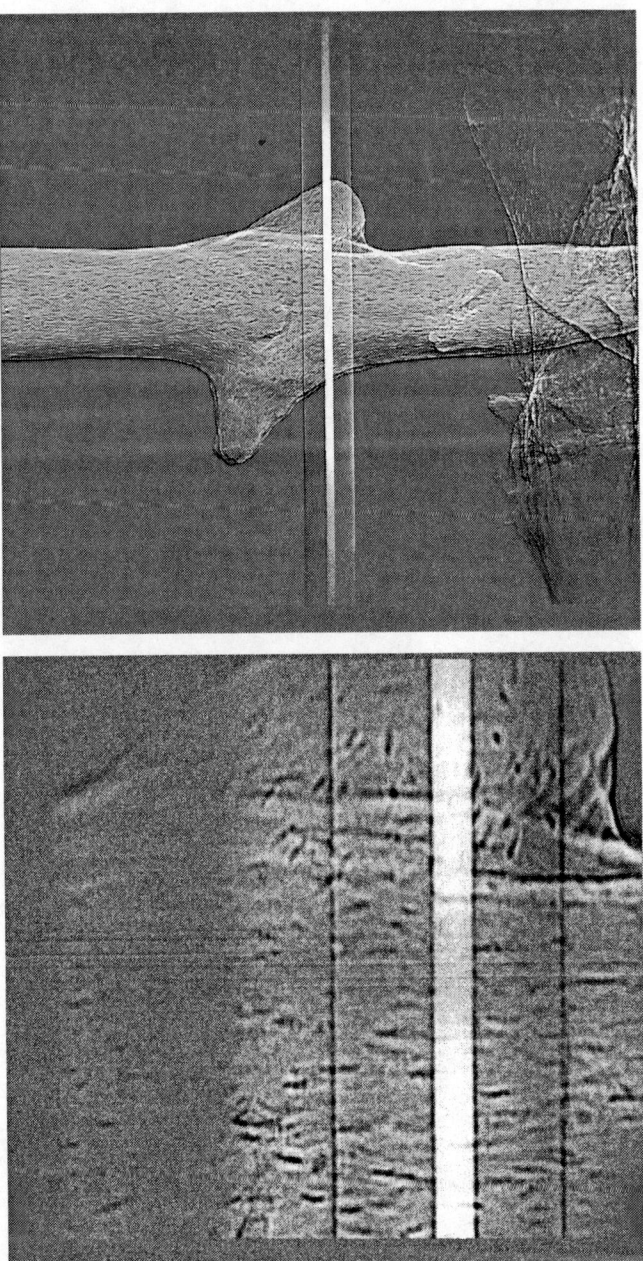

FIGURE 6. A human bone trabecula seen at a magnification of 20x20 (upper part) and 100x100 (lower part). In both cases a boron fiber with a tungsten core was superimposed for calibration. On the right side (up) the glue for trabecula fixation is visible. The shadow on the left side (low) indicates an inhomogeneous illumination of the field of view. Clearly visible becomes the cellular structure of the trabecula.

RESULTS

The Bragg Magnifier was operated at the Materials Science Beamline of the SLS at energies between 21.1 keV and 22.75 keV. By adjusting the Si crystals accordingly, magnification factors between 20x20 and 100x100 have been achieved. As a first sample we investigated a human bone trabecula. For the calibration of the imaging properties we included in the target station a 100 μm boron fiber with a 15μm tungsten core. Two typical images of this arrangement are shown in fig. 5. The magnification factor for the upper picture is 20x20, for the lower one it is 100x100, which yield an effective pixel size from 700x700nm^2 to 140x140nm^2. The corresponding field of view is 1440x1440 μm^2 and 280x280 μm^2, respectively.

CONCLUSIONS

We have developed a Bragg Magnifier that operates with high efficiency at energies above 20 keV. Magnifications up to 100x100 have been reached so far, which yield resolutions down to 140nm. First tomographic pictures have been taken, the analysis of which is on the way. For improving the homogeneity of the image illumination at large magnifications we expert a step forward when the beam intensity of the SLS will be increased. An inspection of the present results indicate that edge-enhanced or phase-contrast investigations are feasible. We will study this feature in the next future.

ACKNOWLEDGMENTS

We thank B. Patterson, D. Vermeulen and M. Lange for their generous help in setting up the beamline. We appreciated the efficient collaboration with B. Lux and Dr. V. Alex from the Institut für Kristallzüchtung, Berlin. We thank Prof. H. Grimmer and Dr. D. Clemens from PSI for their help with the preliminary characterization of the crystals. This work has been supported by the ETH Board.

REFERENCES

1.   U. Bonse and F. Bush, Prog. Biophys. Molec. Biol., Vol. **65**, 133 (1996)
2.   R. Lee *et al.*, Proc. of SPIE Vol. 3149, 257 (1997)
3.   B. Dowd *et al.*, Proc. of SPIE Vol. 3772, 224 (1999)
4.   T. Weitkamp *et al.*, Proc. of SPIE Vol. 3772, 311 (1999)
5.   A. Koch *et al.*, J. Opt. Soc. Am. A, **15**, 1940 (1998)
6.   B. Lai *et al.*, Rev. Sci. Instruments **66**, 2287 (1995)
7.   C. Schroer *et al.*, Proc. of SPIE Vol. 4503, 23 (2002)

8. W. H. Zachariasen, "Theory of X-Ray Diffraction in Crystals", John Wiley and Sons, Inc., New York, 1945.
9. R. W. James, "The Optical Principles of the Diffraction of X-Rays", Oxbow Press, Woodbridge, 1962
10. R. Caciuffo et al., Physics Reports 152, 1 (1987)
11 S. Brennan and P. Cowan, Rev. Sci. Instrum. 63, 850 (1992)
12. Computations performed with the in-house developed software XDF. To be published.
13. M. Kuriyama et al., J. Res. Natl. Inst. Stand. Technol. 95, 559 (1990)
14. The crystals were produced by the company Holm, Thann, and oriented by the IKZ, Berlin , Germany
15. Polishing and optical contacting have been performed by the optical company Carl Zeiss, Oberkochen, Germany.
16. The components have been produced by the company Kohzu Precision Co. Ltd., Japan.
17. The optics system has been produced by the company Optique Peter, Lyon, France
18. The CCD camera was purchased from Pixel Vision of Oregon, Inc. Tigard, USA
19. B. D. Patterson et al., Chimia 55, 534, (2001)
20. M. Stampanoni et al., Acta Physica Polonica B, Vol. 33, 463 (2002)

IV. PHOTOIONIZATION PROCESSES AND HIGHLY CHARGED IONS

High Energy Limit of the Two-Electron Photoionization Processes in the Helium-Like Systems

M. Ya. Amusia*†, E. G. Drukarev**, R. Krivec‡ and V. B. Mandelzweig*

*Racah Institute of Physics, The Hebrew University, Jerusalem 91904, Israel
†A. F. Ioffe Physical-Technical Institute St. Petersburg, 194021, Russia
**Petersburg Nuclear Physics Institute, Gatchina 18830, Russia
‡Department of Theoretical Physics, J. Stefan Institute, PO Box 3000, 1001 Ljubljana, Slovenia

Abstract. We present the analysis of double photoionization and of ionization accompanied by excitation of the helium atom and helium-like ions. The high energy limit corresponding to the ultrarelativistic energies transferred to the atom is considered. The cross-sections ratios are expressed through certain integral parameters of the wave function of the initial state. The calculations with the wave function obtained earlier by Correlation Function Hyperspherical Harmonic Method are carried out. The dependence of the relative role of the quasifree mechanism on the nuclear charge Z in ionization of the ground state and of the excited $n's$ states is analyzed.

INTRODUCTION

The atom of helium and the helium-like ions attract the interest of the physicist as the simplest many-body systems bound by the electromagnetic interactions. The double photoionization of such systems takes place due to the interaction between the electrons. Thus, investigation of this process provides the information about the correlations inside the system. The double photoionization of helium has been studied since the sixties of the last century. The upper limit of the photon energies ω values of about several hundreds eV was available for the experimental investigations at that time. The progress of the experimental technique in nineties made the measurements at the energies up to 20 keV available nowadays.

At the photon energies exceeding 6 keV ionization accompanied by radiation of the photon (Compton scattering) is known to provide the largest contribution among the processes in which one or both electrons are eliminated from the atom of helium. However it is possible to separate experimentally the contributions to the two-electron ionization and ionization with excitation caused by either photon absorption or by photon scattering. This provides motivation for the investigation of the double photoionization and of ionization with excitation at high values of the photon energy.

From the point of view of the theoretical physics the investigation of the high energy region is attractive since the description of the final state can be simplified and the asymptotics of the amplitude of the processes can be expressed through certain characteristics of the ground state only. This makes the information extracted from the experimental data somewhat more simple.

CP652, X-Ray and Inner-Shell Processes: 19th International Conference on X-Ray and Inner-Shell Processes
edited by A. Bianconi, A. Marcelli, and N. L. Saini
© 2003 American Institute of Physics 0-7354-0111-X/03/$20.00

In this talk we present the results of the calculations of the high energy limits of the cross sections of the double photoionizations and of the ionization with excitation of the ground state of helium and helium-like ions and of the excited n^1S and n^3S states of these systems. We use the form which is traditional for the double photoionization studies, i.e. we present the double-to-single cross section ratios

$$R_n^{(k)}(\omega) = \frac{\sigma_n^{(k)++}(\omega)}{\sigma_n^{(k)+}(\omega)} ; \quad R_n^{(k)*}(\omega) = \frac{\sigma_n^{(k)|*}(\omega)}{\sigma_n^{(k)+}(\omega} . \tag{1}$$

Here the notation $+, ++, +*$ corresponds to single ionization, to double ionization and to the ionization with excitation. The upper index (k) corresponds to the excited n^kS initial states, $n = k = 1$ for the ground state. We denote $R_1^{(1)} = R$; $R_1^{(1)*} = R^*$. The wording "high energy limit" means that the photon energy exceeds strongly the energy of the electron at rest. In the system of units with $\hbar = c = 1$ this means that

$$\omega \gg m \tag{2}$$

with $m \approx 511 \,\text{keV}$ being the electron rest mass. We calculate the leading term of the expansion of the ratios $R^{(k)}$ and $R^{(k)*}$ in powers of ω^{-1} in this limit.

We consider the case of not very large values Z of the charge of the nucleus

$$(\alpha Z)^2 \ll 1 \tag{3}$$

with $\alpha = 1/137$ being the fine structure constant. This enables us to treat the interactions of the outgoing electrons with the nucleus in the lowest order of perturbative series. Thus the ratios defined by Eq. (1) depend on the characteristics of the initial state only.

It is instructive to recall the mechanisms of the process in the "high energy nonrelativistic limit" (HENL), i.e. in the case when the photon energy exceeds strongly the binding energy I of the helium atom but the ratio ω/m is still considered as a small parameter. It was shown in [1] that the process can be viewed as the absorption of the photon energy by one of electrons followed by the shake-off of another one. The electron which interacts with the photon directly carries the momentum $p \sim (2m\omega)^{1/2}$ transferred by the nucleus. Thus, the amplitude of the single photoionization is separated as a factor in that of the double photoionization. It contains the main dependence of the cross section σ^{++} on the photon energy, leading to the HENL behavior $R = \text{const}$. Of course, the leading HENL terms are determined by the dipole approximation of the interaction between the photon and electron. The quadrupole terms provide the corrections $(\omega/p)^2 \sim \omega/m$ to the amplitudes of both single and double photoionizations. They cancel in the ratio $R(\omega)$, as well as the relativistic corrections. It was shown in [2] (see also [3]) that there is another mechanism contributing in the next to leading order of HENL. In this quasifree mechanism (QF) both electrons exchange by the large momentum without the participation of the nucleus. The process cannot take place in the dipole approximation [2] thus requiring the quadrupole terms of the photon–electron interaction and producing the correction of the order ω/m to the HENL value of R. The latter becomes quite clear if one notices that the mechanism of transferring of the large momentum between the two electrons is quite similar to that between the electron and the atom [4].

Now we shall see what happens at $\omega \gtrsim m$.

SHAKE-OFF CONTRIBUTION

The HENL [1] can be obtained most easily in the "velocity form" of the photon–electron interaction. In this case the fast outgoing electron can be described by the plane wave, providing

$$R_n^* = \sum_v \frac{J_v}{J_1} \tag{4}$$

with

$$J_v = \left| \int \Psi(0, r_2) \varphi_v(r_2) d^3 r_2 \right|^2 \tag{5}$$

for the HENL of the ionization of the ground state of the helium-like ion accompanied by the excitation of the state of the discrete spectrum of the residual single-electron ion with the quantum numbers v. Thus φ_v is just the Coulomb field wave function while $\Psi(r_1, r_2)$ is the wave function of the initial state.

The HENL of the double photoionization can be presented in the same way

$$R_{s.o.} = \left(\int J_v \frac{d^3 v}{(2\pi)^3} \right) \Big/ J_1 = C_{s.o.} \tag{6}$$

with "v" standing for the asymptotic momenta of the outgoing secondary electron, while φ_v is the properly normalized Coulomb continuum wave function. Here J_1 corresponds to the set of the quantum numbers $v = \{n, \ell, m\} = \{1, 0, 0\}$ of the discrete spectrum.

The distances r_2 in the integrand of the right-hand side (rhs) of Eq. (5) are of the order of the size of the atom r_a. Also, the integral in the rhs of Eq. (6) is saturated by small momenta of the order of that of the bound state. Thus all the dynamics connected with the variable "r_2" in the function $\Psi(0, r_2)$ can be treated in the nonrelativistic way even at $\omega \gtrsim m$. However the function $\Psi(0, r_2)$ is the limiting value of the function $\Psi(r_1 \sim p^{-1}, r_2)$ with p being the momentum of the outgoing electron. At $\omega \gtrsim m$ we find $p \gtrsim m$ and thus the dynamics connected with the variable "r_1" requires the relativistic treatment. This refers to the amplitude of the single ionization as well. Another modification, which emerges in the relativistic case is that one cannot use the plane wave for the description of the fast outgoing electron any more with the lowest order Born correction providing the value of the same order of magnitude [5].

In the considered case the relativistic wave function $\Psi(r_1 \sim p^{-1}, r_2)$ can be expressed through the nonrelativistic one using the lowest order iteration of the Bethe–Salpeter equation

$$\Psi = \Psi_0 + GV\Psi \tag{7}$$

with V standing for the interactions inside the system while G is the relativistic Green function of two free electrons. This equation connects the wave function $\Psi(r_1 \sim p_1^{-1}, r_2 \sim r_a)$ with the wave function $\Psi(r_1 \sim r_a, r_2 \sim r_a)$. It becomes quite explicit in the momentum space

$$\tilde{\Psi}(\mathbf{p}_1, \mathbf{p}_2) = \frac{8\pi\alpha Z}{p_1^4} \left(1 + \frac{(\alpha p_1)}{2m} \right) \int \frac{d^3 f}{(2\pi)^3} \tilde{\Psi}(\mathbf{f}, \mathbf{p}_2) \tag{8}$$

with $\tilde{\Psi}(p_1,p_2)$ being the Fourier transform of the function $\Psi(r_1,r_2)$ while α are the standard Dirac α-matrices. The integral in the rhs of Eq. (8) is saturated by momenta f of the order of the binding momentum of the atom. Thus the function $\tilde{\Psi}(f,p_2)$ in the right-hand side of Eq. (8) can be treated nonrelativistically. Hence, Eq. (8) expresses the relativistic wave function through the nonrelativistic one. All the relativistic effects are contained in the factor GV in the rhs of Eqs. (7) and (8). Note that the result is very much alike the one obtained in [6] by using the Furry–Sommerfeld–Maue wave functions for the analysis of the high energy photoeffect in the Coulomb field.

Similar treatment of the final state wave functions also provide the factorization of the additional dependence on ω with respect to the nonrelativistic treatment of the initial state. Hence, although one needs the relativistic functions for the description of the cross sections $\sigma^{(k)+}$, $\sigma^{(k)++}$, $\sigma^{(k)*}$ the relativistic effects in the initial wave functions cancel in the ratios defined by Eq. (1). Thus, their HENL calculated with the nonrelativistic wave functions $\Psi(r,r_2)$ can be expanded to the energies $\omega \gg m$ [7].

CONTRIBUTION OF THE QUASIFREE MECHANISM

Following the discussion in the Introduction the mechanism (QF) takes place due to the exchange of the large momenta between the outgoing electrons, while the recoil momentum of the nucleus is of the order of the electrons binding momentum. The HENL value of the cross section σ^{++} for the singlet states can be thus expressed through the "cross section" $\sigma_0(\omega)$ of ionization of the system of the two free electrons at rest, i.e. $\sigma^{++}(\omega) = C\sigma_0(\omega)$ with [8]

$$C = \int |\Psi(r_1,r_2)|^2 d^3 r_1 . \tag{9}$$

Since the dependence of the cross section $\sigma^+(\omega)$ on the properties of the initial state manifests itself through the value J_1 defined by Eq. (5) at $v = 1,0,0$, i.e. $\sigma^+(\omega) = J_1 S(\omega)$, we find [8]

$$R_{QF}(\omega) = C_{QF} \frac{\sigma_0(\omega)}{S(\omega)} \tag{10}$$

with the factor

$$C_{QF} = \frac{C}{J_1} \tag{11}$$

containing the dependence of HENL of $R_{QF}(\omega)$ on the properties of the initial state. The function $\Psi(\bar{r}_1, \bar{r}_1)$, which determines the factor C, is the limiting value of the function $\Psi(\bar{r}_1, \bar{r}_2)$ at $|r_1 - r_2| \sim p^{-1} \ll r_a$. At $\omega \ll m$ the nonrelativistic wave function $\Psi(r_1,r_2)$ can be used. However, at $\omega \gtrsim m$ relativistic treatment of the relative motion of the electrons is needed, while their motion with respect to the nucleus remains to be a nonrelativistic one. As in the previous Section, the Bethe–Salpeter equation ties the relativistic function $\Psi(r_1,r_1)$ with a nonrelativistic one. The relativistic effects manifest themselves through the factors which depend on the photon energy but not on the structure of the initial state. Thus, for the energies $\omega \gtrsim m$ corresponding to the relativistic energies of the outgoing electrons we can write

$$R_{QF}^{rel}(\omega) = C_{QF} \cdot \frac{\sigma_0^{rel}(\omega)}{S^{rel}(\omega)} \tag{12}$$

with the factor C_{QF} defined by Eq. (11) containing all the dependence on the structure of the initial state being calculated with the nonrelativistic wave functions $\Psi(r_1, r_2)$.

The ratio being of the order ω/m at $\omega \ll m$ [2] reaches a constant value at ultrarelativistic energies $\omega \gg m$

$$R_{QF}^{urel} = \frac{4}{Z^2} C_{QF} \tag{13}$$

for the singlet states. For the triplet states the value R_{QF}^{rel} is quenched by the additional factor of the order ω^{-2}. This is because the wave function of the initial state is antisymmetric in the position space and the probability for the electrons to approach each other at the distances of the order p^{-1} is small [7].

THE HIGH ENERGY LIMIT

To summarize the results of the previous sections, the HENL values of ionization with excitation to single ionization ratio $R^{(k)*}$ can be expanded to the region $\omega \gtrsim m$. They are determined by the shake-off mechanism. The same refers to the double-to-single cross section ionization ratios of the triplet states. However for the double ionization of the singlet states we find the contributions of both shake-off and quasifree mechanism. For $\omega \gtrsim m$

$$R^{(1)}(\omega) = R_{s.o.}^{(1)}(\omega) + R_{QF}^{(1)}(\omega) \tag{14}$$

providing the limiting value

$$R^{(1)} = C_{s.o.} + \frac{4}{Z^2} C_{QF} \tag{15}$$

with $C_{s.o., QF}$ defined by Eqs. (6) and (11), while the factor Z^2 in denominator of the second term comes from $S(\omega)$. In contrast to this we have $R^{(3)} = C_{s.o.}$ for the triplet states at all the values $\omega \gg I$.

THE WAVE FUNCTION OF THE INITIAL STATE

As we have seen, the ratios we are looking for can be expressed through the nonrelativistic wave function of the initial state $\Psi(\vec{r}_1, \vec{r}_2)$. There are many types of the functions considered as "very accurate". Often the wording "very accurate" is identified with "providing the value of the binding energy very accurately". But this is not enough for the problem which we are investigating now. As we have seen, the values of the ratios are determined by the wave function at $r_1 \to 0$ and $|\vec{r}_1 - \vec{r}_2| \to 0$. The wave function of the initial state (which can be the ground state or the excited state) has singularities at these points due to the square root singularity of the Coulomb interactions

$$V_{eN} = \frac{-\alpha Z}{(r^2)^{1/2}} \quad \text{and} \quad V_{ee} = \frac{\alpha}{[(\vec{r}_1 - \vec{r}_2)^2]^{1/2}}$$

[9]. Behavior of the wave functions near these points is determined by the Kato conditions. For the ground state

$$r_0 \frac{\partial \Psi(\vec{r}_1, \vec{r}_2)}{\partial r_1} = -Z\Psi(\vec{r}_1, \vec{r}_2) \tag{16}$$

at $r_1 \to 0$ and

$$r_0 \frac{\partial \Psi(\vec{r}_1, \vec{r}_2)}{\partial \rho} = \frac{1}{2} \Psi(\vec{r}_1, \vec{r}_2) \tag{17}$$

at $\rho = |\vec{r}_1 - \vec{r}_2| \to 0$.

Here r_0 is the Bohr radius. Equations (16) and (17) can be viewed as the conditions of the cancellation of the singularities in the Schrödinger equation for the atom of helium at $r_1 \to 0$ and at $\rho \to 0$.

Evaluation of the amplitude of the double ionization in the original formalism with the "accurate" wave functions which, however, do not satisfy the Kato cusp conditions may lead to erroneous results. It was shown in [10] that calculation of the shake-off contribution to the ratio $R(\omega)$ in the length form provides the spurious behavior of HENL $R(\omega) \sim \omega$ instead of the true one $R(\omega) = $ const. However, the coefficient appeared to be proportional to the combination $[r_0(\partial/\partial r_1) + Z]\Psi(\vec{r}_1, \vec{r}_2)$, which vanishes if Eq. (16) is fulfilled for the approximate function, used in the computations. It was demonstrated in [11] that the numerical calculations carried out with the functions, which did not satisfy Eq. (17) provided the erroneous results for the distribution of the fast electrons in the double photoionization of helium and for the high energy behavior of the cross section.

However, even if the small distance behavior is treated properly, the various wave functions still provide the numbers which differ rather considerably [8,11,12]. This increases the necessity to have a proper wave function of the initial state.

We use the functions obtained in [13–15] for the description of the initial state. They were found by using the Correlation Function Hyperspherical Harmonic Method. The initial wave function is decomposed as

$$\Psi = e^f \Phi, \tag{18}$$

where f is the correlation function describing the singularities of Ψ and Φ is a smooth remainder which can be expanded in a fast converging hyperspherical harmonic (HH) expansion. The function f depends on the interparticle distances, which is necessary and sufficient to take into account analytically the two- and three-body Coulomb singularities (cusps) in the wave function, i..e, it satisfies the Kato cusp conditions [9] exactly. Furthermore, Ψ is obtained by a direct solution of the three body Schrödinger equation, which guarantees local correctness [13–15] of Ψ because the convergence of Ψ across the configurations space is uniform.

RESULTS

The calculations of the contribution of the quasifree mechanism to the ionization of the ground state of helium with the function defined by Eq. (18) provide the value $C_{QF} = 0.060$. Thus the purely Coulomb calculations [8] providing the value $C_{QF} = 1/8$ overestimate the contribution of QF mechanism by a factor of about two.

Now we follow the relative role of QF mechanism in ionization of $n'S$ states as a function of the nuclear charge Z. While $C_{s.o.}$ is known to drop as Z^{-2} at large Z, the parameter C_{QF} does not depend on Z in this limit. Thus, both shake-off and quasifree mechanisms have the same Z-dependence in the limit $Z \gg 1$. The role of QF mechanism is determined by the function $\zeta(n,Z)$ defined in such a way that

$$R_n^{(1)} = R_{n,s.o.}^{(1)} \cdot \zeta(n,Z) \tag{19}$$

for the limiting values of $R_n^{(1)}$. For the ground state of helium we find $\zeta(1,2) \approx 4.6$, growing slowly to the limiting value $\zeta(1, Z \gg 1) \approx 6.3$. For $n = 2$, corresponding to ionization of 2^1S states the function $\zeta(2,Z)$ reaches its minimum value at $Z = 5$ ($\zeta \approx 1.49$) while $\zeta(2,2) \approx 1$ and $\zeta(2, Z \gg 1) \approx 1.52$. For the higher excited states the function $\zeta(n,Z)$ drops monotonically with Z changing by about 50% between $Z = 2$ and the limiting values $Z \gg 1$ at least for $n \leq 5$.

Turning to the ionization of the triplet states the QF mechanism is quenched and the high energy limit is determined by shake-off only. We mention the data for $R_n^{(3)*}(Z)$. This values are larger than unity, thus demonstrating that the probability to find the residual ion in the ground state in such a process is small. For example, the value $R_2^{(3)*}(2) \approx 28.5$ for the ionization with excitation of the metastable state 2^3S of helium. Also the double-to-single ionization ratio increases with n, since the change of the effective charge is more sensitive for the highly excited states. Say, for helium the ratio $R_2^{(3)} \approx 0.09$ while $R_5^{(3)} \approx 0.34$.

ACKNOWLEDGMENTS

The research was supported by the Hebrew University Intramural Fund (MYA), by the Bilateral Cooperation Program at the Ministry of education, science and sport of Slovenia (RK), and by the Israeli Science Foundation grant 131/00 (VBM)

REFERENCES

1. Kabir, P. K., and Salpeter, E. E., *Phys. Rev.* **108**, 1256 (1957).
2. Amusia, M. Ya., Drukarev E. G., Gorshkov, V.G., and Kazachkov, M. P., *J. Phys. B* **8**, 1248 (1975).
3. Amusia, M. Ya., *Atomic Photoeffect*, Plenum Press, New York and London, 1990.
4. Drukarev, E.G., *Phys. Rev. A* **52**, 3910 (1995).

5. Bethe, H. A., and Salpeter, E. E., *Quantum Mechanics of One- and Two-Electron Atoms*, Plenum/Rosetta, New York, 1977.

6. Pratt, R. H., *Phys. Rev.* **117**, 1017 (1960).

7. Drukarev, E. G., and Karpeshin, F. F., *J. Phys. B* **9**, 913 (1976).

8. Drukarev, E. G., *Phys. Rev. A* **51**, R2684 (1995).

9. Kato, T., *Comm. Pure Appl. Math.* **10**, 151 (1957).

10. Åberg, T., *Phys. Rev. A* **2**, 1726 (1970).

11. Drukarev, E. G., Avdonina, N. B., and Pratt, R. H., *J. Phys. B: At. Mol. Opt. Phys.* **34**, 1 (2001).

12. Krivec, R., Amusia, M. Ya., and Mandelzweig, V. B., *Phys. Rev. A* **64**, 012713 (2001).

13. Haftel, M., and Mandelzweig, V. B., *Phys. Rev. A* **42**, 6342 (1990).

14. Haftel, M., Krivec, R., and Mandelzweig, V. B., *J. Comp. Phys.* **123**, 149 (1996).

15. Krivec, R., Mandelzweig, V. B., and Vagra, K., *Phys. Rev. A* **61**, 062503 (2000).

X-Ray Emission from Highly Charged Ions Colliding with a Relativistic Electron Beam in the SuperEBIT Electron Beam Ion Trap

P. Beiersdorfer*, B. Beck*, J. A. Becker*, J. K. Lepson† and K. J. Reed*

*Lawrence Livermore National Laboratory, Livermore, CA
†University of California, Berkeley, CA

Abstract.
  The high-energy electron beam ion trap SuperEBIT at the Lawrence Livermore National Laboratory allows the study of the x-ray emission from highly charged ions interacting with electrons with energy in excess of 200 keV. Radiation from ions as highly charged as Cf^{96+} has been produced this way. The facility is being used to investigate the contributions from quantum electrodynamics in heavy ions. Here the focus is lithiumlike ions, especially U^{89+}, which provide the opportunity for the most accurate test of QED in highly charged ions. We have also used the facility to measure the degree of x-ray line polarization as a function of the energy of the electron collision energy. For example, we have studied the linear polarization of the K-shell emission lines of Fe^{24+} for electron-impact energies high as 120 keV. A new area of research is the investigation of nuclear excitation by electronic transitions. This is the inverse process of internal conversion, where an atomic x ray is absorbed by the nucleus resulting in an excited nuclear state. We are planning to study this process in ^{189}Os using 217 keV atomic x rays generated in the interaction with a 196 keV electron beam.

INTRODUCTION

The electron beam ion trap is a modified electron beam ion source built to study the interaction of highly charged ions with an electron beam by looking directly into the trap.

  The first electron beam ion trap, dubbed EBIT, was put into operation at the Lawrence Livermore National Laboratory in 1986 and is described in detail by Levine *et al.* [1, 2]. The machine initially was operated with a beam current of about 100 mA and energy of about 10 keV. This provided an electron density in the 10^{12} cm^{-3} range.

  Because the first electron beam ion trap, dubbed EBIT, was designed as an x-ray source, spectroscopic instrumentation centered on analyzing the x-ray emission with broad-band germanium detectors and with high-resolution crystal spectrometers, notably flat-crystal spectrometers. This allowed the first measurement of electron-impact excitation cross sections of a highly charged ion [3], followed by measurements of dielectronic recombination rates [4, 5] and resonance excitation cross sections [6].

  When compared to other x-ray sources at the time, e.g., tokamaks, beam-foil setups at heavy-ion accelerators, vacuum sparks, or laser-produced plasmas, EBIT was a relatively weak x-ray source. Focussing x-ray instrumentation was developed to collect more photons [7], resulting in high-resolution spectra useful for accurate wavelength determinations and QED studies [8] as well as measurements of innershell ionization

CP652, *X-Ray and Inner-Shell Processes: 19ᵗʰ International Conference on X-Ray and Inner-Shell Processes*
edited by A. Bianconi, A. Marcelli, and N. L. Saini
© 2003 American Institute of Physics 0-7354-0111-X/03/$20.00

cross sections [9] and energy-dependent electron-impact excitation cross sections [10].

The success of EBIT resulted in the construction of the second electron beam ion trap. This machine, dubbed EBIT-II, became operational at the Lawrence Livermore National Laboratory in January 1990. EBIT-II could operate with beam energies as high as 26 keV and beam currents as high as 250 mA.

Many high-resolution x-ray spectrometers were developed for this machine [11, 12, 13], which allowed many new atomic measurements, such as measurements of level-specific dielectronic recombination resonance strengths [14, 15], identification of magnetic octupole decay [16], and determination of radiative branching ratios [17]. A crystal spectrometer with resolving power $\lambda/\Delta\lambda = 68,000$ was employed to determine the ion temperature and measure the femto-second natural line width of electric dipole transitions of highly charged ions [18, 19]. In addition, absolutely calibrated monolithic crystals were implemented to make QED measurements of hydrogenic ions [20].

Fast-switching of the electron beam allowed to make the first measurement of the radiative lifetime of electric-dipole forbidden x-ray transitions in highly charged ions in the microsecond regime [21]. Development of the magnetic trapping mode, in which EBIT-II was operated without an electron beam [22], extended radiative lifetime measurements to electric dipole-forbidden x-ray transitions to many other highly charged ions [23, 24, 25].

In 2000, the 36-pixel array x-ray microcalorimeter devoloped by the Goddard Space Flight Center for the ASTRO-E space mission was added to the suite of x-ray instrumentation of EBIT-II [26]. It provided broadband x-ray detection capabilites coupled with a 10-eV spectral resolution and was used for various laboratory x-ray astrophysics measurements [27].

The successful operation of EBIT-II enabled the shutdown of EBIT in order to use parts for building a new, high-energy electron beam ion trap. This high-voltage machine commenced routine operation in January 1992 and was named SuperEBIT. The machine was designed for electron beam energies as high as 250 keV [28], and energies in excess of 200 keV were indeed achieved. With it any ion of essentially any element up to uranium could be produced, including bare uranium [29, 30]. The highest charge state to date is heliumlike Cf^{96+}, as discussed below. In other words, SuperEBIT allowed the production of highly charged ions that were heretofore only accessible by heavy-ion accelerators.

Because the ions were at rest in SuperEBIT (ignoring the small thermal motion of the ions), x-ray measurements were greatly simplified compared to similar measurements on accelerators. Moreover, measurements could be made, such as those of electron-impact excitation, that were impossible to do on accelerators.

X-ray studies on SuperEBIT included determinations of the $2s$ Lamb shift in lithiumlike thorium and uranium [31, 32], and measurements of the variation of the nuclear radii of ^{233}U, ^{235}U, and ^{233}U [33, 34]. These studies have culminated in the most accurate QED measurement of any highly charged ion [35]. This measurement was accurate enough to be sensitive to the two-loop self energy contribution [36].

Dielectronic recombination measurements were performed on U^{90+} and neighboring ions [37]. These measurements provided the first experimental evidence of the quantum mechanical interference between dielectronic recombination and radiative recombination.

The magnetic trapping mode, mentioned earlier, was developed on SuperEBIT [38, 39]. On SuperEBIT it was used to study the x-ray emission from very highly charged ions produced by charge exchange recombination [40]. Using pulsed gas-injection, charge-exchange-induced x-ray spectra were obtained with the magnetic mode for ions as highly charged as heliumlike U^{90+} [41].

Following the successful operation of EBIT, EBIT-II, and SuperEBIT, electron beam ion traps were constructed outside Livermore. The first two were built at Oxford, England, using the Livermore designs of EBIT-II. Some changes, however, were made. One of these machines was delivered to the National Institute of Standards in Gaithersburg, Maryland. A close copy of EBIT-II was built in the United States and delivered to the Institute for Plasma Physics in Berlin. Higher-energy machines were built at the University of Electro-Communications, Tokyo, and at the Albert-Ludwigs-Universität Freiburg, Germany (now moved to Heidelberg). Although all electron beam ion trap devices share common design principles, it should be noted that the performance characteristics of each device is unique. The performance characteristics vary from machine to machine in a similar fashion as, for example, one tokamak device differs from another.

As many of the new electron beam ion traps have been called "EBIT", we now refer to the orginal EBIT electron beam ion trap as EBIT-I to avoid confusion. EBIT-I has been put again in service at the Lawrence Livermore National Laboratory upon an internal move of the facility. At the same time, EBIT-II was moved to the Lawrence Berkeley Laboratory.

In the following we describe experiments performed with SuperEBIT. The experiments were carried out with relativistic electron beams in excess of 100 keV. This is an energy regime still unsurpassed at other ion traps. These energies have allowed us to study the x-ray emission of highly charged actinide ions and test predictions of quantum electrodynamics in novel ways. Our investigations are now shifting beyond uranium to elements as high as curium, berkelium, and californium. We have used the device to measure the linear polarization of x rays well into the relativistic regime. We are also preparing measurements to study nuclear excitation by atomic x rays.

Ten years have passed since SuperEBIT was first operated. This overview shows that many unique measurements at the forefront of highly charged ion research continue to be performed with this machine.

ACTINIDE SPECTROSCOPY

Very heavy elements are of great interest to fundamental atomic physics because they are associated with the strongest electric and magnetic fields found in nature. Uranium (element 92) has been of particular interest because it is the heaviest naturally occuring element. The strong nuclear fields and the finite nuclear extent of their origin affect the atomic levels considerably. The $1s$ Lamb shift in uranium is about 460 eV. In californium (element 98) it is close to 700 eV. Therefore, studying x-ray transitions from highly charged ions beyond uranium is one of the research areas we are pursuing on SuperEBIT.

The actinides are radioactive. Using large quantities of material would, therefore, contaminate our machine. However, only about 10^7 ions are needed to filled the SuperEBIT

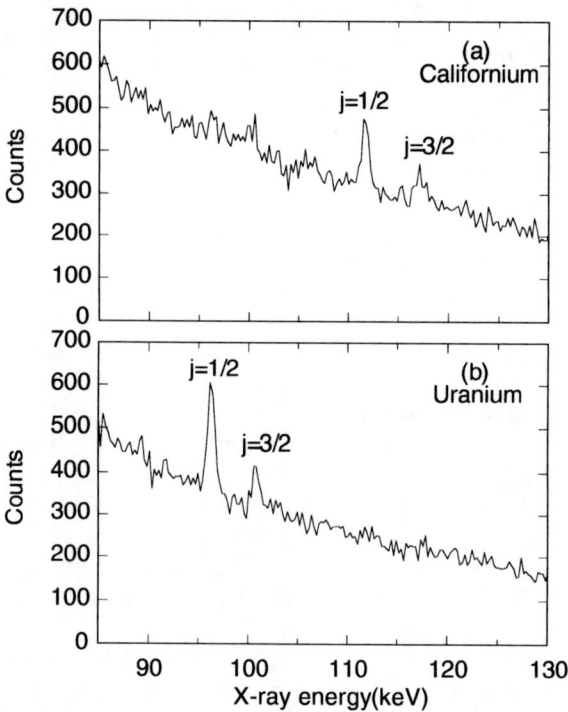

FIGURE 1. K-shell emission (a) of highly charged ^{249}Cf and (b) of highly charged ^{238}U recorded with a high-purity Ge detector. The label refers to the total angular momentum of the excited electron.

trap. This is such a small quantity that even radioactive elements can, in principle, be used in the device without appreciable contamination of the facility. Unfortunately, it is not easy to inject and trap only the amount of material needed to exactly fill the trap.

A method to inject trace amounts of radioactive material was described by Elliott and Marrs [42]. It relies on plating a small amount of radioactive material on the end of a thin wire. The wire is brought near the electron beam, where ion sputtering slowly transfers the material to the trap.

We have employed the wire method for injecting californium into SuperEBIT. A platinum wire was electrolytically coated with 5 ng (20 nC) of ^{249}Cf, and inserted into the SuperEBIT trap [43]. The trap was operated with a 250-mA electron beam at about 140–150 keV. A minimum of about 15 minutes was required before californium x rays were seen.

A K-shell x-ray spectrum collected within 30 minutes after insertion of the wire is shown in Fig. 1(a). The spectrum was recorded with a 5-cm diameter, 2-cm thick Ge detector. The spectrum shows one feature at 110 and a second at 114 keV. The lower-energy feature corresponds to $2s_{1/2} \rightarrow 1s_{1/2}$ and $2p_{1/2} \rightarrow 1s_{1/2}$ transitions; the higher-energy feature corresponds to $2p_{3/2} \rightarrow 1s_{1/2}$ transitions. For comparison, we show a spectrum of highly charged uranium in Fig. 1(b). In addition, we show the K-shell x-ray

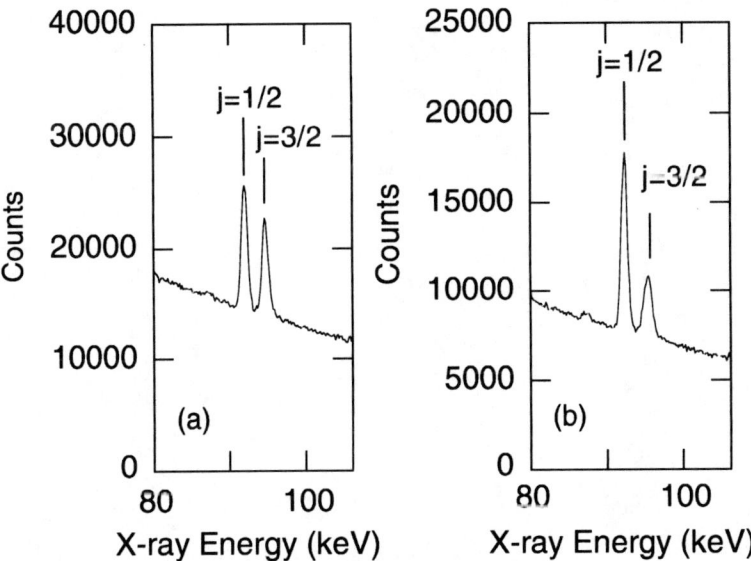

FIGURE 2. K-shell emission of highly charged thorium: (a) average charge around Th^{84+} and (b) average charge reaching Th^{88+}. The label refers to the total angular momentum of the excited electron.

of highly charged thorium in Fig. 2.

The intensity of the lower-energy feature is clearly higher than that of the higher-energy feature. This intensity ratio is very different from that observed in an x-ray tube where a neutral element is bombarded with an electron beam and the higher-energy feature, the so-called $K\alpha_1$ line, is about twice the size as the lower-energy feature, the so-called $K\alpha_2$ line. Only highly charged ions emit K-shell x rays where the lower-energy feature is as large or larger than the higher-energy feature.

The $j = 1/2$ feature dominates the intensity ratio of the two californium K-shell features in Fig. 1(a). This is a definitive signature that the californium was ionized as highly as heliumlike Cf^{96+}.

The thorium emission in Fig. 2 illustrates the dependence of this ratio on the charge balance. The charge balance in (a) was centered around C-like and N-like thorium, that in (b) was centered around He-like through Be-like thorium. The lower-energy feature clearly becomes larger as the charge balance shifts to higher charge states.

2S QED STUDIES

The $2p_{3/2} \rightarrow 1s_{1/2}$ K-shell transitions in the actinides have energies around 100 keV. By contrast, the corresponding $2p_{3/2} \rightarrow 2s_{1/2}$ L-shell transitions, have energies of a few keV. Since the QED contribution to the transition energy drops by a lesser amount than the overall transition energy, the fraction due to QED is actually higher in the

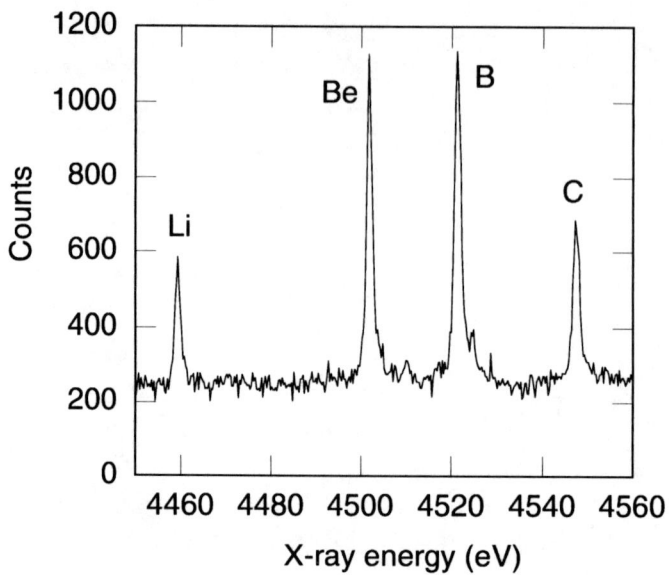

FIGURE 3. Crystal-spectrometer spectrum of the $2p_{3/2} \to 2s_{1/2}$ transitions from highly charged uranium ions in the energy region 4450 to 4560 eV. Transitions in lithiumlike, berylliumlike, boronlike, and carbonlike uranium are labeled by *Li*, *Be*, *B*, and *C*, respectively.

$2p_{3/2} \to 2s_{1/2}$ transitions. Moreover, the lower transition energies make it easier to measure these lines with current techniques. As a result, we have concentrated on measuring the $2p_{3/2} \to 2s_{1/2}$ transitions.

A spectrum of $2p_{3/2} \to 2s_{1/2}$ transitions in uranium near 4500 eV is shown in Fig. 3. An electron beam well in excess of 100 keV was required to produce the appropriate charge states. The spectrum not only shows the $2p_{3/2} \to 2s_{1/2}$ line in lithiumlike U^{89+}, it also shows the equivalent transitions in the neighboring charge states, berylliumlike U^{88+}, boronlike U^{87+}, and carbonlike U^{86+}. All of the $2p_{3/2} \to 2s_{1/2}$ transitions are affected by QED, which accounts for about 40 eV of the energy of these transitions.

Our measurements of lithiumlike $2p_{3/2} \to 2s_{1/2}$ transitions have been extended to bismuth, thorium, and uranium, as summarized in Table I. In uranium, we have made measurements for three different isotopes. The transition energy systematically increases for uranium isotopes with fewer neutrons and thus smaller nuclear radii. This clearly shows the strong influence of the nuclear size on the transition energy.

POLARIZATION SPECTROSCOPY

X-ray radiation generated by an electron beam is inherently polarized. We use this property to make explicit tests of polarization calculations. Because the degree of polarization depends on the relative strength of the populations of the magnetic sublevels, polariza-

TABLE 1. Measured $2p_{3/2} \rightarrow 2s_{1/2}$ transitions energies from different experiments.

Ion	Energy
$^{238}U^{89+}$	4459.37 ± 0.21 eV
$^{235}U^{89+}$	4459.43 ± 0.22 eV
$^{233}U^{89+}$	4459.63 ± 0.24 eV
Th^{87+}	4025.23 ± 0.14 eV
Bi^{80+}	2788.139 ± 0.039 eV

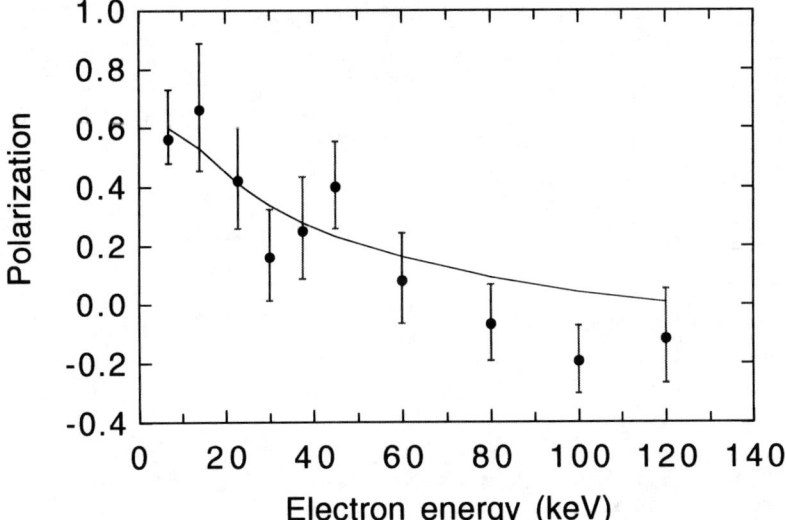

FIGURE 4. Polarization of the $1s2p\ ^1P_1 \rightarrow 1s^2\ ^1S_0$ resonance line in heliumlike Fe^{24+}. All data were obtained on SuperEBIT except the lowest-energy datum, which was obtained on EBIT-II. The solid line represents calculations obtained with the code of Zhang, Sampson, and Clark [48].

tion measurements test electron-impact cross section calculations on a more fundamental level than measurements of the total excitation cross section.

Calculations of the degree of linear polarization have been tested in several measurements with highly charged heliumlike ions on EBIT-I and EBIT-II [44, 45, 46]. These measurements tested the calculations near threshold for polarization.

The high electron collision energies achievable in SuperEBIT allow us to test the calculations in the fully relativistic regime. In Fig. 4 we show our measurement of the $1s2p\ ^1P_1 \rightarrow 1s^2\ ^1S_0$ resonance line in heliumlike Fe^{24+}.

Our measurements made use of the two-crystal technique for measuring polarization [47]. We employed two LiF crystals cut to the (220) and (200) planes. Measurements were made on SuperEBIT with beam energies between 14 keV and 120 keV. The figure

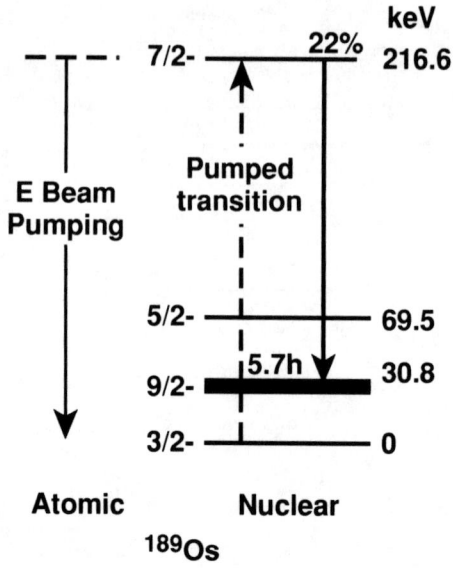

FIGURE 5. Energetics of the ^{189}Os^{75+} system.

also shows the value measured near threshold at 6.8 keV obtained on EBIT-II [45]. The results are compared to predictions obtained with the relativistic code from Zhang, Sampson, and Clark [48], and good agreement is found.

NUCLEAR EXCITATION BY ATOMIC X RAYS IN ^{189}OS

The relativistic electron beam energies achieved in SuperEBIT can also be used for novel nuclear physics measurements. We are beginning experiments to investigate nuclear excitation by absorbing atomic x rays. The aim of such atomic-nuclear interaction experiments is to determine the strength of the matrix element associated with the process dubbed NEET (nuclear excitation by electronic transition). NEET can be thought of as inverse internal conversion, i.e., the nuclear decay process in which a nuclear excited state decays by electron emission as opposed to gamma-ray emission.

Our experiments are centered on ^{189}Os. Here, the required degeneracy between the energy of an atomic transition and the nuclear energy level is accessible with our machine. The degeneracy is seen in Fig. 5, which shows the energy level diagrams for atomic and nuclear ^{189}Os. A beam electron is captured into an L-shell atomic level in heliumlike Os^{75+}. This capture releases an atomic x ray with energy equal to the sum of the electron beam energy and the binding energy of the captured electron. By tuning the electron beam to the right energy, the energy of the atomic x ray can be made to equal that of the 216.663-keV nuclear level indicated in Fig. 5. If the proper energy degeneracy is attained, the atomic x ray may excite the nuclear level via virtual photon exchange.

The 217-keV level gamma decays to the 31-keV metastable state (cf. Fig. 5) with a decay half life of 5.7 hours. Detection of NEET is then achieved by detecting the decay of the 31-keV metastable state.

NEET has never been detected, and there are great uncertainties in the predictions of the matrix element. This makes these experiments a daring challenge.

ACKNOWLEDGMENTS

This work was supported in part by DOE/NERI and performed by the University of California Lawrence Livermore National Laboratory under the auspices of the Department of Energy under Contract No. W-7405-Eng-48.

REFERENCES

1. Levine, M. A., Marrs, R. E., Henderson, J. R., Knapp, D. A., and Schneider, M. B., *Phys. Scripta*, **T22**, 157 (1988).
2. Levine, M. A., Marrs, R. E., Bardsley, J. N., Beiersdorfer, P., Bennett, C. L., Chen, M. H., Cowan, T., Dietrich, D., Henderson, J. R., Knapp, D. A., Osterheld, A., Penetrante, B. M., Schneider, M. B., and Scofield, J. H., *Nucl. Instrum. Methods*, **B43**, 431 (1989).
3. Marrs, R. E., Levine, M. A., Knapp, D. A., and Henderson, J. R., *Phys. Rev. Lett.*, **60**, 1715 (1988).
4. Knapp, D. A., Marrs, R. E., Levine, M. A., Bennett, C. L., Chen, M. H., Henderson, J. R., Schneider, M. B., and Scofield, J. H., *Phys. Rev. Lett.*, **62**, 2104 (1989).
5. Schneider, M. B., Knapp, D. A., Chen, M. H., Scofield, J. H., Beiersdorfer, P., Bennett, C., Henderson, J. R., Levine, M. A., and Marrs, R. E., *Phys. Rev. A*, **45**, R1291 (1992).
6. Beiersdorfer, P., Osterheld, A. L., Chen, M. H., Henderson, J. R., Knapp, D. A., Levine, M. A., Marrs, R. E., Reed, K. J., Schneider, M. B., and Vogel, D. A., *Phys. Rev. Lett.*, **65**, 1995 (1990).
7. Beiersdorfer, P., Marrs, R. E., Henderson, J. R., Knapp, D. A., Levine, M. A., Platt, D. B., Schneider, M. B., Vogel, D. A., and Wong, K. L., *Rev. Sci. Instrum.*, **61**, 2338 (1990).
8. Beiersdorfer, P., Chen, M. H., Marrs, R. E., and Levine, M. A., *Phys. Rev. A*, **41**, 3453 (1990).
9. Vogel, D. A., Beiersdorfer, P., Marrs, R., Wong, K., and Zasadzinski, R., *Z. Phys. D*, **21**, S193 (1991).
10. Chantrenne, S., Beiersdorfer, P., Cauble, R., and Schneider, M. B., *Phys. Rev. Lett.*, **69**, 265 (1992).
11. Beiersdorfer, P., and Wargelin, B. J., *Rev. Sci. Instrum.*, **65**, 13 (1994).
12. Beiersdorfer, P., Crespo-López Urrutia, J. R., Förster, E., Mahiri, J., and Widmann, K., *Rev. Sci. Instrum.*, **68**, 1077 (1997).
13. Brown, G. V., Beiersdorfer, P., and Widmann, K., *Rev. Sci. Instrum.*, **70**, 280 (1999).
14. Beiersdorfer, P., Phillips, T. W., Wong, K. L., Marrs, R. E., and Vogel, D. A., *Phys. Rev. A*, **46**, 3812 (1992).
15. Beiersdorfer, P., Schneider, M. B., Bitter, M., and von Goeler, S., *Rev. Sci. Instrum.*, **63**, 5029 (1992).
16. Beiersdorfer, P., Osterheld, A. L., Scofield, J., Wargelin, B., , and Marrs, R. E., *Phys. Rev. Lett.*, **67**, 2272 (1991).
17. Beiersdorfer, P., Chen, M. H., MacLaren, S., Marrs, R. E., Vogel, D. A., Wong, K., and Zasadzinski, R., *Phys. Rev. A*, **44**, 4730 (1991).
18. Beiersdorfer, P., Osterheld, A. L., Decaux, V., and Widmann, K., *Phys. Rev. Lett.*, **77**, 5353 (1996).
19. Beiersdorfer, P., in *AIP Conference Proceedings 389, X-Ray and Innershell Processes*, edited by R. L. J. ane H. Schmidt-Böcking and B. F. Sonntag, Woodbury, NY, 1997, p. 121.
20. Hölzer, G., Förster, E., Klöpfel, D., Beiersdorfer, P., Brown, G. V., Crespo López-Urrutia, J. R., and Widmann, K., *Phys. Rev. A*, **57**, 945 (1998).
21. Wargelin, B. J., Beiersdorfer, P., and Kahn, S. M., *Phys. Rev. Lett.*, **71**, 2196 (1993).
22. Beiersdorfer, P., Schweikhard, L., Crespo López-Urrutia, J., and Widmann, K., *Rev. Sci. Instrum.*, **67**, 3818 (1996).
23. López-Urrutia, J. R. C., Beiersdorfer, P., Savin, D. W., and Widmann, K., *Phys. Rev. A*, **58**, 238 (1998).

24. Neill, P., Beiersdorfer, P., Brown, G., Harris, C., Träbert, E., Utter, S. B., and Wong, K. L., *Phys. Rev. A*, **62**, 141 (2000).
25. Träbert, E., Beiersdorfer, P., Brown, G. V., Smith, A. J., Utter, S. B., Gu, M. F., and Savin, D. W., *Phys. Rev. A*, **60**, 2034 (1999).
26. Porter, F. S., Audley, M. D., Beiersdorfer, P., Boyce, K. R., Brekosky, R. P., Brown, G. V., Gendreau, K. C., Gygax, J., Kahn, S., Kelley, R. L., Stahle, C. K., and Szymkowiak, A. E., *Proc. SPIE*, **4140**, 407 (2000).
27. Chen, H., Beiersdorfer, P., Scofield, J. H., Gendreau, K. C., Boyce, K. R., Brown, G. V., Kelley, R. L., Porter, F. S., Stahle, C. K., Szymkowiak, A. E., and Kahn, S. M., *Astrophys. J. (Lett.)*, **567**, L169 (2002).
28. Knapp, D. A., Marrs, R. E., Elliott, S. R., Magee, E. W., and Zasadzinski, R., *Nucl. Instrum. Methods*, **A334**, 305 (1993).
29. Marrs, R. E., Elliott, S. R., and Knapp, D. A., *Phys. Rev. Lett.*, **72**, 4082 (1994).
30. Marrs, R. E., Beiersdorfer, P., and Schneider, D., *Phys. Today*, **47**, 27 (1994).
31. Beiersdorfer, P., Knapp, D., Marrs, R. E., Elliott, S. R., and Chen, M. H., *Phys. Rev. Lett.*, **71**, 3939 (1993).
32. Beiersdorfer, P., Osterheld, A., Elliott, S. R., Chen, M. H., Knapp, D., and Reed, K., *Phys. Rev. A*, **52**, 2693 (1995).
33. Elliott, S. R., Beiersdorfer, P., and Chen, M. H., *Phys. Rev. Lett.*, **76**, 1031 (1996).
34. Elliott, S. R., Beiersdorfer, P., Chen, M. H., Decaux, V., and Knapp, D. A., *Phys. Rev. A*, **57**, 583 (1998).
35. Beiersdorfer, P., Osterheld, A., Scofield, J., Crespo López-Urrutia, J., and Widmann, K., *Phys. Rev. Lett.*, **80**, 3022 (1998).
36. Sapirstein, J., and Cheng, K. T., *Phys. Rev. A*, **64**, 022502 (2001).
37. Knapp, D. A., Beiersdorfer, P., Chen, M. H., Schneider, D., and Scofield, J. H., *Phys. Rev. Lett.*, **74**, 54 (1995).
38. Beiersdorfer, P., Beck, B., Marrs, R. E., Elliott, S. R., and Schweikhard, L., *Rapid Commun. Mass Spectrom.*, **8**, 141 (1994).
39. Beiersdorfer, P., Beck, B., Becker, S., and Schweikhard, L., *Int. J. Mass Spectrom. Ion Proc.*, **157/158**, 149 (1996).
40. Beiersdorfer, P., Olson, R. E., Brown, G. V., Chen, H., Harris, C. L., Neill, P. A., Schweikhard, L., Utter, S. B., and Widmann, K., *Phys. Rev. Lett.*, **85**, 5090 (2000).
41. Schweikhard, L., Beiersdorfer, P., Brown, G. V., Crespo López-Urrutia, J. R., Utter, S. B., and Widmann, K., *Nucl. Instrum. Methods*, **B 142**, 245 (1998).
42. Elliott, S. R., and Marrs, R. E., *Nucl. Instrum. Methods*, **B100**, 529 (1995).
43. Beiersdorfer, P., Elliott, S. R., Crespo López-Urrutia, J., and Widmann, K., *Nucl. Phys.*, **A626**, 357 (1997).
44. Henderson, J. R., Beiersdorfer, P., Bennett, C. L., Chantrenne, S., Knapp, D. A., Marrs, R. E., Schneider, M. B., Wong, K. L., Doschek, G. A., Seely, J. F., Brown, C. M., LaVilla, R. E., Dubau, J., and Levine, M. A., *Phys. Rev. Lett.*, **65**, 705 (1990).
45. Beiersdorfer, P., Vogel, D. A., Reed, K. J., Decaux, V., Scofield, J. H., Widmann, K., Hölzer, G., Förster, E., Wehrhan, O., Savin, D. W., and Schweikhard, L., *Phys. Rev. A*, **53**, 3974 (1996).
46. Shlyaptseva, A. S., Mancini, R. C., Neill, P., Beiersdorfer, P., Crespo López-Urrutia, J., and Widmann, K., *Phys. Rev. A*, **57**, 888 (1998).
47. Beiersdorfer, P., Crespo López-Urrutia, J., Decaux, V., Widmann, K., and Neill, P., *Rev. Sci. Instrum.*, **68**, 1073 (1997).
48. Zhang, H. L., Sampson, D. H., and Clark, R. E. H., *Phys. Rev. A*, **41**, 198 (1990).

Absolute Measurements and MCDF Calculation of the Photoionization Cross Section of O^{3+} Ion

J.-P. Champeaux[1], J.-M. Bizau[2], D. Cubaynes[2], C. Blancard[1], D. Hitz[4], J. Bruneau[1], and F.J. Wuilleumier[2]

[1]*Departement de Physique Théorique et Appliquée, CEA-DAM/Ile-de-France, BP 12, 91680 Bruyères-le-Châtel, France*
[2]*Laboratoire d'Interaction des Rayons X avec La Matière (LIXAM), UMR 8624 du CNRS, University Paris-Sud, B. 350, and Laboratoire pour l'Utilisation du Rayonnement Electromagnétique (LURE), 91405 Orsay*
[4]*CEA-Grenoble, Département de Recherche Fondamentale sur la Matière Condensée, 17 rue des Martyrs, 38054 Grenoble Cedex 9, France*

Abstract. In this paper, we present the results of our investigation on the photoionization of the O^{3+} ion in the $1s^2 2s^2 2p$ ^2P ground state and the $1s^2 2s 2p^2$ ^4P metastable states. The experimental results are compared to a 72-states Multi Configurational Dirac Fock (MCDF) calculation.

INTRODUCTION

Photoionization processes play a significant role in the behavior of low density plasmas such as laboratory and astrophysical plasmas. Many plasma modeling codes are used to calculate photoionization data (Opacity Project, Iron Project) from different theoretical approximations such as R-Matrix or Multi Configurationnal Dirac-Fock approximation [1].

Due to the difficulty to obtain a high enough density of ions in the source volume of the spectrometer, the first direct comparison between theoretical and experimental data was achieved only in 1987- 88 when Peart and Lyon measured a photoionization cross section in singly-charged potassium ions [2], and were able to compare it to the prediction of the one electron model [3] and of RRPA calculation [4]. Recently, a significant step was made in photoionization of multiply charged ions by the use of Electron Cyclotron Resonant Ion Sources (ECRIS) [5] and third generation synchrotron radiation center. But today only a few experimental data have been measured on an absolute scale, to allow a valuable test of the theoretical calculations.

In addition to the fundamental aspect of atomic physics, oxygen ions are of importance in many processes occurring in the earth atmosphere as well as in stellar

CP652, *X-Ray and Inner-Shell Processes: 19th International Conference on X-Ray and Inner-Shell Processes*
edited by A. Bianconi, A. Marcelli, and N. L. Saini
© 2003 American Institute of Physics 0-7354-0111-X/03/$20.00

corona or interstellar plasmas. Experimental cross sections have been already measured for singly charged oxygen ions [6,7] and R-Matrix calculations were performed for oxygen ions [8,9]. In the work described here, we have been able to measure the single photoionization cross section of O^{3+} ion over a photon energy range from 65 eV up to 100 eV.

EXPERIMENT AND THEORY

The oxygen ion beam is produced using an Electron Cyclotoron Resonance Ion Source (ECRIS). The ions are extracted from the ECRIS with an electrostatic field and charge selected using a Wien filter. Then the O^{3+} ion beam is merged with the photon beam emitted from the SU6 ondulator of Super-ACO synchrotron radiation source of LURE, after monochromatization by a toroidal grating monochromator. The relative variation of the single photoionization cross section of O^{3+} was obtained from the measurement of the O^{4+} signal produced by the photoionization process according to: O^{3+} + hv --> O^{4+} + εl. This relative cross section was put on an absolute scale by measuring the absolute cross section at a photon energy of 79.1 eV, i.e. in the continuum well above the ionization threshold of the ground state of O^{3+}. This measurement yielded a value of 2.0 ± 0.4 Mb.

In order to interpret our experimental results, we have performed a MCDF calculation. The photoexcitation cross section was calculated introducing 985 levels issued from 84 configurations to describe the initial and final states. The initial state was described using $2s^2 2p$, $2s^2$ ns np nd (n = 3 to n = 11), 2s $2p^2$ and 2s2p ns np nd (n = 3 to n = 11) electronic configurations and the final state by $2s^2$, 2s2p and $2p^2$. The results of our calculations for the ionization thresholds of O^{3+} are 76.36 eV and 67.0 eV for the $1s^2 2s^2 2p$ $^2P_{1/2,3/2}$ and the $1s^2 2s 2p^2$ $^4P_{1/2,3/2,5/2}$ states, respectively. This calculated value of the ground state threshold is in fairly good agreement with our measured value of 77.3 eV. The direct photoionization cross section was calculated for several photon energies every 10 eV above the energy of the ground state threshold with a 12 electron configurations calculation and fit using a power low $A*E(eV)^B$.

RESULTS

Figure 1 shows the experimental spectrum we have measured for the photoionization cross section of O^{3+} (upper panel) and the results of our MCDF calculations (lower panel). The theoretical spectrum was obtained by summing the photoexcitation cross section (followed by autoionization into the O^{4+} continuum), and the direct ionization cross section obtained in the 12 configurations MCDF calculation. This sum was then convolved with a 0.25eV Gaussian profile in order to reproduce the experimental resolution.

In the experimental spectrum, we observe several groups of structures. They are due to different photoionization processes. The step around 77.3 eV is due to the

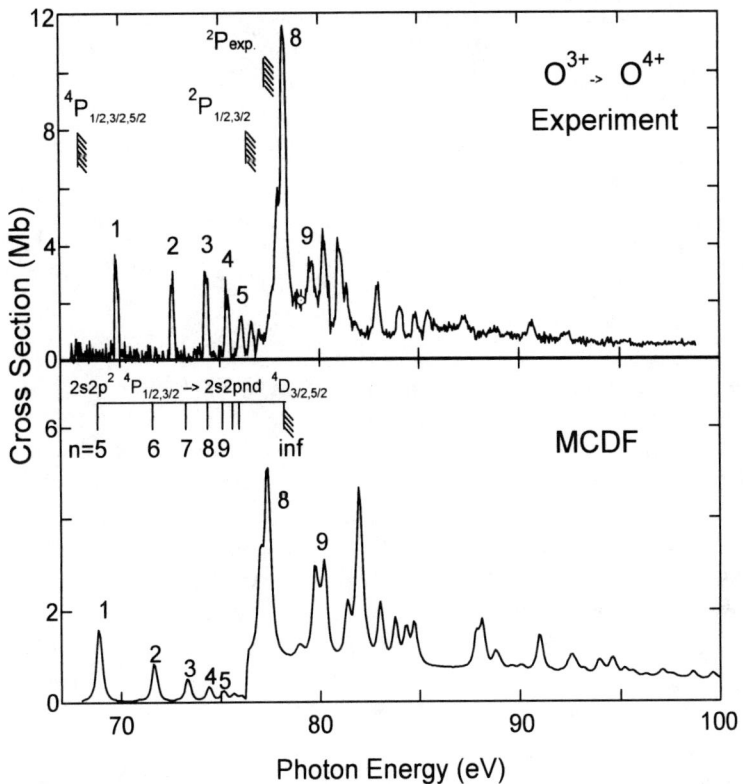

FIGURE 1. Comparison between experimental and calculated spectrum of the photoionization cross section of O^{3+} ion.

opening of the 2p direct ionization channel of the O^{3+} ion in the $1s^2 2s^2 2p$ $^2P_{1/2,3/2}$ ground states; whereas the lines observed for discrete photon energies below and above this threshold correspond to resonant photoexcitation processes followed by autoionization decays.

Below the $^2P_{1/2,3/2}$ thresholds, we observe a very clear Rydberg series (lines noted 1 to 7 in figure 1). This Rydberg series is due to excitation of an outer 2p-electron to a nd (n>4) excited orbital belonging to the first metastable states of O^{3+} ion, followed by autoionization decay into the O^{4+} $1s^2 2s^2$ $^1S_0$ ground state, according to:

$$O^{3+*} 1s^2 2s 2p^2 \; ^4P_{1/2,3/2,5/2} \rightarrow O^{3+**} 1s^2 2s 2pnd \; (n > 4) \; ^4D \rightarrow O^{4+} 1s^2 2s^2 \; ^1S_0 + \varepsilon l$$

Most of the lines observed above 77.3 eV are due to resonant excitation of O^{3+} in the $1s^2 2s^2 2p$ $^2P_{1/2,3/2}$ ground state.

The Rydberg series (1-7) below the ground state threshold is well reproduced by our MCDF calculations with an almost constant shift of 0.7 eV. We note that this series can be reproduced only by relativistic calculation introducing spin-orbit interaction. The agreement with our MCDF calculations is not as good for the resonances located above the ground state threshold. We can nevertheless identified

TABLE 1. Comparison between MCDF and experimental values of most intense lines.

N°	MCDF position (eV)	Experimental position (eV)	MCDF Identification Transitions ^{4}P---^{4}D (1 to 6), ^{2}P --- ^{2}D for line 8 and 9.
1	69,18	69.84	2s 2p^2 -> 2s 2p 5d
2	71,94	72.63	2s 2p^2 -> 2s 2p 6d
3	73,60	74.32	2s 2p^2 -> 2s 2p 7d
4	74,68	75.36	2s 2p^2 -> 2s 2p 8d
5	75,41	76.06	2s 2p^2 -> 2s 2p 9d
6	-	76,58	2s 2p^2 -> 2s 2p 10d
8	77,05	78,22	2s^2 2p -> 2s 2p 5p
9	80,08	80,24	2s^2 2p -> 2s 2p 6p

some of them by comparison with their calculated oscillator strength and energy position (lines **8** and **9**).

The determination of the percentage of metastable states in the beam produced by the ECR source in the interaction region is an important point for the experimental determination of the absolute cross section. To achieve this determination, we compared the relative measurement of well identified resonances (lines **1** and **8**) to their calculated oscillator strengths. Line **1** is due to the photoexcitation of the metastable state while line **8** arises from photoexcitation of ground state. We present in Table 1 the positions of most intense lines and their corresponding assignment.

CONCLUSION

To conclude, we have measured the absolute photoionization cross section of the O^{3+} ion. We have also calculated this cross section using the MCDF approximation. The results of our calculations are in good agreement with the experimental measurements. The transition corresponding to the most intense lines are also identified. The difficulty in obtaining accurate value of the relative population of ions in the metastable ^{4}P states is a factor limiting a more accurate determination of the absolute photoionization cross section.

REFERENCES

1. Bruneau, J., *J. Phys. B* **17**, 3009 (1984)
2. Peart B., and Lyon I.C., *J. Phys. B* **20**, L673 (1987)
3. Reilmann, R.F., and Manson, S.T., *Astrophys. Journal*, Suppl. Series **10**, 815 (1979)
4. Nasreen, G., Deshmukh, P. C., and Manson, S. T., *J. Phys B* **21**, L281 (1988)
5. Bizau et al., *Phys. Rev. Letters* **84**, 435 (2000); ibid. **87**, 273002 (2001)
6. Covington et al., *Phys. Rev. Letters* **87**, 24002(2001)
7. Kjeldsen, H. and Kristensen, B., *Astrophys. Journal*, Suppl. Series **138**, 219(2002)
8. Nahar, S. N., *Phys. Rev. A* **58**, 3766 (1998)
9. Olalla at al., E., *Mon. Not. R. Astron. Society* **332**, 1005 (2002)

Nondipolar photoionization of atoms

B. Krässig, R. W. Dunford, E. P. Kanter, S. H. Southworth, and L. Young

Chemistry Division, Argonne National Laboratory, Argonne, IL 60439, USA

Abstract. Results of the nondipolar angular distribution parameter γ are presented for two cases of photoionization with substantially different photon energies, Kr 1s in the hard x-ray regime, and He 1s in the vuv regime. The nondipolar asymmetries are apparent even at low energies. The presence of dipole and quadrupole resonances strongly modifies the energy dependence of the nondipolar asymmetry parameter γ. Measurements of γ can be used as a spectroscopic tool to characterize quadrupole resonances that are not accessible in photoabsorption measurements.

INTRODUCTION

The vast majority of the phenomena observed in the absorption of photons in matter can be understood on basis of the dipole approximation. In a multipole expansion of the interaction the dipole part is always the dominant one, but at high energies, when the x-ray wavelength approaches the length scale of the atom, the dipole approximation alone may not be sufficient. Yet, the presence of higher multipoles in the photon-atom interaction can be felt even at low photon energies if photoelectron angular distributions are being studied. In particular, the odd-parity dipole interaction and the even-parity quadrupole interaction interfere with each other giving rise to a term in the angular distribution that is odd under inversion. The interference term causes an asymmetry of the angular distribution with respect to the propagation direction of the photon. The interference term vanishes in the integral cross section and it also vanishes in the plane perpendicular to the photon beam where photoelectron angular distributions are most commonly measured. In the following we will describe measurements of the forward-backward asymmetry of the photoelectron angular distribution as a means of accessing information on the nondipolar contents in the interaction. The apparatus used in these measurements exploits special symmetry properties of the photoelectron angular distribution to eliminate the dependence on the polarization characteristics of the photon beam and to reduce instrumental anisotropies. We report results of the nondipolar angular distribution parameter γ for Kr 1s from threshold at 14.3 keV up to 22 keV [1]. As a second example, we present results on the nondipolar asymmetries of He 1s photoionization with ∼ 60 eV photon energy in the region of the doubly excited autoionization resonances [2].

CP652, *X-Ray and Inner-Shell Processes: 19th International Conference on X-Ray and Inner-Shell Processes*
edited by A. Bianconi, A. Marcelli, and N. L. Saini
© 2003 American Institute of Physics 0-7354-0111-X/03/$20.00

THEORETICAL FORMALISM

In the quantum mechanical description of the interaction of a photon with an atom the transition operator contains the oscillatory factor $e^{ik \cdot r}$. Here k is the momentum vector of the photon in atomic units and r is the position vector of the ionized atomic electron. In the dipole approximation the exponential is approximated by unity, assuming $kr \ll 1$. Going beyond the dipole approximation, the plane wave can be expanded in partial waves [3] or in multipoles $E1, M1, E2, M2, \ldots$ [4, 5]. Grouping the expansion terms of orders of k is referred to as the retardation expansion. The leading term in the retardation expansion is unity and thus represents the dipole approximation. It corresponds to the long-wavelength limit of the dipole term, $E1$, in the multipole expansion. Going to the second term in the retardation series amounts to including the long-wavelength limits of the magnetic dipole, $M1$, and electric quadrupole, $E2$. At higher orders the correspondence between terms of the retardation and multipole expansions cannot be made as easily. The multipoles have alternating parities with increasing order j, $\pi(Ej) = (-1)^j$ and $\pi(Mj) = (-1)^{j-1}$, and transfer j units of angular momentum in the interaction. Between electric quadrupole and magnetic dipole the latter is expected to be very weak as it vanishes within the framework of central field calculations [6]. In the present context we will be mainly concerned with electric dipole and electric quadrupole components and the resulting first order retardation correction.

The angular distribution of the emitted photoelectrons depends on the square modulus of the transition matrix element. If the transition operator is written as a coherent sum of two terms, e.g. $T = E1 + E2$, then the square modulus of the transition matrix element $|\langle T \rangle|^2$ will contain cross terms $\langle E1 \rangle \langle E2 \rangle$ in addition to the squares of the pure terms $|\langle E1 \rangle|^2, |\langle E2 \rangle|^2$. The leading retardation correction in the differential cross section beyond the standard dipole approximation contains the $E1$–$E2$ interference term in the short-wavelength limit. This term is of order k and has odd parity which causes a forward-backward asymmetry in the angular distribution. The differential cross section with the first order correction, for linearly polarized photons, can be written as [6]

$$\frac{d\sigma}{d\Omega} = \frac{\sigma}{4\pi} \left(1 + \beta P_2(\hat{\boldsymbol{\varepsilon}} \cdot \hat{\boldsymbol{p}}) + (\gamma(\hat{\boldsymbol{\varepsilon}} \cdot \hat{\boldsymbol{p}})^2 + \delta)\hat{\boldsymbol{k}} \cdot \hat{\boldsymbol{p}} \right). \tag{1}$$

The quantities $\hat{\boldsymbol{\varepsilon}}, \hat{\boldsymbol{k}}, \hat{\boldsymbol{p}}$ are the unit vectors of the polarization, beam, and electron emission directions, respectively. In addition to the usual anisotropy parameter β the differential cross section contains two nondipolar asymmetry parameters, γ and δ. In the way these two parameters were defined, $\delta = 0$ for the ionization of s electrons.

For ionization of s electrons the asymmetry parameter γ depends on the photon energy ω, the radial matrix elements for electric dipole and electric quadrupole transitions R_1, R_2, and the phase differences of the ougoing partial waves from electric dipolar and electric quadrupolar photoionization δ_1, δ_2 [3, 6]

$$\gamma(ns = 3\alpha\omega \frac{R_2}{R_1} \cos(\delta_2 - \delta_1) \tag{2}$$

with α being the fine structure constant.

The second order retardation correction, order k^2, has even parity and is therefore forward-backward symmetric. The form of the differential cross section including the

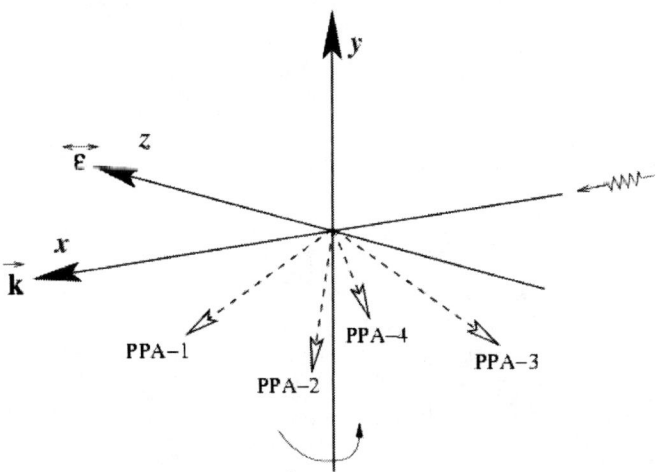

FIGURE 1. Schematic of the positions of the four PPAs relative to the x-ray beam direction *k* and the polarization direction **ε**.

second order correction can be found in [7]. The second order correction adds two more independent angular distribution parameters.

EXPERIMENTAL TECHNIQUE

The apparatus used in the present experiments is a considerably improved version of the one used in our previous experiments [8]. The setup was specifically designed for the measurement of the nondipolar forward-backward asymmetries. The new apparatus utilizes four parallel plate electron analyzers (PPAs) mounted on a common rotation stage. The four PPAs are mounted such that they are aligned along the space diagonals in the cartesian coordinate frame, viewing the interaction region in the center (Fig. 1). In this position the analyzers are at the magic angle $\theta_m = 54.7°$ with respect to all three axes. When the signals of the two analyzers in the forward hemisphere are added together, and the signals of the two backward analyzers are added, then the sum signals are independent of the degree of polarization or the orientation of the polarization vector in the plane perpendicular to the beam axis [9]. The signals are also independent of the anisotropy parameter β. Furthermore, because the geometry of each analyzer with respect to the interaction region is the same for all four analyzers, the instrumental anisotropies are small. With the rotation stage the individual analyzers can be repositioned so that during a measurement each analyzer collects data at all four positions. In this way the efficiency differences between the analyzers are compensated.

Denoting the signals of the four PPAs at the angular positions shown in Fig. 1 as I_1, I_2, I_3, I_4, the measured forward-backward asymmetry according to Eq. (1) is proportional to $\gamma + 3\delta$

$$\frac{I_1 + I_2 - I_3 - I_4}{I_1 + I_2 + I_3 + I_4} = \frac{\gamma + 3\delta}{\sqrt{27}}. \tag{3}$$

If the second order retardation correction is taken into account the normalization constant $I_1 + I_2 + I_3 + I_4$ in the denominator becomes dependent on one of the second order parameters, ξ [7]

$$\frac{I_1 + I_2 - I_3 - I_4}{I_1 + I_2 + I_3 + I_4} = \frac{\gamma + 3\delta}{\sqrt{27}(1 - \frac{7\xi}{18})}. \tag{4}$$

The presence of a nonzero ξ therefore affects the result of $\gamma + 3\delta$ if it is extracted from the observed asymmetry using Eq. (3). The values of ξ calculated by Derevianko are generally $\ll 1$ so that the correction in equation (4) is mostly negligible. Also, $\delta = 0$ in the two cases of 1s ionization studied in this report.

RESULTS FOR Kr 1s

The measurements for the nondipolar asymmetry of 1s photoionization in krypton were carried out at the BESSRC-CAT 12-ID undulator beam line of the Advanced Photon Source at Argonne National Laboratory. The measurements covered the energy range from 11 eV above the Kr-K edge to 8000 eV above (x-ray energy $\sim$14.3 keV $-$ 22.3 keV). A complete account of these measurements will be given elsewhere [1].

Fig. 2 shows the energy dependence of the parameter γ(Kr1s). The data points below 1 keV have negative γ indicating preferential backward emission of the photoelectrons. Near 1 keV the nondipolar asymmetry vanishes with $\gamma = 0$ and above 1 keV the asymmetry is positive and increasing with energy.

Fig. 2 contains also the prediction of relativistic IPA calculations [10] for this case. The calculation is able to include retardation to all orders. However, in order to demonstrate the degree by which the measurement of γ is affected by the omission of the second order contribution we show the result of the first order retardation correction as a full line, and include for comparison the asymmetry resulting from first plus second order retardation corrections. Clearly, the first order calculation reproduces the experimental data very well. The distortion by the presence of the second order correction is less than the experimental error and apparent only at the the highest energies in this measurement.

The energy dependence of the nondipolar asymmetries in Kr is reminiscent of the ones observed in Ar [8], however with Kr having much larger negative values in the low energy region and the zero crossing of the γ parameter being at a much higher energy for Kr. The energy dependences of γ(1s) for different Z smoothly follow this trend of more negative values and shifting of the zero crossing to higher energies for increasing Z. On the low Z end the curves converge on the positive-definite square-root-like energy dependence of the point-Coulombic case of hydrogen.

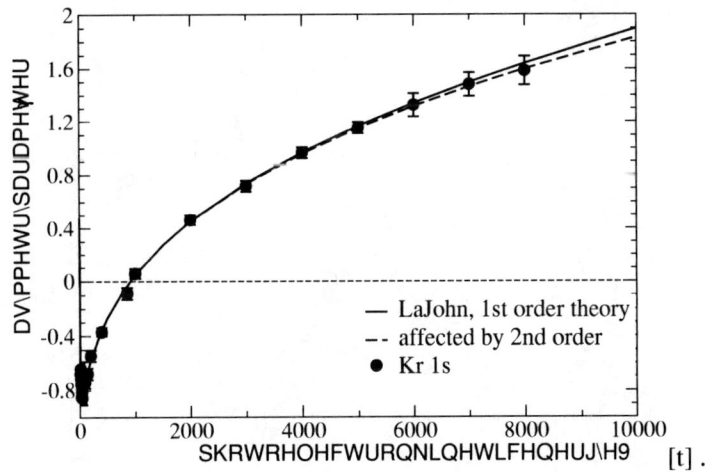

FIGURE 2. Energy dependence of the nondipolar parameter γ for Kr 1s as obtained from Eq. (3). Experimental data, solid circles; relativistic IPA calculation up to 1st order retardation correction [10], full line; calculation up to second order, broken line. See text

RESULTS FOR He 1s

In helium already the lowest doubly-excited state lies energetically above the ionization energy. The resonantly excited neutral states decay rapidly via autoionization. The interference of the sequential process of excitation followed by decay with the direct process leads to the well known Fano-resonance structure in the energy range between 60.1 eV and the double ionization limit at 79.0 eV. The dipole-allowed $^1P^o$ series have been studied and characterized with great precision and detail in photoabsorption experiments [11]. The resonance series with couplings and parity other than $^1P^o$ are observed in electron scattering experiments [12].

The intent of this measurement was to measure the nondipolar asymmetry at low photon energies and to study the influence of the dipole and (dipole-forbidden) quadrupole autoionization resonances on the nondipolar asymmetry (cf. [13, 14]. The low-energy measurement for helium were performed at the PGM Undulator beam line at the Synchrotron Radiation Center (SRC) of the University of Wisconsin. The complete results of this measurement and a more detailed description can be found in [2].

Fig. 3 shows in the lower panel the signal of an ion detector when scanning the monochromator across the region of the lowest-lying $^1D^e$ and $^1P^o$ resonances. Only the dipole resonance is observed in the ion signal. A fit to these data, with convolution to the bandpass function of the monochromator, results in values for the dipole resonance parameters that are in excellent agreement with those from the state-of-the-art measurements of this resonance series [11].

In the upper panel of Fig. 3 the nondipolar asymmetry parameter γ is shown for the same region. Outside the resonances γ has a value of about 0.08, but γ displays pronounced variations at both the dipole and quadrupole resonances. In the maximum of

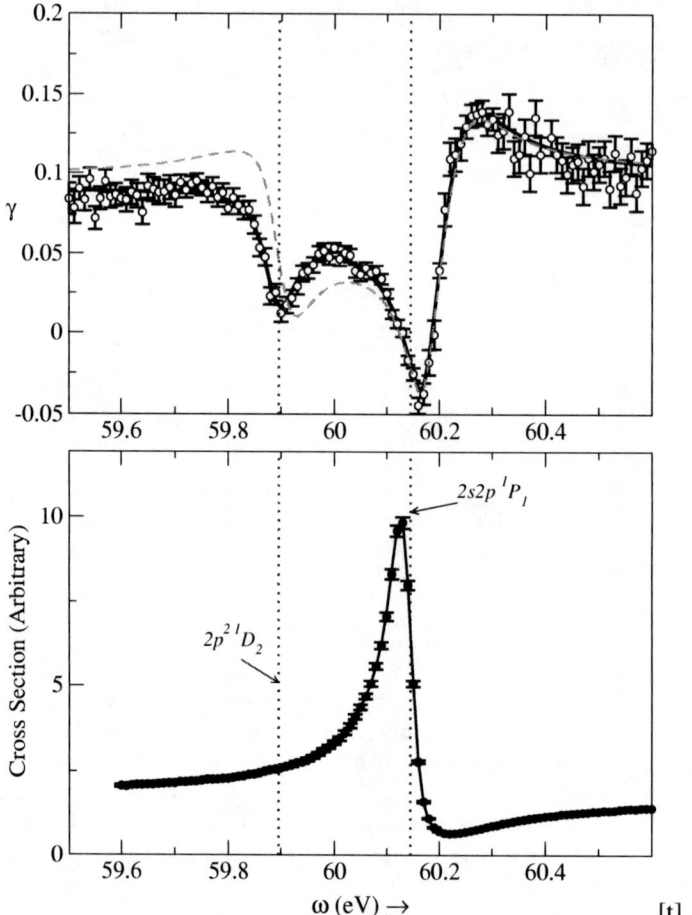

FIGURE 3. Upper panel: Energy dependence of the nondipolar parameter γ for He 1s in the region of the lowest-lying quadrupole and dipole resonances. Experimental data, solid circles; fit to data according to Eq. (5), solid line; curve from Eq. (5) using theoretical values of $q_2, \delta_2 - \delta_1$, broken line. Lower panel: energy dependence of the ion signal recorded when scanning across the same energy region. See text

the dipole resonance γ is close to zero because of the predominance of dipole interaction at the maximum. In the minimum of the dipole resonance γ has a transient from negative values to positive values. The minimum would be exactly zero for a δ-function-like bandwidth of the monochromator and this, following from Eq. (2), would result in a divergence of γ at exactly this energy. With the experimental bandwith of the experiment this divergence is attenuated to the observed transient.

The functional form of the nondipolar asymmetry in the region of the lowest series of autoionization resonances (one continuum) has been derived by Martin [15]. Starting from Eq. (2) and using the well-established formalism of Fano [16] for both dipole and quadrupole,

$$\gamma^r = \gamma^0 \left\{ \frac{\cos(\delta_2 + \Delta_2 - \delta_1 - \Delta_1)}{\cos(\delta_2 - \delta_1)} \right\} \times \left\{ \frac{q_2 + \varepsilon_2}{\sqrt{1 + \varepsilon_2^2}} \right\} \times \left\{ \frac{\sqrt{1 + \varepsilon_1^2}}{q_1 + \varepsilon_1} \right\}, \tag{5}$$

where $\varepsilon_\ell = \frac{\omega - \omega_\ell}{\Gamma_\ell/2}$ and $\cot \Delta_\ell = -\varepsilon_\ell$. The superscript r on γ stands for resonant and γ^0 is the nondipolar asymmetry in the absence of the resonances. The quantities with indices $\ell = 1, 2$ refer to the electric dipole and electric quadrupole interactions, respectively, and concern the Fano q_ℓ parameter, the resonance widths Γ_ℓ, and resonance energies ω_ℓ. Using Eq. (5) as a fit function to the observed nondipolar asymmetries, further requires proper accounting for the finite bandwith of the monochromator.

By fitting the observed nondipolar asymmetry we obtained the Fano resonance parameters for both the dipole and quadrupole resonances, the off-resonance asymmetry γ^0 and also the off-resonant phase shift difference $\cos(\delta_2 - \delta_1)$ [2]. It is the presence of the resonances with teir well-defined resonance phase shifts that allows the separation of γ^0 into the radial matrix-element ratio and the phase shift difference, which otherwise would not be possible.

This measurement is the first experimental determination of the Fano resonance parameter $q_2(2p^2 {}^1D^e) = -0.25(7)$. This value differs significantly from the theoretical predictions $q_2 = -1$ [17, 18]. This quantity is not accessible in scattering experiments. Also the resonance position and width of the quadrupole resonance were determined in this experiment to higher precision than before [2]. The measurement of the nondipolar asymmetry establishes a new spectroscopic tool to characterize accurately the dipole-forbidden quadrupole resonances.

CONCLUSION

Nondipolar effects are noticeably present in the angular distribution of photoelectrons. The influence of higher electric quadrupole components in the photon beam manifests itself as a forward-backward asymmetry in the angular distribution with respect to the beam direction, leaving the plane perpendicular to the photon beam and angle-integrated cross sections unaffected. In general the nondipolar asymmetries increase with energy, but the individual energy dependences can be strongly influenced by variation of the dipole or quadrupole amplitudes and the phase difference between outgoing partial waves for dipole and quadrupole interactions. As an example we presented data for the energy dependence of Kr 1s photoelectron up to 8 keV above threshold. The variations of the observed nondipolar asymmetry become very pronounced in the region of resonances. The variation of the nondipolar asymmetry can be used as a spectroscopic tool to characterize quadrupole resonances that are too weak in intensity to be observed directly in photoabsorption. This was shown in the case of helium. Even at photon energies as low as 60 eV the nondipolar asymmetries and the influence of autoionization resonances are distinctly observable. We observed similarly strong variations in the region of the cross section minimum in Xe 5s photoionization near 20 eV photon energy, which will be the subject of a separate publication.

ACKNOWLEDGEMENTS

We are grateful to N. L. S. Martin, R. Guillemin, O. Hemmers, and D. W. Lindle for their assistance during the experiments at SRC. We thank the BESSRC and SRC staff for excellent research facilities. The University of Wisconsin SRC is funded by the National Science Foundation Grant No. DMR-0084402. This work is supported by the Chemical Sciences, Geosciences, and Biosciences Division of the Office of Basic Energy Sciences, Office of Science, U.S. Department of Energy under contract W-31-109-ENG-38.

REFERENCES

1. Krässig, B., Bilheux, J., Dunford, R. W., Gemmell, D. S., Hasegawa, S., Kanter, E. P., Southworth, S. H., Young, L., LaJohn, L. A., and Pratt, R. H., *Phys. Rev. A* (2002), to be published.
2. Krässig, B., Kanter, E. P., Southworth, S. H., Guillemin, R., Hemmers, O., Lindle, D. W., Wehlitz, R., and Martin, N. L. S., *Phys. Rev. Lett.*, **88**, 203002 (2002).
3. Bechler, A., and Pratt, R. H., *Phys. Rev. A*, **39**, 1774 (1989).
4. Peshkin, M., "Angular distributions of photoelectrons: Consequences of symmetry", in *Advances in Chemical Physics, Vol. XVIII*, edited by I. Prigogine and S. A. Rice, John Wiley & Sons, Inc., 1970.
5. Scofield, J. H., *Phys. Rev. A*, **40**, 3054 (1989).
6. Cooper, J. W., *Phys. Rev. A*, **47**, 1841 (1993).
7. Derevianko, A., Johnson, W. R., and Cheng, K. T., *At. Data Nucl. Data Tables*, **73**, 153 (1999), the parameter ξ corresponds to ν used in this reference, see [19].
8. Jung, M., Krässig, B., Gemmell, D. S., Kanter, E. P., LeBrun, T., Southworth, S. H., and Young, L., *Phys. Rev. A*, **54**, 2127 (1996).
9. Peshkin, M., "Photon beam polarization and nondipolar angular distributions", in *Atomic Physics with Hard X-Rays from High Brilliance Synchrotron Light Sources*, ANL/APS/TM-16, Argonne National Laboratory, 1996, p. 207.
10. LaJohn, L. A., and Pratt, R. H. (2002), private communication.
11. Schulz, K., Kaindl, G., Domke, M., Bozek, J. D., Heiman, P. A., Schlachter, A. S., and Rost, J. M., *Phys. Rev. Lett.*, **77**, 3086 (1996), and references therein.
12. van den Brink, J., Nienhuis, G., van Eck, J., and Heideman, H., *J. Phys. B*, **22**, 3501 (1989), and references therein.
13. Amusia, M. Y., Dolamatov, V. K., and Ivanov, V. K., *Sov. Phys. Tech. Phys.*, **31**, 4 (1986).
14. Martin, N. L. S., Thompson, D. B., Baumann, R. P., Caldwell, C. D., Krause, M. O., Frigo, S. P., and Wilson, M., *Phys. Rev. Lett.*, **81**, 1199 (1999).
15. Martin, N. L. S. (2001), private communication.
16. Fano, U., *Phys. Rev.*, **124**, 1866 (1961).
17. Kheifets, A. S., *J. Phys. B*, **26**, 2053 (1993).
18. Lhagva, O., and Henmedeh, L., *J. Phys. B*, **27**, 4623 (1994).
19. Derevianko, A., Hemmers, O., Oblad, S., Glans, P., Wang, H., Whitfield, S. B., Wehlitz, R., Sellin, I. A., Johnson, W. R., and Lindle, D. W., *Phys. Rev. Lett.*, **84**, 2116 (2000).

2p photoionization of atomic cobalt and nickel

M. Martins*, K. Godehusen†, T. Richter** and P. Zimmermann††

*Universität Hamburg, Institut für Experimentalphysik, Luruper Chaussee 149, D-22761 Hamburg
†BESSY GmbH, Albert-Einstein-Str. 15, D-12489 Berlin
**Technische Universität Berlin, Institut für Atomare Physik und Fachdidaktik, Hardenbergstr. 36, D-10623 Berlin

Abstract. The 2p photoionization of free atoms of the late transition metals cobalt and nickel has been investigated using synchrotron radiation and is compared to CI calculations. Good agreement is found between experiment and theory. Furthermore a remarkable agreement between the atomic nickel 2p photoemission and recent data on nickel compounds is found.

INTRODUCTION

The core-level photoionization is a widely used method to study the electronic and magnetic properties of the 3d transition metals (TM) due to there element specifity. Therefore the 2p absorption spectra of the 3d metals and compounds have been studied for a long time [1]. The interpretation of this data is often done using atomic single electron models. However, multiplet splitting and valence shell re-coupling in the core excited state can give rise to numerous main and satellite lines not taken into account in these models [2, 3, 4]. Unfortunately only sparse atomic data on the 2p photoionization of the TM is available for a critical comparison with the corresponding spectra in the solid phase. First absorption experiments have been made by Arp et al. on chromium, manganese, and copper [5, 6, 7], which are relatively easy to evaporate. Photoemission experiments on these elements have recently performed by Wernet et al. [8, 9, 4]. All other 3d elements are much harder to evaporate due to the high temperatures required (up to 2000 K) and their aggressive melts. In this paper we report on the first 2p photoabsorption and photoemission studies on atomic cobalt and nickel. The experimental data are compared to configuration interaction (CI) calculations.

EXPERIMENT

Figure 1 shows the basic experimental setup. All experiments were performed at the undulator beamlines U49/1-SGM and U49/2-PGM1 at the BESSY II storage ring. Only an undulator beamline can deliver the high flux needed for these experiments. By using electron bombardment nickel and cobalt metal was heated to about 1800K forming an effusive atomic beam. The linearly polarized synchrotron radiation intersected the atomic beam and either the resulting photoelectrons or photoions were recorded.

CP652, X-Ray and Inner-Shell Processes: 19th International Conference on X-Ray and Inner-Shell Processes
edited by A. Bianconi, A. Marcelli, and N. L. Saini
© 2003 American Institute of Physics 0-7354-0111-X/03/$20.00

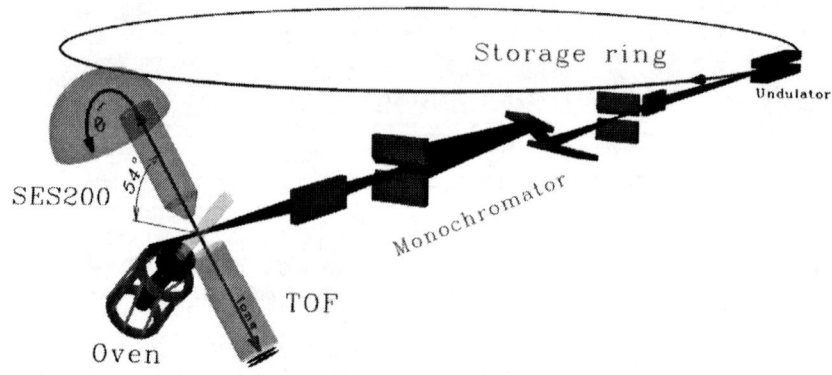

FIGURE 1. Experimental setup used for the experiments at the BESSY II storage ring.

For the detection of the photoelectrons a Scienta SES200 hemispherical electron analyzer mounted under the magic angle was used. For the detection of the photoions our experimental chamber also houses a time-of-flight (TOF) ion spectrometer. Therefore we were able to measure the ion yield at the 2p-3d resonances by scanning the photon energy and recording the intensity for the different ionic states of nickel and cobalt. Neglecting fluorescence decays into neutral species, this ion yield is directly proportional to the absorption spectrum.

Due to the very low target density of the atomic beam one has to cope with, we could not use the high resolving power of the beamline and the electron analyzer to its full extend. In order to achieve a usable count rate to resolution ratio we settled on a total resolution of ≈ 1.7 eV for the electron spectrum and ≈ 0.5 eV for the absorption spectrum.

THEORY

The theoretical spectra were calculated using the configuration interaction (CI) method starting with Hartree-Fock (HF) wavefunctions from the Cowan Code [10] as a zero order approximation including relativistic corrections. Strong CI have been included for the $3d^84s^2$, $3d^94s$, $3d^{10}$, and $3d^84p^2$ nickel valence configurations in the initial, excited, and final state. For cobalt the corresponding configurations are included. To take into account the weak configuration interaction with the large number of high-lying configurations, all Slater integrals were scaled by 80% of their ab initio values [11]. For the spin-orbit parameters, the ab initio values have been used. Matrix elements were calculated in an intermediate coupling scheme with the standard Slater-Condon superposition of configuration method and the cross section was calculated using the Fano and extended Fano method [12]. The linewidth is derived from the subsequent Auger decay of the $2p$ hole states. However, due to the large number of possible final states only the main decay channels have been included.

Due to the high evaporation temperatures for cobalt excitations from $3d^74s^2$

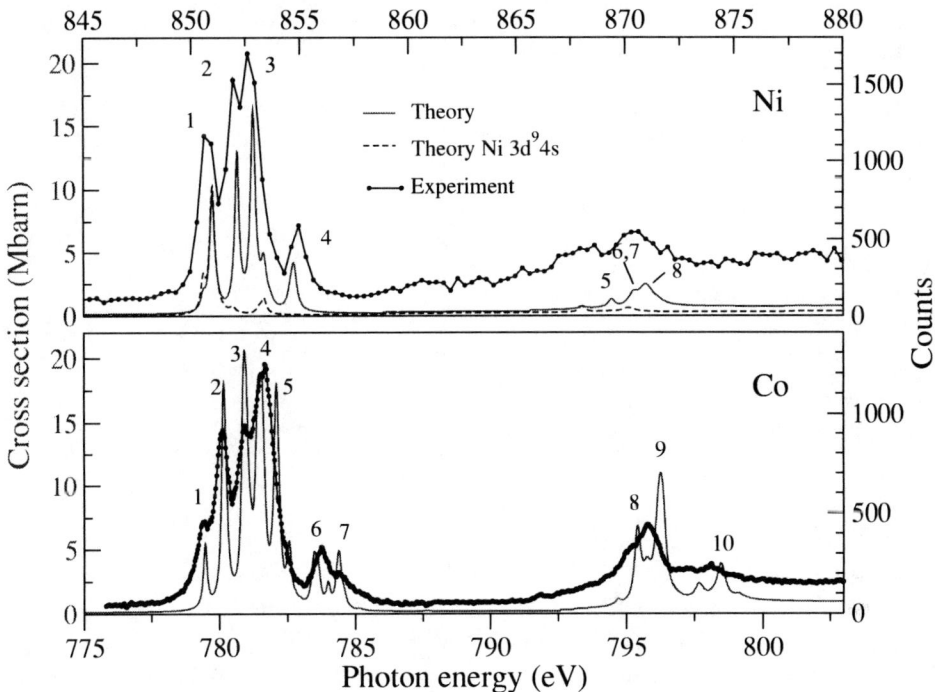

FIGURE 2. Experimental and theoretical photoionization spectra of atomic cobalt and nickel. For the theoretical nickel spectrum the weighted sum for all populated initial states and the cross section for the excitation from the $3d^9 4s$ configuration is shown.

$^4F_{5/2,7/2,9/2}$ initial states have to be included. For nickel the lowest states $3d^8 4s^2$ $^3F_{4,3,2}$ and $3d^9 4s$ $^3D_{3,2,1}$ of *both configurations* $3d^8 4s^2$ and $3d^9 4s$ are populated.

RESULTS

In figure 2 the experimental and theoretical photoabsorption spectra for free cobalt and nickel atoms are shown. The theoretical spectra have been shifted by 3 eV and 2.5 eV to lower energies to match the experimental energy positions for cobalt and nickel, respectively. The main splitting is due to the large 2p spin-orbit splitting of about 20 eV. However in contrast to the solid state data [1, 13] a rich fine structure due to the core valence interaction is observed. In table 1 the assignment for the main resonances are summarized. The Ni spectrum is simpler, even when there are two different configurations populated in the initial state. For the configuration $3d^9 4s$ the states $^3D_3$ has the highest population (28%). From this state only one excitation into the $2p^5(^2P_{3/2})3d^{10}4s$ $^3P_2$ final state (line 1) is possible (table 1), resulting in the simple spectrum from the $^3D_3$ state shown in figure 2. The other $2p_{3/2}$ main resonances 2-4

TABLE 1. Assignment of the main photoabsorption lines for cobalt and nickel. For cobalt only the most prominent excitations into the $2p^5 3d^8 4s^2$ states are given.

		Cobalt				Nickel	
			State				**State**
No.	hν (eV)	initial	final	No.	hν (eV)	initial	final
1	779.5	$^4F_{9/2}$	$(^3F)^4G_{11/2}$	1	851.0	$3d^9 4s\,^3D_3$	$2p^5 3d^{10} 4s\,^3P_2$
2	780.2	$^4F_{9/2}, ^4F_{7/2}$	$(^3F)^4G_{9/2},(^3F)^2F_{7/2}$	2	852.1	$3d^8 4s^2\,^3F_4$	$2p^5 3d^9 4s^2\,^3F_4$
3	781.0	$^4F_{9/2}, ^4F_{7/2}$	$(^3F)^4F_{7/2},(^3F)^4F_{5/2}$	3	852.8	$3d^8 4s^2\,^3F_4$	$2p^5 3d^9 4s^2\,^3D_3$
4	781.4	$^4F_{9/2}, ^4F_{7/2}$	$(^3F)^4F_{9/2},(^3F)^4D_{5/2}$	4	854.7	$3d^8 4s^2\,^3F_4$	$2p^5 3d^9 4s^2\,^1F_3$
5	782.1	$^4F_{9/2}$	$(^3P)^4D_{7/2}$	5	869.3	$3d^8 4s^2\,^3F_2$	$2p^5 3d^9 4s^2\,^3F_2$
6	783.6	$^4F_{9/2}$	$(^1G)^2G_{9/2}$	6	870.4	$3d^8 4s^2\,^3F_3$	$2p^5 3d^9 4s^2\,^3P_2$
7	784.4	$^4F_{7/2}$	$(^3P)^4D_{5/2}$	7	870.8	$3d^8 4s^2\,^3F_3$	$2p^5 3d^9 4s^2\,^3F_3$
8	795.4	$^4F_{7/2}$	$(^3F)^4G_{7/2}$	8	871.0	$3d^8 4s^2\,^3F_4$	$2p^5 3d^9 4s^2\,^3F_3$
9	796.3	$^4F_{9/2}, ^4F_{7/2}$	$(^3F)^2G_{9/2},(^3F)^4D_{7/2}$				
10	798.5	$^4F_{7/2}$	$(^3P)^2D_{5/2}$				

in the nickel spectrum results from the $3d^8 4s^2\,^3F_4$ with 48% population in the initial state. The larger line width of the $2p_{1/2}$ lines (5-8) is due to an additional spin flip decay into the $2p^5(^2P_{3/2})(3d4s)^{10}\varepsilon\ell$ continua. The cobalt spectrum is more complex due to the larger number of possible final states of the $2p^5 3d^8 4s^2$ excited state configuration. Furthermore for the initial states in cobalt the spectra look very similar and therefore no exact assignment of the different lines is possible.

The line width due to the subsequent Auger decay for cobalt and nickel are similar. The $2p_{3/2}$ lines have a width in the order of 0.2 eV and the $2p_{1/2}$ have a larger width in the range of 0.4-1.0 eV due to the spin flip decay.

In figure 3 the experimental $2p$ photoemission spectrum of the Ni atom is shown. Both $2p$ lines show an additional splitting of 8 eV, which can be assigned to the $2p^5 3d^8 4s^2$ and $2p^5 3d^9 4s$ configurations [14]. Therefore, the degeneracy of the $3d^8 4s^2$ and $3d^9 4s$ configuration in the ground state is removed by the influence of the core hole. This can be understood by the better shielding of the $4s$ electrons, which are therefore less influenced by the $2p$ core hole, this results in a stronger binding of the $3d^8 4s^2$ configuration. From a comparison to the $2p$ photoabsorption it can be easily seen, that the $2p$ photoemission is much more sensitive to this final state splitting. The experimental spectrum is compared to two different CI models. In model **A** the mixing with the $4p$ electrons is neglected. In this model the fine structure is no well reproduced. The model gives two additional shoulders for the $2p_{1/2}$ lines, which are not observed in the experiment. Furthermore for the $2p_{3/2}$ lines the overall shape of the structures does not fit well. Model **B** includes also the interaction with the $4p^2$ configuration, which results in a much better agreement for the fine structure. However in this model the intensities for the $2p^5 3d^9 4s$ photoemission lines are underestimated. Here photoemission experiments with higher resolution are highly desirable, to resolve the fine structure better. However, with the current generation of XUV sources this is almost impossible due to the low counting rates.

The atomic data is in very good agreement with a recent photoemission study of NiO [15] showing the importance of atomic multiplet splittings on the fine structure of

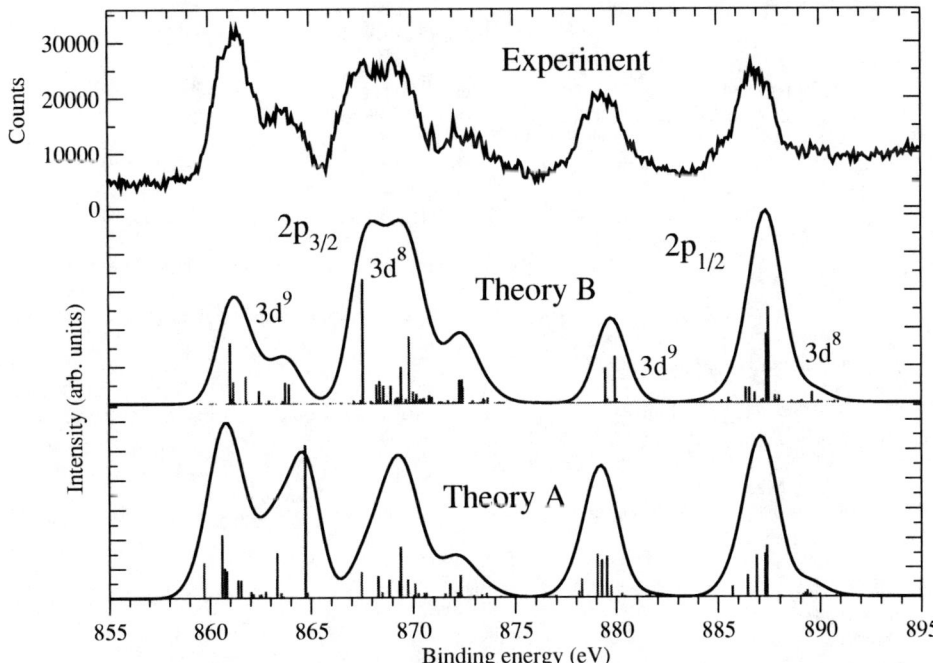

FIGURE 3. Experimental $2p$ photoemission spectra of free nickel (upper graph) in comparison to two different theoretical CI models. The theoretical spectra have been convoluted with a 1.7 eV Gaussian to take into account the experimental broadening.

compounds [14].

CONCLUSIONS

To summarize, we have presented the first experimental $2p$ photoabsorption data on the late $3d$ TM cobalt and nickel. The experimental finding is in good agreement with CI calculations. The nickel $2p$ photoemission is influenced by correlations of the valence electrons. The atomic photoabsorption data shows a strong fine structure not or only marginal resolved in the solid state spectra, which is of great importance for the understanding of the electronic structure and magnetic effects. Therefore experiments on the free atom are an excellent method to get informations about the fine structure not visible in the condensed phase.

Finally we would like to thank the BESSY staff for the excellent support during the beamtimes and the Deutsche Forschungsgemeinschaft (DFG) for the financial support.

REFERENCES

1. Fink, J., Müller-Heinzerling, T., Scheerer, B., Speier, W., Hillebrecht, F., Fuggle, J., Zaanen, J., and Sawatzky, G., *Phys. Rev. B*, **32**, 4899–4904 (1985).
2. Okada, K., and Kotani, A., *J. Phys. Soc. Jpn.*, **61**, 4619–4637 (1992).
3. Bagus, P. S., Broer, R., de Jong, W., Nieuwpoort, W., Parmigiani, F., and Sangaletti, L., *Phys. Rev. Lett.*, **84**, 2259–2262 (2000).
4. Wernet, P., Sonntag, B., Martins, M., Glatzel, P., Obst, B., and Zimmermann, P., *Phys. Rev. A*, **63**, R050702 (2001).
5. Arp, U., Federmann, F., Kallne, E., Sonntag, B., and Sorensen, S. L., *J. Phys. B*, **25**, 3747–3755 (1992).
6. Arp, U., Iemura, K., Kutluk, G., Meyer, M., Nagata, T., Sacchi, M., Sonntag, B., Yagi, S., and Yagishita, A., *J. Phys. B*, **27**, 3389–3398 (1994).
7. Arp, U., Iemura, K., Kutluk, G., Nagata, T., Yagi, S., and Yagishita, A., *J. Phys. B*, **28**, 225–232 (1995).
8. Wernet, P., Schulz, J., Sonntag, B., Godehusen, K., Zimmermann, P., Martins, M., C.Bethke, and Hillebrecht, F., *Phys. Rev. B*, **62**, 14331–14336 (2000).
9. Wernet, P., Schulz, J., Sonntag, B., Godehusen, K., Zimmermann, P., Grum-Grzhimailo, A., Kabachnik, N., and Martins, M., *Phys. Rev. A*, **64**, 042707 (2001).
10. Cowan, R., *The theory of atomic structure and spectra*, University of California Press, Berkeley, Los Angeles, London, 1981.
11. von dem Borne, A., Johnson, R. L., Sonntag, B., Talkenberg, M., Verweyen, A., Wernet, P., , Schulz, J., Tiedtke, K., Gerth, C., Obst, B., , Zimmermann, P., and Hansen, J. E., *Phys. Rev. A*, **62**, 052703 (2000).
12. Martins, M., *J. Phys. B*, **34**, 1321–1335 (2001).
13. van der Laan, G., Dhesi, S. S., and Dudzik, E., *Phys. Rev. B*, **61**, 12277–12284 (2000).
14. Godehusen, K., Richter, T., Zimmermann, P., and Martins, M., *Phys. Rev. Lett.*, **88**, 217601 (2002).
15. Parmigiani, F., and Sangaletti, L., *J. Electron Spectrosc. Relat. Phenom.*, **98-99**, 287 (1999).

Photoionization of the Fe Ions: Structure of the K-Edge

P. Palmeri[*], C. Mendoza[*], T. Kallman[*] and M. Bautista[†]

[*]*Laboratory for High Energy Astrophysics, NASA Goddard Space Flight Center, Greenbelt MD 20771, USA*
[†]*Centro de Física, IVIC, Caracas 1020A, Venezuela*

Abstract. X-ray absorption and emission features arising from the inner-shell transitions in iron are of practical importance in astrophysics due to the Fe cosmic abundance and to the absence of traits from other elements in the nearby spectrum. As a result, the strengths and energies of such features can constrain the ionization stage, elemental abundance, and column density of the gas in the vicinity of the exotic cosmic objects, e.g. active galactic nuclei (AGN) and galactic black hole candidates. Although the observational technology in X-ray astronomy is still evolving and currently lacks high spectroscopic resolution, the astrophysical models have been based on atomic calculations that predict a sudden and high step-like increase of the cross section at the K-shell threshold (see for instance Ref. [1]). New Breit-Pauli R-matrix calculations of the photoionization cross section of the ground states of Fe XVII in the region near the K threshold are presented. They strongly support the view that the previously assumed sharp edge behaviour is not correct. The latter has been caused by the neglect of spectator Auger channels in the decay of the resonances converging to the K threshold. These decay channels include the dominant KLL channels and give rise to constant widths (independent of n). As a consequence, these series display damped Lorentzian components that rapidly blend to impose continuity at threshold, thus reformatting the previously held picture of the edge. Apparent broadened iron edges detected in the spectra of AGN and galactic black hole candidates seem [2, 3] to indicate that these quantum effects may be at least partially responsible for the observed broadening.

INTRODUCTION

X-ray absorption and emission features due to inner shell transitions in iron are of practical importance in astronomy, owing to the cosmic abundance of iron and to the absence of features from other elements nearby in the spectrum. As a result, the strengths and energies of iron K shell features can constrain the ionization state, elemental abundance, and column density of gas in the vicinity of exotic objects such as active galactic nuclei (AGN) and galactic black hole candidates. Although the observational technology in X-ray astronomy is still evolving, many such features have been observed. Emission lines have been used to constrain the geometrical distribution of gas in compact binary stars [4], and to detect black holes in AGN [5]. Iron absorption features have been detected from AGN [6], from compact binaries [7], and from galactic black hole candidates [8, 3]. These divide naturally into bound-bound and bound-free transitions, and much of the interpretation of the bound-free absorption features has been based on atomic calculations such as those of Verner and Yakovlev [9] or Berrington and Ballance [1] which predict a sudden increase in the cross section at the binding energy of the K-shell electron. When

CP652, *X-Ray and Inner-Shell Processes: 19th International Conference on X-Ray and Inner-Shell Processes*
edited by A. Bianconi, A. Marcelli, and N. L. Saini
© 2003 American Institute of Physics 0-7354-0111-X/03/$20.00

a photon is sufficiently energetic to promote a K-shell electron to a higher Rydberg state, the resulting excited state can decay via Auger channels. In the case of Fe XVII, we have these possible decay pathways:

$$h\nu + 2p^6 \rightarrow 1s^{-1}2p^6 np$$

$$\rightarrow \left\{ \begin{array}{c} 2p^5 + e^- \\ 2s^{-1}2p^6 + e^- \end{array} \right\} \tag{1}$$

$$\rightarrow \left\{ \begin{array}{c} 2p^4 np + e^- \\ 2s^{-1}2p^5 np + e^- \\ 2s^{-2}2p^6 np + e^- \end{array} \right\} \tag{2}$$

Eq. (1) represents the participator (KLn) channels where the np Rydberg electron is involved in the decay while it is not in the spectator (KLL) channels (Eq. (2)). R-matrix calculations have been carried out to analyze the role of the spectator channels.

In the following sections, we will detail the methodology and discuss the results.

THEORY

The Breit-Pauli R-matrix method (BPRM) is widely used in electron–ion scattering and in radiative bound–bound and bound–free calculations. It is based of the close-coupling approximation [10] whereby the wavefunctions for states of an N-electron target and a colliding electron with total agular momentum and parity $J\pi$ are expanded in terms of the target eigenfunctions

$$\Psi^{J\pi} = A \sum_i \chi_i \frac{F_i(r)}{r} + \sum_j c_j \Phi_j . \tag{3}$$

The functions χ_i are vector coupled products of the target eigenfunctions and the angular part of the incident-electron functions, $F_i(r)$ are the radial part of the latter and A is an antisymmetrization operator. The functions Φ_j are bound-type functions of the total system constructed with target orbitals; they are introduced to compensate for orthogonality conditions imposed on the $F_i(r)$ and to improve short-range correlations. The Kohn variational gives rise to a set of coupled integro-differential equations that are solved by R-matrix techniques [11, 12] within a box of radius, say, $r \leq a$. In the asymptotic region ($r > a$), solutions are found using the MQDT method [15, 16]. Breit–Pauli relativistic corrections have been introduced in the R-matrix suite by Scott and Burke [13], Scott and Taylor [14]. Inter-channel coupling is equivalent to configuration interaction in the atomic structure context, and presents a formal and unified approach to study the decay properties of both bound states and resonances.

STRATEGY

In order to display the different effects of these two types of Auger decays, two calculations of the photoionization cross section have been carried out (hereafter referred to as Calculations A and B). In Calculation A, the BPRM method have been used to solve the $e^- + Fe^{1/+}$ system. The following target configurations were considered: $2p^5$, $2s^{-1}2p^6$, and $1s^{-1}2p^6$. Only the participator Auger channels described in Eq.(1) were therefore included. The orbitals of the target ion were optimized using the code SUPERSTRUC-TURE [17]. As it was obviously impossible to include all the $2p^4np$, $2s^{-1}2p^5np$, and $2s^{-2}2p^6np$ target configurations in our R-matrix model, Calculation B was carried out in two steps. Firstly, an R-matrix calculation was carried out adding the contributions of KLL Auger decays of the $1s^{-1}2p^63p$ resonances (Eq.(2)). The next step consisted in fitting Lorentzian profiles to these 3p resonances in order to obtain the widths and areas. These symmetric profiles are due to the dominance of the KLL channels over the KLn ones. Actually, these spectator Auger channels produce continuum states that cannot be reached by direct photoionization of the ground state causing the Fano q parameter to go to infinity (see Nayandin et al. [18] and references therein). As the Auger rates of the spectator channels are constant along the Rydberg series, the cross section was extrapolated for the higher members of the series using Lorentzian profiles with a constant width and areas decreasing as the third power of the effective quantum number, n^*. The resonance positions were deduced by the usual Ritz formula using the data of Calculation A.

RESULTS AND DISCUSSION

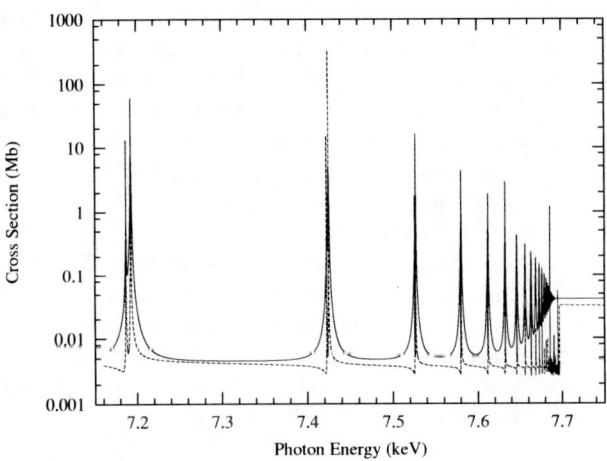

FIGURE 1. Total photoionization cross section in Mb versus photon energy in keV. Solid and dash lines are Calculations B and A respectively (see the text).The sharp K-edge near 7.7 keV in Calculation A has disappeared in Calculation B.

In Fig. 1, a comparison between Calculation A and B is shown. The sharp K-edge near 7.7 keV has disappeared, being filled up by the highest members of the Rydberg series which are blended together. Note also the differences in the resonance shapes and intensities. Because of the constant width, the intensities decrease as $1/(n^*)^3$ in Calculation B. In Table 1, a comparison between widths obtained by two R-matrix calculations (with and without the spectator Auger channels) is presented for the 3p and 4p resonances. One can see the significant effect of the spectator (KLL) Auger decays on the widths. As the KLn widths (Γ^{KLn}) are going down as $(n^*)^{-3}$ and the KLL ones are independent of n^*, the resulting widths ($\Gamma^{KLn+KLL}$) become constant for the highest members of the Rydberg series. Actually, see Table 1, we can conclude that they are almost constant for the first few members. Fig. 2 depicts the cross section obtained in

TABLE 1. Comparison between widths obtained without (Γ^{KLn}) and with the KLL (spectator) Auger decays ($\Gamma^{KLn+KLL}$) using the BPRM method for the 3p and 4p resonances.

Resonance	E (keV)	Γ^{KLn} (eV)	$\Gamma^{KLn+KLL}$ (eV)
$1s^{-1}2p^63p\,^3P^o_1$	7.1865	3.802×10^{-2}	6.495×10^{-1}
$1s^{-1}2p^63p\,^1P^o_1$	7.1923	2.165×10^{-2}	6.335×10^{-1}
$1s^{-1}2p^64p\,^3P^o_1$	7.4224	1.386×10^{-2}	6.146×10^{-1}
$1s^{-1}2p^64p\,^1P^o_1$	7.4246	7.952×10^{-3}	6.082×10^{-1}

Calculation B convoluted with an instrumental (Gaussian) profile of 10 eV. The 'edge' is now shifted down to 7.6 keV. The 9p resonances are the highest resolved member of the Rydberg series. Unfortunately, we have found no experimental data in Fe XVII to compare with our calculations. Nevertheless, Farhat et al. [19] have measured the total electron yield of the $h\nu + 3p^6 \rightarrow 2p^{-1}3p^6 + e^-$ photoionization process in neutral argon. The $2p^{-1}3p^6n\ell$ resonances can decay via dominant spectator Auger channels to the $3p^4n\ell$, $3p^{-1}3p^5n\ell$, and $3s^{-2}3p^6n\ell$ Ar II configurations. Fig. 3 is the reprint of Fig. 2 in Farhat et al. [19] which presents the total electron yield versus photon energy (the resolution is 160 meV). The $2p_{3/2}$ threshold is indicated at 248.6 eV but no sharp edge is seen at this energy. Note the similar behaviour near threshold with Fig. 2. Actually, Gorczyca and Robicheaux [20] used an optical potential within the MQDT method to mimic the effect of the spectator Auger channels on the L-shell photoionization cross section of neutral argon. This technique has been also compared to measurements of the total electron yield near the K-shell threshold in neutral neon [21] and oxygen [22].

The calculations presented here show that the K-shell photoabsorption cross section for Fe XVII does not exhibit a sharp edge at the the binding energy of the K-shell electron, but it is shifted and broadened towards lower energies. It is worth pointing out that apparently broadened iron edges have been detected in the spectra of AGN and galactic black hole candidates [2, 3], and the same physical processes we identify here may be at least partially responsible for the observed broadening.

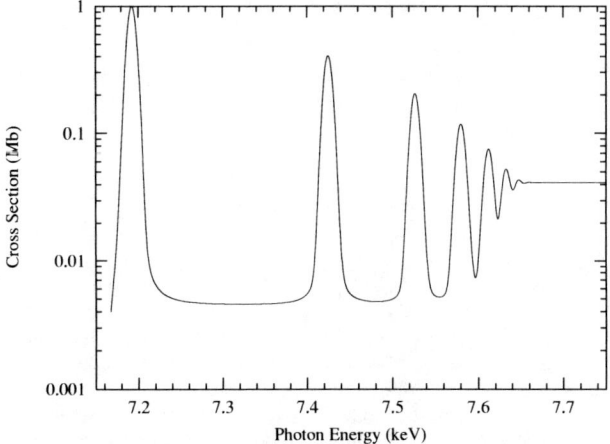

FIGURE 2. Total cross section obtained in Calculation B (see the text) convoluted with a Gaussian profile with a width of 10 eV.

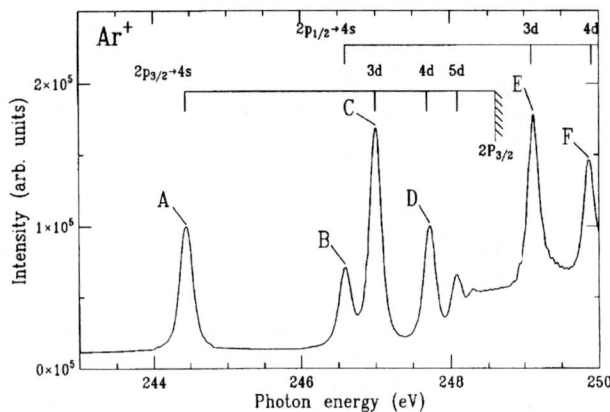

FIGURE 3. Total electron yield versus photon energy in eV as measured in Ar I by Farhat et al. [19]. The energy scan was taken over the $2p^{-1}n\ell$ resonances.

CONCLUSION

As an important conclusion, this work shows that the sharp edge structures at thresholds disappear when the Rydberg series before threshold are affected by spectator Auger decays. These inner-shell Auger channels cause constant-width Lorentzian shape resonances that blend near the threshold changing drastically the structure of the edge. Although this effect has been measured for the L-edge in neutral argon and for the K-shell in neutral neon and oxygen, experimental data are needed for the K-edge in Fe XVII.

163

ACKNOWLEDGMENTS

CM and PP kindly acknowledge a Senior Research Associateship from the National Research Council and a Research Associateship from University of Maryland respectively.

REFERENCES

1. Berrington, K. A., and Ballance, C. P., *J. Phys. B* **34**, 2697-2705 (2001).
2. Done, C., and Zycki, P., *Mon. Not. R. Astron. Soc.* **305**, 457-468 (1999).
3. Miller, J. M., Fabian, A. C. , Wijnands, R., Remillard, R. A., Wojdowski, P., Schulz, N. S. , Di Matteo, T., Marshall, H. L., Canizares, C. R., Pooley, D., and Lewin, W. H. G., *Astrophys. J.*, in press (2002).
4. Ebisawa, K., Day, C. S. R., Kallman, T. R., Nagase, F., Kotani, T., Kawashima, K., Kitamoto, S., and Woo, J. W., *Publ. Astron. Soc. Jpn* **48**, 425-440 (1996).
5. Tanaka, Y., Nandra, K., Fabian, A. C., Inoue, H., Otani, C., Dotani, T., Hayashida, K., Iwasawa, K., Kii, T., Kunieda, H., Makino, F., and Matsuoka, M., *Nature* **375**, 659-661 (1995).
6. Nandra, K., Pounds, K. A., Stewart, G. C., Fabian, A. C., and Rees, M. J., *Mon. Not. R. Astron. Soc.* **236**, 39P-46P (1989).
7. Shulz, N., and Brandt, N., *Astrophys. J.*, in press (2002).
8. Kotani, T., Ebisawa, K., Dotani, T., Inoue, H., Nagase, F., Tanaka, Y., and Ueda, Y., *Astrophys. J.* **539**, 413-423 (2000).
9. Verner, D., and Yakovlev, D. G., *Astron. Astrophys. Suppl. Ser.* **109**, 125-133 (1995).
10. Burke, P. G., and Seaton, M. J., *Meth. Comput. Phys.* **10**, 1 (1971).
11. Burke, P. G., Hibbert, A., and Robb, D., *J. Phys. B* **4**, 153-161 (1971).
12. Berrington, K. A., Burke, P. G., Butler, K., Seaton, M. J., Storey, P. J, Taylor, K. T., and Yan, Yu., *J. Phys. B* **20**, 6379-6397 (1987).
13. Scott., N. S., and Burke, P. G., *J. Phys. B* **12**, 4299-4314 (1980).
14. Scott, N. S., and Taylor, K. T., *Comput. Phys. Commun.* **25**, 347-387 (1982).
15. Seaton, M. J., *Proc. Phys. Soc. London* **88**, 801-814 ; 815-832 (1966).
16. Seaton, M. J., *Rep. Prog. Phys.* **46**, 167-257 (1983).
17. Eissner, W., Jones, M., and Nussbaumer, N., *Comput. Phys. Commun.* **8**, 270-306 (1974).
18. Nayandin, O., Gorczyca, T. W., Wills, A. A., Langer, B., Bozek, J. D., and Berrah, N., *Phys. Rev. A* **64**, 022505 (2001).
19. Farhat, A., Humphrey, M., Langer, B., Berrah, N., Bozek, J. D., and Cubaynes, D., *Phys. Rev. A* **56**, 501-513 (1997).
20. Gorczyca, T. W, and Robicheaux, F., *Phys. Rev. A* **60**, 1216-1225 (1999).
21. Gorczyca, T. W, *Phys. Rev. A* **61**, 024702 (2000).
22. Gorczyca, T. W, and McLaughlin, B. M., *J. Phys. B* **33**, L859-L863 (2000).

Photoionization Cross Sections of Kr and Xe from Threshold up to 1000 eV

[1]M. Richter, [1]G. Ulm, [2]Chr. Gerth, [2]K. Tiedtke, [2]J. Feldhaus,
[3]A.A. Sorokin, [3]L.A. Shmaenok, [3]S.V. Bobashev.

[1] *Physikalisch-Technische Bundesanstalt, Berlin, Germany*
[2] *Deutsches Electronensynchrotron, Hamburg, Germany*
[3] *Ioffe Physico-Technical Institute RAS, St.Petersburg, Russia.*

Abstract. Updated recommended data for absolute photoionization cross sections of Kr and Xe with uncertainties of ± 3 % are presented. The data are the result of a comprehensive analysis of reliable experimental measurements in absolute terms taking preceding recommendations into account. Significant corrections of the cross section values for certain photon-energy regimes have been derived from the comparison of gas ionization by photons and electrons using a new experimental approach. The values presented serve as a background for new techniques to be used for the quantitative characterization of VUV-FEL photon beams.

INTRODUCTION

In atomic and molecular physics, photoionization (PI) is a fundamental process. The knowledge of absolute PI cross sections with low uncertainty is crucial for understanding the interaction between photons and atoms. Moreover, accurate PI cross sections of rare gas atoms are needed in many fields of applied research, for example, when the absolute photon intensity of sources of VUV and soft X-ray radiation is measured using an ionization chamber.

Reviews of total PI cross sections were given by Marr and West [1], West and Morton [2], and Henke, Gullikson, and Davis [3] who have evaluated published experimental data then available and presented recommended values for PI cross sections of He, Ne, Ar, Kr and Xe. Later, Bizau and Wuilleumier [4] proposed and tabulated recommended values for PI cross sections of He and Ne taking into account previously published data including those of Samson and co-workers [5-7] which had not been taken into consideration in earlier reviews [1-3]. In our previous papers [8,9] we found that total PI cross sections of Ne and Ar obtained

CP652, *X-Ray and Inner-Shell Processes: 19th International Conference on X-Ray and Inner-Shell Processes*
edited by A. Bianconi, A. Marcelli, and N. L. Saini
© 2003 American Institute of Physics 0-7354-0111-X/03/$20.00

by different experimental groups agree fairly well within a wide spectral range, while significant discrepancies between different sets of data occur for Kr and Xe in selected photon energy regimes. Moreover, it was found that the data obtained by Samson and co-workers [5,6] are the most accurate and reliable for all rare gas atoms.

In this paper we propose recommended data for absolute total PI cross sections of Kr and Xe with a standard relative uncertainty of 3 %. The data were obtained by Samson and co-workers [5,6] by accurate measurements carried out using the double-ionization chamber technique as well as by our own measurements [9] using a new experimental approach that will be described in the next chapter.

EXPERIMENTAL

The new experimental approach used to derive total PI cross sections of Kr and Xe is based on the accurate measurement of ratios of total cross sections for PI and electron-impact ionization (EI) followed by normalization to the well-known total PI cross section reported in [5]. The apparatus used for cross-section ratios measurements (Fig. 1) as well as the experimental procedure were discussed in detail previously [8,9]. Briefly, the apparatus consists of an chamber, an ion detector, an electron gun, a Faraday cup for electron current measurements, and a photodiode calibrated against an electrical substitution radiometer (ESR), for photon flux measurements [see, e.g., Ref. 10 and references therein]. The target rare gas of 99.99 % purity homogeneously fills the ionization chamber. During the measurements its pressure is maintained at certain levels in the range between 2×10^{-4} Pa and 3×10^{-3} Pa.

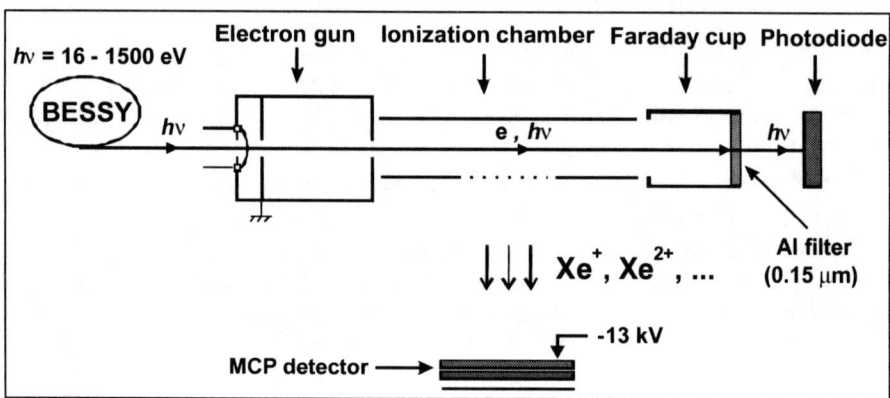

FIGURE 1. Scheme of the apparatus for accurate measurements of ratios of total cross sections for photoionization and electron-impact ionization in rare gases.

The operation of the apparatus is based on the successive ionization of the target gas by electrons with known energy and by monochromatized synchrotron radiation. Ions produced in these processes are collected and recorded by a microchannel plate (MCP) detector operated in the counting mode. In order to obtain the same collection and detection efficiency for differently charged ions, the latter are accelerated in a constant electric field up to kinetic energies corresponding to an energy of single charged ions of 13 keV. Essential features of the apparatus are the hollow-axis electron gun and the thin aluminum filter of known transmittance at the bottom of the Faraday cup. This design ensures coincidence of the photon and electron-beam trajectories in the ionization chamber, and allows PI and EI measurements to be performed without any change in the position of the electron gun and the Faraday cup. This guarantees equivalent conditions for PI and EI as regards the gas density and the electric field potential distribution within the ionization chamber. Under these conditions the ratio of the total EI cross section $\sigma_e(E)$ to the total PI cross section $\sigma_{ph}(h\nu)$ can be expressed as [9]:

$$\frac{\sigma_e(E)}{\sigma_{ph}(h\nu)} = \frac{1}{\tau_{ph}(h\nu)} \cdot \frac{1}{\eta_{ph}(h\nu)} \cdot \frac{f_e/I_e}{f_{ph}/I_{ph}}. \tag{1}$$

I_e and I_{ph} denote the currents of Faraday cup and photodiode, respectively, f_e and f_{ph} are the MCP detector count rates for ions formed by electron impact and by photon impact, respectively, $\tau_{ph}(h\nu)$ is the transmittance of the aluminum filter, and $\eta_{ph}(h\nu)$ is the quantum efficiency of the photodiode determined by calibration against the ESR.

The measurements of the cross-section ratios were performed at the SX700 and NIM beam lines in the Radiometry Laboratory of the Physikalisch-Technische Bundesanstalt at the Electron Storage Ring BESSY I [see, e.g., Ref. 10 and references therein].

RESULTS

Ratios $\sigma_e(E)/\sigma_{ph}(h\nu)$ of total cross sections for EI and PI of Kr and Xe were measured at an electron energy E of 1000 eV and selected photon energies $h\nu$ between 16 eV and 1012 eV. To avoid unexpected errors in our measurements, we chose the energies of photons for each target rare gas outside the intervals containing absorption edges and resonance structures in the photoionization spectrum. Therefore, total relative standard uncertainties as low as 1.3 % to 1.9 % for the cross-section ratios were achieved [9]. The measurements yield a common scale of total cross sections for EI and PI that allows both total EI cross sections [9] and total PI cross sections to be deduced as described below.

Total PI cross sections of Kr and Xe were determined in a two-step procedure. In the first step values for the total EI cross sections at 1000-eV electron energy were derived from the measured ratios by normalization to absolute total PI cross sections as reported in the literature by different experimental groups [5,6,11-13] with the relative uncertainties quoted being better than 7 % (Fig. 2). The relative uncertainties for these values arise from the relative uncertainties of the measured ratios and the relative uncertainties of the absolute total PI cross sections claimed by the groups. This step enables us to examine different sets of PI cross-section

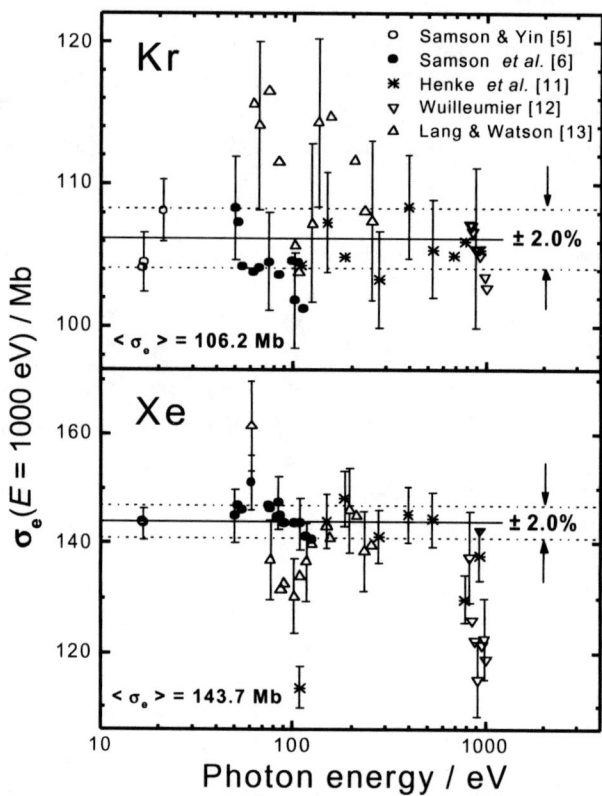

FIGURE 2. Total electron-impact ionization cross section of Kr and Xe at an electron energy of 1000 eV obtained from measured cross-section ratios and normalization to total photoionization cross sections at different photon energies. The symbols stand for data obtained by normalization to different sets of photoionization cross sections [5,6,11-13]. The continuous line represents the average value over the data obtained in the VUV spectral range below 25 eV photon energy by normalization to the photoionization cross sections reported in [5].

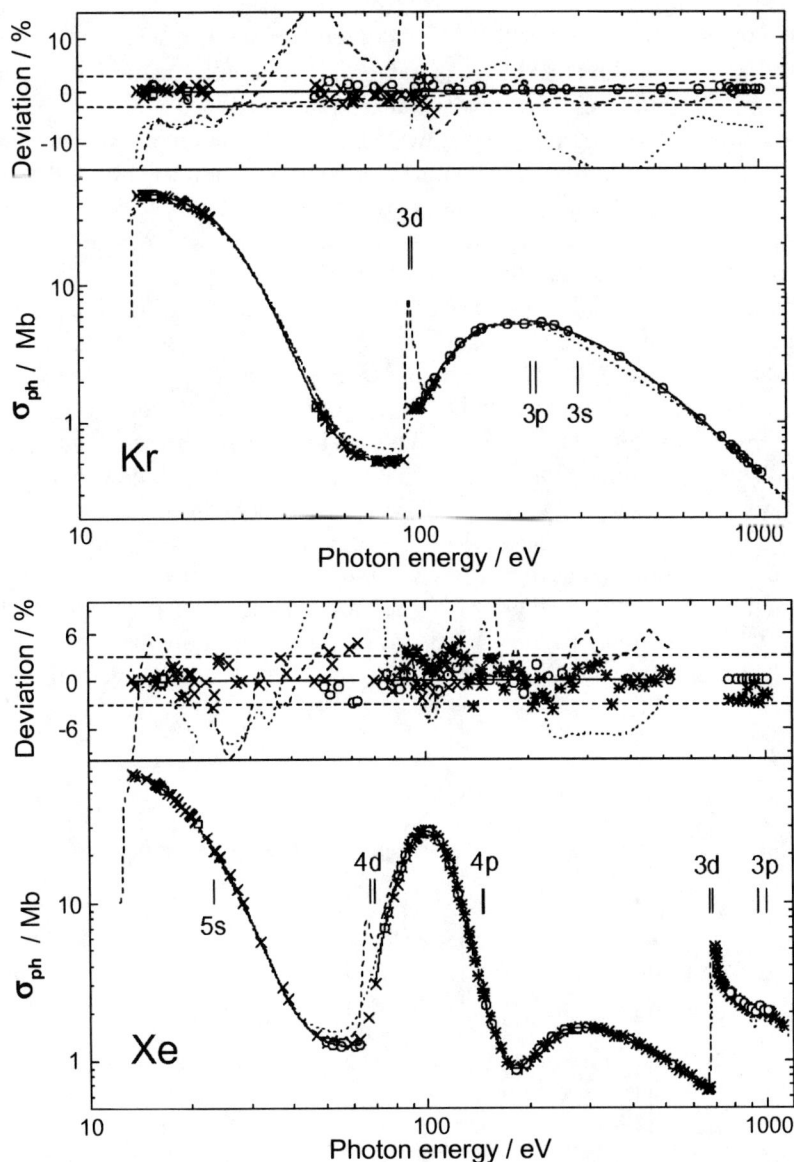

FIGURE 3. Total photoionization cross sections of Kr and Xe: present experimental data (o), Samson and Yin [5] (□), Samson *et al.*[6] (×), Saito and Suzuki [14] (✳), present recommended data (——), Henke *et al.* [3] (----), Marr and West [1] and West and Morton [2] (······). The upper plots show the fractional deviation of the experimental and preceding recommended data from the present recommended data. In the case of Xe, in the spectral range above 3d thresholds, the fractional deviation of the experimental data [14] from the present experimental data are shown.

data, to choose the most reliable of them and to derive an average value $\sigma_e(E = 1000 \text{ eV})$ for the total EI cross section [9]. The latter value for Kr and Xe was determined with a relative standard uncertainty of 2 % by averaging the data obtained in the VUV spectral range below 25 eV photon energy by normalization to the PI cross sections reported in [5] with a quoted relative uncertainty of 0.8 %.

In the second step, taking the measured values for the cross-section ratios and the values for $\sigma_e(E = 1000 \text{ eV})$, we derived total PI cross sections for Kr and Xe in the photon energy range from 50 eV to 1012 eV with relative standard uncertainties of 2.4 % to 2.7 %, i.e. with uncertainties smaller than those reported in the literature for this spectral range. The present total PI cross sections of Kr and Xe are shown in Fig. 3 together with the experimental data of Samson and co-workers obtained with quoted relative uncertainties of 0.8 % to 3 % [5,6]. Also shown are recommended values for absolute total PI cross sections proposed in this paper with a relative uncertainty of 3 % as well preceding recommended data [1-3] and absolute total PI cross sections of Xe [14] recently measured with a quoted relative uncertainty of 1 %. The present recommended values were obtained for Kr and Xe in the spectral range below the 2p and 3d threshold, respectively, from a polynomial fit to our experimental values and to the experimental values reported in [5,6] by the least-squares method, taking into consideration the relative cross-section dependence of preceding recommended data [1-3]. In the spectral range

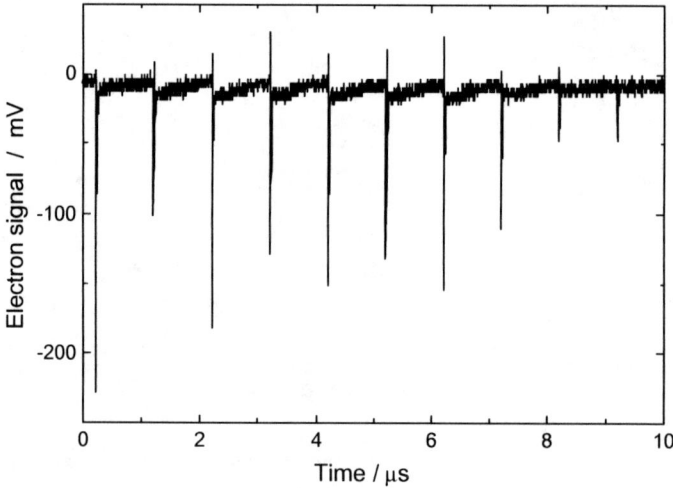

FIGURE 4. Electron signal of VUV radiation at 14.3 eV photon energy from a bunch train filled with 10 bunches as measured at the TTF-FEL at DESY with a recently developed gas-monitor detector.

where the two data sets overlap, averaging of the two polynomial fits was performed. No polynomial fit was performed in the spectral intervals containing absorption edges and resonance structures in the photoionization spectrum. The present experimental and recommended data for Xe are in good agreement with the data obtained in [14], while significant discrepancies occur between our recommended data and those published in [1-3].

The present total PI cross sections will serve as a background for absolute calibration of a recently developed gas-monitor detector. The detector, which is based on atomic photoionization at low target gas density, allows to perform a quantitative characterization of sources of extremely intense and pulsed VUV and soft X-ray radiation, like the VUV free electron laser (FEL) at the TESLA Test Facility (TTF) at DESY which emits up to about 10^{14} photons within a pulse of 100 fs duration [15]. First promising results from the gas-monitor detector were recently obtained at the TTF-FEL at DESY. Fig. 4 shows an example of a pulse measurement at a photon energy of 14.3 eV about one order of magnitude below saturation of the FEL. Single-pulse readout was realized by photoelectron detection.

ACKNOWLEDGMENTS

This work has been performed within the scope of a project jointly funded by DFG and RFBR.

REFERENCES

1. Marr G.V., West J.B., *At. Data Nucl. Data Tables* **18**, 497-508 (1976).
2. West J.B., Morton J., *At. Data Nucl. Data Tables* **22**, 103-107 (1978).
3. Henke B.L., Gullikson E.M., Davis J.C., *At. Data Nucl. Data Tables* **54**, 181-342 (1993); (see also Gullikson E., http://www-cxro.lbl.gov/optical_constants/).
4. Bizau J.M., Wuilleumier F.J., *J. Electron Spectrosc. Relat. Phenom.* **71**, 205-224 (1995).
5. Samson J.A.R., Yin L., *J. Opt. Soc. Am. B* **6**, 2326-2333 (1989).
6. Samson J.A.R., Lyn L., Haddad G.N., Angel G.C., *J. Phys. IV* **C1**, 99-107 (1991).
7. Samson J.A.R., He Z.X., Yin L., Haddad G.N., *J. Phys. B* **27**, 887-898 (1994).
8. Sorokin A.A., Shmaenok L.A., Bobashev S.V., Möbus B., Ulm G., *Phys. Rev. A.* **58**, 2900-2910 (1998).
9. Sorokin A.A., Shmaenok L.A., Bobashev S.V., Möbus B., Richter M., Ulm G., *Phys. Rev. A.* **61**, 022723 (2000).
10. Richter M., Ulm G., *J. Electr. Spectrosc. Relat. Phenom.*, **101-103**, 1013-1018 (1999).
11. Henke B.L., Elgin R.L., Lent R.E., Ledingham R.B., *Norelco Reporter* **14**, 112-134 (1967).
12. Wuilleumier F., *Ph. D. Thesis*, Université de Paris, 1969 (unpublished).
13. Lang J., Watson W.S., *J. Phys. B* **8**, L339-343 (1975).
14. Saito N., Suzuki I.H., *Nucl. Inst. Meth. A* **467-468**, 1577-1580 (2001).
15. Ayvazyan V. *et al.*, *Phys. Rev. Lett.* **88**, 104802 (2002).

Dynamics of Photoionization and Photoexcitation of Molecules Probed by Multiple Coincidence Momentum Imaging

Norio Saito*, Kiyoshi Ueda[¶], and Inosuke Koyano[$]

*National Metrology Institute of Japan, AIST, Tsukuba 305-8568, Japan
[¶]IMRAM, Tohoku University, Sendai 980-8577, Japan
[$]Department of Material Science, Himeji Institute of Technology, Kamigori, Hyogo 678-1297, Japan

Abstract. Recent studies on dynamics of photoexcitation and photoionization of molecules probed by multiple coincidence momentum imaging technique are reviewed. This technique consists of electron and ion time-of-flight analyzers with multi-hit two-dimensional position sensitive detectors and a supersonic jet. As specific examples we discuss nuclear motion caused by the C $1s \rightarrow 2\pi_u$ excitation in CO_2 and the C $1s$ photoelectron angular distributions from fixed-in-space CO_2.

INTRODUCTION

A multiple coincidence momentum imaging apparatus, which consists of electron and ion time-of-flight analyzers with multi-hit two-dimensional position sensitive detectors, is a powerful tool to investigate dynamics of photoexcitation and photoionization of molecules [1-10]. This technique provides us the complete information on the momenta of the particles and the vector correlations among the particles. We review here our recent investigations for the nuclear motion in core-excited states [3-5] and the behavior of photoelectrons in the molecular flame [9,10] using this apparatus.

When a core electron of an atom in a molecule is excited, in general, the core hole relaxes by the Auger electron emission and ionic fragmentation proceeds along the dissociation path on the potential energy surface of the Auger-final state. The lifetime of the core-excited state for light atoms such as C, N and O is of the order of 10^{-15} s. Within this time scale, however, nuclear motion of the molecule proceeds in the core-excited state before the Auger decay [11-19]. We are particularly interested in core excitation of polyatomic molecules because of the following reasons. In case of polyatomic molecules the symmetry of the stable geometry of the core-excited state is often different from that of the ground state [20-27] and as a result nuclear motion takes place towards new stable geometry with the different symmetry. The non-totally symmetric nuclear motion thus induced is of particular interest because it may open up a new dynamical pathway of molecular dissociation [28,29].

CP652, X-Ray and Inner-Shell Processes: 19th International Conference on X-Ray and Inner-Shell Processes
edited by A. Bianconi, A. Marcelli, and N. L. Saini
© 2003 American Institute of Physics 0-7354-0111-X/03/$20.00

Photoemission from $1s$ orbital in small molecules often presents broad resonances in the energy region of several electron volts above threshold. In these resonances, the photoelectron wave experiences scattering due to the non-isotropic molecular environment, before it is detected by a distant macroscopic detector. This effect results in the photoelectron wave, which has a predominant $l=1$ character just after emission, to be a mixture of partial waves where high angular momentum components can have large weights [30]. If the direction of the photoelectron with respect to the molecular frame can be measured, the photoelectron angular distribution (PAD) from fixed-in-space molecules often reflects the angular momentum composition of the photoelectron wave. Therefore, such measurements provide a specific fingerprint of the scattering effect experienced by the photoelectron within the molecular potential, often referred to as the shape resonance effect [6-8,31-36].

We have been investigating nuclear motion in core-excited states and subsequent dissociation dynamics using multiple coincidence momentum imaging technique [3-5], as well as excitation [37-42], and de-excitation [16,19] spectroscopies, and also photoionization dynamics using multiple coincidence momentum imaging technique [9,10]. This paper reviews a selection of the relevant studies we have performed very recently.

The next section describes the experimental methods for investigating nuclear dynamics of core-excited molecules and photoionization dynamics. Following that, the RESULTS AND DISCUSSION section describes the studies on nuclear motion of CO_2 in the C $1s^{-1}2\pi_u$ excited states [3-5], the studies on the C $1s^{-1}2\pi_u$ Renner-Teller pair states [42], and the studies on photoionization dynamics by measuring PADs from fixed-in-space CO_2 molecules [9,10].

EXPERIMENT

The experiments are carried out on the c branch of the soft X-ray photochemistry beamline 27SU at SPring-8 [43]. Multiple coincidence momentum imaging technique is based on the electron and ion time-of-flight (TOF) measurements using two position sensitive TOF spectrometers (longer and shorter) (Fig. 1). The TOF and position on

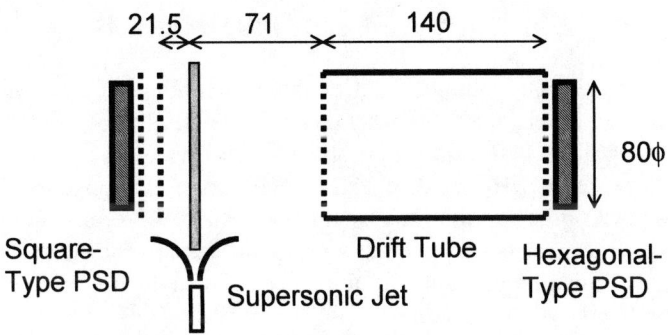

FIGURE 1. A schematic illustration of the multiple coincidence momentum imaging system.

the detector of the particle allow one to extract the complete information about the linear momentum (p_x, p_y, p_z) for each particle without ambiguity. The lengths of the acceleration region and the drift region for the longer TOF spectrometer are 71 mm and 140 mm, respectively. For the shorter TOF spectrometer, the length of the acceleration region is 22.5 mm and no drift region is provided. Both of the TOF spectrometers are equipped with multi-channel plates of 80 mm effective diameter, followed by a 2-dimensional (2D) multi-hit readout delay-line anode (RD-80 for the short TOF spectrometer and HEX-80 for the long TOF spectrometer) manufactured by Roentdek [44].

In the case of the study for nuclear motion in core-excited molecules, ions are measured using the long TOF spectrometer and electrons are measured using the short TOF spectrometer. The electron signals are used as the start timing for the TOF measurement of ions without analyzing momenta of electrons.

In the case of the study for the PAD of fixed-in-space molecules, a uniform magnetic field (10.2 G) helps collect all electrons. The TOF of electrons are measured with respect to the bunch marker of the synchrotron radiation source. We record only the events in which at least two ions and one electron are detected in coincidence. The direction of the molecular axis at the time of photoemission is defined by the momentum vector of the O^+ and CO^+ fragments, created by Coulomb dissociation of CO_2^{2+} after rapid Auger decay.

The angle-resolved yield curves for the energetic ions are measured by use two energetic-ion detectors mounted at $0°$ and $90°$ with respect to the photon polarization vector E of the incident light [38,40]. The retarding voltage used is typically 6 V and thus ions with kinetic energies higher than 6 eV were detected. The measurements are carried out for both horizontal and vertical directions of the E vector. In this way the energetic-ion yield spectra measured at $0°$ and $90°$ compensated for the difference in the detection efficiency of the two detectors are directly obtained. Total ion yield was measured simultaneously with the energetic-ion yield spectra with a different detector.

RESULT AND DISCUSSION

C $1s^{-1}2\pi_u$ Core-Excited States in CO_2

A doubly degenerate Π state of a linear molecule splits into two states along the bending coordinate due to the vibronic interaction, referred to as static Renner-Teller (RT) effect [45,46], and the molecular structure of one of them changes from the linear to bent geometry [16,17,20,24,25,28]. Here we focus on the C $1s^{-1}2\pi_u$ core-excited states in CO_2 that split into two states due to static RT effect. The lower branch of the RT pair states, A_1, becomes bent whereas the upper branch, B_1, remains linear [17]. The dipole transition moments from the ground state to the bent A_1 and linear B_1 states are parallel and perpendicular to the direction of the bending motion, respectively. Thus these two states can be separately observed if the direction of the bending motion can be specified relative to the transition dipole moment. To realize

174

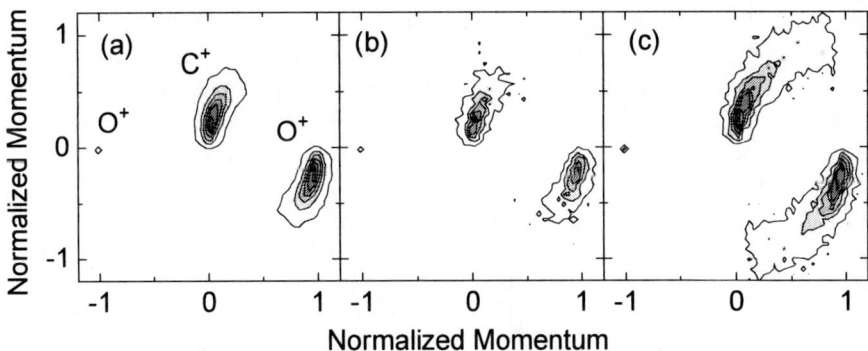

FIGURE 2. Newton diagrams for the three-body break-up $CO_2^{3+} \rightarrow C^+ + O^+ + O^+$ at the photon energies of 312 eV (a) and 290.8 eV (b, c). (b) B_1 and (c) A_1 in C_{2v}.

such a measurement, we employ the triple-ion-coincidence momentum imaging technique [3].

We focus on the vector correlation among the linear momenta of the three fragment ions produced in the three-body break-up $CO_2^{3+} \rightarrow C^+ + O^+ + O^+$. This correlation can be well represented by the Newton diagrams. In the Newton diagrams in Fig. 2, the amplitude of the linear momentum of the first O^+ is normalized to unity and the normalized vector is placed on the negative x-axis. The linear momentum of C^+ is then plotted in the positive x and y direction, while the linear momentum of the second O^+ in the positive x and negative y direction. The intensity is given in the form of contour plots in linear scale.

As references for later discussion, we first show diagram (a) recorded at the C $1s \rightarrow \sigma^*$ shape resonance at 312 eV. Here, we have not applied the procedure taken in (b) and (c) below to separate the two RT pair states. It can be seen that the C^+ ion

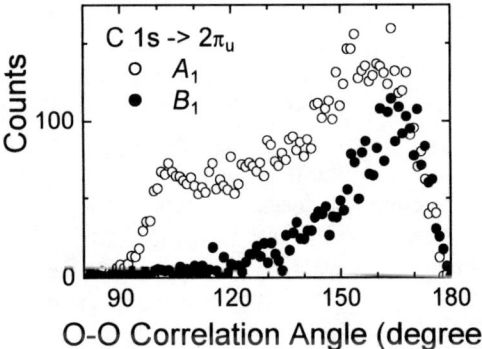

FIGURE 3. O-O correlation angle distributions for the excitation to the state with the polarization direction parallel to the bending motion (A_1: open circle) and to the state with the polarization direction perpendicular to the bending motion (B_1: closed circle) measured at 290.8 eV. Here, the O-O correlation angle is the angle between the linear momenta of the two O^+ ions obtained by the triple-ion-coincidence momentum imaging technique for the process $CO_2^{3+} \rightarrow C^+ + O^+ + O^+$.

goes off in the direction slightly transverse to the two O^+ vectors on the average even at the $C1s \rightarrow \sigma^*$ shape resonance, despite the fact that the molecule in this core-excited state is believed to remain linear until ionic dissociation. In diagrams (b) and (c), the excitation energy is set at the $C\ 1s \rightarrow 2\pi_u$ excitation peak (290.8 eV). In the construction of diagram (b), however, we selected the dissociation events from the B_1 state, i.e., the events in which the vector product of the two linear momenta of the two O^+ ions is parallel to E. In constructing diagram (c), on the other hand, we selected only the events from the A_1 state, i.e., the events in which the linear momentum of C^+ is parallel to E. It can be seen that diagram (b) coincides with diagram (a) demonstrating that the B_1 state has a linear geometry. In diagram (c), we can clearly see that the very long tail appears for each island, demonstrating that the C^+ ion goes off with a significant momentum transverse to the two O^+ vectors. This is a direct proof that the bending motion proceeds significantly in the A_1 state.

We now focus on the distribution of the O-O correlation angle between the two linear momenta of the two O^+ ions detected in coincidence with C^+. We plot in Fig. 3 the O-O correlation angle distributions for all solid angles for the excitation to $C\ 1s \rightarrow 2\pi_u A_1$ and B_1. It has been found that the O-O correlation angle distribution for $C\ 1s \rightarrow 2\pi_u B_1$ coincides with those for $C\ 1s^{-1}\sigma^*$ (not shown here), exhibiting a peak at $\sim 165°$: the peak at $\sim 165°$ corresponds to the Franck-Condon point where the excitation takes place in a linear geometry. The O-O correlation angle distribution for $C\ 1s^{-1}2\pi_u A_1$ on the other hand, exhibits a peak at $\sim 160°$ and extends to lower angles, forming a second peak at $\sim 100°$. The second peak corresponds to the classical cut-off. Detailed discussion is presented in separate papers [3,5].

Symmetry-Resolved Spectra

In the angle-resolved energetic ion spectroscopy, we detect fragment ions in the directions parallel ($0°$) and perpendicular ($90°$) to the E vector of the incident light. The $C\ 1s \rightarrow 2\pi_u$ excitation takes place preferentially for the molecules with molecular axis aligned in the direction perpendicular to the E vector. The fragment ions from the linear CO_2 molecule are detected only by the detector at $90°$. As a result, the detection of the fragment ions by the detector at $0°$ directly reflects the bending motion of the molecule. The bending motion (the displacement of the C atom) is in the direction parallel to the dipole moment for the A_1 state and perpendicular to the dipole moment for the B_1 state. Thus it is clear that the ions originated from the B_1 excitation cannot be detected at $0°$, as long as the axial recoil approximation is valid.

We introduce here the quantities $I(0)$ and $I(90)$ for ion yields recorded by the $0°$ and $90°$ detectors, respectively, and $I(A_1)$ and $I(B_1)$ for ion yields originated from the A_1 and B_1 excitations, respectively. These quantities are normalized in such a way as to satisfy the following relation:

$$I(A_1) + I(B_1) = I(0) + 2 \times I(90) \qquad (1)$$

Because $I(0)$ does not include the contribution from the B_1 excitation, we can express $I(0) = \alpha \times I(A_1)$ with α being a branching ratio of $I(A_1)$ to $I(0)$. Then the value of α can be related to the ratio $p \equiv I(B_1)/I(A_1)$ using eq. (1).;

$$\alpha = \frac{I(0)}{I(0) + 2 \times I(90)} \times (1 + p) \tag{2}$$

The values of p have been measured at several photon energies across the C $1s \to 2\pi_u$ resonance by means of the triple-ion-coincidence momentum imaging technique [3,4]. Thus we can extract $I(A_1)$ and $I(B_1)$ from the measurements of $I(0)$ and $I(90)$.

Figure 4 (a) presents the angle-resolved energetic-ion (≥ 6 eV) yield spectra of CO_2 measured across the C $1s \to 2\pi_u$ resonance, after subtraction of the baselines which include contributions from the valence ionization and the ionization by the second order light. Vibrational structure can be seen in both $I(0)$ and $I(90)$. Careful inspections however reveal that the vibrational structures observed in $I(0)$ and $I(90)$ are slightly different from each other.

We obtained the values of α from the values of p given in Refs. [4,5] and the measured spectra $I(0)$ and $I(90)$ using Eq. (2) and plotted the results in Fig. 4 (b). The values of α exhibit weak energy dependence. We thus carried out a linear fitting to the data points. Finally we extracted symmetry resolved excitation spectra $I(A_1)$ and $I(B_1)$ from α, $I(0)$ and $I(90)$.

Figure 4 (c) shows the symmetry-resolved excitation spectra for A_1 and B_1. The $I(B_1)$ spectrum in Fig. 4 (c) is well described by overlapping seven Voigt profiles whose Gaussian and Lorentzian widths are 30 meV and 170 meV, respectively. The $I(A_1)$ spectrum in Fig. 4 (c), on the other hand, is well described by one broad component, whose Gaussian and Lorentian widths are 30 meV and 440 meV, respectively, and five narrow lines whose Gaussian and Lorentian widths are 30 meV and 170 meV, respectively.

In order to identify the vibrational structure, we have simulated the excitation

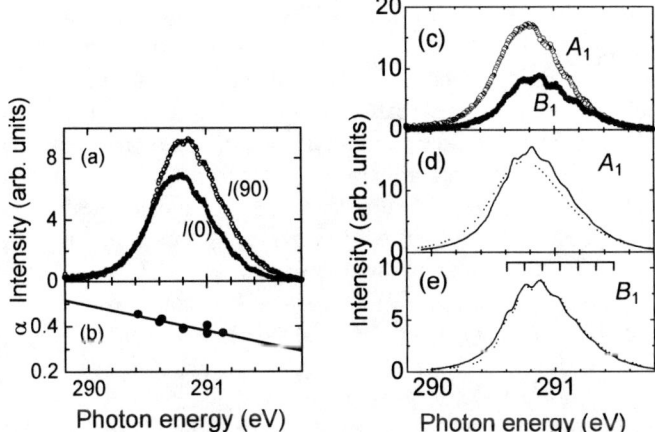

FIGURE 4. (a): Angle-resolved energetic-ion yield spectra of CO_2 measured across the C $1s \to 2\pi_u$ resonance. Closed and open circles, $I(0)$ and $I(90)$, respectively. (b):Photon-energy dependence of the α. (c): Experimental symmetry-resolved A_1 and B_1 excitation spectra. (d) and (e): theoretical excitation spectra $I(A_1)$ and $I(B_1)$, respectively. Dotted lines denote the spectra calculated in the adiabatic representation. Solid lines are the spectra calculated taking account of the non-adiabatic effect using a simple approximation. See the text for details.

spectra by using reliable *ab initio* potential energy surfaces (PESs) of the core-excited state. Figures 4 (d) and 4 (e) show the theoretical spectra. The dotted lines are calculated in the adiababic representation. The detailed vibrational structures in the dotted line for the B_1 state are in good agreement with the experimental spectrum. However, the theoretical calculation for the A_1 state only reproduces the global broad feature well and does not reproduce the vibrational structure. This is due to the dense vibrational components of the highly excited bending mode, as expected.

In order to explain the vibrational structure observed in the experimental $I(A_1)$ spectrum and large Lorentzian width in both $I(A_1)$ and $I(B_1)$ spectra, one should take account of the non-adiabatic coupling between the A_1 and B_1 states neglected in the present theoretical calculation. This non-adiabatic effect, often referred to as the dynamical Renner-Teller effect [45], is caused by the coupling between the bending nuclear motion and the electronic motion. Then, the excitation spectrum may be approximated by a weighted sum of the $I(A_1)$ and $I(B_1)$ spectra calculated in the adiabatic representation, by shifting the energy slightly with respect to each other. The solid lines in Figs. 4 (d) and 4 (e) are calculated as the weighted sum of the adiabatic $I(A_1)$ and $I(B_1)$ spectra. The solid lines reproduce the experimental $I(A_1)$ spectrum well and illustrates that the vibrational structure in the $I(A_1)$ spectrum originates from the symmetric stretching mode in the B_1 state. Detailed discussion is presented in a separate paper [42].

C1s PADs From Fixed-In-Space CO$_2$ Molecules

The PADs from fixed-in-space molecules provide a specific fingerprint of the scattering effect experienced by the photoelectron within the molecular potential, often referred to as the shape resonance effect. The experimental techniques have been developed rapidly and PADs from fixed-in-space molecules have been studied extensively for N$_2$ and CO [6-8,31-34]. Both the calculation based on the random phase approximation with the nonspherical relaxed-core Hartree-Fock potential and the one based on the multiple scattering theory in non-spherical self-consistent potentials succeeded to reproduce the experimental features reasonably well for N$_2$ and CO [8,34,35]. Thus it is a natural consequence to consider extensions of this kind of experimental investigations to triatomic molecules and to see how the theoretical predictions can reproduce the measured PADs. Here we present a discussion based on comparison of recent measurements and calculations for C 1s PAD from fixed-in-space CO$_2$ molecules near the C $1s^{-1}4\sigma_u^*$ shape resonance at ~ 312 eV. The O 1s PADs from fixed-in-space CO$_2$ were reported for the fixed molecular axis in parallel to the polarization vector E, i.e., for the $\Sigma \rightarrow \Sigma$ transitions [36].

The upper panel of Fig.5 shows the experimental three-dimensional C 1s PAD from fixed-in-space CO$_2$ molecules measured at the photon energy of 312 eV, which corresponds to the electron energy of 14.6 eV. The x-axis shows the angle of the molecular axis with respect to the E vector. The y-axis indicates the angle of the electron emission direction with respect to the E vector. The scale is a linear gray scale. The lower panel of Fig. 5 shows C 1s PADs from fixed-in-space CO$_2$ molecules sliced at the molecular angles of 0°, 45°, and 90° in the upper panel of Fig. 5, which correspond to the molecular direction of 0°, 45°, and 90° with respect to the photon polarization directions from left to right.

The direction of polarization, E, is horizontal. The dots show the experimental results and the solid curves the calculated results using the partially relaxed-core Hartree-Fock approximation (PRCHF). In order to allow a better comparison between experimental and theoretical data, the calculated PADs presented in Fig.5 have been convoluted with a Gaussian distribution with width (FWHM) of 15°, corresponding to the experimental angular resolution. The PADs for different angles of the molecular axis (0°, 45° and 90°) are in relative scales: the intensity reflects real data.

The 3D PAD represents the global feature of photoionization. The 3D PAD in Fig. 5 shows that ions are preferentially aligned parallel to the E vector of the incident light. One might think that C $1s$ photoelectrons are preferentially ejected in the direction along the E vector. However, the feature is tilted by 45° in Fig. 5, illustrating that photoelectrons *see* the molecular field and that they are ejected preferentially in the direction along the molecular axis.

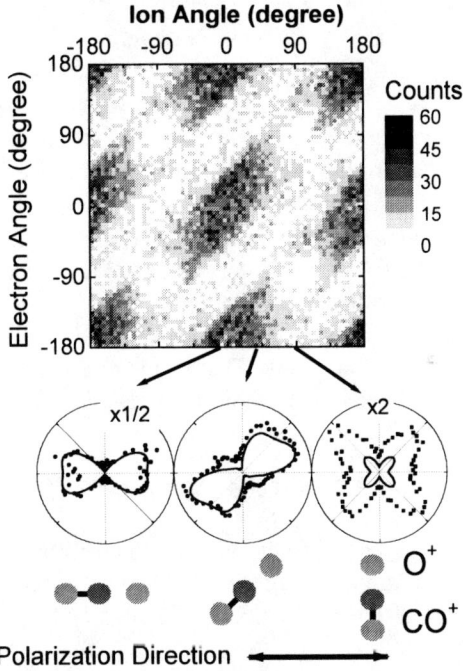

FIGURE 5. C $1s$ PADs from fixed-in-space CO_2 molecules measured at near the C $1s$ shape resonance (the photon energy of 312 eV). Upper panel shows the experimental three-dimensional C $1s$ PAD from fixed-in-space CO_2 molecules. The x-axis shows the angle of the molecular axis with respect to the photon E vector. The y-axis indicates the angle of the electron emission direction with respect to the E vector. Lower panel shows C $1s$ PADs from fixed-in-space CO_2 molecules at the molecular axis of 0°, 45° and 90° with respect to E as shown the illustration below. The dots and curves correspond to measurements and calculations, respectively. The direction of polarization, E, is horizontal.

To see PADs quantitatively, the lower panel of Fig.5 is useful. Generally, the calculated PADs are in qualitative agreement with the measured ones. Focus on the PAD for the Σ channel. The experimental PAD for the molecular axis of $0°$ is slightly fat compared to the well-known shape of the p wave and slightly hollow along the molecular axis. Compared with the PAD of N_2 and CO [6-8,31-33], however, the observed PAD of CO_2 is less structured. This feature of the PAD can be attributed mainly to the change in the relative intensities of different partial waves. The dipole photoionization promotes the C $1s$ σ_g electron to the l=1, 3, 5,... of σ_u continua, within the independent particle approximation. The calculation using the PRCHF approximation shows that at near the shape resonance (14.2 eV) all the partial waves with l=1, 3, 5 are enhanced. The l=1 wave is most intense but both l=3 and l=5 waves are also significant. A relatively fat PAD at 14.2 eV is interpreted as a result of interference among these three waves. This situation is in sharp contrast with the PAD of a homonuclear diatomic molecule N_2 of $D_{\infty h}$ symmetry, which exhibits a rich structure dominated by the l=3 wave.

We now focus on the PAD for the Π channel. We notice that the intensities of the calculated PAD are smaller than the measured ones. This is because we normalized the intensity for both experimental and theoretical PAD in the Σ channel: an overestimate of the calculation for the enhancement due to the shape resonance in the Σ channel can be the reason for this apparent disagreement. The dipole photoionization promotes the C $1s$ σ_g electron to the l=1, 3, 5,... of π_u continua, within an independent particle approximation. No shape resonance exists in the Π channel.

The PADs at $45°$ are determined by a coherent mixture of Σ and Π channels: both the ratios of the amplitudes and the phase correlation between the two channels play significant roles. Thus the PADs at $45°$ can be additional benchmarks for the stringent test of the theory. The general agreement is reasonable, suggesting that the calculated phase correlations between Σ and Π are reasonable. The intensities of the small lobes are underestimated by the calculations. This is due to the overestimate of the intensity in Σ relative to that in Π. Detailed discussion is presented in a separate paper [9].

ACKNOWLEDGMENTS

The experiments were carried out with the approval of the SPring-8 program advisory committee and supported in part by Grants-in-Aid for Scientific Research from the Japan Society of the Promotion of Science and by the Matsuo Foundation. The authors are grateful to M. Machida, A. De Fanis, K. Kubozuka, H. Chiba, Y. Muramatsu, M. Takahashi, H. Yoshida, K. Okada, I. H. Suzuki, K. Nobusada, H. Schmidt-Böcking, R. Dörner, A. Czasch, Th. Weber, M. Hattass, O. Jagutzki, L. Schmidt, R. Moshammer, A. Cassimi, V. McKoy, K. Wang, B. Zimmermann, M. Lavollée and U. Becker for fruitful collaborations.

REFERENCES

1. Saito, N., Heiser, F., Hemmers, O., *Phys. Rev.* A **54**, 2004 (1996).
2. Lavollée, M. and Brems, V., *J. Chem. Phys.* **110**, 918 (1999).
3. Muramatsu, Y., Ueda, K., Saito, N., *et al.*, *Phys. Rev. Lett.* **88**, 083001 (2002)
4. Muramatsu, Y., Ueda, K., Chiba, H., *et al.*, *Surf. Rev. Lett.* (in press).
5. Muramatsu, Y., Saito, N., Lavollée, M., , *et al.*, to be published.
6. Weber, Th., Jagutzki, O., Hattass, M., *et al.*, *J. Phys.* B **34**, 3669 (2001).
7. Landers, A., Weber, Th., Ali, I., *et al.*, *Phys. Rev. Lett.* **87**, 013002 (2001)
8. Jahnke, T., Weber, Th., Landers, A. L., *et al.*, *Phys. Rev. Lett.* **88**, 073002 (2002).
9. Saito, N., De Fanis, A., Kubozuka, K., *et al.*, submitted to *J. Phys.* B.
10. De Fanis, A., Saito, N., Pavlychev, A. A., *et al.*, *Phys. Rev. Lett.* (in press).
11. Neeb, M., Rubensson, J. E., Biermann, M., and Eberhardt, W., *J. Electron Spectrosc. Relat. Phenom.* **67**, 261 (1994).
12. Becker, U. and Menzel, A., *Nucl. Instr. Methods* B **99**, 68 (1995).
13. Kukk, E., Aksela, H., Aksela, S., *et al.*, *Phys. Rev. Lett.* **76**, 3100 (1996).
14. Sundin, S., Gel'Mukhanov, F. K., Ågren, H., *et al.*, *Phys. Rev. Lett.* **79**, 1451 (1997).
15. Björneholm, O., Sundin, S., Svensson, S., *et al.*, *Phys. Rev. Lett.* **79**, 3150 (1997).
16. Muramatsu, Y., Shimizu, Y., Yoshida, H., *et al.*, *Chem. Phys. Lett* **330**, 91 (2000).
17. Kukk, E., Bozek, J. D., and Berrah, N., *Phys. Rev.* A **62**, 032708 (2000).
18. Hjelte, I., Piancastelli, M. N., Fink, R. F., *et al.*, *Chem. Phys. Lett.* **334**, 151 (2001).
19. De Fanis, A., Nobusada, K., Hjelte, I., *et al.*, *J. Phys.* B **35**, L23 (2002)
20. Wight, G. R. and Brion, C. E., *J. Electron Spectrosc. Relat. Phenom.* **3**, 191 (1973).
21. Lebrun, T., Lavollée, M., Simon, M., and Morin, P., *J. Chem. Phys.* **98**, 2534 (1993).
22. Simon, M., Morin, P., Lablanquie, P., *et al.*, *Chem. Phys. Lett.* **238**, 42 (1995).
23. Ueda, K., Ohmori, K., Okunishi, M., *et al.*, *Phys. Rev.* A **52**, R1815 (1995).
24. Adachi, J., Kosugi, N., Shigemasa, E., and Yagishita, A., *J. Chem. Phys.* **102**, 7369 (1995).
25. Adachi, J., N. Kosugi, Shigemasa, E., and Yagishita, A., *J. Chem. Phys.* **107**, 4919 (1997).
26. Simon, M., Miron, C., Leclercq, N., *et al.*, *Phys. Rev. Lett.* **79**, 3857 (1997).
27. Ueda, K., Tanaka, S., Shimizu, Y., *et al.*, *Phys. Rev. Lett.* **85**, 3129 (2000).
28. Ueda, K., Simon, M., Miron, C., *et al.*, *Phys. Rev. Lett.* **83**, 3800 (1999).
29. Morin, P., Simon, M., Miron, C., *et al.*, *Phys. Rev.* A **61**, 050701 (2000).
30. Dehmer, J. L. and Dill, D., *Phys. Rev. Lett.* **35**, 213 (1975).
31. Shigemasa, E., Adachi, J., Oura, M., and Yagishita, A., *Phys. Rev. Lett.* **74**, 359 (1995).
32. Heiser, F., Gesner, G., Viefhaus, J., *et al.*, *Phys. Rev. Lett.* **79**, 2435 (1997).
33. Motoki, S., Adachi, J., Hikosaka, Y., *et al.*, *J. Phys.* B **33**, 4193 (2000).
34. Cherepkov, N. A., Raseev, G., Adachi, J., *et al.*, *J. Phys.* B **33**, 4213 (2000).
35. Cherepkov, N. A., Semenov, S. K., Hikosaka, Y., *et al.*, *Phys. Rev. Lett.* **84**, 250 (2000).
36. Watanabe, N, Adachi, J., Soejima, K., *et al.*, *Phys. Rev. Lett.* **78**, 4910 (1997).
37. Okada, K., Ueda, K., Tokushima, T., *et al.*, *Chem. Phys. Lett.* **326**, 314 (2000).
38. Saito, N., Ueda, K., Simon, M., *et al.*, *Phys. Rev.* A **62**, 042503 (2000).
39. Hiraya, A., Nobusada, K., Simon, M., *et al.*, *Phys. Rev.* A **63**, 042705 (2001).
40. Ueda, K., Yoshida, H., Senba, Y., *et al.*, *Nucl. Instrum. Methods* A **467-468**, 1502 (2001).
41. Okada, K., Tamenori, Y, Koyano, I., and Ueda, K., *Surf. Rev. Lett.* (in press).
42. Yoshida, H., Nobusada, K., Okada, K., *et al.*, *Phys. Rev. Lett.* **88**, 083001 (2002).
43. Ohashi, H., Ishiguro, E., Tamenori, Y., *et al.*, *Nucl. Instrum. Methods* A **467-468**, 533 (2001).; *Nucl. Instrum. Methods* A **467-468**, 533 (2001).
44. see http://roentdek.com for details on the detectors
45. Herzberg, G., *Infrared and Raman Spectra of Polyatomic Molecules*, New York, Van Nostrand, 1973.
46. Köppel, H., Domcke, W., and Cederbaum, L. S., *Adv. Chem. Phys.* **57**, 59 (1984).

Multiphoton Excitation of Plasmons in Clusters.

Jean-Patrick Connerade* and Andrey V.Solov'yov†

*The Blackett Laboratory, Imperial College, London SW7 2BW, UK
†A.F.Ioffe Physical-Technical Institute of the Academy of Sciences of Russia, Polytechnicheskaya 26, St. Petersburg 194021, Russia

Abstract. We present a theoretical framework for the multiphoton excitation of plasmons. We show that, in addition to dipole plasmon excitations, multipole plasmons (quadrupole, octupole, etc) are excited in a metallic cluster by multiphoton absorption processes, resulting in a significant difference between plasmon resonance profiles in multiphoton and single-photon absorption. The method is quite general, and applies to any system with delocalised electrons, of which the simplest are spherical metallic clusters.

INTRODUCTION

Plasmons are characteristic of systems containing many delocalised electrons. They occur from the quantum to the classical limit. At the quantum end, atoms do not exhibit conspicuous plasmon behaviour, because of the absence of a clear 'surface'. Metallic clusters provide excellent examples of plasmons in quantum systems, appearing for as few as eight atoms. They persist right through to very large cluster sizes, which can be considered as the solid state limit. Metallic clusters allow one to study the evolution of plasmons from quantum to classical regimes.

A feature of plasmons is their presence both in the bulk and on the surface. They possess many oscillatory modes. Dipole excitation from the ground state using a single photon has been the traditional way to explore their spectroscopy, but provides limited information on plasmon dynamics. Our purpose is to demonstrate that much more detail is accessible by multiphoton spectroscopy, and that the full dynamics of the plasmon, by coupling with more than one photon, induces a richer spectrum from which much more information can be gained.

We have developed two simple models, leaving out inessential detail to concentrate on the mechanisms by which multiphoton excitation of metallic clusters occurs. These two models are (i) a quantum and (ii) a classical picture. The first is based on the jellium approximation, in which delocalised electrons are confined within a spherical cluster, and the second treats forced oscillations in the Mie picture. We omit molecular vibrations or phonons, and consider merely collective motions of conduction electrons. This approach brings out essential features common to many systems to which jellium picture can be applied. The main conclusions about multiphoton excitation are similar in the quantum and in the semi-classical limits, so that a smooth transition from one to the other occurs. This theoretical formalism is not confined to photons. It can be used to describe any kind of higher order plasmon excitation processes, for example multiple scattering of electrons within a cluster.

CP652, X-Ray and Inner-Shell Processes: 19th International Conference on X-Ray and Inner-Shell Processes
edited by A. Bianconi, A. Marcelli, and N. L. Saini
© 2003 American Institute of Physics 0-7354-0111-X/03/$20.00

Recently, a number of papers have discussed metallic clusters [1, 2] and fullerenes [3] in strong laser fields. Our prime interest is in lower laser powers, for which the integrity of clusters is preserved, and multiphoton excitation just begins to intrude. In our semi-classical model, the collective flow of charge is driven by a periodic field. The results can be related to the multiphoton absorption cross section of the cluster, which takes account of quantum mechanics. In principle we could include the turn-on and turn off of laser pulses for various power levels and initial charge distributions. However, we concentrate on a novel feature, which arises even for an infinite wavetrain interacting with a cluster (the simplest and most fundamental problem): multiple plasmon excitations driven by multi-photon excitations.

Surface plasmons are well-known in atomic clusters. Dipole surface plasmons are responsible for the formation of giant resonances in photoabsorption spectra of metal clusters (see e.g. [4, 5, 7, 6]). They determine inelastic collisions of charged particles with metal clusters (see [7] and references therein), where it was demonstrated that collective excitations contribute to the electron energy loss spectrum near the surface plasmon resonance. In the energy range above the ionization threshold, volume plasmons dominate the differential cross section, resulting in resonance behaviour [7]. The role of the polarization interaction and of plasmon excitations in electron attachment to metal clusters has been examined both theoretically [7] and experimentally [8]. Plasmon excitations induce resonance enhancement of the electron attachment cross section.

PLASMON RESONANCE APPROXIMATION

Our quantum analysis is based on the plasmon resonance approximation. We consider the simplest example, namely the cross section for single-photon absorption, viz:

$$\sigma_1 = \frac{4\pi^2 e^2}{c} \omega \sum_n |z_{on}|^2 \delta(\omega_{no} - \hbar\omega) \tag{1}$$

Here, e is the charge of electron, c the velocity of light, $\hbar$ is Planck's constant, $\omega_{no} = \varepsilon_n - \varepsilon_0$ the electronic excitation energy, ω the photon frequency and ez_{on} the matrix element of the z- component of the cluster dipole moment. The summation includes all final states of the excited electron, belonging to both discrete and continuous spectra.

In the jellium picture, which holds for metal clusters and to some extent for fullerenes, the main contribution to the cross section (1) arises from a small group of excited states or sometimes even a single transition of frequency close to the classical Mie resonance – also known as the plasmon resonance. For a spherical metal cluster, it is given by $\omega_l^2 = \frac{4\pi N e^2}{mV} \cdot \frac{l}{(2l+1)}$ (see e.g. [4, 5, 7]) Here $V = 4\pi R^3/3$ is the cluster volume, $R = r_o N^{1/3}$ the cluster radius, r_o the Wigner-Seitz radius; N the number of delocalized electrons, l the angular momentum of the plasmon mode, m the electron's mass. Note that a single $1 - 4eV$ photon can, in practice, excite only $l = 1$ dipole plasmon oscillations.

The plasmon resonance approximation assumes that excitations near a plasmon resonance exhaust the sum rule almost completely (see [4, 5, 7, 6]), which means that $\sum_n \omega_{no}|z_{on}|^2 = \frac{N\hbar^2}{2m}$ (see e.g. [9]).

Assuming a Lorentzian distribution of width Γ_1 for the plasmon resonance states and replacing the delta function in (1) by the profile (see e.g. [9]), one recovers the well-known expression for the single photon absorption cross section (see e.g. [4, 7])

$$\sigma_1 = \frac{\pi N e^2}{mc} \frac{\Gamma_1}{(\omega_1 - \omega)^2 + \frac{\Gamma_1^2}{4}} \approx \frac{4\pi N e^2}{mc} \frac{\omega^2 \Gamma_1}{(\omega_1^2 - \omega^2)^2 + \omega^2 \Gamma_1^2} \tag{2}$$

The width Γ_1 is due to Landau damping. Its calculation is performed in [7]. The cross section (2) reproduces plasmon resonances in single- photon absorption spectra of metal clusters. In the dipole approximation, the two-photon cross section is

$$\sigma_2 = \frac{32\pi^3 e^4 \hbar}{c^2} \omega^2 \sum_n \left| \sum_m \frac{z_{nm} z_{mo}}{\hbar\omega - \omega_{mo} + i\delta} \right|^2 \delta(\omega_{no} - 2\hbar\omega) \tag{3}$$

We evaluate it in the same way as for the single-photon case. The main contribution to the sum over intermediate states m arises from virtual dipole plasmon excitations. With the use of the sum rule, it reduces to $|r_{10}|^2 \approx N\hbar/2m\omega_1$. The remaining matrix elements z_{n1} in (3) describe dipole transitions from the dipole plasmon resonance state to other excited states. Thus, for the final state, $l = 0$ or $l = 2$ only. However, there is no surface plasmon excitation with $l = 0$. Thus, only transitions to states with $l = 2$ are of interest. These arguments show that, by using two photons simultaneously, one can excite the quadrupole plasmon resonance in a metal cluster or in a fullerene. When calculating the cross section (3) near the quadrupole plasmon resonance excitation, i.e. at $2\omega \sim \omega_2$, we need consider only transitions to the resonance final state, i.e. put $\sum_n |z_{n1}|^2 \approx |z_{21}|^2$ (we use indices 1 and 2 for dipole and quadrupole plasmon resonance states) and replace the delta function, $\delta(\omega_{no} - 2\hbar\omega)$ by a Lorentzian distribution of width Γ_2, as for the single photon case.

The transition matrix element z_{21} describes the electronic transition between dipole and quadrupole plasmon resonance states. It occurs as a single electron transition rather than a collective one. Therefore, calculation of z_{21} on the basis of the sum rule overestimates this matrix element. Instead, we have calculated it by assuming that transition electron densities of both dipole and quadrupole plasmon modes are localised in a narrow layer of width ΔR near the cluster surface. This assumption is consistent with ab initio transition density analysis in [7] for the Na_{40} and Na_{92} clusters. For details, we refer to our forthcoming paper [10] and present only the final result $z_{21} = -\frac{8}{3} \left(\frac{6}{5}\right)^{1/4} \frac{\hbar}{m\omega_1 \Delta R}$.

Finally, we obtain the two-photon absorption cross section:

$$\sigma_2 = \frac{8\pi^2 A^2 N e^4 \hbar}{m^3 c^2 \Delta R^2 \omega_1^3} \frac{\omega^2}{(\omega - \omega_1)^2 + \frac{\Gamma_1^2}{4}} \cdot \frac{\Gamma_2}{(\omega_2 - 2\omega)^2 + \frac{\Gamma_2^2}{4}} \tag{4}$$

where $A = \frac{8}{3} \left(\frac{6}{5}\right)^{1/4} \approx 2.79$.

In figure 1 we plot cross section profiles per unit atom for single-photon (dashed line) and two-photon (solid line) absorption calculated according to (2) and (4). These profiles do not depend on the number of atoms in the cluster.

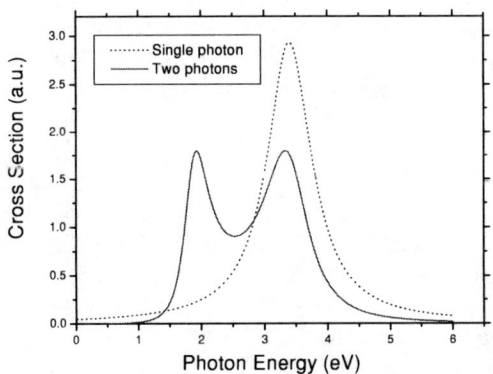

FIGURE 1. The profiles of single-photon (dotted line) and two-photon (solid line) absorption calculated according to (2) and (4) and normalised per unit atom. The two-photon absorption profile is scaled by a factor 1/100. The scales are not identical for the two curves for reasons of definition of the cross sections in the single- and two-photon cases, but both are given in atomic units.

The peak in the single-photon plot gives the location of the dipole resonance. The other peak in the two-photon plot is the quadrupole resonance. This figure illustrates the contribution due to quadrupole plasmon excitation in the two-photon spectrum. In this calculation we used $r_0 = 4.0$ and $\Gamma_1 = \omega_1/4$, $\Gamma_2 = \omega_2/4$, $\Delta R = r_0$. The choice can be different for different clusters, but should always lead to qualitatively similar single- and two-photon absorption profiles. Accurate parameters are obtained only on the basis of *ab initio* calculations.

Our formalism can be used for a larger number of photons. In this case the cross section can be analysed in a similar way. This leads to the conclusion that plasmon resonances with larger angular momenta (octupole, etc) can be excited.

HYDRODYNAMIC DESCRIPTION

We now turn to the semiclassical or hydrodynamic picture, i.e. the opposite extreme to the quantum model. A classical description of the electron density variation $\rho(\mathbf{r},t)$ is appropriate when plasmon excitations dominate over the single-particle spectrum, since plasmons are essentially classical. We describe collective motion of $\rho(\mathbf{r},t)$ using Euler's equation and the equation of continuity.

Euler's equation couples the acceleration of the density $d\mathbf{v}/dt$ with the total local electric field $\mathbf{E}$ at the point $(\mathbf{r},t)$. It has the form:

$$\frac{\partial \mathbf{v}(\mathbf{r},t)}{\partial t} + \{\mathbf{v}(\mathbf{r},t) \cdot \nabla\}\mathbf{v}(\mathbf{r},t) = -\frac{e}{m}\nabla\varphi(\mathbf{r},t) - \frac{e}{m}\nabla \int d\mathbf{r}' \frac{\delta\rho(\mathbf{r}',t)}{|\mathbf{r}-\mathbf{r}'|} \qquad (5)$$

Here $\varphi(\mathbf{r},t)$ is the potential of the external field. The density $\rho(\mathbf{r},t) = \rho_o(\mathbf{r}) + \delta\rho(\mathbf{r},t)$, where $\rho_o(\mathbf{r})$ is the density in a free cluster without external fields and $\delta\rho(\mathbf{r},t)$ is the variation of density caused by the external field and polarization force acting together.

The motion of $\rho(\mathbf{r},t)$ obeys the equation of continuity:

$$\frac{\partial\rho(\mathbf{r},t)}{\partial t} + \nabla \cdot \{\rho(\mathbf{r},t)\mathbf{v}(\mathbf{r},t)\} = 0 \tag{6}$$

The simultaneous solution of (5), and (6) with appropriate initial conditions, and the initial distribution $\rho_o(\mathbf{r})$ determine $\delta\rho(\mathbf{r},t)$ as well as its velocity $\mathbf{v}(\mathbf{r},t)$. The full solution of this problem will be given in a forthcoming paper [10]. By solving a set of equations with $\varphi(\mathbf{r})$ describing the dipole electron-photon interaction up to n-th order, one can calculate $\rho(\mathbf{r},t)$ due to the field of n photons. The resulting equations describe volume and surface eigen-oscillations of $\rho(\mathbf{r},t)$ characterised by l. The surface plasmon resonance frequency ω_l is the same as above. The volume plasmon resonance frequency is $\omega_p = \sqrt{\frac{4\pi e^2 N}{mV}}$. Equations for the volume and surface plasmon oscillations can be separated from each other and solved iteratively. The solutions $\delta\rho_{lm}^{v(n)}(r)$ (for volume plasmons) and $\delta\rho_{lm}^{s(n)}$ (for surface plasmons) can be found for arbitrary large order of perturbation theory n, although the formulae become more and more tedious with increasing n. They demonstrate that, in higher orders, plasmon resonances with angular momenta larger than the angular momentum of the external field can be excited. The equations reveal a significant shift of the plasmon resonance profiles in the highest orders towards lower frequencies.

We have also applied our theory to the description of multiphoton absorption. We focus on the analysis of plasmon excitations. If surface or volume plasmon resonances are excited by photons, i.e. $\omega \sim \omega_p$, the following condition is fulfilled $\omega R/c \sim \omega_p R/c \ll 1$, which implies the dipole approximation. In this limit, volume plasmon excitations play a negligible role, resulting in a considerable simplification. In this case, one derives

$$\left((\omega n)^2 - \omega_l^2\right)\delta\rho_{l,m}^{s(n)} = -\sqrt{\frac{4\pi}{3}\frac{Ne^2E}{mV}}\delta_{n,1}\delta_{l,1}\delta_{m,0} +$$

$$+\sqrt{\frac{4\pi}{3}\frac{eE}{mR}}\sum_{l_1,m_1}I_2(l,m|l_1,m_1,|1,0)\delta\rho_{l_1,m_1}^{s(n-1)} -$$

$$-\frac{4\pi eE}{mR}\sum_{j_1=1}^{n-1}\frac{1}{n-j}\sum_{\substack{l_1,m_1\\l_2,m_2}}I_2(l,m|l_1,m_1,|l_2,m_2)\frac{\delta\rho_{l_1,m_1}^{s(j)}\delta\rho_{l_2,m_2}^{s(n-j)}}{2l_2+1} \tag{7}$$

where $E = \sqrt{2\pi\hbar\omega/V_o}$ is the strength of the linearly polarized electric field of the photon and V_o is the normalization volume of the photon mode. The angular integral is $I_2(l,m|l_1,m_1|1,0) = \sqrt{2l_1(l_1+1)}\int d\Omega_{\mathbf{n}_r}Y_{l,m}^*(\mathbf{n}_r)\mathbf{Y}_{l_1,m_1}^{(1)}(\mathbf{n}_r)\mathbf{Y}_{1,0}^{(1)}(\mathbf{n}_r)$.

For details, we refer to [10]. Equation (7) should be solved iteratively starting from $n=1$. For $n=1,2$, the non-trivial solutions read as $\delta\rho_{1,0}^{s(1)} = -\sqrt{\frac{4\pi}{3}\frac{Ne^2E}{mV(\omega^2-\omega_1^2)}}$, $\delta\rho_{0,0}^{s(2)} =$

$-\frac{\pi^{1/2}}{3m^2RV}\cdot\frac{Ne^3E^2}{(\omega^2-\omega_1^2)^2}$, $\delta\rho_{2,0}^{s(2)} = -\frac{4\pi^{1/2}}{3\sqrt{5}m^2RV}\cdot\frac{Ne^3E^2\omega^2}{(\omega^2-\omega_1^2)^2((2\omega)^2-\omega_2^2)}$

We now calculate multipole moments of the cluster induced by an external radiation field: $Q_{l,m}^{(n)} = \sqrt{\frac{4\pi}{2l+1}} R^{l+2} \delta\rho_{l,m}^{s(n)}$. Substituting here $\delta\rho_{1,0}^{s(n)}$, one obtains the dipole moment of the cluster, $D^{(1)}(\omega) = Q_{1,0}^{(1)}$ induced in the single-photon absorption process

$$D^{(1)}(\omega) = -\frac{Ne^2 E}{m(\omega^2 - \omega_1^2 + i\omega\Gamma_1)} \tag{8}$$

Subsituting $\delta\rho_{2,0}^{s(2)}$, into $Q_{l,m}^{(n)}$, one finds the quadrupole moment of the cluster, $Q^{(2)}(\omega) = Q_{2,0}^{(2)}$ induced in the two-photon regime:

$$Q^{(2)}(\omega) = Q_{2,0}^{(2)} = -\frac{2}{5} \frac{Ne^3 E^2 \omega^2}{m^2((\omega^2 - \omega_1^2)^2 + \omega^2\Gamma_1^2)((2\omega)^2 - \omega_2^2 + i2\omega\Gamma_2)} \tag{9}$$

Here, we have introduced the plasmon resonance widths Γ_1 and Γ_2 which take into account Landau damping of the dipole and quadrupole surface plasmon resonances. They must be determined separately, e.g. by an *ab initio* approach (see [7] and references therein).

Three photons can induce dipole and octupole moments in the cluster, and we have derived explicit expressions for them [10]. They demonstrate that multipole moments induced in the cluster by multiphoton absorption processes possess a prominent plasmon resonance structure. The connection between $D^{(1)}(\omega)$ from (8) and the cross section σ_1 found in (2) is straightforward: $\sigma_1 = \frac{4\pi\omega}{cE} Im D^{(1)}(\omega)$.

ACKNOWLEDGMENTS

The authors acknowledge support from the Royal Society of London and INTAS.

REFERENCES

1. L. Köller, M. Schumacher, J. Köhn, Tiggesbäumker, and K.H.Meiwes-Broer, Phys. Rev. Lett **82**, 3783 (1999)
2. C.A. Ullrich, P.-G. Reinhard and E. Suraud, J. Phys. B **30**, 5043 (1997)
3. S. Hunsche, T. Starczewski, A. l'Huillier, A. Persson, A. Wahlström, C-G. van Linden, B. van der Heuvell, and S. Svanberg, Phys. Rev. Lett. **77**, 1966 (1996)
4. W. A. de Heer, *Rev.Mod.Phys.* **65**, 611 (1993)
5. C. Brechignac and J.P.Connerade, *J.Phys.B: At. Mol. Opt. Phys.* **27**, 3795 (1994)
6. V.K.Ivanov, G.Yu.Kashenock, R.G.Polozkov and A.V.Solov'yov, *J.Phys.B: At.Mol.Opt.Phys.* **34**, L669-L677 (2001)
7. A.V. Solov'yov, in NATO Advanced Study Institute, Les Houches, Session LXXIII, Summer School "Atomic Clusters and Nanoparticles", Edited by C.Guet, P.Hobza, F.Spiegelman and F.David, EDP Sciences and Springer Verlag, Berlin, Heidelberg, New York (2001).
8. S.Sentürk, J.P.Connerade, D.D.Burgess and N.J.Mason, J.Phys.B: At.Mol.Opt.Phys. **33**, 2763 (2000)
9. L.D. Landau and E.M. Lifshitz, *Quantim Mechanics*, Pergamon, London (1965)
10. J.P. Connerade, A.V. Solov'yov, Phys.Rev. **A66**, 013207-(1-16) (2002)

Double K-Vacancy Production by X-Ray Photoionization

S. H. Southworth*, R. W. Dunford*, E. P. Kanter*, B. Krässig *, L. Young*,
G. B. Armen†, J. C. Levin†, M. H. Chen** and D. L. Ederer‡

*Argonne National Laboratory, Argonne, IL 60439, USA
†University of Tennessee, Knoxville, TN 37996, USA
**Lawrence Livermore National Laboratory, Livermore, CA 94550, USA
‡Tulane University, New Orleans, LA 70118, USA

Abstract.
    We have studied double K-shell photoionization of Ne and Mo (Z = 10 and 42) at the Advanced Photon Source. Double K-vacancy production in Ne was observed by recording the *KK-KLL* Auger hypersatellite spectrum. Comparison is made with calculations using the multiconfiguration Dirac-Fock method. For Mo, double K-vacancy production was observed by recording the $K\alpha, \beta$ fluorescence hypersatellite and satellite x rays in coincidence. From the intensities of the Auger or x-ray hypersatellites relative to diagram lines, the probabilities for double K-vacancy production relative to single K-vacancies were determined. These results, along with reported measurements on other atoms, are compared with Z-scaling calculations of the high-energy limits of the double-to-single K-shell photoionization ratio.

Introduction

    Double photoionization of He has been extensively studied because it is sensitive to electron correlation in the simplest case of a two-electron system [1]. Calculation of the double-to-single photoionization ratio R as a function of photon energy provides a testing ground for the development of theories that incorporate electron correlation in the ground state and in the final continuum states [2, 3]. Among several theoretical approaches [3], the many-body perturbation theory (MBPT) [4] provides a description of the double-photoionization process in terms of three first-order amplitudes in the electron-electron interaction (see [2, 3] for discussions of the MBPT). The three lowest-order MBPT diagrams are referred to as GSC (ground state correlation), SO (shake off), and TS1 (two-step one). The GSC accounts for radial and angular correlation between the two equivalent 1s electrons in the ground state (see, e.g., [5]). SO accounts for relaxation in the final ionic state upon ejection of one of the electrons into the continuum. The TS1 is also a final state interaction and describes photoemission of one electron that knocks out the second electron by binary collision.

    The MBPT amplitudes are added coherently, so describing the double-photoionization process in terms of their relative importance is not strictly valid. The amplitudes are also gauge dependent, i.e., they depend on the form of the dipole operator used (length, velocity, or acceleration), a topic that has been discussed in the theoretical literature ([2, 3, 6]). Nevertheless, it is informative to consider the

CP652, *X-Ray and Inner-Shell Processes: 19th International Conference on X-Ray and Inner-Shell Processes*
edited by A. Bianconi, A. Marcelli, and N. L. Saini
© 2003 American Institute of Physics 0-7354-0111-X/03/$20.00

energy dependencies of the three amplitudes in the acceleration form of the dipole operator [4] which gives better agreement with the measurements of [1] than the length and velocity results. All three amplitudes increase strongly from threshold, but the GSC and SO amplitudes level out asymptotically, while the TS1 amplitude passes through a maximum and decreases monotonically to high energy. The picture that emerges is that R increases from threshold to a maximum, then falls off slowly to an asymptotic ratio determined by the GSC and SO terms. This general picture is supported by Z-scaling calculations of double K-shell photoionization of two-electron systems [7].

High-energy-limit calculations [8, 9] are interesting and useful, because they are considered to be asymptotically exact and provide reference values for experiments on atoms other than He. In the high energy limit, final-state correlation is negligible and only the ground-state correlation needs to be accurately treated. Very accurate $1s^2$ ground state wavefunctions can be constructed for He and He-like ions [8, 9]. The results of those calculations provide reference values for experiments that measure R of many-electron targets under the assumption that interaction with outer-shell electrons is negligible. However, scaling shows a Z^2 dependence on the energy required to reach the asymptotic region [7] and may be difficult to study experimentally for higher Z atoms. Good agreement is obtained between theory and the high-energy measurement (5.4-9.1 keV) on He using the ion-recoil momentum imaging method to distinguish photoabsorption from Compton scattering [10]. The experiments on higher-Z atoms described here and in other publications [1, 11, 12, 13] were measured using photon energies close to the expected maxima of R [7]. Calculations that treat energy-dependent correlation interactions for this ratio are lacking for atoms other than He. The experiments may motivate more detailed calculations on higher-Z atoms and may require treatment of relativistic interactions as well as electron correlation.

Ne[KK] at 5 keV

The threshold for double-K ionization of Ne is 1863 eV [14] compared with 870.21 eV for single-K ionization [5]. Ne K vacancies decay predominantly by emission of K-LL Auger electrons rather than by fluorescence [15]. The energies and relative intensities of diagram Auger lines and the satellites resulting from additional excitation or ionization of the L shell have been determined experimentally [16, 17]. Using the multi-configuration Dirac-Fock method, Chen [18] calculated the energies and transition rates of the KK-KLL Auger hypersatellite spectrum. The hypersatellites were used to determine R for Ne by electron impact [14], and the same method was used here. For the present experiment, double-K vacancies were produced in Ne by photoionization with a 5 keV photon beam at Argonne's Advanced Photon Source (APS), and the resulting K-LL and KK-KLL Auger spectra were recorded with a cylindrical-mirror electron analyzer (CMA). The CMA was positioned with its symmetry axis in the plane perpendicular to the x-ray beam and at the magic angle (54.7^o) with respect to the linear-polarization direction to eliminate the dependence of measured intensities on angular anisotropies. The electron spectra were corrected for energy variations of the collection efficiency of the CMA and for variations of the x-ray beam intensity. The spectrum in Fig. 1 shows the

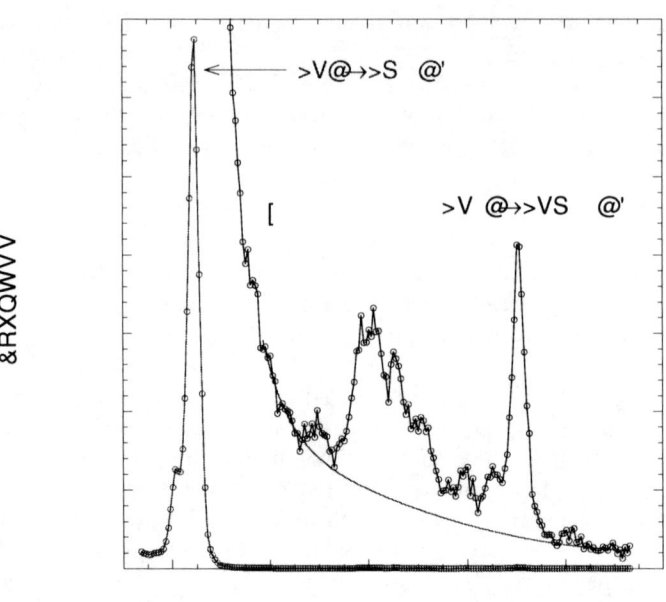

FIGURE 1. Auger electron spectrum of Ne excited by absorption of 5 keV x rays. The $K - L_{2,3}L_{2,3}$ ^{1}D diagram line at 804.5 eV and the $KK - KL_{2,3}L_{2,3}$ ^{2}D hypersatellite line at 870.5 eV are indicated. The structure between 835-855 eV is discussed in the text.

$K - L_{2,3}L_{2,3}$ ^{1}D diagram line at 804.5 eV and the $KK - KL_{2,3}L_{2,3}$ ^{2}D hypersatellite line at 870.5 eV. The structure between 835-855 eV is complicated due to contributions from several initial multi-vacancy states. This energy region includes the $KK - KL_1L_{2,3}$ $^{2}P^{(+)}$ hypersatellite and satellites from the decay of initial states such as [1s2p]np and [1s2s]ns, i.e., shakeup states observed in the 1s photoelectron spectrum [16, 19]. Additional multi-vacancy states apparently also contribute to this region of the Auger spectrum. We are attempting to identify these states using shake calculations and calculations of their associated *K-LL* Auger spectra. The $KK - KL_{2,3}L_{2,3}$ ^{2}D hypersatellite peak at 870.5 eV appears to be essentially free of contributions from other initial vacancy states, so we use its relative intensity to determine the relative number of double-*K* vacancies produced by photoionization at 5 keV.

The area of the $K - L_{2,3}L_{2,3}$ ^{1}D diagram peak is proportional to the number of single *K*-vacancies produced times the branching ratio into that particular final state. The branching ratio was deduced from experimental results [15, 17]. Similarly, the area of the $KK - KL_{2,3}L_{2,3}$ ^{2}D peak is proportional to the number of double *K*-vacancies produced times its branching ratio. The branching ratio was determined from the calculated Auger rates [18] and estimates of the probabilities of fluorescence and multiple-Auger final states. From this analysis, the ratio R for Ne at 5 keV was determined and is plotted in Fig. 3 below with results for other atoms. We stress that the result reported here is

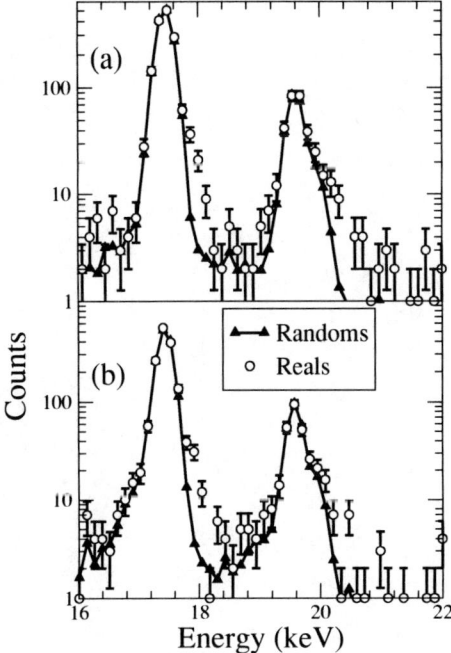

FIGURE 2. X ray/x ray fluorescence coincidence spectra from Mo excited by absorption of 50 keV x rays. The energy spectra in each detector are shown for coincidences with $K\alpha, \beta$ x rays in the other detector. The spectrum for detector "1" is in (a) while that for "2" is in (b). The data plotted as open circles are for real coincidences while the filled triangles with solid curve are random coincidences.

preliminary due to the need to fully account for weak Auger satellites from other multi-vacancy initial states.

Mo[KK] at 50 keV

A full report on our measurement of R for Mo at 50 keV is given in [13]. The threshold for double-K ionization is 40.654 keV compared with 20 keV for single-K ionization. Mo K vacancies decay predominantly by x-ray fluorescence, and double K vacancies emit a $KK - KL$ hypersatellite x ray shifted up in energy from the $K\alpha, \beta$ diagram lines followed by a $KL - L^2$ satellite. Our experiment at the APS used x-ray beams of 40.2 and 50 keV passing through a thin Mo target. Two Si(Li) detectors faced each other with the target inbetween and were normal to the beam and parallel to the polarization direction to minimize detection of scattered x rays. Coincidence electronics recorded the x-ray energies deposited in each detector and their time difference. The energy spectra recorded in each detector for coincidences with a $K\alpha, \beta$ x-ray (encompassing also the $KL - L^2$ satellite) are plotted in Fig. 2. The real coincidences show weak shoulders on the high-energy sides of the diagram lines due to the hypersatellites. These shoulders

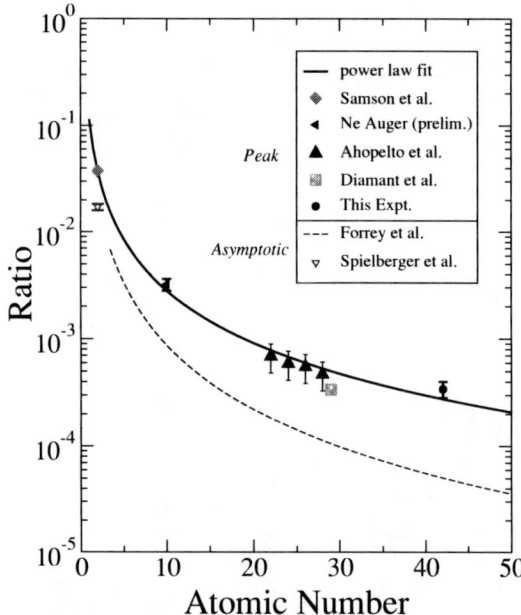

FIGURE 3. The ratio of double-to-single K-shell photoionization as a function of atomic number. Measured results are for He at 200 eV [1] (diamond), He at 5.4-9.1 keV [10] (open triangle), Ne (this work; left triangle), Ti, Cr, Fe, and Ni [11] (up triangle), Cu [12] (square), and Mo [13] (solid circle). The solid curve is a power law fit to the measurements. The dashed curve shows the Z-scaling law calculated for He-like ions in the high-energy limit [8].

were absent in the coincidence spectra recorded at the subthreshold energy of 40.2 keV. The ratio R was determined from the relative intensities of hypersatellites and diagram transitions after accounting for solid angles, detection efficiencies, and fluorescence yields [13].

Z-dependence of double-to-single K-shell photoionization

Our results for the double-to-single K-shell photoionization ratio R for Ne at 5 keV and Mo at 50 keV are plotted in Fig. 3 along with results for other atoms [1, 10, 11, 12]. For reference, the calculated asymptotic Z-scaling law of [8] is also shown. Except for the high-energy measurement on He [10], the measured ratios are significantly higher than the asymptotic limit, as expected. With the exception of [10], the measurements were made at photon energies within the broad maximum of R predicted by Z-scaling calculations [7]. The solid line through the measurements is a power-law fit with the form $1/Z^{1.61}$ [13]. This dependence on Z for values of R measured near their maxima is somewhat weaker than that calculated for the asymptotic limits [8].

With the availability of intense sources of synchrotron x-rays, additional experimental studies are expected in the near future. Our group has used the x ray/x ray fluorescence coincidence method to study double-K photoionization of Ag over 50-90 keV and observed the rise of the ratio R from near threshold to the expected maximum. Oura *et al.* [20] used a crystal spectrometer to study x-ray hypersatellite production in Ca, Ti, and V. Huotari *et al.* [21] also used high-resolution x-ray fluorescence spectroscopy to study hypersatellites in the transition metal elements from V to Zn. We hope that these experimental results will stimulate further theoretical work on the Z dependence of double K-vacancy production.

ACKNOWLEDGMENTS

We are grateful to the staff of the Basic Energy Sciences Synchrotron Radiation Center at the Advanced Photon Source for their assistance in performing the experiments. The Argonne group is supported by the Chemical Sciences, Geosciences, and Biosciences Division of the Office of Basic Energy Sciences, Office of Science, U.S. Department of Energy, under Contract No. W-31-109-Eng-38. Use of the Advanced Photon Source is supported by the U. S. Department of Energy, Basic Energy Sciences, Office of Science, under Contract No. W-31-109-Eng-38.

REFERENCES

1. Samson, J. A. R., Stolte, W. C., He, Z.-X., Cutler, J. N., Lu, Y., and Bartlett, R. J., *Phys. Rev. A*, **57**, 1906 (1998).
2. McGuire, J. H., Berrah, N., Bartlett, R. J., Samson, J. A. R., Tanis, J. A., Cocke, C. L., and Schlachter, A. S., *J. Phys. B*, **28**, 913 (1995).
3. Sadeghpour, H. R., *Can J. Phys.*, **74**, 727 (1996).
4. Hino, K., Ishihara, T., Shimizu, F., Toshima, N., and McGuire, J. H., *Phys. Rev. A*, **48**, 1271 (1993).
5. Schmidt, V., *Electron Spectrometry of Atoms using Synchrotron Radiation*, Cambridge University, Cambridge, 1997.
6. Forrey, R. C., Yan, Z.-C., Sadeghpour, H. R., and Dalgarno, A., *Phys. Rev. Lett.*, **78**, 3662 (1997).
7. Kornberg, M. A., and Miraglia, J. E., *Phys. Rev. A*, **49**, 5120 (1994).
8. Forrey, R. C., Sadeghpour, H. R., Baker, J. D., MorganIII, J. D., and Dalgarno, A., *Phys. Rev. A*, **51**, 2112 (1995).
9. Krivec, R., Amusia, M. Y., and Mandelzweig, V. B., *Phys. Rev. A*, **63**, 052708 (2001).
10. Spielberger, L., Jagutzki, O., Dörner, R., Ullrich, J., Meyer, U., Mergel, V., Unverzagt, M., Damrau, M., Vogt, T., Ali, I., Khayyat, K., Bahr, D., Schmidt, H. G., Frahm, R., and Schmidt-Böcking, H., *Phys. Rev. Lett.*, **74**, 4615 (1995).
11. Ahopelto, J., Rantavuori, E., and Keski-Rahkonen, O., *Phys. Scr.*, **20**, 71 (1979).
12. Diamant, R., Huotari, S., Hämäläinen, K., Kao, C. C., and Deutsch, M., *Phys. Rev. A*, **62**, 052519 (2000).
13. Kanter, E. P., Dunford, R. W., Krässig, B., and Southworth, S. H., *Phys. Rev. Lett.*, **83**, 508 (1999).
14. Pelicon, P., Čadež, I., Žitnik, M., Ž. Šmit, Dolenc, S., Mühleisen, A., and Hall, R. I., *Phys. Rev. A*, **62**, 022704 (2000).
15. Kanngießer, B., Jainz, M., Brünken, S., Benten, W., Gerth, C., Godehusen, K., Tiedtke, K., van Kampen, P., Tutay, A., Zimmermann, P., Demekhin, V. F., and Kochur, A. G., *Phys. Rev. A*, **62**, 014702 (2000).
16. Krause, M. O., Carlson, T. A., and Moddeman, W. E., *J. Phys. (Paris)*, **32**, C4–139 (1971).

17. Albiez, A., Thoma, M., Weber, W., and Mehlhorn, W., *Z. Phys. D*, **16**, 97 (1990).
18. Chen, M. H., *Phys. Rev. A*, **44**, 239 (1991).
19. Svensson, S., Eriksson, B., Mårtensson, N., Wendin, G., and Gelius, U., *J. Electron Spectrosc.*, **47**, 327 (1988).
20. Oura, M., Yamaoka, H., Kawatsura, K., Takahiro, K., Takeshima, N., Zou, Y., Hutton, R., Ito, S., Awaya, Y., Terasawa, M., Sekioka, T., and Mukoyama, T., *J. Phys. B*, **35**, 3847 (2002).
21. Huotari, S., Hämäläinen, K., Kao, C.-C., Diamant, R., Sharon, R., and Deutsch, M., *NSLS Activity Report*, Brookhaven National Laboratory, 2001, p. 2-59.

Multiple Electron Scattering in Ion-Atom Collisions: Fermi-Shuttle Acceleration in Ionization

B. Sulik[1,2], Cs. Koncz[1], K. Tőkési[1], A. Orbán[1], Á Kövér[1], S. Ricz[1],
N. Stolterfoht[2], R. Hellhammer[2], J.-Y. Chesnel[2,3], P. Richard[4],
H. Tawara[4], H. Aliabadi[4], and D. Berényi[1]

[1]*Institute of Nuclear Research (ATOMKI), H--4001 Debrecen, P.O.Box 51, Hungary*
[2]*Hahn-Meitner Institute, Glienickerstr. 100, D-14109 Berlin, Germany*
[3]*Centre Interdisciplinaire de Recherches Ions Lasers, Unité Mixte CEA-CNRS-ISMRA-Université Caen, France*
[4]*J. R. Macdonald Laboratory, Manhattan, Kansas, U.S.A. 66506*

Abstract. We present experimental evidences for consecutive multiple projectile-target-projectile-... (P-T-P-... or T-P-T-...) scattering of the electrons liberated in ion-atom collisions. The highest order of the observed multiple scattering sequences is a quadruple P-T-P-T scattering. We observed the P-T scattering in different collisions, and found strong indications for the accelerating T-P scattering. Distinct signatures of the P-T, P-T-P and P-T-P-T multiple electron scattering contributions to the high-energy part (300 - 3400 eV) of the double differential electron spectra have been separated and identified with the help of reference measurements and auxiliary calculations in single C^+ + Xe collisions at 150 and 233 keV/u impact energies. In the collisions of few keV energy ions with inert gas atoms, preliminary results indicate the presence of even higher order scattering contributions.

INTRODUCTION

Ionization process in atomic collisions can be associated with multiple electron scattering by the projectile and the target cores. This kind of mostly forward-backward scattering can accelerate the light electrons up to relatively high energies due to the repeated collisions with the incoming heavy projectile ion. It is usually referred to as Fermi-shuttle acceleration. In recent years, several laboratories have started to study that process [1-17] in ion impact ionization. Originally, Fermi [18] had proposed the mechanism as a possible origin of cosmic rays. Giant magnetic fields, moving against each other in space, can accelerate charged particles to very high energies in long sequences of reflections. It was later shown that this kind of ``ping-pong" game can

CP652, X-Ray and Inner-Shell Processes: 19th International Conference on X-Ray and Inner-Shell Processes
edited by A. Bianconi, A. Marcelli, and N. L. Saini
© 2003 American Institute of Physics 0-7354-0111-X/03/$20.00

also be played with other ``paddles", such as the microscopic fields of atoms, molecules or clusters [19-21] where even a short sequence of reflections or scattering events might be of great interest.

Let us consider the sequence of backscatterings of a liberated electron between the incoming projectile ion (moving with a velocity V) and the target core. The velocity of the electron is increased by approximately $2V$, in every $180°$ elastic scattering with the incoming projectile, while only the direction of the electronic motion is changed by the target field. This follows directly from the kinematics of small particles (m) elastically scattered by heavy centers (M) with mass ratios $M/m \gg 1$. In Ref. [16], we introduced the shorthand P and T to denote the electron-projectile and electron-target scatterings, respectively. With this notation, the so-called binary encounter (BE) ionization of the target [22,23] is denoted by P, while projectile ionization (electron loss [24]) by T. Longer sequences can be referred to as for example P-T-P or T-P-T-P.

Target ionization sequences start with a P process, and may emit electrons up to the velocity $2nV$ in both forward and backward directions relative to the projectile motion. Here, n is the number of encounters with the projectile. For electron loss sequences (starting with T), the corresponding velocity is $(2n+1)V$. The observation of such hot electrons is of fundamental importance for basic research in collision physics. They are emitted in particular ionization processes, which can be associated with specific three-body states. Moreover, since fast electrons form a long-range secondary particle radiation, this kind of acceleration process may be important in ion-matter interactions, and therefore, relevant in many applications such as cancer therapy, ion track formation, or the modification of material properties (see e.g., [7, 9, 10, 25]).

The first theoretical prediction of the Fermi-shuttle acceleration of electrons was published by Wang *et al.* [21]. They applied a quantum mechanical model with zero-range potentials to two-center collisions. Subsequently, both classical [9, 14, 26, 27] and quantum mechanical calculations [27, 28] have been applied to the ``ping-pong" scenario. Some qualitative features can be well described by non-perturbative classical models, e.g., classical trajectory Monte Carlo calculations (see in [9,14-17] or other classical scenarios [26,27]. Since the free electron diffraction patterns play a significant role in the formation of Fermi-shuttle sequences [6-9, 14-17, 26-31], classical models can provide only qualitative predictions. No attempt has been made for a full theoretical treatment of Fermi shuttle acceleration in atomic collisions yet, but some simple properties can be derived from general considerations (see below).

In collisions of free atomic species, observable yield of hot electrons were reported in slow ion-Ar collisions as early as 1979 [32]. Fermi-shuttle acceleration was systematically searched for [5], and first identified by Suarez *et al.* [6], as the T-P process in H + He collisions. The first observation of the P-T process in gaseous targets was reported later by Bechtold *et al.* [7, 8] in the significantly faster collisions with 5.9 MeV/u U^{27+} ions (see also [9]). Larger yields of fast electrons have been observed in ion-solid collisions [1-4, 10-12] in a wide (keV-GeV) range of impact energies. This can be explained by the high density of atomic centers in solids [9-11, 33]. Recently, Fermi-shuttle acceleration was successfully invoked to explain the high-velocity tail of the spectrum of electrons emitted in the remarkably fast 45

MeV/u Ni + Au collisions by Lanzano *et al.* [12]. Single collisions, however, are rather rare events in ion-solid experiments. To investigate the mechanism in more detail, it looks most promising to proceed with thin gaseous targets.

In this work, we present an overview of our observations on the P-T process [14, 15], the search and indications [16] for the T-P and P-T processes in different collision systems, and the first observation of the triple (P-T-P), and the quadruple process P-T-P-T [17] in single collisions of ions with free atoms. In the following, we start with a short overview of the physical picture and the theoretical methods and models, which can be used to estimate the quantitative features of the processes. Only a brief description of the experimental arrangements is given, where we present unpublished results, or where such details are considered to provide relevant information.

GENERAL CONSIDERATIONS

The backscattering of the electrons within a narrow cone around 180° is a necessary condition of forming longer sequences. Therefore, one can consider the effective area of an "ion-core paddle" as roughly proportional to the cross section for 180° elastic electron scattering by the core. Since, in a first approximation, it is proportional to the square of the atomic number Z^2 [7], one might expect that the probability of longer sequences strongly increases with increasing atomic number of both the projectile and the target. The obeservability of a P-T-P process on the "background" of a single scattering P process, e.g., can be characterized by the ratio $\sigma_{P\text{-}T\text{-}P} / \sigma_P \sim Z^2_{proj}$ [17].

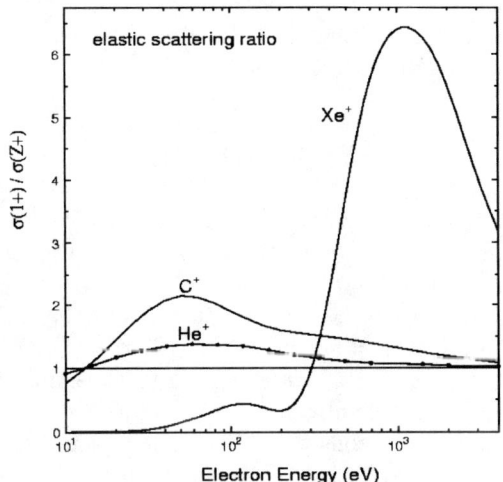

FIGURE 1. The role of the core potentials in increasing the probability of the Fermi-shuttle process. Single differential free-electron elastic scattering cross sections at 180° divided by the corresponding Rutherford cross sections calculated for the *bare* nuclei. Depending on the electron velocity, both screening and enhancement processes may undergo.

However, as it was widely recognized [6-9, 14-17, 26-31], the screened fields of the collision partners may enhance the forward-backward focusing, compared to the Coulomb fields of bare ions. In Fig. 1, we compare the elastic free-electron scattering cross sections on three different single ionized ionic cores with the corresponding Rutherford cross sections calculated for the *nuclear* charges. It is clearly seen that in wide ranges of the electron velocity, the presence of the electron cloud strongly enhances the backscattering probabilities, making the formation of Fermi-shuttle sequences significantly more probable.

In the experimental spectra, one can search for distinct structures at the mean velocity of the electrons emitted e.g., in P_n-T_m (n-1 $\leq m \leq n$) sequences, where a general $v(\theta)$ formula [17] can be derived from simple kinematics [17]:

$$v = \begin{cases} V(\cos\theta + \sqrt{\cos^2\theta + 4m + 4m^2}) & \text{if } m < n \\ 2nV & \text{if } m = n \end{cases} \tag{1}$$

Since there is no quantitative theory exists for the relative yields, the search for even small enhancements, which follow the above equation, is a reasonable way to study the process. We note that the mean velocity given by Eq. (1) corresponds to a start with a target electron initially ``at rest''. Since a backscattering needs a finite time, the length of possible sequences is limited for such electrons. Therefore, the high-momentum tail of the initial state distribution (Compton profile) could be favored, and a shift of the observed mean velocity to higher values can be expected for longer sequences.

Multiple scattering sequences are more likely to be formed in small impact parameter collisions. This intuitive picture has been confirmed by CTMC calculations [16, 17]. We also have found by individual analysis of the CTMC trajectories, that a small impact parameter backscattering can be easily continued by a second scattering, if the enhancement factor for the next collision (see Fig. 1) is large.

EXPERIMENTS

In the experiments, double differential cross sections for electron emission were measured. In most of the measurements, we took electron electron spectra for energies of 20-3400 eV and for observation angles θ=0-180°. The experimental method is given in more details in Refs. [14-17], so only a brief description is given here. In the majority of the measurements, performed at ATOMKI, Debrecen, beams of He^+ ions with the lowest available energy (233-keV/u), and C^+ ions with 150- and 233-keV/u energies were directed onto He, Ne, Ar and Xe gas jet targets. The electron spectra were collected simultaneously in 13 angular channels with a triple-pass electrostatic spectrometer (ESA-21) [34]. For cross sections $\sigma < 10^{-23}$ cm^2/eV/sr, the absolute experimental uncertainty was less than 40 %. This has been estimated from reference data [35], and the reproducibility of the spectra (which was better than 15

%). For testing single collision conditions, the effective target density was typically varied between 10^{12} and 10^{13} atom/cm^3, and no difference between the spectra were found within the statistical uncertainties.

Our first experiments [14,15] have been performed in a limited range of electron energies (20-550 eV), but we already could observe an enhanced intensity for the P-T process at 180° in 150 keV/u C$^+$ + Ne and Ar collisions. We concluded that a third scattering should be likely in this case. The extension of the measured electron energy range up to 2500 eV, however, did not allow to see any further significant structures in the above spectra. A classical trajectory Monte-Carlo (CTMC) analysis [16] provided a possible explanation for the structureless continuous spectra found in the measurements. The results of the analysis are illustrated in Fig. 2, where we compare the calculated CTMC contribution separately for the ionization of the target and the projectile, for C$^+$ + Ne collisions. Since CTMC is a classical (and therefore strongly approximate) but non-perturbative theory, it provides the expected double scattering structures for both ionization components. They are clearly seen at V and 3V for the projectile ionization (electron loss) process, and at 2V and 4V for the target ionization contribution. The sum of the two contributions, however, appears to be a smooth curve without any significant structure. It compares very well with the experimental data, providing only an indication but not an evidence for the presence of the multiple scattering components.

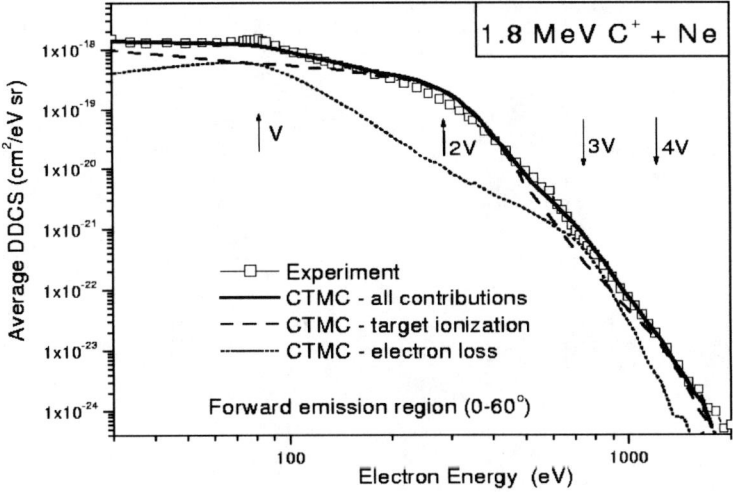

FIGURE 2. Analysis of the CTMC contributions to the forward electron emission and comparison with experiment. The projectile ionization (electron loss) part shows shoulders at the projectile velocity V and at 3V, while the target ionization contribution at 2V and 4V. The sum is almost structureless, in agreement with experiment (see [16]).

Searching for the way out of the above limitation, we measured collision systems at higher projectile velocities. The effective width of above shoulders and peaks are practically determined by the Compton profiles of the electronic orbitals playing role in the process. Accordingly, one might expect that at higher velocities (similarly to the binary encounter peak at zero degree), the studied structures become sharper, and they separate better. Therefore, we performed measurements in the J. R. Macdonald Laboratory at Kansas State University, at higher projectile energy (0.5 MeV/u) to study the T-P process in B^{q+} + He, Ne collisions. The projectile ions were provided by the KSU tandem Van de Graaff accelerator, and directed to a gas cell, where the target gas pressure was varied in a vide range to check single collision conditions. The spectra of the emitted electrons has been collected at zero degree relative to the beam direction by a two-stage tandem electron spectrometer (see e.g. [23]).

The results provided an indication for the presence of a T-P scattering component at 3V. However, clear evidence could also not be provided for these collision systems, for two reasons. One was that the spectra were taken at rather high electron energies (up to 12 keV), where we could not ensure satisfying background conditions. The other reason was that we could compare only with theories. All the ionization components have been calculated within the 1st Born approximation framework, except the binary encounter region, where we applied an improved impulse approximation [31]. All the calculations have been performed by using realistic (Hartree-Fock-Slater) wave functions (CTMC calculations are still running). We illustrate the results in Fig. 3, where the measured total cross electron emission section is compared to the sum of the calculated cross section.

There are two distinct peak-like structures observed in Fig. 3. The peak at 2V belongs

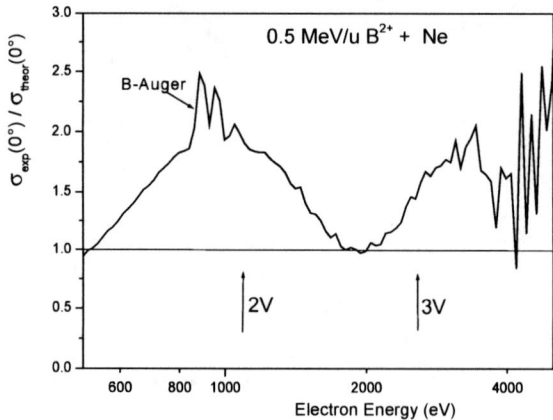

FIGURE 3. The ratio of the experimental ant theoretical cross sections at zero degree relative to the beam direction in 0.5 MeV/u B^{2+} + Ne collisions. The peak at and above $3V$ may be associated with the T-P process (projectile ionization by the target core followed by a 180° scattering on the projectile core).

to the excess intensity in the binary encounter peak due to the dispersion in the screened projectile field (see Fig. 1), which are not properly treated by 1st order theories. The other peak at and above 3V may be associated with the T-P process.

The other way to search for higher orders of the Fermi-shuttle process was to focus the interest in the intermediate velocity region, but turn to heavier collision partners. This way, we finally could find the expected evidences. In the following, we briefly report this study [17]. To increase the length of the sequences, xenon was selected as a relatively heavy inert gas target. It was important to find an independent reference system, where we can get data without significant multiple scattering contributions. Therefore, we used both He$^+$ and C$^+$ projectiles, to get less and more multiple scattering contribution, respectively (the estimated σ_{P-T-P} / σ_P ratio was 11 times larger for the carbon than for the helium ion).

Double differential electron emission cross sections measured for 150 keV/u (V=2.45 au) C$^+$ + Xe collisions are shown in Fig. 4. We would like to note that more

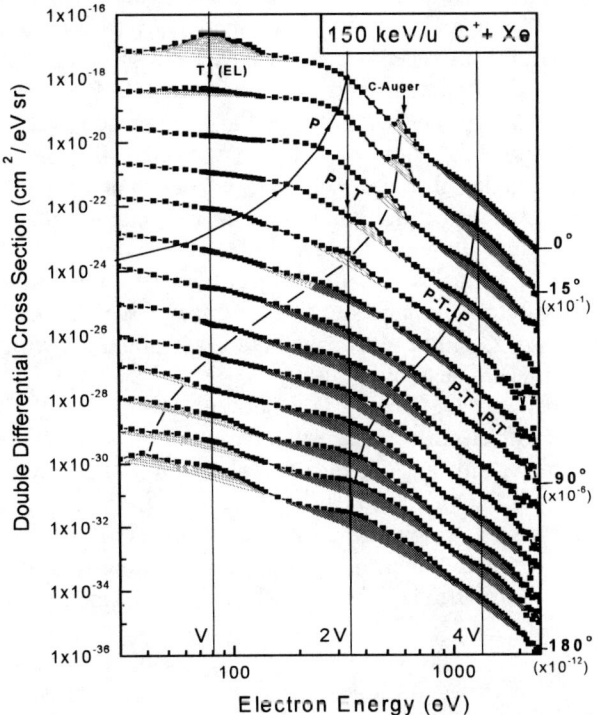

FIGURE 4. Experimental double differential cross section for electron emission at 0, 15,...165, 180°, in 150 keV/u C$^+$ +Xe collisions. Lines with arrows indicate the expected location of the T, P, P-T, P-T-P, and P-T-P-T ridges. The arrows are drawn to represent the ``direction of acceleration''. Dominant single scattering target ionization yields (including the P process) are not shaded. Proposed multiple scattering contributions are marked by dark gray, while the C K-Auger and T components by light gray shading [17].

than six orders of magnitude variation in the cross sections was measured without background limitations. Low energy electron emission is dominated by target ionization, while the electron loss (EL) peak (T process) appears at the so-called cusp energy, 81.8 eV. The carbon K-Auger group originating from the moving projectile is clearly seen shifted in energy with observation angle as expected. The xenon MNN Auger group in the energy region of about 500-550 eV was found to be at the limit of observability in the reference He^+ + Xe spectra, and its relative contribution for C^+ is negligible. The energy position of the BE peak follows from Eq. 1 (with $n=1$, $m=0$), forming the single scattering binary ridge, indicated by the P curve. The double scattering P-T ridge ($n=1$, $m=1$) at the constant energy of 328 eV is clearly observable at backward angles. The next acceleration phase ($n=2$, $m=1$) forms the P-T-P ridge, which is rather pronounced at forward angles at the expected velocity of $4V$. A closer inspection at backward angles also shows an enhancement around $4V$ which can be associated with the P-T-P-T process.

In Fig. 4, in agreement with our consideration, a shift toward higher energies can be seen for the proposed P-T and P-T-P-T components at backward angles. This is just the opposite case than the observed decrease of peak energies in the first scattering due to finite binding energies (e.g., for the 0° BE peak [36]). In Fig. 4, multiple scattering processes starting with electron loss (T) might also be present, but they do not show up as any distinct features. We note, that a T-P structure should be very close to the C K-Auger group at all angles, thus decreasing the chance to observe it.

To identify the signatures of multiple scattering, one should separate them from the single scattering contributions. In order to estimate the latter components, we measured reference spectra with He^+ projectile, where only a negligible P-T-P contribution was expected compared to the C^+ impact. We also performed 1st-Born calculations for the ionization of all subshells of the xenon atom with both projectiles, according to the formalism described in Ref. [37].

An excellent agreement has been found between the measured He^+ data and the 1st-Born calculations at electron energies above the 0° binary encounter peak (~500 eV). Even the calculated subshell effects were observed in the experimental data [17]. Hence, instead of using the measured He^+ spectra as a direct reference, we used the 1st Born data to represent the single scattering target ionization contribution. This also worked well for C^+ impact, where only the carbon K-Auger group and the marked multiple scattering contributions (see Fig. 4) deviated from the theory in the high electron energy region.

It is seen in Fig. 4, that the cross sections vary with electron energy by several orders of magnitude, whereas multiple scattering seems to increase the yields only by a factor of 2-4. Therefore, to observe the multiple scattering contributions, we removed the strong energy variations of the cross sections by displaying the ratio of the experimental cross sections and the 1st-Born results calculated for the target. For the 150 keV/u C^+ + Xe collision system, this ratio is shown in Fig. 5. The energy and angular coordinates of Fig. 4 have been transformed to electron velocity components parallel (v_z) and perpendicular (v_x) to the beam direction (similarly to Refs. [5,7,33], and both v-components have been divided by the the projectile velocity V. In this normalized velocity space, an elastic scattering by the target and the projectile center is represented by a circle centered at the $t(0,0)$ and the $p(1,0)$ points, respectively.

The T process in Fig. 5 is strong at forward $(1,0)$ and rather weak at backward $(-1,0)$ directions. The C K-Auger electrons show up as a circle centered at p. The P process shows only a moderate enhancement at forward angles. This deviation from the 1st-Born results is due to the non-Coulomb projectile field [29, 30]. The P-T process (which is only partially reproduced by the 1st-Born theory [31]) is also clearly observable at backward angles. The most significant structure in Fig. 6 is the circular P-T-P ridge centered at p, with an intense forward component at $v_z=4V$ Finally, the P-T-P-T peak is also clearly identifiable at backward angles, $v_z=-4V$. In summary, the contour plot provides clear evidence for double, triple and quadruple scattering of the target electrons. We note that the corresponding plot for the faster (233 keV/u) C^+ ion (not shown) gives an entirely similar pattern.

The integrated cross sections (see Fig. 2) over the forward (0-60°) and backward (120-180°) observation angle regions have also been analyzed in Ref. [17], where the experimental cross sections are compared to CTMC results. In both data sets, distinct structures appeared at $2V$ in backward angles (P-T), and close to $4V$ in both forward (P-T-P) and backward angles (P-T-P-T). Surprisingly large fractions seem to be backscattered by the xenon core in the fourth scattering, as a possible indication for a trapping mechanism [26,27].

Finally, we would like mention here that our preliminary data taken in collisions of 45 keV N^{3+} ions with Ne and Ar target atoms show a strong enhancement of the

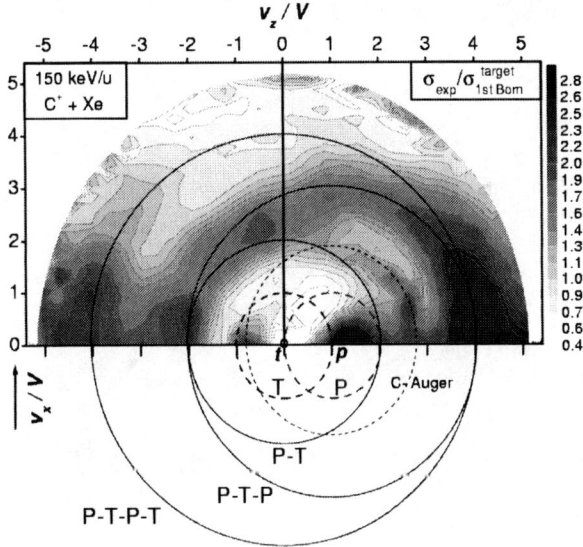

FIGURE 5. Contour plot for the ratio of experiment and 1st-Born target ionization theory. The normalized electron velocity components v_x and v_z are parallel and perpendicular to the beam direction, respectively [17].

electron intensities in the electron velocity region of $6V$-$10V$ at forward observation angles. The measurements have been performed at the beamline of an electron cyclotron resonance (ECR) ion source at the Ionen Strahl Labor (ISL), of Hahn-Meitner Institut Berlin. If confirmed, these data would indicate the existence of longer multiple scattering sequences at lower collision velocities. This way, we could enter the "matching" region for multiple electron scattering and highly excited molecular states.

In conclusion, the evidence for longer electron scattering sequences between the projectile and the target has been found in single ion-atom collisions. Till now, triple and quadruple scattering of the electrons ejected in the ionization of the target has been observed in intermediate velocity single collisions of ions with atoms. Evidence for the above processes has been supported with the help of reference measurements using He^+ impact, and different calculations. We also got some indication that longer scattering sequences could be significant at lower (0.3-1 au) projectile velocities.

ACKNOWLEDGMENTS

We acknowledge support from the Hungarian Scientific Research Fund (OTKA, Grant No: T032942, in part T032306, and the Infrastructure Grant M27839), the Hungarian-German Intergovernmental S&T Collaboration (D17/99), and the US Department of Energy. B. S. expresses thanks to Hahn-Meitner Institute for the excellent conditions during his sabbatical period in Berlin.

REFERENCES

1. R. A. Baragiola et al., *Phys. Rev. A* **45**, 5286 (1992).
2. D. Schneider, G. Schiwietz and D. DeWitt, *Phys. Rev. A* **47**, 3945 (1993).
3. Y. Yamazaki et al., RIKEN *Accel. Prog. Rep.* **24**, 54 (1990), & **26**, 66 (1992). & **27**, 60 (1993).
4. R. A. Sparrow, R. E. Olson and D. Schneider, *J. Phys. B* **28**, 3427 (1995).
5. S. Suarez, R. O. Barrachina, W. Meckbach, *Phys. Rev. Lett.* **77**, 474 (1996).
6. S. Suarez et al., *Nucl. Instr. Meth. B* **124**, 358 (1997).
7. U. Bechthold, et al., *Phys. Rev. Lett.* **79**, 2034 (1997).
8. U. Bechtold, et al., *Nucl. Instr. Meth. B* **143**, 441 (1998).
9. C. O. Reinhold et al., *Phys. Rev. A* **58**, 2671, (1998).
10. H. Rothard, *Nucl. Instr. Meth. B* **146**, 1 (1998).
11..H. Rothard, D. H. Jakubassa-Amundsen and A. Billebaud, *J. Phys. B* **31**, 1563 (1998).
12..G. Lanzano et al., *Phys. Rev. Lett.* **83**, 4518 (1999).
13. G. Lanzano et.al., *Phys. Rev. A.* **63**, 032702 (2001).
14. B. Sulik et al. *Nucl. Inst. and Meth.* **B** 154, 281 (1999).
15. B. Sulik et al. *Physica Scripta* T **80**, 338 (1999).
16. B. Sulik et.al. *Physica Scripta* T **92**, 463 (1999).
17. B. Sulik et al. *Phys. Rev. Lett.* **88**, 073201 (2002).
18. E. Fermi, *Phys. Rev.* **75**, 1169 (1949).
19. R. J. Beuler, G. Friedlander and L. Friedman, Phys. Rev. Lett. 53, 1292 (1989).
20. J. Burgdörfer, J. Wang and R. H. Ritchie, Physica Scripta 44, 391 (1991).

21. J. Wang, J. Burgdörfer and A. Bárány, Phys. Rev. A 43, 4036 (1991).
22. D. H. Lee, et al., Phys. Rev. A 41, 4816 (1990).
23 N. Stolterfoht, R.D. DuBois and R.D. Rivarola, *Electron Emission in Heavy Ion-Atom Collisions*, (Springer, Berlin, 1997).
24. F. Drepper and J. S. Briggs, J. Phys. B 9, 2063 (1976).
25. R. Baragiola (Ed.), *Ionization of Solids by Heavy Particles*, NATO ASI Ser. B 306, (Plenum, New York, 1993).
26. Mario M. Jakas, Phys. Rev. A 52, 866 (1995).
27. Mario M. Jakas, Nucl. Instr. Meth. B 111, 255 (1996).
28. S. Yu. Ovchinnikov and J. H. Macek, Nucl. Instr. Meth. B 154, 41 (1999).
29. P. Richard, et al., J. Phys. B 23, L213 (1990).
30. T. J. M. Zouros, et al., Phys. Rev. A 53, 2272 (1996).
31. Cs. Koncz, B. Sulik, J. Phys B 32, 5009 (1999).
32. N. Stolterfoht and D. Schneider, IEEE Trans. Nucl. Sci. 26, 1130 (1979).
33. C.O. Reinhold and J. Burgdörfer, Phys. Rev. A 55, 450 (1997).
34. D. Varga, et al., Nucl. Inst. and Meth. A 313, 163 (1992).
35. M. E. Rudd, L. H. Toburen, and N. Stolterfoht, Atomic Data and Nuclear Data Tables 23, 405 (1979).
36. H. I. Hidmi et al., Phys. Rev. A 48, 4421 (1993).
37. R. D. DuBois and S. T. Manson, Phys. Rev. A 42, 1222 (1990).

X-ray Emission from Highly Charged Heavy Ions Studied at Storage Rings

X. Ma [1], Th. Stöhlker[2], F. Bosch[2], A. Gumberidze[2,3], C. Kozhuharov[2],
A. Muthig[2], P.H. Mokler[2], A.Warczak[4]

[1]Institute of Modern Physics, Lanzhou, 730000, China
[2]Gesellschaft für Schwerionenforschung, 64291, Darmstadt, Germany
[3]Tbilisi State University, 380028 Chavchavadze Avenue, Tbilisi, Georgia
[4]Institue of Physics, Jagiellonian University, Cracow, Poland

Abstract. Radiative electron capture at low projectile energies is studied via angular differential cross sections for collisions of bare uranium with low-Z target atoms. Our results show that for high-Z systems relativistic effects such as spin-flip transitions show up in an unambiguous fashion which still persist even in the low-energy domain. Moreover, following REC into the $2p_{3/2}$ state a strong alignment of this level was observed by measuring the angular distribution of the Lyα_1 transition in H-like uranium. Here, an interference between the leading **E1** decay channel and the much weaker **M2** multipole transition gives rise to a remarkable modified angular distribution of the emitted photons. For the particular case of hydrogen-like uranium the former variance of the experimental data with theoretical findings is removed when this **E1/M2** multipole mixing is taken into account. Finally, with respect to atomic structure studies, a very recent experiment will be discussed aiming on a precise determination of the electron-electron QED contribution to the groundstate ionization potential in He-like uranium.

INTRODUCTION

X-ray transitions of highly charged heavy ions produced in encounters with low-Z atoms are an important tool for atomic structure and collision studies in the realm of strong Coulomb fields [1,2]. In the past ten years collision investigations for highest projectile charges (e.g. U^{92+}) conducted at the ESR storage ring addressed in particular radiative electron capture (REC), the time-reversed photoionization process in ion atom collisions[2,3]. In general, the experimental results obtained for total as well as angular differential cross-sections are found to be well described by rigorous relativistic calculations [4]. Most remarkably, the experiments achieved to identify unambiguously spin flip contributions to the angular differential cross sections for REC into the ground state of bare uranium [3-6]. The latter emphasizes the sensitivity of such studies to magnetic and retardation effects appearing in relativistic ion-atom collisions. Beside the REC transitions, the angular differential photon distributions at high-Z allowed us even to identify the admixtures of higher order multipole transitions to the leading dipole term for transitions between bound states [7].

Heavy ion storage rings also provide favorable experimental conditions for a test of quantum electrodynamics (QED) in strong Coulomb field. This has been demonstrated

CP652, *X-Ray and Inner-Shell Processes: 19th International Conference on X-Ray and Inner-Shell Processes*
edited by A. Bianconi, A. Marcelli, and N. L. Saini
© 2003 American Institute of Physics 0-7354-0111-X/03/$20.00

within an series of experiments performed at the ESR. For the case of the 1s Lamb shift in hydrogenic uranium, the achieved precision of ±13eV [8] manifests a substantial improvement by almost one order of magnitude compared to a former experiment conducted at the BEVALAC accelerator [9]. Beside the one-electron systems, the two-electron ions are of particular interest for atomic structure studies as they represent the simplest multi-electron systems. Very recently, substantial progress in the theory of two-electron systems has been achieved and the predictions for the two-electron ground state binding energies are expected to be as precise as the one for the one-electron systems. These predictions can now be probed uniquely at the ESR storage ring by measuring the Radiative Recombination transitions (RR) into H-like ions relative to the one into the bare species [10]. This technique introduced at the Super-EBIT at Livermore isolates the pure two-electron contributions and the one-electron contributions such as the nuclear size effect cancel out almost completely [11]. Here, only a relative precision of about 10^{-3} is required to probe sensitively higher-order contributions such as the screened Lamb shift.

In this paper we will report and summarize the results of the latest x-ray related experiments performed at the ESR storage ring. In addition the project of a new heavy-ion cooler-ring which is presently under construction in Lanzhou, China will be discussed.

EXPERIMENTAL ARRANGEMENT AT THE ESR

The experimental studies reported in this review were all carried out at the heavy ion storage ring ESR at GSI in Darmstadt. The heavy ions are accelerated up to a few hundred MeV/u in the synchrotron SIS, then extracted and transferred to the storage ring. Before injected into the storage ring, the ions pass through a copper foil with

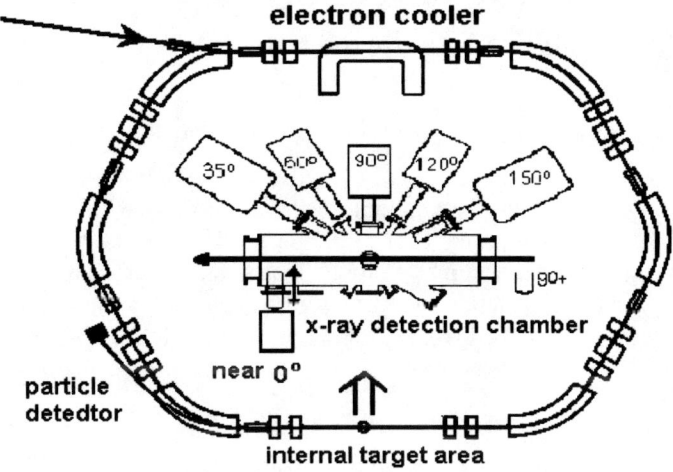

FIGURE 1. Layout of the ESR storage ring. The scattering chamber used for atomic physics experiments is depicted in addition.

proper thickness, and the desired charge state is selected. In the electron cooler the ions are cooled by a copropagating cold electron beam of about 200 mA. Up to 10^8 ions are stored and cooled forming a DC ion beam with an approximate diameter of 2 mm and a longitudinal momentum spread of about 5×10^{-5} [12]. After cooling, if needed, the ions can be decelerated down to an energy of close to ten MeV/u [13]. The layout of the storage ring ESR is depicted in figure 1.

At the internal target area of the ring [14], the scattering chamber used for atomic physics experiments is installed which allows us to record the x-rays emitted from the beam-target interaction volume at different observation angles, ranging from close to 0° to 150° with respect to the beam direction [5]. In the experiments planar germanium detectors are used, isolated from the ultra-high vacuum environment of the ring either via 50 μm stainless steel or 100μm beryllium foils. Except for the near 0° detector, each detector is equipped with an x-ray collimator of a narrow angular acceptance in order to reduce the Doppler broadening. The emitted x-rays are registered in coincidence with charge-changed projectiles. They are collected in the plastic scintillator and/or gas-filled multiwire chamber installed down stream of the target chamber behind the next dipole magnet. In figure 2, a typical x-ray spectrum recorded for $U^{92+} \rightarrow N_2$ collisions at 88 MeV/u is displayed. The spectrum was measured at the observation angle of 150° in coincidence with the down-charged U^{92+} ions. For details of the spectrum see the next section in this paper.

At the electron cooler the x-rays are produced by radiative recombination transitions of free electrons into the ground state and the excited states of the stored uranium ions [15]. In our measurement a high purity germanium detector with four segments was used at an observation angle close to zero degree. Again, a coincidence technique is applied between the x-rays and the down-charged ions.

RADIATIVE ELECTRON CAPTURE IN LOW ENERGY COLLLISIONS BETWEEN BARE URANIUM IONS AND LOW-Z TARGET

Radiative electron capture is considered as the time-reversed process of photoionization [12], and has been used to study the phoionization phenomena for the heaviest system of hydrogen like uranium ions by Stöhlker and collaborators [5,16]. It turns out that REC is a unique tool for precision studies of photoionization process in high-Z domain. These studies already proved that at high energies (310 MeV/u) spin-flip transitions are an important contribution to the deviation from a $\sin^2\theta$ distribution at forward angles. Due to the inverse kinematics, the origin of the observed spin-flip transitions can be identified as events related to large angle backward scattering in photoionization.

Most important, no corrections due to electron scattering occurring in solid targets are required, leading in conventional photoionization studies for high-Z elements to a considerable broadening of the electron emission angle [17]. Therefore in the high-Z regime, almost none of the few fine structure resolved photoionization data can be

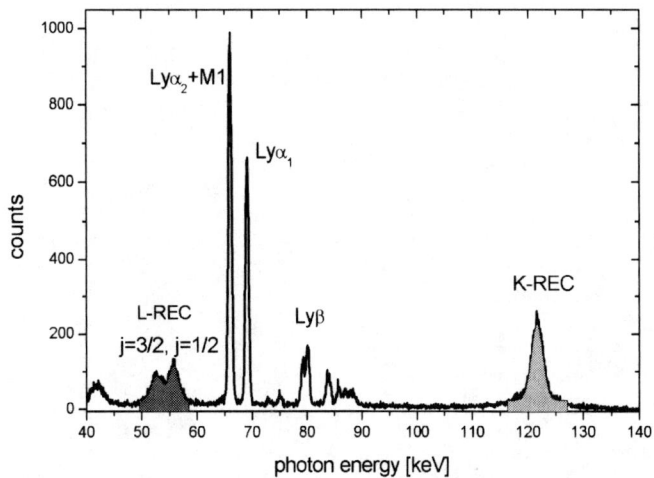

FIGURE 2. X-ray spectrum recorded with a planar germanium detector at an observation angle of 150° (88MeV/u $U^{92+} \rightarrow N_2$) [16].

compared with theory since no corrections due to multiple scattering were performed. The REC experiment has no such restrictions and in particular, it allows us to extend the photoionization studies to much lower photon energies than available for the direct channel. In the recent experiment we studied the photoionization process in the low-energy domain by the angular-differential study of its time-reversed process at the jet-target using decelerated bare uranium ions [16]. For this purpose, bare uranium ions were actively decelerated in the ESR to the energy of 88 MeV/u. For the direct channel this corresponds to a photoelectron energy of 48 keV, which is small compared to the ground state ionization potential of H-like uranium of about 132 keV. The typical x-ray spectrum (recorded at 150° in the collision) is given in Fig. 2.

In the spectrum the REC lines of interest show up as broad peaks due to target electron Compton profile (see the shaded area in figure 2), whereas the characteristic Lyman transitions are the most intense lines in the spectrum. The REC distribution is normalized to the simultaneously observed **Lyα_2 + M1** transition because it is exactly known to follow an isotropic emission pattern in the emitter frame. This technique greatly facilitated the angular distribution analysis and almost all possible systematic uncertainties cancel out by normalization technique. It also allows us to obtain directly angular-differential cross-sections for the emitter frame. The experimental angular differential cross sections for K-REC in the emitter frame are presented in Fig. 3 (solid circles).

Although we are dealing with the low-energy domain, the distribution still shows a considerable backward peaking corresponding to a strong forward peaking of the electron emission in photoionization (see upper x-axis in Fig. 3). This points out the importance of strong retardation corrections and of the presence of strong higher-order multipole contributions. Therefore our data support the theoretical predictions for the

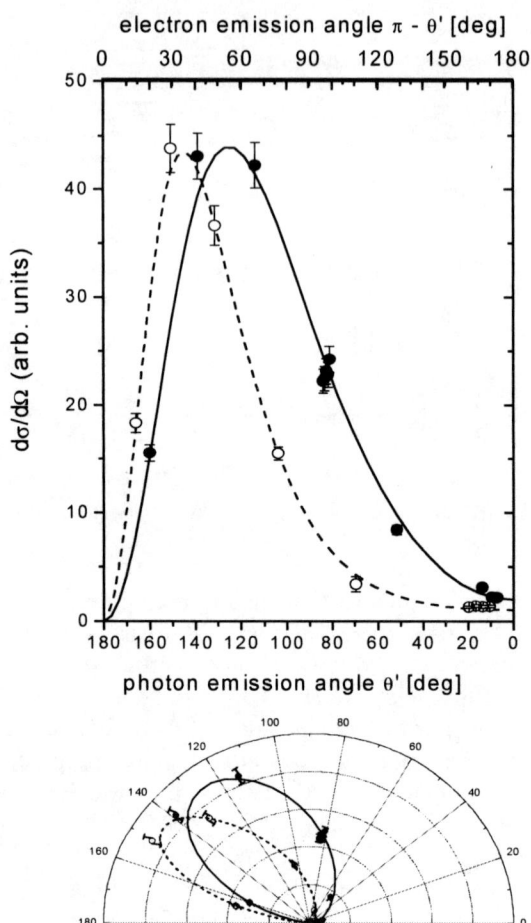

FIGURE 3. K-REC angular distribution (solid circles) versus the emission angle θ' (bottom x-axis) in the emitter frame. The data are from the collision of $U^{92+} \rightarrow N_2$ at 88 MeV/u [16]. In the upper plot, the x-axis at the top refers to the corresponding photoelectron emission angle in photoionization of H-like uranium (photon electron energy 48 keV). The full line: complete relativistic calculations, open circles and dashed line: the results of experiment and theory for 310 MeV/u beam energy, respectively. Lower part of the figure is a polar presentation of the angular distribution.

high-Z regime that the relevance of higher-multipole contributions even persist close to the threshold for photoionization [18].

Indeed, the data are in good agreement with rigorous relativistic calculations [19] (see full line in Fig. 3). Most remarkably, both theory and experiment show a non-vanishing cross-section close to 0°, which proofs that magnetic contributions are still present in the low-energy domain. This also emphasizes the sensitivity of the applied method since magnetic transitions contribute to the total K-REC cross-section by 3% only. To facilitate a comparison with an angular distribution for high-energies, the

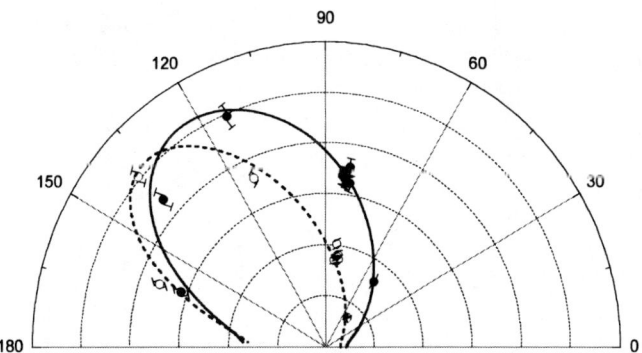

FIGURE 4. Experimental angular distribution for REC into the first excited states of H-like uranium as measured for $U^{92+} \rightarrow N_2$ collisions at 88 MeV/u, corresponding kinetic photoelectron energy of 48 keV [16]. The angles refer to the emitter frame. Solid points: REC into $2s_{1/2}$ and $2p_{1/2}$; open circles: $2p_{3/2}$. The lines depict the result of rigorous relativistic calculations. Note that in the case of the $2p_{3/2}$ state, the experimental and theoretical results were multiplied by a factor of 1.6.

results obtained recently for 310 MeV/u for $U^{92+} \rightarrow N_2$ collisions [3] are displayed in addition (dashed line). This comparison illustrates the enhanced importance of higher multipole contributions (retardation) at high energies leading here to a very strong backward peaking of the distribution. This is shown by the polar presentation in figure 3.

Furthermore, by means of REC, photoionization can be studied even for the excited states in high-Z H-like ions. An important aspect of the current investigation is that the enhanced experimental resolution - due to the narrow Compton profiles at low beam energies- allowed us to derive even j-subshell selective differential cross-sections for photoionization of the first excited states in hydrogenlike uranium by measuring REC into the L-shell. These data are given in Fig. 4. There, the solid points refer to capture into the two j=1/2 fine structure components of the L shell ($2s_{1/2}$ and $2p_{1/2}$) and the open circles give the data obtained for the $2p_{3/2}$ state. The corresponding result of rigorous relativistic calculations are displayed in addition in the figure (full line: $2s_{1/2}$ and $2p_{1/2}$; dotted line: $2p_{3/2}$). Note, that for comparison of the angular distributions for the different j-fine structure components, the $2p_{3/2}$ distribution was multiplied by a factor of 1.6. For the case of the j=1/2 distribution, the $2s_{1/2}$ gives the strongest contribution and the final shape is quite similar to the one observed for the ground state. In contrast, the differential cross-section observed for the experimentally isolated $2p_{3/2}$ level exhibits a much more pronounced backward shift. Its maximum shows up at angles quite similar as it is observed for the K-REC distribution measured at the much higher energy of 310 MeV/u. This illustrates that retardation corrections depend crucially on the angular momentum of the final state and that they are much more pronounced for p- than for s-states.

ALIGNMENT STUDIES: MULTIPOLE MIXING OBSERVED FOR THE DECAY OF THE $2p_{3/2}$ STATE IN HEAVY H-LIKE IONS

The angular distribution of photons emitted in collisions of bare uranium with light target atoms has been studied intensively at the ESR storage ring. It is proved that the angular distribution is sensitive to the population of magnetic substates by the electron capture process. If the electron capture populates the magnetic substates statistically, alignment will not occur and one should expect an isotropic emission pattern for the photon decay of these states. For the $2p_{3/2}$ state with the magnetic quantum numbers of $\mu=\pm1/2, \pm3/2$, however, our experimental data for high-Z ions show a strong anisotropic emission characteristic of the decay photons (Lyα_1 transition) for the case that the $2p_{3/2}$ state is populated by REC [5,6]. The angular distribution W(θ) is characterized by the so called anisotropy parameter β_{20}:

$$W(\theta(\propto 1 + \beta_{20}(1 - \tfrac{3}{2}\sin^2\theta) \tag{1}$$

For the case of the $2p_{3/2}$ level, the anisotropy parameter can be expressed by the population $\sigma(3/2,\mu)$ of the magnetic sublevels μ as

$$\beta_{20} = \frac{1}{2}\frac{\sigma(\tfrac{3}{2},\pm\tfrac{3}{2})-\sigma(\tfrac{3}{2},\pm\tfrac{1}{2})}{\sigma(\tfrac{3}{2},\pm\tfrac{3}{2})+\sigma(\tfrac{3}{2},\pm\tfrac{1}{2})} = \frac{1}{2}A_2 \tag{2}$$

where A_2 denotes the alignment parameter. Note, for Eq. 2 only the dominant electric dipole transition (**E1**) is considered. When the weak magnetic quadrupole transition (**M2**) is taken into account in a coherent way, the angular distribution is changed remarkably [7]. Within density matrix theory, the anisotropy parameter β_{20} is modified to $\beta_A = \tfrac{1}{2}A_2 \cdot f(E1,M2)$ where $f(E1,M2)$ is a structure function related to the interference effect. For H-like uranium ion it reaches a value of 1.28 [7] although the **M2** contribution to the total decay rate of the $2p_{3/2}$ state amounts 1% only. Here, it is important to note that the function $f(E1,M2)$ exhibits a moderate scaling as function of the nuclear charge Z ($f(E1,M2) \propto Z^2$) so that even for medium heavy ions such as Xe^{53+} a 10% correction persists.

In Fig. 5 we compare the experimental results for H-like uranium (solid points) with the corresponding theoretical findings (full line) and the results obtained by assuming $f(E1,M2)=1$ (dotted line), i.e. neglecting the interference term [7]. From the figure it is evident that the former deviation of the theoretical results from the experimental values is removed when taking the interference term into account. This proves the importance of the interference between the **E1** and **M2** decay branches for the decay of $2p_{3/2}$ state.

TWO ELECTRON CONTRIBUTIONS TO THE GROUND STATE OF HE-LIKE URANIUM IONS

The He-like system is the simplest multi-electron system for testing many-body theories in various aspects. In the case of heavy ions, the study of two-electron systems gives insight into the electron–electron interaction in the presence of the strong

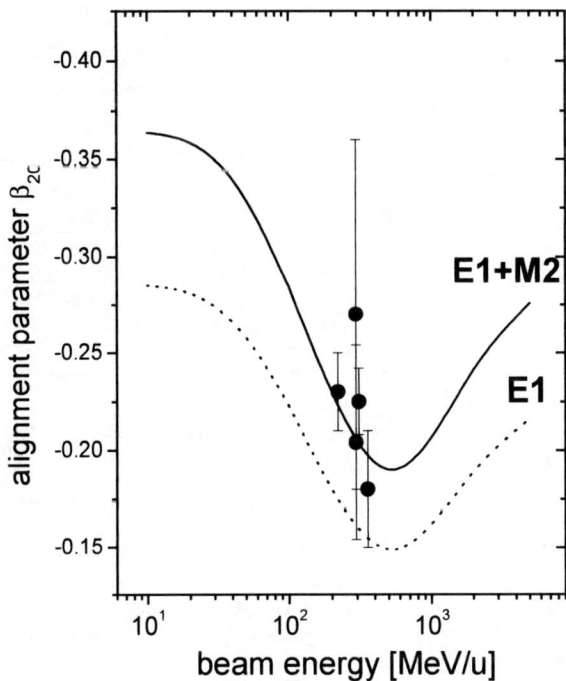

FIGURE 5. Comparison of effective anisotropy parameter between theory and experimental values (solid circles) for the $Ly\alpha_1$ transition of U^{91+} produced in collisions between bare U-ions with N_2 as a function of projectile energy [7]. Solid line: Inclusion of the multipole mixing effect. Dotted line: theoretical results obtained by considering only the decay by **E1** transitions

Coulomb field of the nucleus. Here the electron–electron interaction is strongly influenced by the Breit interaction, as well as by higher order radiative corrections. A new method to address these effects experimentally has been introduced at Super-EBIT [11]. This method exploits RR transitions into the groundstate of stationary bare and H-like ions. Here the difference in the photon energy for RR into the two ion species refers directly to their different ionization potential.

Therefore the measurement of this energy difference allows for a direct experimental determination of the two-electron contribution to the ionization potentials in the He-like ions [11,20,21]. The principle of the measurement and the Feynman diagram for higher order QED contributions are illustrated in figure 6 a and b, respectively. At Super-EBIT this technique was applied to various ions in the Z-range between 32 and 83 and the results obtained approached already an accuracy comparable to the size of the predicted screened Lamb shift corrections [11]. However, due to counting statistics the experiment could not be performed for nuclear charges larger than 83.

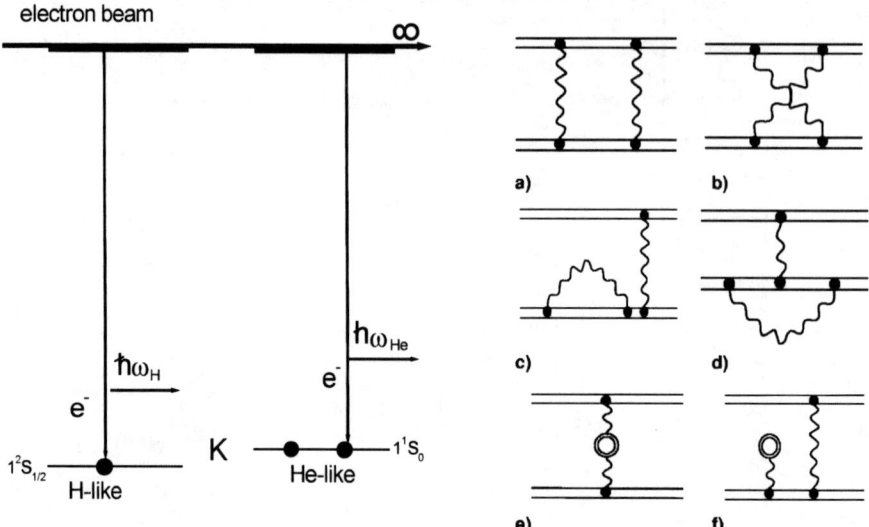

FIGURE 6a. Schematic presentation of radiative recombination transitions for the groundstate of H- and He-like ions. The energy difference $\hbar\omega_H - \hbar\omega_{He}$ gives directly the two-electron contributions to the ionization potentials in He-like ions.

FIGURE 6b. Feynman diagrams for the non-radiative QED corrections (a,b), the two-electron self energy (c,d) and the two-electron vacuum polarization (e,f) [21].

Recently using the same concept as introduced at the Super-EBIT, an experiment was performed at the ESR electron cooler device for bare and hydrogen-like uranium projectiles, aiming to achieve a precision of better than 5 eV for the determination of the two-electron contribution (about 2.2 keV) to the ground state in He-like uranium ions [10]. Note that for uranium the size of the screened Lamb shift is predicted to amount –9.4 eV [20,21].

At the electron cooler a solid state germanium x-ray detector was placed at an observation angle close to 0°. This geometry has the advantage that the Doppler corrections are not sensitive to slight variations of the detector/beam geometry [15]. We chose a low beam energy of 43 MeV/u by employing the deceleration technique. At such low beam energies, the uncertainties associated with the Doppler effect are much less severe compared to high beam energies. Furthermore, background events are almost completely eliminated by a coincidence measurement of the x-rays with the down charged projectiles.

In figure 7 the x-ray spectrum of K-RR recorded for bare and H-like uranium ions is given. From the preliminary analysis, a statistical accuracy for the separation of K-RR lines can be estimated better than 9 eV. Along with a second run, which is already scheduled, an accuracy of better than 4 eV can be anticipated. This aim of the

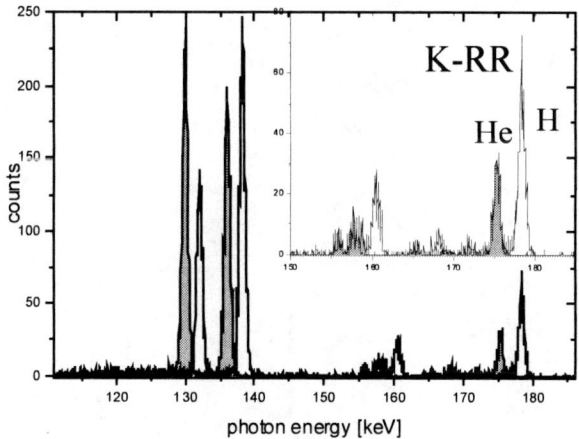

FIGURE 7. X-ray spectra recorded at the FSR electron cooler for decelerated bare and H-like uranium ions [10]. The most prominent features of the spectra are L→K transitions following recombination into the excited states. The solid line refers to initial bare ions and the line with a shaded area refers to initially H-like ions. The recombination lines for capture into the K-shell (K-RR) are enlarged in the inset.

experiment must be compared with the predicted two-electron self-energy contribution of –9.7 eV.

LANZHOU COOLER STORAGE RING PROJECTS

Presently a double-ring system is under construction in Lanzhou. As shown in figure 8 the present cyclotron system (SFC, SSC) will be used as an injector. The two rings have the circumstance of 161.0 and 128.8 meters and the maximum magnetic rigidity of 10.64 and 8.4 Tm, respectively. In the first ring, the CSRm, ions can be accelerated up to 900 MeV/u for carbon and to 400 MeV/u for uranium. In the experimental storage ring CSRe an internal gas-jet target system will be installed and will provide normal cluster targets as well as polarized target beams. One of the main research topics will be atomic physics with highly charged heavy ions [22,23]. The planned research will cover high precision x-ray spectroscopy measurements aiming for probing higher order QED effects in strong Coulomb fields, relativistic effects, as well as the structure and dynamics of few-electron systems. In addition laser cooling, laser induced/assistant ion-electron and ion/atom collision experiments will be possible. The construction phase will be completed in the second half of the year 2004, and the first test run for atomic physics experiment is expected for 2005.

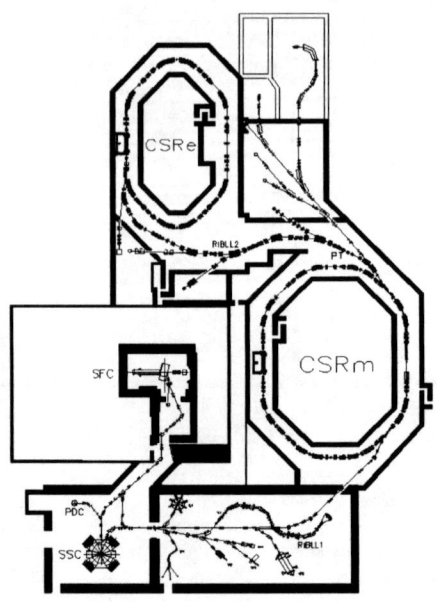

FIGURE 8. Layout of the heavy ion facility at Lanzhou and the ongoing cooler storage ring projects [22]. The present facility: SFC: Sector Focusing Cyclotron with energy constant K=69 and SSC: Separated Sector Cyclotron with energy constant K=450.

SUMMARY

In summary, we measured the angular-distributions for REC into the ground state and the first excited states of decelerated bare uranium ions. This allowed us to obtain for the very first time differential data for photoionization of a high-Z atom in the low-energy regime. The results demonstrate that the applied method is a powerful tool for the study of photon matter interaction in regimes not accessible in direct photoionization experiments. A strong alignment is observed for the radiative decay of the $2p_{3/2}$ state following radiative electron capture. Here, for the experimentally deduced anisotropy parameter the discrepancy between theory and experiment is removed when the interference between the leading **E1** transition and the magnetic quadrupole contribution is taken into account. Finally a very recent experiment was discussed where we aim on a precise determination of the electron-electron QED contribution to the groundstate ionization potential in high-Z ions. In contrast to the former Super-EBIT experiment, a preliminary data analysis already shows that at the ESR storage ring such investigations can now be extended up to the heaviest ions such as uranium.

ACKNOWLEDGMENTS

The authors would like to thank the close collaboration with J. Eichler, S. Fritzsche, and A. Surzhykov. We would also like to thank the ESR team for providing high quality beams and O. Klepper for the preparation of the particle detectors. X. Ma would like to acknowledge the financial support from GSI, BR-Plan (2000, CAS), and NSFC (10210201077, 10134010).

REFERENCES

1. J. Eichler and W.E. Meyerhof, *Relativistic Atomic Collisions* (Academic Press, San Diego, 1995).
2. P. H. Mokler and Th. Stöhlker, Adv. At. Mol. Opt. Phys. **37**, 297 (1996).
3. Th. Stöhlker, T. Ludziejewski, F. Bosch, R. W. Dunford, C. Kozhuharov, P. H. Mokler, H. F. Beyer, O. Brizanescu, B. Franzke, J. Eichler, A. Griegal, S. Hagmann, A. Ichihara, A. Krämer, J. Lekki, D. Liesen, F. Nolden, H. Reich, P. Rymuza, Z. Stachura, M. Steck, P. Swiat and A. Warczak , Phys. Rev. Lett. **82**, 3232(1999); Phys. Rev. Lett. **84**, 1360 (2000).
4. A. Ichihara, T. Shirai, and J. Eichler, Phys. Rev. A **49**, 1875 (1994).
5. Th. Stöhlker, O. Brizanescu, A. Krämer, T. Ludziejewski, X. Ma, P. Swiat and A. Warczak; in X-Ray and Inner Shell Processes, AIP Conf. Proc. No.506, edited by D. S. Gemmel; E. P. Kanter; L. Young, (AIP New York, 1999) 389.
6. Th. Stöhlker, F. Bosch, A. Gallus, C. Kozhuharov, G. Menzel, P. H. Mokler, H. T. Prinz, J. Eichler, A. Ichihara, T. Shirai, R. W. Dunford, T. Ludziejewski, P. Rymuza, Z. Stachura, P. Swiat, and A. Warczak, Phys. Rev. Lett. **79**, 3270 (1997).
7. A. Surzhykov, S. Frizsche, A. Gumberidze, and Th. Stöhlker, Phys. Rev. Lett. **88**, 153001 (2002).
8. Th. Stöhlker, P. H. Mokler, F. Bosch, R. W. Dunford, F. Franzke, C. Kozhuharov, T. Ludziejewski, F. Nolden, H. Reich, P. Rymuza, Z. Stachura, M. Steck, P. Swiat and A. Warczak, Phys. Rev. Lett. **85**, 3109 (2000).
9. J. P. Briand, P. Chevallier, P. Indelicato, K. P. Ziock, and D. D. Dietrich, Phys. Rev. Lett. **65**, 2761 (1990).
10. A. Gumberidze, et. al., to be published .
11. R. E. Marrs, S. R. Elliott, and Th. Stöhlker, Phys. Rev. A **52**, 3577 (1995) .
12. K. Blasche, and B. Franzke, in proceedings of the 4th European Particle Conference, edited by V. Suller and Ch. Petit-Jean-Genaz (World Scientific, Singapore), 133 (1994).
13. M. Steck, K. Beckert, F. Bosch, H. Eickhoff, B. Franzke, O. Klepper, F. Nolden, H. Reich, B. Schlitt, P. Spädtke, T. Winkler, Nuclear Physics A **626**, 495c (1997).
14. H. Reich, W. Bourgeois, B. Franzke, A. Kritzer, and V. Varentsov, Nucl. Phys. A **626**, 417c(1997).
15. H.F. Beyer, G. Menzel, D. Liesen, A. Gallus, F. Bosch, R. Deslattes, P. Indelicato, Th. Stöhlker, O. Klepper, R. Moshammer, F. Nolden, H. Eickhoff, B. Franzke, M. Steck, Z.Phys. **D35**, 169 (1995)
16. Th. Stöhlker, X. Ma, T. Ludziejewski, H. F. Beyer, F. Bosch, O. Brizanescu, R. W. Dunford, J. Eichler, S. Hagmann, A. Ichihara, C. Kozhuharov, A. Krämer, D. Liesen, P. H. Mokler, Z. Stachura, P. Swiat and A. Warczak, Phys. Rev. Lett. **86**, 983(2001).
17. S J Blakeway, W Gelletly, H R Faust and K Schreckenbach, J. Phys. B **16**, 3751(1983).
18. J. Eichler, A. Ichihara, and T. Shirai, Phys. Rev. A **51**, 3027 (1995).
19. A. Ichihara, T. Shirai, and J. Eichler, Phys. Rev. A **54**, 4954 (1996).
20. V.A. Yerokhin, A.N. Artemyev, V.M. Shabaev, Phys. Lett. A **234**, 361 (1997).
21. H. Persson, S. Salomonson, P. Sunnergren, and I. Lindgren, Phys. Rev. Lett. **76**, 204 (1996).
22. J. Xia, Proposal of the HIRFL-CSR Project (1998).
23. X. Ma, Hyperfine Interactions **115**, 107 (1998).

V.   ATOMIC AND NUCLEAR
X-RAY PROCESSES

Nuclear internal conversion between bound atomic states

J.F. Chemin[1], M.R. Harston[1], F.F. Karpeshin[1,5], J. Carreyre[1],
F. Attallah[2], M.M. Aleonard[1], J.N. Scheurer[1], G. Boggaert[3],
J.R. Grandin[4], M.B.Trzhaskovskaya[5]

[1]CENBG, BP 120, F33175, Gradignan Cedex, France
[2]GSI, D-6100 Darmstad, Germany
[3]LPNHE, Ecole Polytechnique, 91128 Palaiseau, France
[4]CIRIL, BP 5133, F-14040 Caen, France
[5]Saint Petersbourg, Nuclear Physics Institute, Gatchira, Saint-Petersbourg, Russia

Abstract : We present experimental and theoretical results for rate of decay of the $(3/2)^+$ isomeric state in ^{125}Te versus the ionic charge state. For charge state larger than 44 the nuclear transition lies below the threshold for emission of a K-shell electron into the continuum with the result that normal internal conversion is energetically forbidden. Rather surprisingly, for the charge 45 and 46 the lifetime of the level was found to have a value close to that in neutral atoms. We present direct evidence that the nuclear transition could still be converted but without the emission of the electron into the continuum, the electron being promoted from the K-shell to an other empty bound state lying close to the continuum. We called this process BIC. The experimental results agree awhith theoretical calculations if BIC resonances are taken into account. This leads to a nuclear decay constant that is extremely sensitive to the precise initial state and simple specification of the charge state is no longer appropriate. The contribution to decay of the nucleus of BIC has recently been extended to the situation in which the electron is promoted to an intermediate filled bound state (PFBIC) with an apparent violation of the Pauli principle. Numerical results of the expected dependence of PFBIC on the charge state will be presented for the decay of the 77.351 keV level in ^{197}Au.

INTRODUCTION

Although the dependence of the nuclear lifetime ($T_{1/2}$) or decay constant (λ) on the extranuclear environment had been investigated for a long time, only very small variations ($\Delta\lambda/\lambda \sim 10^{-3}$) were found [1]. The revival of interest in this field in recent years is due to accelerator developments which make it possible to induce drastic changes in the extranuclear environment by ionizing the inneratomic shells. Such processes, as β decay, electron capture, and internal conversion (IC), all of great interest for astrophysics [2], are very sensitive to the electronic density in the vicinity of the nucleus.

CP652, X-Ray and Inner-Shell Processes: 19th International Conference on X-Ray and Inner-Shell Processes
edited by A. Bianconi, A. Marcelli, and N. L. Saini
© 2003 American Institute of Physics 0-7354-0111-X/03/$20.00

Recently, it was demonstrated that in highly ionized atoms the electronic configurations can have a drastic effect on the nuclear decay. Jung et al [3] found a spectacular decrease of the ^{163}Dy half-life from $T_{1/2} = \infty$ (stable) in a neutral atom to $T_{1/2} = 50d$ in the bare nucleus which is the experimental proof of bound β decay a process previously suggested by Daudel et al [4]. In this paper we report on strong variation of an other nuclear decay process on the electronic environment [5]. In normal internal conversion decay a nucleus deexcites by ejection of a bound electron into the continuum. If the nuclear excitation energy is less than the binding energy of a given electron shell then the conversion probability to continuum states from this shell should become zero. However a possible significant contribution from a subthreshold conversion process in which the nuclear deexcitaion occurs with excitation of a bound electron to a higher-lying bound orbital is expected even when the nuclear deexcitation energy does not exactly match the electron excitation energy owing to the finite width of the atomic states, particularly the hole state produced by the excitation [6].

Such a process, called BIC (Bound Internal Conversion) can play an important role in cases where the binding energies of the atomic electrons are close to the nuclear transition energy. For the particular case of electronic transitions from the K shell and transitions connecting one excited state to the ground nuclear state, this situation is encountered in several nuclei ^{125}Te, ^{191}Ir, ^{196}Au, ^{183}Ta, ^{177}Ta, ^{187}Os. Many more cases can be found if transitions between excited states are considered.

Thus far, investigations of BIC have focussed on the first excited $3/2^+$ level of ^{125m}Te at 35.4917 keV which undergoes an M1 transition to the $1/2^+$ ground nuclear level. In the neutral atom the excited level has a half-life of 1.486 ns and the transition is associated with a total internal conversion coefficient of 13.9. The transition lies just above the K-shell threshold in the neutral atom (31,8 keV). The removal of outer-shell electrons has two different effects on the IC decay process. (i) It modifies the electron number in the different shells ; the corresponding ICC diminishes and vanishes at the limit of empty shells, (ii) It increases the effective Coulomb field seen by the outer electrons and increases the binding energy of the remaining electrons. The increases of the K-shell ionization energy with the result that K-shell internal conversion lies subthreshold in ions with ionic charge greater than 44^+, referred hereafter in this paper as the critical charge state Q_c. In the usual IC process with the emission of a bound electron into the energy continuum, the K-shell ionization by IC becomes energetically impossible despite the presence of the two K-shell electrons .Consequently the half-life should increase by a factor corresponding to the suppression of the K-shell conversion channel.

The half-life of the first excited nuclear state of ^{125}Te, $T_{1/2}^Q$ as a function of a charge state Q can be related to the neutral-atom half-life $T_{1/2}^0$ and the total internal conversion coefficient α_T by :

$$T_{1/2}^Q = T_{1/2}^0 \frac{1 + \alpha_T^0}{1 + \alpha_T^Q} \tag{1}$$

$$\alpha_T^O = \alpha_K^O + \alpha_L^O + \alpha_M^O \qquad (2)$$

One finds for the half-life of He-like ^{125}Te a value $T_{1/2}^{50^+} = 22.1$ ns compared to $T_{1/2}^o = 1.486$ ns.

In such ions with $Q \sim Q_c$ the nuclear transition energy may therefore lie close to the electronic excitation energy of a 1s electron to a final Rydberg orbital just below the continuum threshold and the BIC process is expected to occur.

EXPERIMENT

The experiment was performed at GANIL. A beam of 25 MeV/nucleon of ^{125}Te ions in the charge state 38$^+$ impinged on a 1 mg/cm^2 thick ^{232}Th target in order to strip the ions and to Coulomb excite the Te nuclei. The beam intensity was of the order of 1 enA. The Te ions exiting from the target had an approximately Gaussian charge distribution centered on <Q> ~ 46.5 with standard deviation $\sigma_Q = 1.6$. Scattered Te ions with Q = 44-48 within the angular range $\theta_{coulex} = 1$-6° were accepted in the SPEG spectrometer, which was rotated to a mean angle of 3.5°. The corresponding time of flight of the ions from the target to the spectrometer was 40 ns, allowing for atomic and nuclear decay before entering the magnetic field. The Te ions were then separated in the magnetic field of SPEG according to their charge state. At the exit of the separation system, the horizontal (X) and vertical (Y) positions of the ^{125}Te ions were detected event by event in two identical drift chambers. From these parameters, the distributions in horizontal angle, θ, and vertical angle, ϕ, of ^{125}Te ions entering the detection system were calculated. The coincidence signal between a parallel plate avalanche counter and the cyclotron high frequency provided a fast trigger signal for the data acquisition system. The energy of the scattered ions was measured in an ionization chamber located downstream of the drift chambers.

In the passage through the target, nuclear and atomic excitation of the Te beam were induced. In the range of scattering angles analyzed by SPEG, below the grazing angle at 6.6°, nuclear Coulomb excitation was the dominant mechanism for the excitation of the Te nuclei. A fraction of ^{125}Te nuclei were Coulomb excited to high spin, short-lived states which populate the first excited, J = 3/2$^+$ state of ^{125}Te at 35.492 keV. The decay in-flight of this state by γ-ray emission or by internal conversion with subsequent emission of Te K x rays, referred to hereafter as IC x rays, was detected by an array of four 10-cm^2 planar Ge detectors.

Pure atomic collisions between the beam and the Th target atoms induced vacancies in the Te K shell which were filled promptly with emission of Te K x rays. The lifetime of the K shell vacancies in Te ions with a half-filled L shell is of the order of 10^{-16}s. The K vacancies created in the Te ions are filled by 2p electrons within 100 µm behind the target with emission of K x rays, the energies of which depend on the charge state and electronic configuration. In Te^{46+} the calculated K x-ray energy is approximately 28.1 keV.

The other collision partner Th is more likely to undergo direct ionization of the L shell. In the following, we refer to x rays produced in pure atomic collisions as atomic x rays (AT x rays).

In order to optimize the discrimination between IC x rays emitted with a mean delay of the order of the nuclear half-life and the AT x rays the target was surrounded by lead and coppper shielding so that the Ge detectors viewed a zone along the beam located at a mean distance of 8.7 cm behind the target. Therefore, only the delayed Te K x rays following the IC process and γ rays at 35.49 keV, were expected to reach the detectors. These two radiations, emitted from Te ions in flight and detected in the backward direction at approximately 120° undergo a large Doppler effect (v/c = 0.227), which induces a negative energy shift, lowering the mean energies of the Te IC K x rays and the γ rays to 24 and 29.6 keV, respectively.

In spite of the shielding, prompt (AT) x rays emitted at the target position can reach the detectors if they are emitted in a forward direction, through the beam aperture in the shielding material surrounding the target, and then are scattered in the direction of the detectors. The Doppler shift associated with these AT Te K x rays is positive and leads to a maximum energy for AT Te K x rays equal to 34.3 keV. Hence, the Doppler effect enables a discrimination between the two different mechanisms producing Te K x rays in the collision.

The Ge spectra from the four detectors were gain-matched off-line and added together.

Ge energy spectra corresponding to the sum of the spectra for all charge states, are shown in Figure 1. Spectra (a) and (b) are conditioned by contours set on the scattering angles corresponding to θ_{coulex} between 2.3° and 6° and between 1.3° and 1.9° degrees, respectively. In the following, the events selected by these two contours, are referred

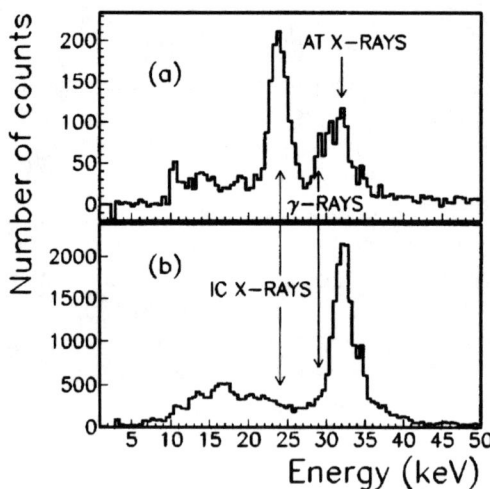

FIGURE 1 Experimental photon spectra in coincidence with Te ions in the charge states ranging between 44^+ and 48^+. Spectrum (a) is gated by the inelastic contour. Spectrum (b) is gated by the elastic contour. The position of AT X rays, IC X rays, and γ rays are indicated by arrows.

to as inelastic and elastic, respectively.

The line at 31.8 keV is due to AT x rays, following the direct Coulomb ionization of the projectiles in the target, emitted in the forward direction with a maximum Doppler shift and then scattered by the surrounding materials in the direction of the detectors. This line dominates in the spectrum (b) associated with the elastic contour for which the nuclear excitations is strongly suppressed. We find that the relative intensity of this Te component remains independent of the ionic charge state.

The line at 24 keV corresponds to IC Te K x rays emitted in the backward direction, at around 130°. This line is not seen in spectrum (b) which is associated with large impact parameter collisions and a small nuclear excitation probability. On the other hand, this line dominates spectrum (a), reflecting the strong dependence of the Coulomb excitation probability of the nucleus on the scattering angle. The intensity of this line is strongly dependent on the Te charge state. The γ transition at 35.49 keV, competing with the IC decay, is also seen in spectrum (a) doppler shifted to an energy of 29.6 keV. This line appears as a shoulder on the low-energy side of the 31.8-keV line. A more detailed account of the experimental arrangement is given in [7].

INTERPRETATION OF THE RESULTS

Below we present the results of calculations of BIC decay in the ions Te^{45+} and Te^{46+} taking into account excited initial states. Details of these calculations are given in Ref. [8]. The total rate for nuclear decay of a Te isomeric nuclear state for an ion in a given initial atomic state i can be written :

$$\lambda_i = \sum_f \lambda_{if}^{BIC} + \lambda^{IC} + \lambda_\gamma \tag{3}$$

where λ_{if}^{BIC} is the BIC decay rate to the final atomic state f, λ_γ is the radiative nuclear decay rate, and λ^{IC} is the rate for internal conversion decay to continuum states, which in the ions under consideration is limited by energy conservation to L-shell conversion.

The BIC decay rate is strongly dependent on the electronic transition energy as shown by the expression for the BIC coefficient R_{if} [6] :

$$R_{if} = \frac{\lambda_{if}^{BIC}}{\lambda_\gamma} = \frac{A_{if}}{2\pi} \frac{\Gamma_{if}}{(\omega_\gamma - \omega_{if})^2 + \Gamma_{if}^2 / 4} \tag{4}$$

where ω_γ is the nuclear transition energy, ω_{if} the atomic transition energy, and the quantity A_{if} is given by the same expression as that for the internal conversion coefficient to continuum states except that the wave function for the continuum electron, normalized on the energy scale, is replaced by that for a bound final state orbital. The quantity Γ_{if} is the total transition width which is dominated by the width ($\sim$ eV) of the 1s hole in the final state following excitation of a 1s electron by BIC.

TABLE 1. Transitions energies and BIC coefficients R_{if} associated with the excitation $1s \rightarrow 8s$ in Te^{46+} for transitions with a resonance defect $|\omega_\gamma - \omega_{if}| < 80$ eV

Initial state i	Final state f	$\omega_\gamma - \omega_{if}$ (eV)	R_{if}
$1s^2 2s^2 2\overline{p}{}^1 2p^2$ J=0	$1s^1 2s^2 2\overline{p}{}^2 8s^1$ J=1	-77.51	0.09
$1s^2 2s^2 2\overline{p}{}^1 2p^2$ J=1	$1s^1 2s^2 2\overline{p}{}^2 8s^1$ J=0	-76.58	0.05
$1s^2 2s^2 2\overline{p}{}^1 2p^1$ J=1	$1s^1 2s^2 2\overline{p}{}^1 2p^1 8s^1$ J=2	-63.75	0.07
$1s^2 2s^2 2\overline{p}{}^1 2p^1$ J=1	$1s^1 2s^2 2\overline{p}{}^1 2p^1 8s^1$ J=0	-62.95	0.04
$1s^2 2s^2 2\overline{p}{}^1 2p^1$ J=1	$1s^1 2s^2 2\overline{p}{}^1 2p^1 8s^1$ J=0	-32.11	0.62
$1s^2 2s^2 2\overline{p}{}^1 2p^1$ J=1	$1s^1 2s^2 2\overline{p}{}^1 2p^1 8s^1$ J=1	-31.98	1.19
$1s^2 2s^2 2\overline{p}{}^1 2p^1$ J=2	$1s^1 2s^2 2\overline{p}{}^1 2p^1 8s^1$ J=3	-66.25	0.06
$1s^2 2s^2 2\overline{p}{}^1 2p^1$ J=2	$1s^1 2s^2 2\overline{p}{}^1 2p^1 8s^1$ J=2	-66.56	0.04
$1s^2 2s^2 2\overline{p}{}^1 2p^1$ J=2	$1s^1 2s^2 2\overline{p}{}^1 2p^1 8s^1$ J=1	-31.55	0.34
$1s^2 2s^2 2\overline{p}{}^1 2p^1$ J=2	$1s^1 2s^2 2\overline{p}{}^1 2p^1 8s^1$ J=2	-31.33	0.51
$1s^2 2s^2 2p^2$ J=2	$1s^1 2s^2 2p^2 8s^1$ J=3	-43.88	0.08
$1s^2 2s^2 2p^2$ J=2	$1s^1 2s^2 2p^2 8s^1$ J=2	-43.13	0.07
$1s^2 2s^2 2p^2$ J=2	$1s^1 2s^2 2p^2 8s^1$ J=1	-4.26	16.39
$1s^2 2s^2 2p^2$ J=2	$1s^1 2s^2 2p^2 8s^1$ J=2	-4.13	25.65
$1s^2 2s^2 2p^2$ J=0	$1s^1 2s^2 2p^2 8s^1$ J=1	-28.23	0.81

Here and below we use the notation $2\overline{p}$ to denote a $2p$ $j = 1/2$ orbital and the notation $2p$ to denote a $2p$ $j = 3/2$ orbital. Calculations [8] of excited atomic radiative decay rates of ionic states of the form Te $1s^2$ $2s^2$ $2\overline{p}{}^x$ $2p^y$, with $x \neq 2$ have revealed that such states have lifetimes of the order of nanoseconds or longer. These are comparable with, or longer than, the time during which IC x rays and γ rays are measured in the present experiment. This has the important consequence that the number of possible BIC resonances is substantially increased relative to the case in which the $2p_{1/2}$ shell has the maximum occupancy.

For the two ionic states, the different possible distributions of electrons in $2\overline{p}$ and $2p$ orbitals give rise to five atomic states. For the M1 transition of ^{125m}Te we have considered final states of the form $1s^1$ $2s^2$ $2\overline{p}{}^x$ $2p^y$ ns^1, created by $1s \rightarrow ns$ BIC transitions. The coupling between the open shell electrons in the final states considered leads to 19 different levels for Te^{45+} and 16 levels for Te^{46+}.

An example of $1s \rightarrow 8s$ BIC transitions in Te^{46+} is shown in Table 1 for those transitions lying within 80 eV of exact resonance. The transition energies, widths and A_{if} values have been calculated using Dirac-Fock electron wave functions calculated with GRASP [9]. In the calculations, the Auger contribution to the hole width and the effect of hyperfine interaction have been neglected.

In Table 2 we compare the half-life for radiative atomic decay with the nuclear decay half-life for BIC decay based on the BIC coefficients calculated above, summed

TABLE 2. Summary of theoretical results obtained here for the lifetimes of atomic and nuclear of highly ionized ions of ^{125}Te

	Te^{45+}			Te^{46+}		
Initial Atomic Level	Atomic half life (ns)	K-shell coefficient	Nuclear half-life (ns)	Atomic half-life (ns)	K-shell coefficient	Nuclear half-life (ns)
1	∞	13	1.4	∞	0.7	6.5
2	5	42	0.5	7	2.6	4.2
3	29	6	2.6	10^5	1.4	5.4
4	5	52	0.4	4	42	0.5
5	3	19	1.0	3	0.9	6.1

over the all final states. For both Te^{45+} and Te^{46+}, the atomic level 4 is seen to possess the shortest nuclear decay half-life. This is particular notably in the case of Te^{46+}.

In order to make the link with experiment, we have calculated the ratio $\rho^Q = N\gamma/N_x$ assuming an initial statistical distribution of the five initial levels, taking account of the population and decay by atomic radiative transitions. After summing over all final levels with n = 8 – 20 for Te^{45+} and n= 6 – 11 for Te^{46+} we obtain $\rho^{45} = 0.14$ and $\rho^{46} = 0.41$. The experimental ρ values from the data shown in Figure 2 are, respectively, equal to $\rho^{45} = 0.08 \pm 0.04$ and $\rho^{46} = 0.35 \pm 0.12$. The good agreement between these values and the experimental data given above confirms the importance of BIC resonances in excited initial states.

In this situation the nuclear decay half-life becomes extremely sensitive to the precise atomic state because electronic level shifts of a few eV can significantly change the contributions arising from different near-resonance electron-nucleus transitions. In the subthreshold case, therefore, one cannot properly speak of a half-life or internal conversion coefficient associated with a given charge state ; it is also necessary to specify the precise electronic state. For ^{125m}Te, the energy shifts associated with the coupling between the open shell electrons in the initial and final states play an important role in determining the contributions from different resonant transitions . In addition, BIC transitions from long-lived excited initial states are found to be important. In this case the necessity to take into account time-dependent atomic state populations has the result that the ratio of the rates for internal conversion decay and γ decay becomes sensitive to the experimental configuration.

INTERNAL CONVERSION BETWEEN BOUND STATES AND THE PAULI EXCLUSION PRINCIPLE

We consider now the case of BIC nuclear deexcitation, when the intermediate state bound orbital is fully occupied. This process, which apparently violates the Pauli principle, is hereafter referred to as Pauli-forbidden bound internal conversion (PFBIC).

We take the M1 decay of the 77.351 $\mp$0.002 keV excited nuclear level of ^{197}Au as an example of the PFBIC process. The main channels for the decay of this excited state proceed by emission of an M1 γ or by normal internal conversion (IC) to final electronic states in the continuum. IC occurs principally in the 2s and 3s shells and corresponds to a total internal conversion coefficient of approximately 4.2. The half-life of the first excited state is 1.91 $\mp$ 0.01 ns [10]. Our aim here is to investigate the contribution of PFBIC to the nuclear decay rate and its dependence on the ionization state. In PFBIC in ^{197}Au, a virtual photon from the nucleus interacts with a 1s electron creating a hole in the K shell. The lifetime of the K-shell vacancy is very short due to the strength of the radiative dipole transition 2p $\rightarrow$ 1s. This lifetime corresponds to a hole width of the order of 52 eV that facilitates energy matching between the nuclear transition energy and the atomic 1s $\rightarrow$ 3s electronic excitation energy (77 300 eV) in neutral atom). An essential feature of PFBIC is the formation of a virtual 3s hole that can be occupied by the electron excited from the 1s shell. Several mechanisms can contribute to the formation of such a virtual hole. One possibility is the double transition 2p $\rightarrow$ 1s and 3s $\rightarrow$ 2p with emission of two photons in the final state. A related possibility is the transition 2p $\rightarrow$ 1 s together with filling of the 2p hole by an Auger transition leading to the excitation of a 3s electron and emission of an Auger electron into the continuum. A third possibility is filling of the initial 1s hole by an Auger transition that excites a 3s electron into the continuum. In the final state for PFBIC the virtual 3s hole has been filled so that the Pauli principle is indeed satisfied in this state. The creation and filling of this intermediate virtual hole necessarily imply that PFBIC involves terms of higher order in the perturbation series than the usual internal conversion process. The total internal conversion coefficient R_{PFBIC} is obtained by adding the partial internal coefficient in each of the channels mentioned above.

$$R_{PFBIC} = \sum_K R_{PFBIC}^K \tag{5}$$

We choose the nucleus ^{197}Au for a first investigation of PFBIC effects because the excitation energy from the 1s shell to the 3s shell lie within 100 eV of the nuclear excitation energy in the neutral atom so that the Pauli-forbidden transition 1s $\rightarrow$ 3s is nearly resonant with the nuclear transition as shown in Figure 2. In addition, we examine the charge state dependence of the PFBIC rate since the atomic binding energies change with the charge state, thereby allowing the possibility that the PFBIC transition lies closer to resonance in ionized states [5].

In order to examine the magnitude of the charge state dependence of PFBIC, we have computed the bound internal conversion coefficient in ions up to 69$^+$. The results are shown in Table 3. For each value of the charge state, we give in Table 3 the electronic configuration, the calculated energy mismatch, Δ, and the contributions to

TABLE 3. Calculated values of the energy of the resonance defect Δ, the discrete conversion coefficient A, and the PFBIC conversion coefficient R for various ions of ^{197}Au. Columns 5,6 and 7 give, respectively, the contribution to R_{PFBIC} of the two photon process, $(R_{PFBIC}^{\gamma_1\gamma_2})$, the one photon and one electron process, $(R_{PFBIC}^{\gamma_1 LM_1})$, and the one step K-shell Auger process, $(R_{PFBIC}^{KXM_1})$. The last column gives the total PFBIC conversion coefficient, R_{PFBIC} All of the values are in eV except the R coefficients that are dimensionless.

Ion	Configuration	Δ	A	$R_{PFBIC}^{\gamma_1\gamma_2}$	$R_{PFBIC}^{\gamma_1 LM_1}$	$R_{PFBIC}^{KXM_1}$	R_{PFBIC}
Au^0	$[Xe]4f^{14}5d^{10}6s$	-60	606 19	0.026	0.053	0.11	0.19
Au^{11+}	$[Xe]4f^{14}$	-58	606 58	0.028	0.056	0.118	0.20
Au^{25+}	$[Xe]$	-102	610 29	0.010	0.020	0.043	0.074
Au^{33+}	$[Kr]4d^{10}$	-65	611 84	0.023	0.046	0.098	0.168
Au^{42+}	$[Kr]4d^1$	13	622 64	0.136	0.272	0.578	0.99
Au^{51+}	$[Ar]3d^{10}$	178	631 58	0.004	0.015	0.026	0.03
Au^{61+}	$[Ar]$	722	698 53	0.000	0.000	0.001	0.00
Au^{68+}	$[Ne]3s$	1258	753 74	0.000	0.000	0.000	0.39
Au^{69+}	$[Ne]$	1306	760 42	0.000	0.000	0.000	0.74

PFBIC from the three processes considered above. The details of the calculations are given in [11].

The principal factor contributing to the charge state dependence of the rate for the PFBIC process arises from the energy mismatch, Δ. As the atom is ionized from the neutral atom (Δ = - 60 eV), the PFBIC transition first moves away from resonance (Δ < 0) until charge state 25^+ is attained and then moves back towards resonance. In ions

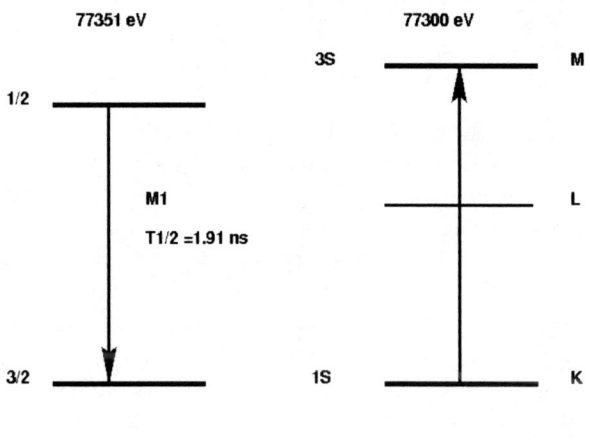

FIGURE 2. Nuclear and electronic levels of ^{197}Au in the neutral atom. The nuclear levels are shown on the left and the electronic levels on the right-hand side.

with $Q \sim 41^+$ the Δ value lies only a few eV from exact resonance. As the charge state is further increased the Δ values becomes more positive and the transition moves rapidly away from resonance. These changes in the proximity to resonance are reflected in the value of R_{PFBIC} shown in the last column of Table 3 that attain a maximum in value of 1.2 in the ion 40^+.

The PFBIC process constitutes a new channel for the decay of the first excited state in Au. Taking into account this channel leads to an effective internal conversion coefficient :

$$\alpha = \langle \alpha \rangle + R_{PFBIC}$$

The half-life of this nuclear transition has been extensively studied. A mean value of 1.91 ns is reported for the half-life of the 77 351 eV nuclear state in the data tables of Ref. [10]. A change of the value of α induces a change in the half-life $T_{1/2}$ of the nuclear level. The value $T_{1/2}^{Q}$ is related to the value $T_{1/2}^{o}$ in the neutral atom by the relation.

$$T_{1/2}^{Q} = T_{1/2}^{o} \frac{1 + \langle \alpha \rangle}{1 + \langle \alpha \rangle + R_{PFBIC}} \tag{6}$$

In the neutral atom, the inclusion of PFBIC changes the nuclear lifetime by only about 3.6 %. In this case the influence of BIC on the nuclear lifetime thus turns out to be small. The situation is rather different in Au ions. As seen in Table 3 the value of R_{PFBIC} reaches a maximum, $R_{PFBIC} \sim 1$ around charge state $Q \sim 40$. In this case, the inclusion of PFBIC would increase the value α by an amount of the order of 20 %.

In ions of ^{197}Au with $Q \sim 40^+$ the half-life is then shortened to $T_{1/2}^{40} \sim 1.6$ ns compared to $T_{1/2}^{o} = 1.9$ ns in the neutral atom.

CONCLUSION

We have presented experimental and theoretical results for a new mode of internal conversion which takes place by excitation of an electron to a bound orbital. When the binding energy of the converted electron becomes larger than the nuclear transition energy. The internal conversion has a resonant behaviour and becomes strongly dependent on the details of the nuclear structure. In particular the coupling between electrons in different open shells produces rich resonance structure which leads to an enhancement of the nuclear decay probability. In this situation it is no longer possible to define a unique internal conversion coefficient or a single lifetime for an ion in a given charge state. Instead the full electronic configuration should be considered in the definition of the BIC coefficient.

It is interesting to note that the BIC process is exactly the reverse process of NEET in which the nucleus is excited by a near-resonant electronic transition [12]. If we assume a prepared excited atomic state which can decay by a transition whose approximate energy matches a nuclear transition in the same atom to be transfered to the nuclear part of the atom. The matrix elements coupling the electronic and nuclear currents are exactly the same in BIC and in NEET if the electronic and nuclear states involved in BIC and NEET are identical. In order to observe the NEET process corresponding to the BIC decay of ^{125}Te observed here, one would have to prepare a ^{125}Te ion in the electronic configuration $1s^1 2s^2 2p^3 ns^1$ (with $n \sim 17$) or in the configuration $1s^1 2s^2 2p^2 ns^1$ (with n equal 8), which is obviously very difficult. Nevertheless suitable situation for the observation of NEET might be encountered in other atoms.

BIC can also take place by excitation of an electron to a bound orbital which is fully occupied. This process, which appears to violate the Pauli exclusion principle, can take place by virtue of the finite widths of the electron transitions that depopulate the Pauli-forbidden state.

Numerical calculations described above for ^{197}Au show that in the neutral atom the Pauli-forbidden internal conversion to bound states corresponds to a decay rate that is 19 % of that for radiative nuclear decay.

REFERENCES

1. Emery, G.T., *Annu. Rev. Part. Sci.* **22**, 165 (1972).
2. Takahashi, K., and Yokoi, K., *At. Data Nucl. Data Tables* **36**, 375 (1987), Phys. Rev. **C 36**, 1522 (1987).
3. Jung, M., Bosch, F., Beckert, K., Eickhoff, H., Folger, H., Franzke, B., Gruber, A., Kienle, P., Klepper,O.,Koenig, W., Kuzhuharov, C., Mann, R., Moshammer, R., Nolden, F., Scharf, U., Soff,G., Spïadtke,P., Steck, M., Stöhlker, Th., Sümmerer, K., *Phys. Rev. Lett.* **69**, 2164 (1992).
4. Daudel, R., Jean. M., and Lecoin, N., *J. Phys. Radium* **8**, 238 (1947).
5. Attallah, F., Aiche, M., Chemin, J.F., Scheurer, J.N., Meyerhof, W.E., Grandin, J.P., Aguer, P., Bogaert, G., Grunberg, C., Kiener, J., Lefebvre, A., and Thibaud, J.P., *Phys. Rev. Lett.* **75**, 1715 (1995) and Phys. Rev. **C 55**, 1665 (1997).
6. Karpeshin, F.F., Harston, M.R., Attallah, F., Chemin, J.F., Scheurer, J.N., Band, I.M., and Trzhaskovskaya, M.B., *Phys. Rev.* **C 53**, 1640 (1996).
7. Carreyre, T., et al, *Phys. Rev.* **C 62**, 24311 (2000).
8. Harston, M.R.., Careyre, T., Chemin, J.F., Karpeshin, F., Trzhaskovskaya, M.B. Nucl. Phys. **A 676**, (143) (2000)
9. Grant, L.P., McKenzie, B.J., Norrington, P.H., Mayers, D.F., and Pyper, N.C., *Comput. Phys. Commun.* **21**, 207 (1980).
10. Firestone, R.B., *Table of Isotopes*, 8th *ed., edited by*, Shirley, V.S., (Wiley, New York, 1996, Vol II.
11. Karpeshin, F.F., et al, *Phys. Rev.* **C 65**, 34303, (2002).
12. Morita, M., *Prog. Theor. Phys.* **41**, 1574 (1973).

L1 Atomic-Level Widths Of Heavy Elements

J.-Cl. Dousse and P.-A. Raboud

Department of Physics, University of Fribourg, Chemin du Musée 3, CH-1700 Fribourg, Switzerland

Abstract. High-resolution measurements of the L_1-$M_{4,5}$ fluorescence x-ray lines of Sm, Ho, Yb, W, Pt, Hg and Bi were performed, using a transmission DuMond-type and a reflection von Hamos-type bent crystal spectrometer. Despite the poor intensity of the dipole-forbidden transitions, precise linewidths were obtained. Assuming for the $M_{4,5}$ level widths the values reported recently by Campbell and Papp, the L_1 level widths of the investigated elements could be determined with a precision of 4-10%. A good agreement with Campbell and Papp's data was observed for the heaviest elements. For Sm and Ho, however, an intriguing discrepancy was found. The latter was explained by a splitting effect of the L_1 subshell resulting from the coupling of the 2s electron spin in the initial excited state with the total spin of the open 4f level. Based on this splitting picture, adjustements of the L_1 level widths recommended by Campbell and Papp were proposed for the lanthanides region.

INTRODUCTION

Several review papers dealing with atomic-level widths can be found in the literature (see, eg., [1-3]. Most of them rely on theory. Recently Campbell and Papp assembled a large number of experimental data from which they derived an internally consistent set of level widths for the K shell to the N_7 subshell for all elements across the periodic table [4]. For the subshell L_1 they used results from XPS (x-ray photoelectron spectroscopy) measurements for the light elements and XES (x-ray emission spectroscopy) data for the elements pertaining to the region $40 \leq Z \leq 51$. To connect the latter data to recent XES results obtained for Th and U [5,6], they chose to employ a set of L_1 Coster-Kronig and relative fluorescence yields of elements in the range $62 \leq Z \leq 79$ (Refs. [7-9]). The alternative of using in the region $62 \leq Z \leq 79$ old XES measurements [10] of the L_1-$N_{2,3}$ linewidths instead of the set of L_1 Coster-Kronig yields was also probed by Campbell and Papp. The results were found to be 0.3-3.3 eV higher than the values derived from Coster-Kronig spectroscopy. Because of the age of these XES data and the relatively big uncertainty on the $N_{2,3}$ widths, they renounced to consider them in their determination of the recommended L_1 widths. They pointed out, however, that there would be merit in performing modern XES measurements of L_1 transitions to ascertain if the trend determined from the Coster-Kronig spectroscopy data is correct. Following this suggestion, we have undertaken a series of high-resolution XES measurements in the Z-region of interest. Although dipole-forbidden L_1-$M_{4,5}$ transitions are very weak, they were preferred to the stronger L_1-$M_{2,3}$ transitions because the single recent source of M_2 and M_3 level widths is the

CP652, *X-Ray and Inner-Shell Processes: 19th International Conference on X-Ray and Inner-Shell Processes*
edited by A. Bianconi, A. Marcelli, and N. L. Saini
© 2003 American Institute of Physics 0-7354-0111-X/03/$20.00

compilation by Campbell and Papp in which, however, the $M_{2,3}$ widths were already determined from L_1-$M_{2,3}$ transitions, assuming their recommended L_1 level widths. L_1-$N_{2,3}$ transitions could neither be employed because nonlifetime broadening effects have been observed for the $N_{2,3}$ levels between $Z = 50$ (Sn) and $Z = 70$ (Yb) [11]. In addition, the uncertainties quoted by Campbell and Papp for the $N_{2,3}$ levels of elements with $Z > 60$ is rather large (0.8 eV). Thus, in the present work the level widths Γ_{L1} were determined from the observed linewidths of the L_1-$M_{4,5}$ transitions, assuming the widths $\Gamma_{M4,5}$ reported in [4] that have errors of only 0.1-0.3 eV.

MEASUREMENTS AND DATA ANALYSIS

The L_1-$M_{4,5}$ x-ray spectra of seven metallic elements in the range $62 \leq Z \leq 83$ were measured at the University of Fribourg by means of high-resolution x-ray spectroscopy, using two different curved-crystal spectrometers. For the elements $_{78}$Pt, $_{80}$Hg and $_{83}$Bi, a transmission DuMond-type crystal spectrometer was employed. As this instrument cannot be used for photon energies below about 10 keV, the L_1-$M_{4,5}$ spectra of $_{62}$Sm, $_{67}$Ho, $_{70}$Yb and $_{74}$W were measured with a reflection von Hamos-type crystal spectrometer. Except for Hg, the targets consisted of 15-22 mg/cm^2 thick metallic foils with chemical purities of 99.9% or more. The liquid Hg was enclosed in a stainless steel reservoir whose front wall was made of a 8-μm-thick Kapton foil. The target L x-ray emission was induced with a commercial 3-kW Coolidge x-ray tube with a Cr anode and a 0.5-mm thick Be window.

The transmission spectrometer [12] was operated in the so-called modified DuMond slit geometry. In this geometry the target is viewed by the bent crystal through a narrow slit which is located on the focal circle. A 0.5 mm thick (110) SiO_2 crystal plate was used for the diffraction of the x rays. The quartz lamina was bent cylindrically to a a radius R=313 cm and the effective reflecting area of the crystal was 12 x 12 cm^2. The x-ray detector was a 5-in.-diameter two-component Phoswich scintillation counter, consisting of a thin (0.25 in.) NaI(Tl) crystal followed by an optically coupled thick (2 in.) Cs(Tl) crystal. This type of detector permits to strongly reduce the Compton noise arising from high-energy photons. A further reduction of the background was achieved by sorting on-line the events of interest as a function of their energy. In the DuMond geometry, the measurements are carried out in a point-by-point way. As a result of the poor intensity of the L_1-$M_{4,5}$ transitions, very long acquisition times were needed. In order to minimize systematic errors related to long-term instabilities of the experimental set-up, each spectrum was measured in 30-40 successive scans, which were then summed off-line. The Gaussian angular instrumental broadening of the DuMond spectrometer was determined from the 2p-1s transition of Sn measured in first, second and third orders of reflection. A value of $\sigma =$ 5.21 arcsec was found, which corresponds to FWHM energy resolution of 3.2, 3.6 and 4.4 eV, respectively for the Pt, Hg and Bi L_1-$M_{4,5}$ x-ray spectra.

The von Hamos spectrometer is described in detail in [13]. It consists mainly of an x-ray source (target viewed through a narrow rectangular slit), a cylindrically bent crystal and a position sensitive detector. In the von Hamos geometry the crystal is bent around an axis that is parallel to the axis of dispersion and provides focusing in the non-dispersive direction. In contrast to the DuMond geometry, the von Hamos geometry permits at one position of the spectrometer components data collection over an energy bandwidth, which is limited primarily by the detector length. For this project the spectrometer was equipped with a 5 cm-wide and 10-cm-high $(2\bar{2}3)$ quartz crystal. The radius of curvature of the crystal was 25.4 cm. Reflected photons were detected by a 27.6-mm-long and 6.9-mm-high deep-depleted (50 μm) CCD (charge-coupled device) position sensitive detector. The latter, which consisted of 1024 pixels in the horizontal direction of dispersion and 256 pixels in the vertical direction, with a pixel resolution of 27 μm, was thermoelectrically cooled down to –60° C. The diffracted x rays hitting the CCD built a two-dimensional pattern on the detector. Each measurement consisted in collecting several thousands of two-dimensional images. For each image, good event pixels were sorted by setting an energy window corresponding to the x rays of interest. The filtered images were then added and their sum projected on the horizontal axis to obtain the final energy spectrum. The instrumental response of the von Hamos spectrometer was determined by measuring the $K\alpha_1$ x-ray lines of several elements ranging between Fe and Ge. It was found that the instrumental response could be well reproduced by a Gaussian profile whose standard deviation σ varied as a function of the x-ray energy between 1.4 and 1.8 eV.

The L_1-$M_{4,5}$ x-ray spectra were analyzed by means of the least-squares-fitting program MINUIT [14]. Voigtian profiles were employed to reproduce the observed x-ray lines because they correspond to the convolution of the Gaussian instrumental broadening with the Lorentz function representing the natural line shape of an x-ray transition. The natural widths of the L_1-$M_{4,5}$ x-ray lines were extracted by keeping

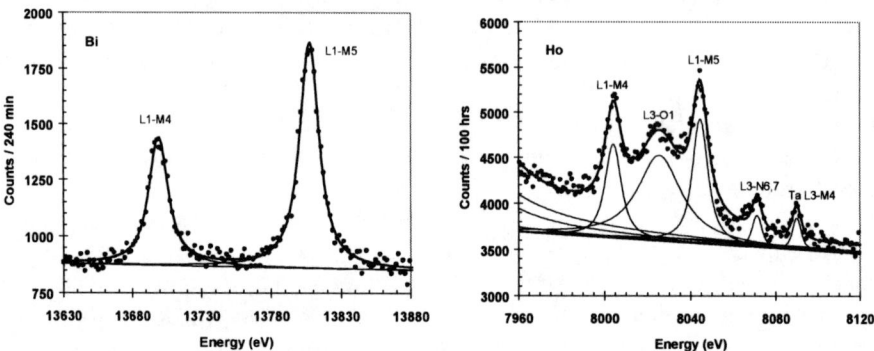

FIGURE 1. Fitted high-resolution L_1-$M_{4,5}$ x-ray spectra of Bi and Ho. The Bi spectrum which was measured with the DuMond spectrometer could be fitted easily. The analysis of the Ho spectrum which was measured with the von Hamos spectrometer was more complicated due to the presence of the L_3-O_1 and L_3-$N_{6,7}$ transitions. The Ta $L\alpha_2$ line observed at 8086 eV originates from trace impurities in the Ho target. The intensity increase appearing on the low-energy side of the spectrum is due to the tail of the strong L_3-$N_{4,5}$ transition. The latter was measured in a separate scan. The values of the fitting parameters obtained from the analysis of the L_3-$N_{6,7}$ x-ray line were then kept fixed in the fit of the L_1- $M_{4,5}$ spectrum.

TABLE 1. Natural linewidths in eV of the measured L_1-$M_{4,5}$ transitions.

| | | L_1-M_4 | | | L_1-M_5 | |
| | | Error | | | Error | |
Z	Linewidth	Fit	Total	Linewidth	Fit	Total
$_{62}$Sm	5.94	0.68	0.76	5.53	0.56	0.62
$_{67}$Ho	7.45	0.60	0.61	7.31	0.42	0.44
$_{70}$Yb	6.60	0.38	0.40	6.91	0.38	0.40
$_{74}$W	7.96	0.60	0.65	8.21	0.50	0.52
$_{78}$Pt	-	-	-	10.63	0.66	0.66
$_{80}$Hg	13.39	1.39	1.39	12.75	0.93	0.93
$_{83}$Bi	15.27	0.79	0.79	15.00	0.47	0.47

fixed in the fit the known Gaussian instrumental broadening. The energies and the intensities of the transitions as well as a linear background were used as additional free fitting parameters. For illustration, the fitted spectra of Ho and Bi are shown in Fig. 1. More details about the measurements and data analysis are given in [15].

RESULTS AND DISCUSSION

The observed linewidths of the L_1-M_4 and L_1-M_5 transitions are presented in Table 1. As a result of the poor intensity of the quadrupole transitions, the principal contribution to the total error originates from the matrix error of the fitting procedure. For Pt, Hg and Bi, which were measured with the DuMond spectrometer, the natural linewidths are 3-4 times bigger than the instrumental broadening. As the latter could be determined with a precision of about 2%, the contribution of the instrumental response uncertainty to the total errors quoted in Table 1 is negligibly small. For Pt, only the L_1-M_5 transition was measured.

The L_1 level widths obtained in the present study are presented in Table 2. They were determined from the weighted average of the linewidths of the L_1-M_4 and L_1-M_5 transitions given in Table 1 and the recommended values of the $M_{4,5}$ level widths quoted by Campbell and Papp [4]. For Pt, the L_1 width was derived from the sole L_1-M_5 transition. The 10% uncertainties assumed in [4] for the $M_{4,5}$ level widths are included in the indicated errors. The latter vary from 4% for the heaviest element (Bi) up to 10% for the lightest one (Sm). Also reported in Table 2 are the L_1 level widths recommended by Campbell and Papp [4] and theoretical predictions based on the independent-particle model from two different sources. The first one corresponds to the Lawrence Livermore National Laboratory Evaluated Atomic Data Library (EADL) [16]. The second set of theoretical predictions is due to McGuire [17]. Here, however, results are available for Ho, W, and Bi only. Our results are also presented graphically in Fig. 2, which gives an overview of existing information about the level width of the subshell L_1 in the range $50 \leq Z \leq 92$.

TABLE 2. Atomic-level widths in eV for the subshell L_1.

Z	Experimental		Theoretical	
	Present	Ref. [4]	Ref. [16]	Ref. [17]
$_{62}$Sm	4.83 ± 0.49	3.3 ± 1.5	4.3	-
$_{67}$Ho	6.23 ± 0.36	4.5 ± 1.5	5.0	5.1
$_{70}$Yb	5.40 ±0.30	5.2 ± 1.5	8.0	-
$_{74}$W	6.41 ± 0.43	6.3 ± 1.5	11.6	6.7
$_{78}$Pt	8.55 ± 0.69	8.8 ± 2.0	13.4	-
$_{80}$Hg	10.67 ± 0.79	10.5 ± 2.0	13.8	-
$_{83}$Bi	12.50 ± 0.45	12.3 ± 2.0	14.0	14.5

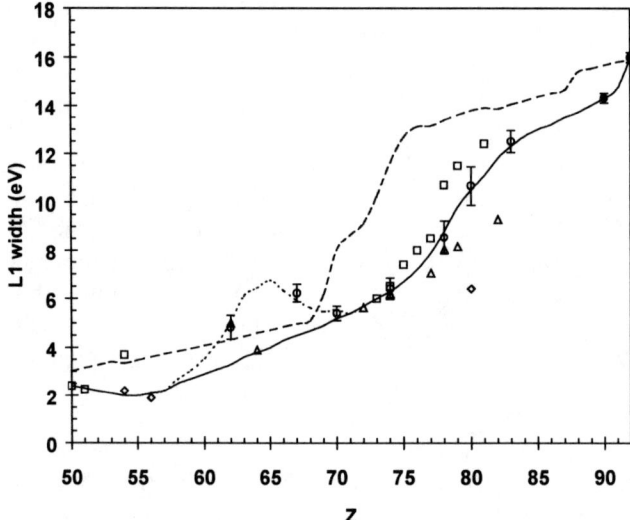

FIGURE 2. Width L_1 versus atomic number Z. Present results are depicted by open circles. Other symbols stand for experimental data from different sources. The solid line represents the recommended values of Campbell and Papp [4], the dashed line results of independent-particle calculations [16], and the dotted line predictions derived from the present work for the rare-earth-metal elements.

As shown in Table 2, an excellent agreement is found between our results and the values quoted in [4] for the elements $70 \leq Z \leq 83$. In contrast to that, the results obtained for Ho and Sm are significantly bigger than the values quoted by Campbell and Papp. We are inclined to explain the observed deviations by a splitting effect of the 2s level because a similar effect was already observed in rare-earth elements for the 5s level [18]. The splitting originates from the energy difference between the s-electron spin-up and spin-down states. Actually, the spin s = ½ of the 2s subshell couples with the spin S of the 4f subshell to yield the total spin S' = S ± ½, so that the coupling results in a doublet. As the intensities of the two components are proportional

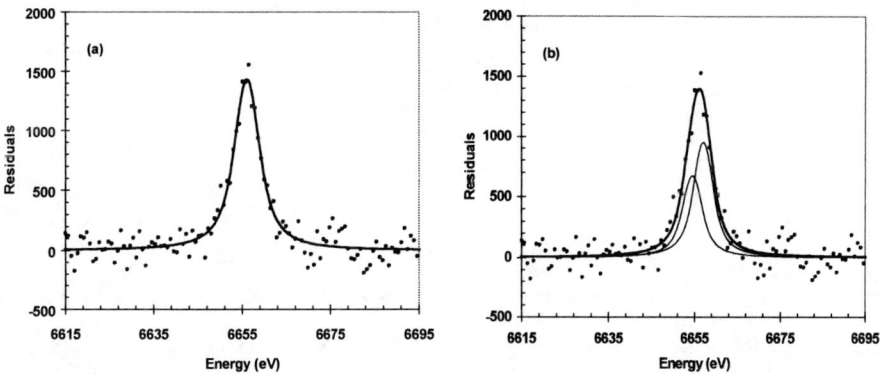

FIGURE 3. One-component (a) and two-component (b) fits of the L_1-M_5 x-ray line of Sm (see text).

to the respective multiplicities $(2S'+1)(2L+1)$, the yield ratio of the two components is given by:

$$\frac{I_1}{I_2} = \frac{[2(S+\frac{1}{2})+1]\cdot(2L+1)}{[2(S-\frac{1}{2})+1]\cdot(2L+1)} = \frac{S+1}{S},$$ (1)

where I_1 stands for the intensity of the higher energy component. The splitting energy δE can be approximated by the following expression:

$$\delta E(Z) = \varepsilon \cdot S(Z) \cdot B_{L_1}(Z).$$ (2)

In (2), B_{L1} represents the binding energy of the L_1 subshell and ε is a scaling parameter. To probe this splitting assumption, the L_1-M_5 transitions of Sm and Ho were fitted with two Lorentzians having the same width, derived from [4], and relative yields according to Eq. (1). The energy separation between the two components was used as a free fitting parameter, leading to an average value $\varepsilon = 1.31$ eV. For illustration, the one- and two-component fits of Sm are shown in Fig. 3. Theoretical profiles of the L_1-M_5 transitions of other rare-earth elements were then constructed by summing two Lorentzians of equal width, the latter being derived again from the values Γ_{L1} and Γ_{M5} reported in [4]. For each element, the relative intensities of the two Lorentzians and their energy difference were deduced from Eqs. (1) and (2), using for ε the above mentioned value. Subtracting finally from the linewidths of the calculated theoretical profiles the $M_{4,5}$ level widths quoted in [4], we were able to determine a new set of L_1 widths for the lanthanides region (dotted line of Fig. 2).

ACKNOWLEDGMENTS

This work was supported by the Swiss National Science Foundation.

REFERENCES

1. Keski-Rahkonen, O., and Krause, M.O., At. Data Nucl. Data Tables **14**, 139-146 (1974).
2. Salem, S.I., and Lee, P.L., At. Data Nucl. Data Tables **18**, 233-241 (1976).
3. Krause, M.O., and Oliver, J.H., J. Phys. Chem. Ref. Data **8**, 329-338 (1979).
4. Campbell, J.L., and Papp, T., At. Data Nucl. Data Tables **77**, 1-56 (2001).
5. Raboud, P.-A., Dousse, J.-Cl., Hoszowska, J., and Savoy, I., Phys. Rev. A **61**, 012507-1-13 (2000).
6. Hoszowska, J., Dousse, J.-Cl., and Rhême, Ch., Phys. Rev. A **50**, 123-131 (1994).
7. Werner U., and Jitschin, W., Phys. Rev. A **38**, 4009-4018 (1988).
8. Stötzel, R., Werner, U., Sarkar, M., and Jitschin, W., J. Phys. B **25**, 2295-2307 (1992).
9. Papp, T., Campbell, J.L., and Raman, S., Phys. Rev. A **58**, 3537-3543 (1998).
10. Cooper, J.N., Phys. Rev. **65**, 155-161 (1944).
11. Kowalczyk, S.P., Ley, L., Martin, R.L., McFeely, F.R., and Shirley, D.A., Faraday Discuss. Chem. Soc. **60**, 7-17 (1975).
12. Perny B., et al., Nucl. Instrum. Methods Phys. Res. A **267**, 120-138 (1988).
13. Hoszowska, J., Dousse, J.-Cl., Kern, J., and Rhême, Ch., Nucl. Instrum. Methods Phys. Res. A **376**, 129-138 (1996).
14. James F., and Roos, M., Comput. Phys. Commun. **10**, 343-367 (1975).
15. Raboud, P.-A., Berset, M., Dousse, J.-Cl., and Maillard, Y.-P., Phys. Rev. A **65**, 022512-1-10 (2002).
16. Perkins, S.T., Cullen, D.E., Chen, M.-H., Hubbel, J.H., Rathkopf, J., and Scofield, J.H., Lawrence Livermore National Laboratory Report No. UCRL-50400, 1991 (unpublished).
17. McGuire, E.J., Phys. Rev. A **3**, 587-594 (1971).
18. Cohen, E.R., Wertheim, G.K., Rosencwaig, A., and Guggenheim, H.J., Phys. Rev. B **5**, 1037-1039 (1972).

Nuclear Transitions Induced by Synchrotron X-rays

Donald S. Gemmell

Physics Division, Argonne National Laboratory, Argonne, Illinois 60439 USA

Abstract. We discuss two rare but interesting processes by which synchrotron x-rays with energies up to about 100 keV may be used to induce nuclear transitions.

In the NEET (Nuclear Excitation by Electronic Transition) process, an intense x-ray beam is employed to make vacancies, e.g. K-holes, in the atoms of a specific nuclear isotope. When a vacancy is filled by an electronic transition from a higher atomic level, there is some probability that instead of the usual x-ray or Auger emission, the nucleus of the atom itself will be excited. This is then followed by a nuclear decay exhibiting characteristic gamma-rays or other types of radiation, with time delays typical of the nuclear states involved. The probability for NEET increases when the energies of the atomic and the nuclear transitions become close. We address some theoretical aspects of the process and describe experimental efforts to observe it in ^{189}Os and ^{197}Au.

The second process to be discussed is the possibility of "triggering" the decay of a nuclear isomer by irradiation with an x-ray beam. We focus on the case of the 31-year, 2.4-MeV, 16+ isomer of ^{178}Hf. There has been speculation that if one could isolate gram quantities, say, of this isomer and then have the capability to accelerate its decay in a controlled way, one would have a powerful triggerable source of enormous energy. This could be used to generate explosions, for rapid irradiations, or for more general energy-storage applications, depending on the rate of energy release. We describe attempts to observe this process.

INTRODUCTION

Transitions between energy levels of atomic nuclei can be induced by photons in various well understood ways. Examples include resonant absorption, Coulomb excitation, and inelastic electron scattering. (The latter two are mediated by virtual photons.) Here we discuss two processes, which occur with low probability, but which nevertheless are of interest from a fundamental point of view and may have significant potential for various applications. These are 1) NEET (Nuclear Excitation by Electronic Transition), which has only recently, after many erroneous claims by several groups over a period of almost thirty years, been definitively observed in the nucleus ^{197}Au [1], and 2) the acceleration of the decay of long-lived nuclear isomers by "triggering" with x-ray irradiation, a process that has been claimed [2 - 4] to have been observed in the second isomeric state (16^+) of the nucleus ^{178}Hf, but that is disputed by other authors [5,6].

CP652, *X-Ray and Inner-Shell Processes: 19th International Conference on X-Ray and Inner-Shell Processes*
edited by A. Bianconi, A. Marcelli, and N. L. Saini
© 2003 American Institute of Physics 0-7354-0111-X/03/$20.00

In this discussion we give particular emphasis to measurements made at the Advanced Photon Source (APS) at Argonne National Laboratory and at SPring-8 in Japan. At the APS we have looked for NEET in [189]Os [7,8] and have attempted to verify the x-ray triggering of the decay of [178 m2]Hf [6]. At SPring-8, Kishimoto et al.[1] have observed NEET in [197]Au, and other experimenters [4] have reported the observation of triggering in [178 m2]Hf. Interestingly, this most recent claim to observe triggering in [178 m2]Hf postulates that the process occurs via NEET.

NEET

The concept of the NEET process was introduced in 1973 by Morita [9], who pointed out that an atomic inner-shell vacancy might have an observable probability to decay by other than the "normal" modes of x-ray or Auger emission if the decay energy were to match closely an excitation to a state of that atom's nucleus. The associated changes in spin and parity of the atomic and nuclear transitions would also have to match. The energy of the atomic state would be transferred via the exchange of a virtual photon into excitation of the nucleus of that same atom, a process akin to the inverse of internal conversion. The NEET probability, P_{NEET}, is defined as the probability that the decay of the initial excited atomic state will result in the excitation of (and subsequent decay from) the corresponding nuclear state. P_{NEET}, although generally small, is expected to be larger in heavier atoms (more compact atomic wave functions), such as the [189]Os and [197]Au considered here. Reference 10 gives a list of heavy atoms where the energy, spin, and parity requirements for atomic and nuclear transitions are simultaneously met and where, therefore, one might possibly detect NEET.

To calculate the magnitude of P_{NEET}, let us consider a heavy atom in which an electronic transition can occur between an initially excited K-vacancy state and a final M-vacancy state (a similar discussion applies to the case of an initial L-vacancy). This situation is illustrated in Fig. 1.

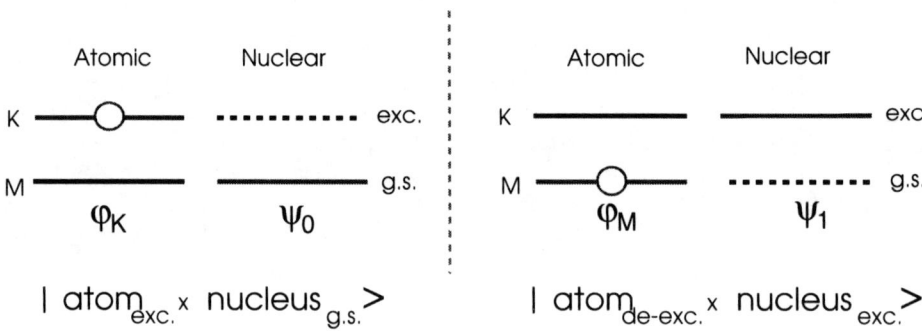

FIGURE 1. Initial and final states involved in a NEET transition.

Let us assume that the nucleus of that atom can undergo an excitation to a level involving the same changes in angular momentum and parity as are involved in the atomic transition. NEET can occur when the two product states are nearly degenerate. They are coupled by a residual interaction, V_{em}, the electromagnetic interaction of the electron hole with the protons in the nucleus. Let φ_i denote the atomic wavefunction and ψ_j the nuclear wavefunction. Following the creation of a K-hole, the initial state has a product wavefunction $|\alpha> = |\varphi_K \psi_0>$. The residual interaction generates an amplitude for the state $|\beta> = |\varphi_M \psi_1>$ and one can detect this component by measuring the nuclear decay. We can write the time evolution of the total wavefunction as

$$|\Phi(t)\rangle = a_\alpha(t)|\alpha\rangle + a_\beta(t)|\beta\rangle \quad , \tag{1}$$

where the amplitudes a_α and a_β have initial ($t=0$) values of 1 and 0, respectively. We determine the two time-dependent amplitudes from the following coupled equations, which include the off-diagonal matrix element, $\kappa = \langle \alpha|V_{em}|\beta\rangle$, and the decay rates of both states explicitly:

$$i\hbar \frac{da_\alpha}{dt} = \left(E_\alpha - i\Gamma_\alpha/2\right)a_\alpha + \kappa a_\beta \quad ,$$

$$\tag{2}$$

$$i\hbar \frac{da_\beta}{dt} = \kappa a_\alpha + \left(E_\beta - i\Gamma_\beta/2\right)a_\beta \quad ,$$

where $(E_\alpha, \Gamma_\alpha)$ and (E_β, Γ_β) are the energies and decay widths of the two product states, $|\alpha>$ and $|\beta>$, respectively. The associated decay probabilities are

$$P_\alpha = \frac{\Gamma_\alpha}{\hbar} \int_0^\infty |a_\alpha(t)|^2 dt , \quad P_\beta = \frac{\Gamma_\beta}{\hbar} \int_0^\infty |a_\beta(t)|^2 dt . \tag{3}$$

The state $|\beta>$ can decay either by a nuclear decay or by an electronic transition from the M-vacancy. An electronic transition will, however, still result in a nuclear decay at a later time. Thus P_β is equal to the "NEET probability."

The coupled equations (Eq. 2) for the coefficients, a_α and a_β, can be solved analytically [7] and this leads to an exact, if somewhat complex, expression for P_{NEET} ($=P_\beta$). For small κ, this expression reduces to

$$P_{NEET}^{\kappa \to 0} = \frac{\Gamma_\alpha + \Gamma_\beta}{\Gamma_\alpha} \frac{\kappa^2}{\left(E_\alpha - E_\beta\right)^2 + \left(\frac{\Gamma_\alpha + \Gamma_\beta}{2}\right)^2} \quad . \tag{4}$$

Estimates of P_{NEET} for various atomic/nuclear systems have been given by several authors [9 - 19], beginning with Morita [9]. Many of the early estimates involved the use of simplifying approximations that led to results at considerable variance with Eq. 4. Also, it was not recognized at first that P_{NEET} tends to be significantly higher for $M1$ transitions than for $E2$ transitions, mainly because of the involvement of atomic s-states, which leads to a stronger coupling. The more recent calculations of Tkalya [14, 16] give values that are close to those obtained from Eq. 4.

Experiments designed to demonstrate NEET face three major problems:

(1) The values of P_{NEET} tend to be small (typically less than 10^{-7}) and therefore one must generate a large number of initial vacancy states. This usually requires the use of very intense beams and highly sensitive detection techniques.

(2) Evidence needs to be established that the NEET-excited nuclear state has, in fact, been populated. This can be done by observing characteristic decay radiation from that state (as in the case of ^{197}Au) or from a lower nuclear state to which the NEET-excited nuclear state decays (as in the case of ^{189}Os). In both cases the decay times of the nuclear states involved can be employed as an additional identification of the nuclear transitions. (Observation of "prompt" radiation is generally infeasible because of difficulties with background radiation.)

(3) One has to be certain that the observed radiation is that which follows NEET transitions and not some other process(es). For example, using particle bombardment to generate the initial atomic vacancy state may also result in the Coulomb or inelastic-scattering excitation of the same state – often much more intensely than that due to NEET. Also, higher-lying nuclear states may be excited, which decay through the "NEET state", mimicking the NEET process. The use of "white" synchrotron or bremsstrahlung photon beams can similarly result in the population of the nuclear "NEET state" or of higher states by resonant nuclear absorption, thereby producing misleading results.

Early attempts to demonstrate NEET suffered from one or more of these problems. The solution [1, 7, 8, 20] was to perform the experiments using an intense monochromatic x-ray beam with an energy just above the ionization energy needed to create the initial atomic vacancies, but with the energy carefully chosen not to coincide with excited nuclear states (due attention also being given to the energies of harmonics of the fundamental beam energy).

Much of the experimental and theoretical work has been dedicated to studies of the third excited state ($5/2^-$) of ^{189}Os, lying at 69.537 keV [21]. This state decays rapidly (1.6 ns) with a partial branch ($\sim 1.2 \times 10^{-3}$) to a lower-lying metastable (6-hr. half-life) $9/2^-$ state at 30.814 keV, which in turn decays primarily by internal conversion and can readily be measured. In the NEET process involving the 69.537-keV state in ^{189}Os an initial atomic K-vacancy decays via an electronic transition from the M-shell. The KM_I (70.822 keV, $M1$), KM_{IV} (71.840 keV, $E2$), and KM_V (71.911 keV, $E2$) atomic transitions can contribute. The corresponding nuclear state at 69.537 keV can be excited via $M1$ or $E2$ transitions from the $3/2^-$ nuclear ground state.

Following the derivation of Ref. [7], the theoretical expression for the electromagnetic ($M1$ or $E2$) coupling matrix element is

$$\kappa^2 = 4\pi e^2 B\left(\Pi L, \frac{3}{2}^- \rightarrow \frac{5}{2}^-\right)\left\langle j_K \frac{1}{2} L 0 \middle| j_M \frac{1}{2}\right\rangle^2 \left(\frac{q^{L+1}|m_{\Pi L}(q)|}{(2L+1)!!}\right)^2 , \qquad (5)$$

where ΠL represents $M1$ or $E2$, and q (=35.24 Å^{-1}) is the wave number of the nuclear transition. The atomic matrix elements $m_{\Pi L}(q)$, defined in Eq. 12 of Ref. [14], were calculated using wavefunctions from the "GRASP2" code [22], and tabulated values [21] of B($M1$) and B($E2$) were used for the nuclear transition. The Clebsch-Gordan coefficient refers to the total angular momenta j_K of the K-vacancy and j_M of the M-vacancy states.

Inserting the calculated values of κ together with the atomic transition energies given above, and the calculated atomic level widths of Ref. [23], we obtain the following predictions for the NEET probabilities:

$$P_{NEET}(M1) = 1.3 \times 10^{-10}, \quad \text{and} \quad P_{NEET}(E2) = 3.8 \times 10^{-13}. \tag{6}$$

Our experiments at ANL on ^{189}Os were performed using x-ray beams from a wiggler operated by the Basic Energy Sciences Synchrotron Radiation Center [24] at the APS. After initial measurements with a white beam, in which the difficulties mentioned above became very apparent, we switched to the use of a monochromatic 98.74-keV x-ray beam to produce the K-vacancy states. This beam (5×10^{11} photons/s) was formed by Bragg diffraction from a single (440) Si crystal placed in the wiggler beam. The diffraction angle was $2\theta = 7.5°$. This particular beam energy was chosen because it lies above the osmium K-edge at 73.9 keV, does not lie near the energy of a nuclear level in ^{189}Os, and corresponds to a convenient and intense diffraction. The energy width of the beam was about 0.1% (100 eV) and the beam-spot size at the target was 0.2-mm wide and 4-mm high. The incident x-ray beam contained a comparably intense component at 49.37 keV (below the osmium K-edge) and also a few-percent component at 148.1 keV. None of the beam components had energies overlapping any of the ^{189}Os nuclear level energies, thereby avoiding problems with nuclear resonant absorption.

The decay of the 30.814-keV metastable nuclear state was measured off-line using a Ge (LEPS) detector to count the L x-ray spectrum associated with the L-conversion electrons. Figure 2(a) shows the L x-ray spectrum obtained in the initial runs using a white beam. The inset shows a corresponding measurement of the decay of this radiation. The half-life of the 30.8-keV state was measured to be 5.65 ± 0.15 h, in good agreement with the tabulated value [21] of 5.8 ± 0.1 h.

The monochromated x-ray beam from the wiggler was incident upon a thin (9.3 mg/cm^2) layer of isotopically separated (95.3%) metallic ^{189}Os electroplated onto a 0.015"-thick Cu disk using the method of Stuchbery [25]. Individual targets were irradiated in this fashion for periods of about 20 hours. Large numbers of K-vacancies were produced, some of which were expected to lead via NEET to the 69.5-keV state and thence to the 30.8-keV metastable state of the nucleus. The number of Os K-vacancies generated was monitored by on-line observation of the K x-rays using a Ge detector.

After irradiation the targets were removed and the L x-rays associated with decays of the metastable state were detected in a low-background shielded underground counting room where counting with a Ge detector proceeded also for about 20 hours. Figure 2(b) shows the results summed for two targets. From a comparison with Fig.2(a), it is apparent that within the sensitivity of this measurement, there was no evidence of the x-rays that accompany the decay of ^{189}Os metastable state (the peak at

10.3 keV in Fig. 2(b) is due to very weak natural background radiation and is only observable in extremely well shielded conditions).

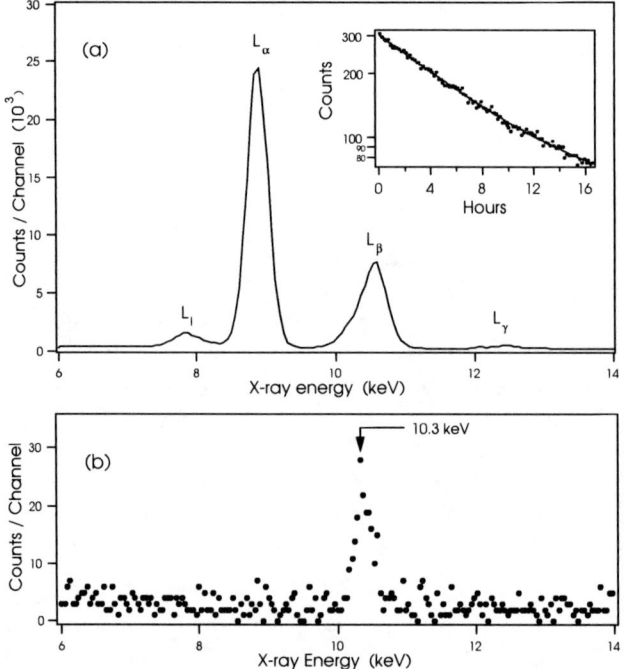

FIGURE 2. a) The osmium L x-ray spectrum obtained in the initial runs after irradiation with a white x-ray beam. (In this case, the excitation of the 30.8-keV state followed resonant nuclear absorption - - not NEET.) The inset shows a corresponding measurement of the 6-hr. decay of this radiation. b) The x-ray spectrum summed for two targets irradiated with monochromatic 98.74-keV x-rays as described in the text. The total irradiation time was 43.9 hours and the total counting time was 37.4 hours.

After taking into account factors such as the number of K-holes created during the irradiation (and their decay), the branching ratio for feeding the metastable state from the 69.5-keV state, geometrical factors, the emission probability for L x-rays in the isomeric decay, self-absorption in the target, etc, we obtained the result [7] $P_{NEET} < 9 \times 10^{-10}$. In a subsequent measurement with multiple targets and improved detection sensitivity [8], we were able to reduce this upper limit even further to $P_{NEET} < 3 \times 10^{-10}$. In more recent work at SPring-8, Aoki et al.[20], using a similar technique found $P_{NEET} < 4.1 \times 10^{-10}$. All of these values are significantly smaller than the various values obtained in previous measurements and predicted by previous calculations (see a summary in Ref. [7]). They are, however, consistent with our calculated value (Eq. 6) of $P_{NEET}(M1) = 1.3 \times 10^{-10}$ and also the recent values calculated by Tkalya [14, 16] ($P_{NEET} = 1.2 \times 10^{-10}$) and by Harston [19] ($P_{NEET} = 1.1 \times 10^{-10}$).

The small value of P_{NEET} in ^{189}Os ($\sim 10^{-10}$), which has made the process undetectable so far, is mainly due to the poor energy matching between the nuclear and atomic transitions. The mismatch is about 1.3 keV, or about 2% of the transition energy. A much more favorable case is found in ^{197}Au, where the energy of the transition from the $3/2^+$ nuclear ground state to the 77.351-keV $1/2^+$ first excited state differs by only 51 eV (0.07%) from that of the $K \rightarrow M_1$ atomic hole transition. Both transitions are $M1/E2$. However, the excited nuclear state has a half-life of only 1.91 ns [26] and so the decay radiation from the state (mainly L internal-conversion electrons) is extremely difficult to measure in a situation where the prompt background radiation from a synchrotron is overwhelming.

In an outstanding technical *tour-de-force*, Kishimoto et al.[1] have succeeded in measuring P_{NEET} in ^{197}Au. At SPring-8 this group exploited the pulsed nature of a monochromatic (80.989-keV) x-ray beam from an undulator. The pulses were of 50 ps duration, spaced 42 ns apart. Using a silicon avalanche diode to detect the internal-conversion electrons from excited nuclei, the group was able to measure time spectra showing the characteristic decay time (2.8 ns) of the 77.351-keV nuclear state. The largest source of background was the very low-level ($\sim 10^{-6}$ of the main bunch intensity) occupation of beam "buckets", spaced about 2 ns apart during the 42-ns periods between the main bunches. The NEET probability was found to be $P_{NEET} = (5.0 \pm 0.6) \times 10^{-8}$, in good agreement with the recently calculated values of Tkalya [27] (3.8×10^{-8}) and Harston [19] (3.6×10^{-8}), and in fair agreement with the results of Ljubicic [28] (7.2×10^{-9}), and Sumi and Tanaka [18] (1.1×10^{-7}).

The work of Kishimoto et al.[1] stands as the only definitive and unequivocal experimental demonstration to date of the phenomenon of NEET.

"TRIGGERING" IN $^{178\,m2}$HF

The possibility of storing energy in nuclear isomers has been discussed for many years. One of the most favorable candidates is considered to be the 2.4-MeV, 16+, 31-year level in ^{178}Hf. This isomeric state, designated $^{178\,m2}$Hf, is produced among the residual activities in nuclear spallation reactions on target materials like tantalum. The attractive notion is that if one could isolate gram quantities, say, of this isomer and then have the capability to trigger its decay so that the nuclei all dump their 2.4-MeV of gamma-ray energy in a fraction of a second rather than gradually over 31 years, one would have a powerful triggerable source of very high energy content (one gram of the isomer would contain a releasable energy of 1.3 Gigajoules, equivalent to about one third of a ton of TNT). This could be used for explosions, for rapid irradiations, or for more general energy-storage applications, depending on the rate of energy release.

This topic remained one of primarily academic interest until 1999 when a group at the University of Texas at Dallas published [2] the results of experiments in which they observed about a 4% increase in the decay rate of the $^{178\,m2}$Hf isomer when it was irradiated by a broad spectrum of x-rays from a dental x-ray machine. Over the next couple of years the same group published further results [3] basically reinforcing those in their original publication. These results generated a great deal of discussion (see, for example, Ref. [29]) and there was speculation on applications in gamma-ray lasers, countering bio-terror attacks, energy storage, etc.

In its normal decay mode, the 16⁺ isomeric state decays primarily through an *E3* transition to the 13⁻ member of the *K*=8 band whose *J*^π = 8⁻ bandhead is itself an isomer (half-life = 4 s.), decaying through a cascade in the ground-state band (Fig. 3).

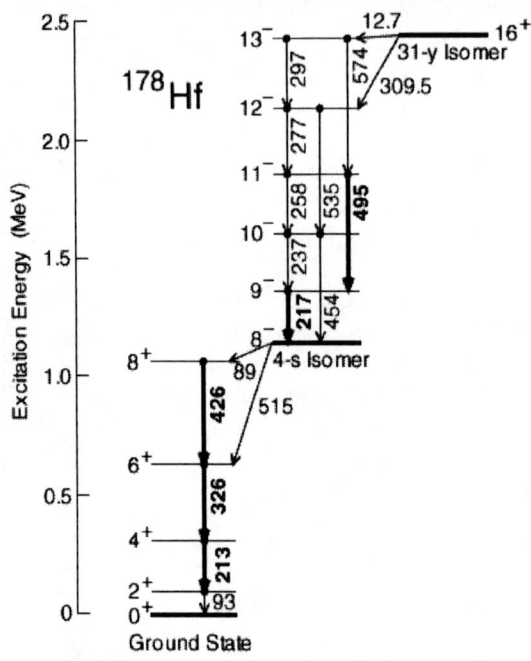

The mechanism proposed by Collins et al.was that x-rays are resonantly absorbed by ¹⁷⁸ m² Hf into a state (or states) of mixed *K* lying about 20 to 60 keV above the isomeric state. Such a state could then decay promptly to a lower-lying level in the *K* = 8 band and thence rapidly to the *J*^π = 8⁻ bandhead, which in turn decays in a well understood cascade to the nuclear ground state. A drawback with this explanation is that the measurement implies that the initial resonant absorption has an integrated cross section (1 x 10⁻²¹ cm² keV), which exceeds the values normally found for nuclear photoabsorption in this mass region by about 7 orders of magnitude [5].

FIGURE 3. Energy level diagram showing the decay of the 31-yr ¹⁷⁸Hf isomer. The transition energies are labelled in keV. Those transitions reported in [2 - 4] to be enhanced are highlighted.

An attempt to confirm this acceleration of the decay rate in ¹⁷⁸ m² Hf was made last year by a collaboration of scientists from Argonne, Los Alamos, and Livermore national laboratories [6]. This group used the APS light source to provide beams of x-rays with intensities that were over 4 orders of magnitude greater than those provided by the dental x-ray machine of Collins et al. A "white" beam from a tapered undulator at the SRI-CAT 1-1D beam line was used. Tapering the undulator gap and using two different average gap settings (15 mm and 20 mm) allowed us to smooth out the otherwise sharp energy structure inherent in undulator radiation and to be sure that there were no "holes" in our beam energy coverage in the range from about 8 keV to over 100 keV. The size of the beam spot at the target, 37 m from the undulator, was 2 x 2 mm. During the measurement, the stored electron beam at the APS was maintained at a steady 100 mA through the use of a continuous "top-up" mode of operation.

We employed three separate targets containing 7.3 x 10¹⁴, 3.0 x 10¹⁵, and 6.4 x 10¹⁵ isomeric ¹⁷⁸Hf nuclei, respectively. The Hf was produced at LANL using the LANSCE accelerator to induce 800-MeV proton spallation reactions on thick Ta targets/beam stops. The Hf was chemically extracted from the Ta target material, purified, precipitated and fired to produce HfO_2 for use in the current experiment. In addition to the 31-year [178]Hf isomer, the target contained some stable Hf from impurities in the beam stop. The spallation reaction also produced other Hf isotopes, including trace amounts of [172]Hf ($t_{1/2}$ =1.89 yr), that proved useful in monitoring the target during the experiment. For each of the three targets, the hafnium oxide was mixed with fine aluminum powder (for heat conduction purposes - - the beam typically deposited ~200 W into the HfO_2) and then sealed by electron-beam welding into a 2-mm diameter by 1.6-mm deep cavity machined into a water-cooled aluminum block. The aluminum entrance window for each target was 0.15 mm thick.

X-rays and gamma rays from the target were measured with two heavily shielded planar Ge detectors at 90° to the beam. The Hf K X-rays were used to monitor the target. Gamma-ray spectra were accumulated and time-sorted as the incident beam was cycled on (for 11 s.) and off (for 22 s., divided into two equal periods of 11 s.). In this way we were able to search for both prompt and delayed (through the $J^\pi = 8^-$ bandhead) enhancements of the [178 m2]Hf decay rate. (A beam-induced decay of the isomer would result in an increased production of the 4-s. isomer and the emission of the subsequent gamma rays in the ground-state band would be enhanced in the first half of the beam-off period as compared to the second.)

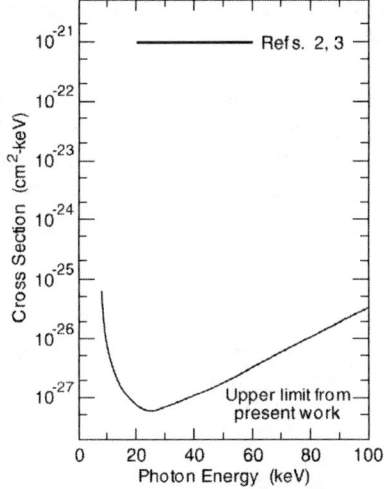

FIGURE 4. Upper limit of the integrated cross section for photon-induced de-excitation of the 31-yr [178]Hf isomer through the 4-s, 8^- isomer. The corresponding value reported by Collins et al.[2, 3] is also shown.

Based on the results of Refs [2, 3] (a ~4% enhancement of the decay rate due to resonant photon absorption in the range 20 - 60 keV), the decay rates observed in our experiment should have been enhanced by a factor of ~20,000. Instead, we saw that the decay rate was constant to within our uncertainty of ± 2%, i.e. about a million-fold smaller than would be consistent with Refs [2, 3]. Our result implies an integrated photon absorption cross section of less than 2×10^{-27} cm^2 keV in the 20 - 60 keV range, a value much more in keeping with the accumulated body of knowledge of gamma-ray transition strengths in this mass and energy region. Figure 4 shows our upper limit as a function of photon energy.

At low energies, the upper limit rises steeply due to (a) the rapid rise of absorption in the target material, and (b) the fact that the incident beam loses flux at energies below about 8 keV (due to absorption effects in various windows in the beam line).

Although our measurement was designed primarily to address the question of resonant absorption in the originally reported 20 - 60 keV range, our data, as indicated in Fig. 4, cover a somewhat wider energy region. At the extreme low-energy portion of this range (~8 keV), the difficulties mentioned above come into play. For this reason, and to obtain more accurate data at energies around 5 - 20 keV, we have recently performed additional measurements [30] using thin targets of pure HfO_2 (containing $^{178\ m2}Hf$, as before) sandwiched between thin Be foils. These HfO_2 targets were thin compared to the mean-free-path of 8-keV photons. Again, no enhancements were observed. A preliminary analysis shows that the upper limit for the integrated cross section in the range 7 - 30 keV is everywhere less than 8×10^{-27} cm^2 keV. At 8 keV, the upper limit has been reduced to 5×10^{-28} cm^2 keV, a factor of 100 lower than obtained in our previous measurement (Fig. 4).

In a recent publication [4], Collins et al.show results obtained by scanning a monochromatic x-ray beam at SPring-8 over the energy range 9 - 13 keV. Here too, they claim to observe x-ray induced acceleration of the decay of $^{178\ m2}Hf$. The enhancements range from 1% to 3%. In this case, the authors postulate that the mechanism is NEET, involving initial L-vacancies created by the incident photons. The enhancements appear to track the L-absorption yields. Just above the L_1 edge, for example, their measurements imply $P_{NEET} = 2 \times 10^{-3}$, which vastly exceeds any other measured or calculated value. For instance, Tkalya [16, 31] calculates that even in the extremely favorable (and unlikely) case in which the atomic and the nuclear energies match exactly, and assuming that the $(E1)$ NEET transition proceeds through the K-mixed state with a reduced transition probability of 0.01 Weisskopf units (the upper limit for transitions in this mass region), then one expects a value $P_{NEET} \approx 10^{-20}$. This includes a factor of $\sim 10^{-14}$ to take into account the admixture of $\Delta K = 8$ in the intermediate state.

In summary, it appears that attractive as the possibility of "triggered" decay of isomers may seem, the accuracy of the measurements of Collins et al.[2 – 4] do not provide convincing evidence that the process has been observed. The measurements made at the APS indicate upper limits in the case of $^{178\ m2}Hf$ that are consistent with known nuclear parameters in this region of mass and energy.

ACKNOWLEDGEMENTS

This work was supported in part by the U.S. Department of Energy (DoE) under contract No. W-31-109-ENG-38 (ANL), in part by the DoE under contract Nos. W-7405-ENG-48 (UC-LLNL), and W-7405-ENG-36 (LANL), and in part by the DoE Nuclear Energy Research Initiative (NERI). We are indebted to J. Greene, T.L. Khoo, C.J. Lister, R. Nelson, W. Patterson, and K. Teh for their help in various aspects of these measurements, to D.R. Haeffner, D.M. Mills, and G. Shenoy for their encouragement of this investigation at the SRI-CAT 1-ID beamline, and to E. V. Tkalya for his theoretical assistance.

REFERENCES

1. Kishimoto, S., Yoda, Y., Seto, M., Kobayashi, Y., Kitao, S., Haruki, R., Kawauchi, T., Fukutani, K., and Okano, T., *Phys. Rev. Lett.* **85**, 1831 (2000).
2. Collins, C.B., Davanloo, F., Iosif, M.C., Dussart, R., Hicks, J.M., Karamian, S.A., Ur, C.A., Popescu, I.I., Kirischuk, V.I., Carroll, J.J., Roberts, H.E., McDaniel, P., and Crist, C.E. *Phys. Rev. Lett.* **82**, 695 (1999).
3. Collins, C.B. et al., *Laser Phys.* **9**, 8 (1999); Collins, C.B. et al., *Phys. Rev.* **C 61**, 054305 (2000); Collins, C.B. et al., *Phys. At. Nucl.* **63**, 2067 (2000).
4. Collins, C.B. et al., *Europhys. Lett.*, **57**, 677 (2002).
5. Olariu, S., and Olariu, A., *Phys. Rev. Lett.*, **84**, 2541 (2000); McNabb, D.P. et al., *Phys Rev. Lett.*, **84**, 2542 (2000); von Neumann-Cosell, P., and Richter, A., *Phys. Rev. Lett.*, **84**, 2543 (2000).
6. Ahmad, I., Banar, J.C., Becker, J.A., Gemmell, D.S., Kraemer, A., Mashayekhi, A., McNabb, D.P., Miller, G.G., Moore, E.F., Pangault, L.N., Rundberg, R.S., Schiffer, J.P., Shastri, S.D., Wang, T.F., and Wilhelmy, J.B., *Phys. Rev. Lett.*, **87**, 072503 (2001).
7. Ahmad, I., Dunford, R.W., Esbensen, H., Gemmell, D.S., Kanter, E.P., Rütt, U., and Southworth, S.H., *Phys. Rev.* **C 61**, 051304(R) (2000).
8. Ahmad, I., Dunford, R.W., Gemmell, D.S., Lister, C.J., Siemssen, R.H., and Southworth, S.H., *(2000, unpublished)*.
9. Morita, M., *Prog. Theor. Phys.* **49**, 1574 (1973)
10. Okamoto, K., *Laser Interactions and Related Phenom.* **4 A**, 283 (1977).
11. Pisk, K., Kaliman, Z., and Logan, B.A., *Nucl. Phys.* **A 504**, 103 (1989).
12. Bondar'kov, M.D., and Kolomietz, V.M., *Izv. Akad. Nauk. SSSR, Ser. Fiz.* **55**, 983 (1991).
13. Ljubicic, A., Kekez, D., and Logan, B.A., *Phys. Lett.* **B 272**, 1 (1991).
14. Tkalya, E.V., *Nucl. Phys.* **A 539**, 209 (1992).
15. Ho, Y-k., et al, *Phys. Rev.* **C 48**, 2277 (1993).
16. Tkalya, E.V., "Nuclear Excitation by Electronic Transition Between Atomic Shells," in *X-Ray and Inner-Shell Processes*, edited by R.W.Dunford et al., AIP Conference Proceedings 506, Melville, New York, 2000, pp.486-495.
17. Ljubicic, A., "Excitations of Nuclear Levels in Atomic Transitions," in Proceedings of the International Meeting on Frontiers of Physics 1998, Kuala Lumpur, 26 – 29 October 1998, edited by S.P. Chia and D.A. Bradley, World Scientific, pp. 133 – 141.
18. Sumi, Y., and Tanaka, S., *Jpn. J. Appl. Phys.* **39**, 1894 (2000).
19. Harston, M.R., *Nucl. Phys.* **A 690**, 447 (2001).
20. Aoki, K., et al., *Phys. Rev.* **C 64**, 044609 (2001).
21. Firestone, R.B., *Nucl. Data Sheets*, **59**, 869 (1990).
22. Parpia, F.A., Fischer, C.F., and Grant, I.P., *Comput. Phys. Commun.* **94**, 249 (1996).
23. See, for example, Leisi, H.J., et al, *Helv. Phys. Acta*, **34**, 161 (1961), and McGuire, E.J., *Phys. Rev.* **A 5**, 1043 (1972).
24. Montano, P.A., et al, *Rev. Sci. Instrum.* **66**, 1839 (1995).
25. Stuchbery, A.E., *Nucl. Instrum. Methods.* **211**, 293 (1983).
26. Chunmei, Z., *Nucl. Data Sheets*, **62**, 433 (1991).
27. Tkalya, E.V., Ref. [16], corrected for the omission of a small numerical factor *(Private Communication)*.
28. Ljubicic, A., Ref. [17], corrected for a printing error *(Private Communication)*.
29. *SCIENCE*, **283**, 769 (1999).
30. Ahmad, I., Banar, J.C., Becker, J.A., Bredeweg, T., Cooper, J., Gemmell, D.S., Mashayekhi, A., McNabb, D.P., Moore, E.F., Palmer, P.D., Rundberg, R.S., Schiffer, J.P., Shastri, S.D., Wang, T.F., and Wilhelmy, J.B., *(to be published)*.
31. Tkalya, E.V., *(Private Communication)*.

Multiple excitations in the K fluorescence emission of Mn, Fe and Ni compounds

P. Glatzel*, U. Bergmann†, F. M. F. de Groot** and S. P. Cramer†*

*University of California at Davis, USA
†Lawrence Berkeley National Laboratory, USA
**Universiteit Utrecht, The Netherlands

Abstract. The creation and the decay of a *1s* vacancy can result in the excitation of a second electron. In this paper, two different modes of *1s* core hole creation as a diagnostic tool to study multi-electron excitations in the K fluorescence emission are compared. The *1s* core hole excited state can be created either by photoionization or by radioactive K capture decay. In the latter case, a *1s* electron reacts with a proton in the nucleus to yield a neutron and an escaping electron neutrino. We report a comparison of Kβ spectra obtained from x-ray excitation in Mn and K capture in 55-Fe in various chemical environments. The K$\beta_{1,3}$ main lines of the photoexcited spectra are broader than the corresponding lines obtained after K capture and the weak satellite lines at higher energies (K$\beta_{2,5}$) differ in shape. A theoretical model on the basis of different electron relaxation depending on the mode of core-hole creation is presented.

INTRODUCTION

Any electronic transition in an atom causes a readjustment of the passive electrons, *i.e.* the electrons that are not directly involved in the transition, to the perturbed potential. A passive electron in an orbital ϕ_μ^i before the perturbation will relax into an orbital $\phi_{\mu'}^f$. For an adiabatic relaxation we have $\mu=\mu'$, *i.e.* the final state orbital is described by the same set of quantum numbers. If the relaxation is non-adiabatic, the electron will occupy an orbital with different quantum numbers ($\mu \neq \mu'$). Those transitions can be referred to as shake or multiple electronic transitions.

In a two step picture that can be used to describe fluorescence emission, the excitation of a second electron can either occur during the creation or during the decay of a *1s* vacancy. Multi-electron transitions upon inner-shell vacancy creation in *3d* transition metals have been studied using photoelectron [1] and absorption spectroscopy [2, 3, 4]. In order to obtain an estimate for the fluorescence energies due to transitions from doubly ionized excited states we can use a simple Z+1 model as shown in Figure 1. For Mn we expect the KLβ lines at about 70 eV above the Kβ main lines. As an example, the KLβ lines in MnO$_2$ are shown. The KLβ peak intensity exhibits a slow rise over several keV as predicted by the Thomas model [5]. The spectra shown in Figure 1 are not corrected for self-absorption and we therefore do not attempt for a detailed analysis. A comprehensive treatment of the KLα lines in Cu can be found in reference [6, 7].

When the core hole is filled the excitation of a second electron can occur via a radiative Auger emission (RAE) where a photon is emitted and an electron is simultanuously

CP652, *X-Ray and Inner-Shell Processes: 19th International Conference on X-Ray and Inner-Shell Processes*
edited by A. Bianconi, A. Marcelli, and N. L. Saini
© 2003 American Institute of Physics 0-7354-0111-X/03/$20.00

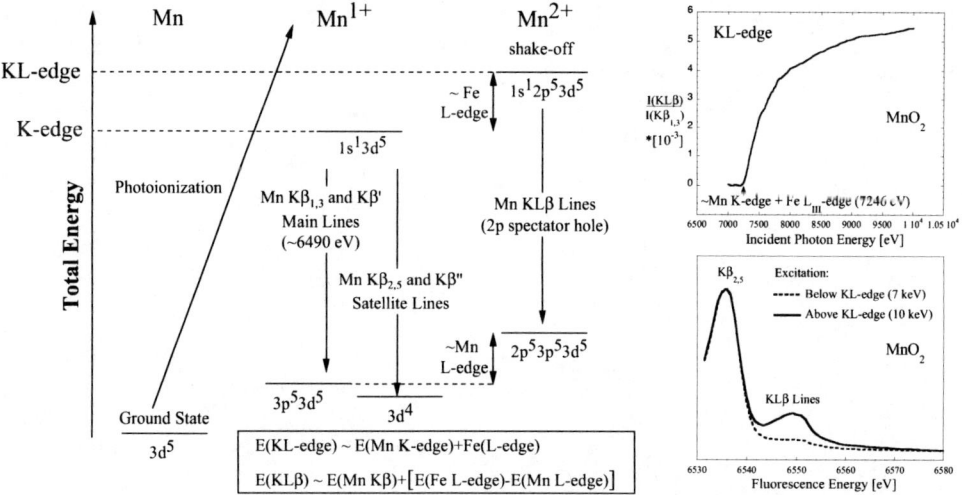

FIGURE 1. *Left: Term scheme for Kβ and KLβ emission in Mn. Atomic configurations are used and only the partly occupied orbitals are given. An estimate for the KL-edge and the KL fluorescence energy is given based on a simple Z+1 model. Right: Kβ main and satellite lines. For the latter, the fluorescence emissions following photoexcitation below (solid line) and above (dashed line) the KL-edge are shown.*

elevated into a higher orbital or into the continuum [8]. Figure 2 shows the KLL RAE in metallic Ni and in K_2NiF_6. The shake probability is related to the overlap integral between the emitting and the receiving orbital and is largest for monopole shake transitions [9]. The KLL-edge therefore resembles the K-edge and not the L-edge because it probes the p-density of unoccupied states. The Ni KLL onset is at about 880 eV below the Ni Kα lines.

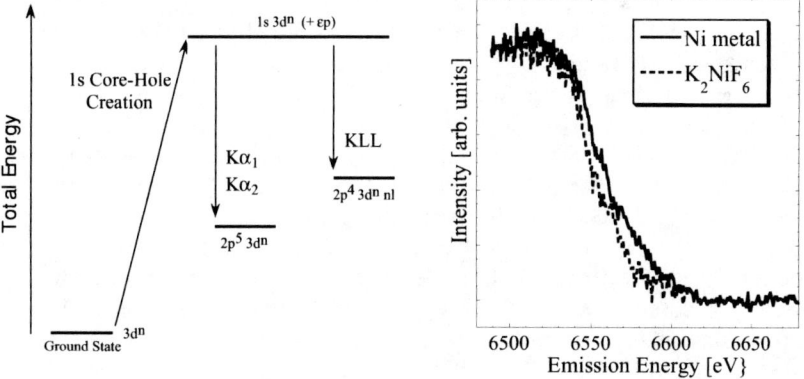

FIGURE 2. *Left: Term scheme of KLL radiative Auger emission in a 3d metal. Right: KLL edge in Ni metal and K_2NiF_6. The KLL edges differ between the metallic Ni and Ni(IV) in an ionic compound, i.e. they show a chemical dependence.*

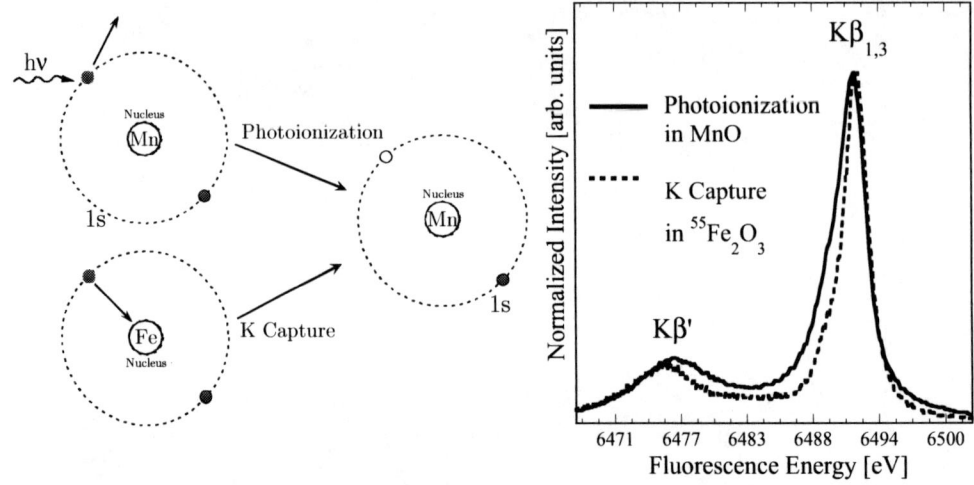

FIGURE 3. *Left: K capture in* ^{55}*Fe and 1s photoionization in Mn. Both processes result in an ion that has a Mn (Z=25) nucleus and a hole in the 1s shell. Right: Experimental Kβ spectra for MnO and* ^{55}*Fe$_2$O$_3$. The instrumental broadening is ≈ 0.8 eV for both spectra. The spectra are normalized to each other in the Kβ$_{1,3}$ peak. The energy scale is the measured fluorescence energy and the spectra were not shifted in energy relative to each other.*

COMPARING DIFFERENT MODES OF 1S CORE HOLE CREATION

Kβ Main Lines

As we have seen in the previous examples, doubly excited intermediate or final states yield sufficiently shifted (*i.e.* can be separated experimentally) K fluorescence lines if the second vacancy occurs in the L or K shell. In contrast, valence electron shake transitions will result in overlapping spectral features between singly and doubly fluorescence final states. In order to tackle this problem one can compare K fluorescence spectra following two different modes of *1s* core hole creation [10, 11]. Provided that the two modes have different valence electron shake probabilities one expects different K fluorescene lines if shake transitions have considerable probability.

On the left side of Figure 3 it is shown that photoionization in Mn and radioactive electron capture decay from the K shell (K capture) lead to the formally identical fluorescence initial state with a Mn-55 nucleus and a hole in the *1s* shell. We compare on the right side of Figure 3 the Kβ main lines, that arise from *3p* to *1s* transitions, for MnO and ^{55}Fe$_2$O$_3$. Both compounds have a metal $3d^5$ configuration in the ionic approximation. The striking difference between the two spectra is that the Kβ$_{1,3}$ peak appears sharper on the low energy side in the ^{55}Fe$_2$O$_3$ than in the MnO spectrum than. A multiplet approach using atomic self-consistent field calculations can account for the Kβ$_{1,3}$ and Kβ′ features[12]. For a more detailed treatment the local symmetry (O$_h$ for both comounds) and orbital hybridization can be taken into account [13, 14]. It is found

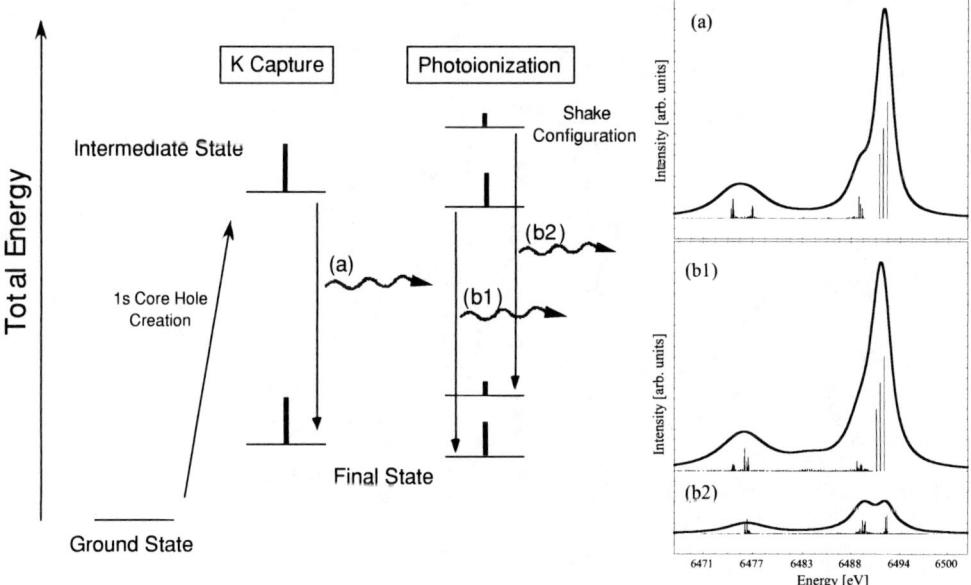

FIGURE 4. *Left: Term scheme for Kβ emission after core hole creation via K capture and photoionization, respectively. The change of effective potential upon photoionization leads to a populated shake configuration in the intermediate state. Assuming that the shake configuration does not relax into the lowest intermediate state we obtain two Kβ emitting transitions. Right: Calculated Kβ main lines after (a) K capture and (b) photoionization. For the latter the two spectra as described in the term scheme on the left are shown and the spectra are scaled according to the calculated intensities of the 1s intermediate states.*

that different hybridization and crystal field splittings between MnO and Fe_2O_3 cannot explain the differences observed in the experimental spectra [15].

Figure 4 shows how the dependence of the spectral features on the mode of $1s$ core hole creation can be incorporated into a theoretical model. The change of effective potential experienced by the valence electrons after $1s$ photoionization causes shake transitions and, in a simplified model, two populated intermediate states. Both states act as fluorescence initial states and give rise to Kβ emission. We therefore assume that the shake configuration in the $1s$ intermediate state does not decay into the lowest $1s$ intermediate state before the decay of the $1s$ core hole, *i.e.* the fluorescence intial state is not fully relaxed. On the other hand, the effective potential experienced by the valence electrons hardly changes in K capture decay because a negative charge (the $1s$ electron) annihilates with a positive charge (a proton in the nucleus) [15].

The calculated spectra on the right side of Figure 4 are based on atomic multiplet calculations including crystal field splitting and ligand-to-metal charge transfer [13, 15]. Two Kβ spectra are shown for MnO after photoionization. They are scaled according to the calculated population of the $1s$ excited states that are reached in a photoionization process. Only one calculated Kβ spectrum is shown in the case of K capture because only one $1s$ excited state is reached. Adding up the two photoionization Kβ spectra

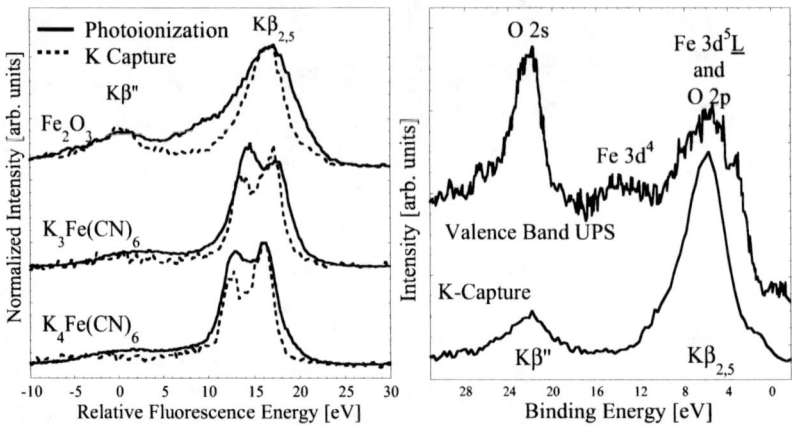

FIGURE 5. *Kβ satellite emission in Fe compounds after K capture (KC) and photoioniziation (PI). To facilitate direct comparison a common energy scale (relative fluorescence energy) is given in the bottom. The feature in Fe₂O₃ at about 0 eV relative fluorescence intensity is the Kβ″ or cross over peak. It was used to align the photoionization and K capture spectra relative to each other.*

will yield a spectrum that is broadened on the low energy side of the $K\beta_{1,3}$ peak in comparison to the K capture spectrum and thus reproduce the experimental observation [15].

Kβ Satellite Lines

Transitions from shells higher than *3p* (Kβ satellite lines) can be interpreted using density functional theory [16, 17]. The Kβ satellite lines in Fe₂O₃, K₄Fe(CN)₆ and K₃Fe(CN)₆ are shown in Figure 5. While the Kβ main lines are dominated by splittings due to intraatomic Coulomb and spin-orbit interactions whose magnitudes depend on the nuclear charge, the Kβ satellite lines are mainly shaped by ligand field effects and band formation. We therefore compared a Mn to an Fe-55 compound in the previous section and now compare identical Fe and Fe-55 compounds for the Kβ satellites. The fluorescence energy scale had to be shifted for a comparison. We used the Kβ″ peak that has been assigned to a ligand *2s* to metal *1s* transition [18] to align the K capture and photoionization spectra.

All spectral features recorded after photoionization are broader than in the data taken after K capture. Furthermore, the intensity ratios of the two bands in the $K\beta_{2,5}$ structure of the cyanides change between the two modes of excitation. The broadening indicates that more final states are populated after photoionization as we already concluded for the Kβ main lines.

On the right of Figure 5 we compare Fe₂O₃ valence band photoemission spectra (taken from reference [19]) to the ^{55}Fe₂O₃ Kβ satellite spectra. Fujimori *et al.* interpreted the valence band UPS spectra using a ligand-to-metal charge transfer model and assigned the strong peak at low binding energies to a screened $3d^5\underline{L}$ configuration and an O $2p$ band and the weaker structure at about 14 eV binding energy to an unscreened $3d^4$ configuration. The $K\beta$ satellite lines are dominated by transitions from orbitals with mainly O $2s$ and $2p$ character. These orbitals exhibit some p-character relative to the metal center and therefore serve for dipole allowed transitions to the metal $1s$ shell.

ACKNOWLEDGMENTS

We thank Sandra Fiskum at PNNL for preparing the [55]Fe compounds. SSRL is funded by the Department of Energy, Office of Basic Energy Sciences. We are indebted to Bernd Sonntag and Jorgen Hansen for fruitful discussions. Use of the Advanced Photon Source was supported by the U.S. Department of Energy, Basic Energy Sciences, Office of Science, under contract No. W-31-109-ENG-38. BioCAT is a National Institutes of Health-supported Research Center RR-08630 This work was supported by the National Institutes of Health GM-44380 and the Department of Energy, Office of Biological and Environmental Research.

REFERENCES

1. Hüfner, S., *Photoelectron spectroscopy : principles and applications*, Springer series in solid-state sciences ; 82., Springer, 1996, 2nd edn.
2. Bianconi, A., Garcia, J., Benfatto, M., Marcelli, A., Natoli, C. R., and Ruiz-Lopez, M. F., *Phys. Rev. B, Condens. Matter (USA)*, **43**, 6885–92 (1991).
3. Bianconi, A., Chenxi, L., Campanella, F., Della Longa, S., Pettiti, I., Pompa, M., Turtu, S., and Udron, D., *Phys. Rev. B, Condens. Matter (USA)*, **44**, 4560–9 (1991).
4. Guo, J., Ellis, D. E., Goodman, G. L., Alp, E. E., Soderholm, L., and Shenoy, G. K., *Phys. Rev. B, Condens. Matter (USA)*, **41**, 82–95 (1990).
5. Thomas, T. D., *Phys. Rev. Lett. (USA)*, **52**, 417–20 (1984).
6. Deutsch, M., Gang, O., Hamalainen, K., and Kao, C. C., *Phys. Rev. Lett. (USA)*, **76**, 2424–7 (1996).
7. Fritsch, M., Kao, C. C., Hamalainen, K., Gang, O., Forster, E., and Deutsch, M., *Phys. Rev. A, At. Mol. Opt. Phys. (USA)*, **57**, 1686–97 (1998).
8. Aberg, T., *Phys. Rev. A, Gen. Phys. (USA)*, **4**, 1735–40 (1971).
9. Fujikawa, T., and Kawai, J., *J. Phys. Soc. Jpn. (Japan)*, **68**, 4032–6 (1999).
10. Mukoyama, T., and Uda, M., *Phys. Rev. A*, **61**, 030501/1–4 (2000).
11. Hansen, P. G., Jonson, B., Borchert, G. L., and Schult, O. W., *Atomic inner-shell physics*, Plenum Press, 1985, pp. 237–267.
12. Meisel, A., Leonhardt, G., and Szargan, R., *X-Ray Spectra and Chemical Binding*, vol. 37 of *Chemical Physics*, Springer-Verlag, 1989.
13. de Groot, F. M. F., Fontaine, A., Kao, C. C., and Krisch, M., *J. Phys.*, **6**, 6875–84 (1994).
14. Kotani, A., *J. Electron Spectrosc. Relat. Phenom (Netherlands)*, **100**, 75–104 (1999).
15. Glatzel, P., Bergmann, U., de Groot, F. M. F., and Cramer, S. P., *Phys. Rev. B, Condens. Matter Mater. Phys. (USA)*, **64**, 045109/1–10 (2001).
16. Bergmann, U., Bendix, J., Glatzel, P., Gray, H. B., and Cramer, S. P., *J Chem Phys*, **116**, 2011–2015 (2002).
17. Drager, G., and Brummer, O., *Phys. Status Solidi B (East Germany)*, **124**, 11–28 (1984).
18. Bergmann, U., Horne, C. R., Collins, T. J., Workman, J. M., and Cramer, S. P., *Chem. Phys. Lett.*, **302**, 119–124 (1999).
19. Fujimori, A., Saeki, M., Kimizuka, N., Taniguchi, M., and Suga, S., *Phys. Rev. B, Condens. Matter*, **34**, 7318–28 (1986).

Cascading Decays of Inner-Shell Vacancies

A. G. Kochur[1], V. L. Sukhorukov[1], V. F. Demekhin[1], C. Gerth[2],
B. Kanngießer[3] and P. Zimmermann[3]

[1]Rostov State University of Transport Communication, Rostov-na-Donu, 344038 Russia
[2]Deutsches Elektronen-Synchrotron DESY, D-22603 Hamburg, Germany
[3]Institut für Atomare Physik und Fachdidaktik, Technische Universität Berlin, D-10623 Berlin, Germany

Abstract. Inner-shell vacancy can decay by consecutive radiative and non-radiative transitions giving rise to multiply ionized ionic states. We propose a theoretical scheme based on the straightforward construction of de-excitation trees allowing one to calculate various characteristics of complex decay cascades. Recent results on the decay of hollow atoms and photoelectron-photoion coincidence measurements are reported. The role of many-electron correlations in multiple processes is discussed.

STATE OF THE PROBLEM

Ionization of an atomic inner shell gives rise to a highly excited short-lived state liable to decay. The vacancy can decay either radiatively with the emission of photon or non-radiatively, *via* Auger or Coster-Kronig transitions, with the emission of electron. Radiative transitions move the vacancy to shallower levels, while upon radiationless transitions one deep vacancy is replaced with two vacancies in outer shells. Most of the vacancies produced after the first decay step can decay further, and a new set of vacancies is produced. The decay process is then a multi-step process of consecutive decays which stops when all the vacancies are in the uppermost shells and can decay no further.

Since in most of the cases radiationless transitions are favoured, the cascading decays produce multiply charged ions. Every transition in a cascade leads to emission of either a photon or an electron, therefore, the cascade-affected emission and electron spectra can be measured. To register those spectra one must excite an inner shell of an atom (say, K) while the spectrometer should be tuned to detect the spectra of some outer shell (say, L or M). Since the low-energy transitions being detected may take place in presence of a variety of additional vacancies produced by the cascade, the spectra of cascades have very complex multi-component satellite structures.

The processes of cascading decay of vacancies, have been studied since sixties. The first measurements of the yields of multiply charged final ions were performed by Krause and Carlson with co-workers [1-5]. They used X-ray tubes to prepare initial deep-initial-vacancy states and magnetic mass-spectrometers to detect the photoions.

CP652, X-Ray and Inner-Shell Processes: 19th International Conference on X-Ray and Inner-Shell Processes
edited by A. Bianconi, A. Marcelli, and N. L. Saini
© 2003 American Institute of Physics 0-7354-0111-X/03/$20.00

They were also the first to perform the calculations on multiply charged ions yields using Monte-Carlo scheme to simulate the decay trees.

A new interest to cascading production of photoions was inspired by the possibility to greatly increase the accuracy of measurements using the synchrotron radiation for the ionization of atoms near thresholds and the time-of-flight mass spectrometers to detect ions [6-13].

The number of works devoted to the cascade-affected electron spectra is not great. The argon $L_{23}MM$ spectrum et exciting photon energies between the K and L_1 edges has been reported by Cooper *et al.* [14]. Southworth *et al.* [15] measured the argon $L_{23}MM$ spectrum at monochromatic excitation just above the K-threshold. Von Busch *et al* [16] measured the same spectrum at broad-band photon excitation above the K-threshold. Calculations on cascade-affected argon $L_{23}MM$ spectra were performed in [14,16] as well. Omar and Hahn [17] calculated Ar $L_{23}MM$ spectrum upon K-ionization, Mirakhmedov and Parilis [18] calculated Auger spectra of various series of krypton upon K-ionization, Kochur and Sukhorukov calculated low-energy cascade-affected electron spectra in argon and krypton [19], and in xenon [20] upon ionization of various inner shells.

The experimental data on cascade-affected emission spectra are very scarce. Xenon $N_{45}-O_{23}$ emission spectra emitted after predominantly M_{45}-ionization by 1keV electron impact were reported by Verkhovtseva et al. [21,22]. Bruhl [23] measured the same spectrum at photon excitation just above the M_{45} threshold. The assignments of the spectra in refs [21-23] were given based mostly on energy positions of specific spectral features, and they differ drastically from each other. The first attempt to assign the Xe $N_{45}-O_{23}$ emission upon decay of the M_{45} vacancy based on calculation of the whole de-excitation tree was made by Mitkina [24].

A number of theoretical works were dealing with emission spectra of cascades [17,24-27]. The difficulty with the cascade-affected emission spectra is that some of the LSJ terms in multi-vacancy configurations produced by a cascade cannot decay non-radiatively due to the symmetry reasons, and the emission from those states increases enormously. One has, in principle, to consider accurately the decays of each specific term of each configuration created by the cascade. This, however, has been possible so far only for very simple cases, like M-emission upon L-ionization in Al [27]. For a more complex case of L-emission upon K-ionization of Ar, a simplified scheme based on modification of configuration-average approach was proposed in ref. [25].

The experimental analysis of the cascade dynamics is faciliated greatly by using coincidence techniques. Auger-electron-photoion coincidence measurements were employed in [12,13] which allowed to study separately the decay of each state produced after the first-step decays of the initial vacancy. One can also use photoelectron-photoion coincidence scheme to separate the initial inner-shell vacancy state [13,28]. This selection is important since upon, say, photoionization a set of various vacancy states is normally produced. A more elaborate recent coincidence experiments [29,30,31] will be discussed in detail in the following sections of this paper.

The experiments on cascade-produced photoions in the cases of less-hollow initial vacancies detect the presence of ions with charges that cannot be reached if only diagram transitions are considered [32,31,28,13]. This is a direct evidence of double Auger processes when two electrons are ejected simultaneously.

It should be noted that already Carlson and Krause with co-workers [1-5] were able to include in their theory one of the double mechanisms – the monopole shake processes, i.e. the processes when due to the sudden change of the core potential, say, upon diagram Auger transition, one of the outer-shell electrons is shaken up or off without change of its orbital momentum. However, with consideration of only monopole double processes it was impossible to get agreement with the experiment, for example, on Ne^{3+} yield upon Ne K ionization [38].

Non-monopole double Auger processes (DAP) upon KLL decays in neon were approached theoretically first by Amuia et al. [34] and then by Kanngiesser et al. [28]. Both works underestimated the total probability of DAP: 4% [34] and 3% [28] with experimental values of about 6% [33,28]. Although total DAP probabilities calculated in [34] and [28] are in reasonable agreement with each other, calculated partial contributions from specific final states differ greatly. Since [34] and [28] are the only theoretical works devoted to non-monopole double processes so far, the need for further research in this direction is obvious.

In the following sections of this paper, after a brief description of the theoretical model, a number of recent results are given. These deal with the cascades in hollow atoms, spin-sensitive coincidence experiments, and many-electron correlations responsible for DAP.

THEORETICAL MODEL

Theoretical approaches to the description of cascading decays of inner shell vacancies fall into two major categories, namely, those based on Monte-Carlo simulations [2,5,18,35,36], and those based on straightforward analytical construction of de-excitation trees [16,17,37,38]. The latter approach had been normally used for the description of comparatively simple cascades, the limit being the decay of Ar K vacancy [16,17]. The decay trees of such cascades have the number of branches of about a hundred, and it is still possible to analyse them 'by hand'. For more complicated cases the number of branches in the decay trees may amount to several million, and automatic analysis of the decay trees is the only way to deal with them.

Kochur et al. [37,38] developed an approach allowing one to straightforwardly construct and automatically analyse very complex cascades and to calculate their various characteristics. A brief description of this approach is given in this section.

For each branching point in the de-excitation tree of a cascade, i.e. for a specific intermediate configuration of a cascade, the branching ratios are expressed through partial (radiative and non-radiative) and total configuration widths. Mean partial widths of non-radiative i-jk and radiative i-j transitions are expressed in a factorised form which makes it possible to separate the dependencies on electron subshell occupation numbers. For any ionic configuration C we have:

$$\Gamma_{ijk}(C) = N_i^v(C)N_{jk}^p(C)\gamma_{ijk}(C) \tag{1}$$

$$\Gamma_{ij}(C) \propto \omega_{ij}^3 N_i^v(C)N_j(C)R_{ij}^2(C) \tag{2}$$

Here i,j,k denote electron subshells involved in a transition, N_i^v is the number of vacancies in the subshell i, N_{jk}^p is the number of electron pairs in the subshells j and k ($N_{jk}^p=N_jN_k$ for $j\neq k$, $N_{jk}^p=N_j(N_j-1)/2$ for $j=k$), N_j, N_k are the occupation numbers, γ_{ijk} is the partial width per one pair of electrons [39], ω_{ij} is the transition energy, R_{ij} is the dipole transition matrix element. The values $\gamma_{ijk}(C)$ and $R_{ij}(C)$ are rather smoothly dependent on the electron configuration C of the decaying state, which allows one to consider the effect of relaxation by introducing simple linear dependencies of γ_{ijk} and R_{ij} on numbers of vacancies in electron subshells [37].

If an intermediate configuration C of a cascade can decay into a number of configurations $\{C_m\}$ then respective branches are characterised by the branching ratios:

$$\chi(C \rightarrow C_m) = \frac{\Gamma(C \rightarrow C_m)}{\sum_m \Gamma(C \rightarrow C_m)} \tag{3}$$

where $\Gamma(C \rightarrow C_m)$ is either (1) or (2).

The change of the core potential during cascade transitions cause the monopole ejection of additional electrons. The ejection of additional electrons in the course of the cascade development leads to the appearance of a great number of additional decay branches in the de-excitation tree making the latter much more complex. The probabilities of monopole ejection of electrons are calculated in the sudden limit [40] using the mean radii of electron subshells to calculate the overlap integrals as described in [37].

Mirakhmetov and Prilis [18] were the first to notice that in highly ionized configurations some transitions are forbidden energetically. However switching those transitions off altogether as it had been done in [18] excludes some transitions that really exist and distorts the decay tree.

The multiplets of initial and final configurations may overlap for some of the low-energy transitions between multivacancy states of a cascade. In those cases the transitions between some multiplet states are forbidden energetically, and the partial configuration-average-approximation transition widths (1,2) should be decreased. On the other hand, the transitions between some of the states of the overlapping multiplets can still be possible even if the centre of gravity of the initial configuration is lower than that of the final one.

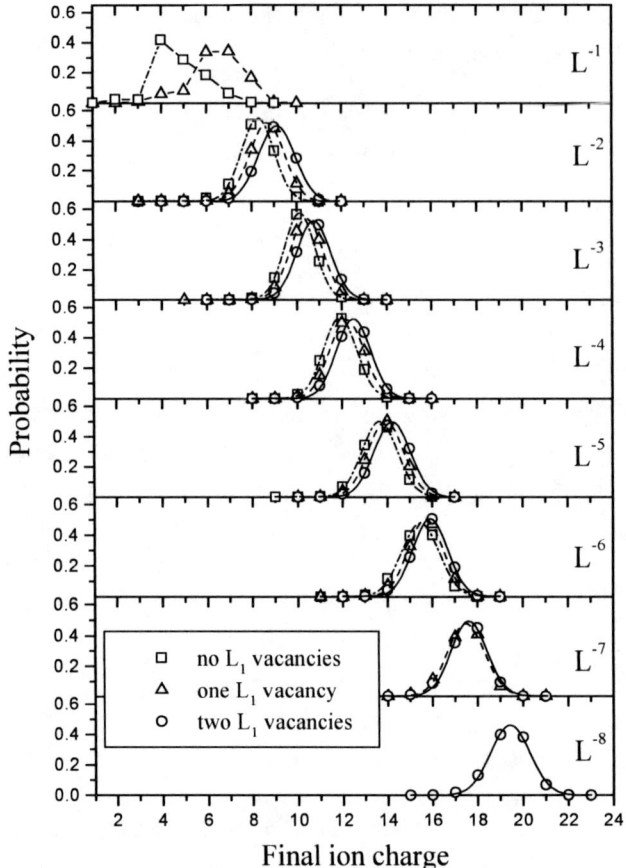

FIGURE 1. Final-ion-charge spectra of krypton produced by the decays of L^{-n} ($n=1$ to 8) states. [45]. Symbols, calculated probabilities; curves, Gaussian fits.

To describe the transition arrays between the states of the overlapping multiplets, the energy spectra of multivacancy configurations are simulated with Gaussian distributions

$$p(E) = \sqrt{\frac{1}{2\pi\sigma^2}} \exp\left(-\frac{1}{2}\frac{(E - E_0^2)^2}{\sigma^2}\right) \qquad (4)$$

centred on centres of gravity E_0 of the multiplets, and with variances σ^2 calculated by the methods described in [41]. With the distributions (4) for initial and final configuration of a transition it is easy to calculate in the case of overlapping multiplets the correcting coefficients to the configuration-average widths (1) and (2) [37].

SELECTED RESULTS

Cascading Decay of Hollow Atoms

A great number of studies on the yields of multiply charged cascade-produced photoions from singly ionized atoms have been performed both experimentally and theoretically (see [38] and references therein).

If a greater number of initial inner-shell vacancies is created then the decay cascades become much more complex, and much higher stages of ionization can be reached. Such hollow states are exotic ones, and the studies on their decays are scarce [42]. However, with coming fourth generation light sources the flux densities can be reached that could produce multiply inner-shell-ionized atoms with reasonable probabilities. Even with present day facilities the hollow states of heavy atoms can be obtained, for example, Kanter *et al.* [43] reported on the observation of the molybdenum K^{-2} states produced by photoionization.

Within the scheme described above, Kochur [44,45] calculated final photoion charge spectra produced by cascading decay of argon and krypton atoms with one to eight vacancies in their K and/or L shells. Figure 2 demonstrates calculated final-ion charge spectra [45] produced by the decay of krypton L^{-n} vacancy states.

For each n all possible combinations of initial L_1 and L_{23} vacancies were considered. One can see that the charge spectra depend on how the vacancies are distributed among L_1 and L_{23} subshells. The most dramatic difference is seen between L_1^{-1} and L_{23}^{-1} ion yields spectra. Starting from $n = 2$, the $L_1^{-k}L_{23}^{-(n-k)}$ ($k=0,1,2$) the ion yields spectra can be fit with Gaussian distributions surprisingly well. The groups of the charge spectra corresponding to specific numbers of vacancies n shift almost linearly with the growth of n. Neighbouring groups of L^{-n} charge spectra are separated by about 1.75. As their widths vary from 1.42 to 1.73 (they slowly grow with the increase of n), there is a hope that they could be resolved experimentally. The split between $L_1^{-k}L_{23}^{-(n-k)}$ spectra for the same n decreases with the increase of n. No simple qualitative explanation is possible for such a behaviour of the L^{-n} final-ion-charge spectra of krypton since the decay cascades in these cases are very complex (several million branches).

One can see that the final-ion-charge spectra produced by the cascading decay of hollow ions are sensitive to the number of initial innner shell vacancies and to their distribution among inner electron subshells. This may allow one to use them to identify hollow states and to study the dynamics of their decay.

Final Ion-Charge Resolved Electron Spectroscopy

As pointed out above, the Auger electron – photoion, and photoelectron – photoion coincidence techniques can give very useful supplementary experimental information allowing one to study the cascading decays in more detail.

A new type of photoelectron – photoion coincidence experimental technique with energy-analysed photoelectrons has been developed recently [29-31]. This technique allows one to split a photoelectron spectrum into partial contributions corresponding to formation of cascade-produced photoions of various charges. Such partial electron spectra were called Final Ion-charge Resolved Electron spectra, or FIRE spectra

Luhmann *et al.* [29] applied FIRE spectroscopy to study the structure of the $4d$-multiplets in Eu and Sm. A strong dependence of the FIRE spectra on the final photoion charge has been reported. Specifically, the low ionization energy part of the $4d$ FIRE(+2) spectra was found to be suppressed. It was suggested in [29] that such a behaviour of the FIRE spectra might be explained by the fact that (otherwise very intense) super-Coster-Kronig $4d$-$4f4f$ transitions are forbidden for the high-spin components of the $4d^{-1}$ multiplets. The high-spin states are those states where the spin orientation of the $4d$ hole and of the $4f$ electrons are the same. Then, for example, in europium the state $4d^9 4f^7$ ($S_{max}= 1/2+7/2=4$) cannot decay into $4d^{10} 4f^5 \varepsilon l$ ($S_{max}= 5/2+2/2=3$). The same consideration stands, of course, for all rare earth elements with less-than-half filled $4f$ subshell. The state $4f^{-2}$ reached by the $4d$-$4f4f$ transition cannot decay further, and 2+ photoions are predominantly formed. Therefore, if the $4d$-$4f4f$ transition is forbidden then the yield of the 2+ ions is suppresses, and the intensities of the FIRE(+2) for high-spin components are low.

The first theoretical description of the $4d$ FIRE spectra in rare earth atoms was given by Kochur *et al.* [46] who combined multiplet structure calculations with the calculations of the yields of the cascade-produced photoion.

For each component $|EJ>$ of the $4d^{-1} 4f^n$ multiplet, the weight of the high-spin states is introduced by summing up the contributions from all high-spin basic states $|4d^9 4f^n \gamma_i L_i S_{max} J >$:

$$\beta(EJ)=\sum_i < 4d^9 4f^n \gamma_i L_i S_{max} J \mid EJ >^2 \qquad (5)$$

Then the cross section of photoabsorption leading to formation of the photoion with the charge q (*i.e.* FIRE(+q) intensity) is

$$\sigma^{+q}(EJ)=\sigma(EJ)[\beta(EJ)P^f(+q)+(1-\beta(EJ))P^a(+q)] \qquad (6)$$

where $P^f(+q)$ and $P^a(+q)$ are the probabilities of formation of the +q photoion upon condition that $4d$-$4f4f$ transitions are either forbidden or allowed.

Gerth *et al.* [30] measured and calculated the $4d$-FIRE spectra in atomic Ce, Pr, Nd, Sm, and Eu. A simple model (5,6) [44] allowed us [30] to explain the variations in the shapes of measured FIRE spectra. The FIRE spectra of Pr [30] are reproduced in Figure 4. as an example.

One can conclude that the FIRE spectroscopy is a new spin-sensitive technique able to distinguish between high-spin and low-spin states within complex multiplets. The spin-sensitivity of FIRE spectroscopy is not induced by an external influence, such as magnetisation or laser pumping, but is an intrinsic property of the cascading decay processes following the creation of an inner-shell vacancy.

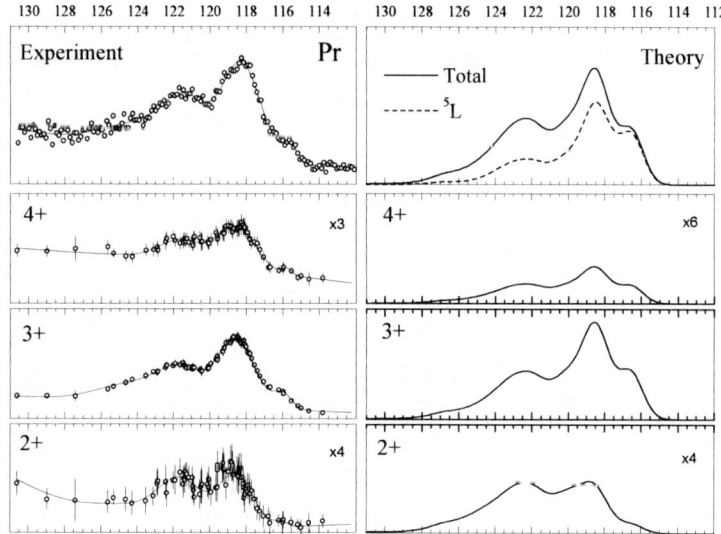

FIGURE 4. Measured and calculated final ion-charge resolved 4d-electron spectra of atomic Pr [30]. Dot line in the right upper panel is the electron spectrum weighted by the fraction of high-spin ^{5}L states β, see (5)

Many-Electron Correlations and Double Auger Process

Multiple processes, such as double Auger processes (DAP) with ejection of two electrons are found to be noticeable in the cascades starting with less-deep initial vacancies in atoms with a lot of valence electrons (K-vacancy in Ne, L-vacancies in Ar, M-vacancies in Kr [13,31,33]).

In this section we discuss DAP upon the decay of the K vacancy in the neon atom.

Total experimental DAP probability (measured by detecting Ne^{+3} photoions) is about 6% [28,33]. Consideration of only monopole shake processes [38] gave just 0.7% of DAP. In ref. [28], together with monopole shake processes we included also non-monopole core-core correlations in final states of Auger transitions and obtained the DAP probability of 3%. Core-core (CC) correlations are the correlations with excitation of only core electrons.

Demekhin and Demekhina [47] considered the contribution to DAP in Ne from the correlations of core and Auger electrons, *i.e.* inelastic scattering of Auger electron by core electrons. They included only radial correlations, i.e. those where both core and Auger electrons are excited into the states with the same orbital quantum numbers. The contribution from radial core-Auger electron (CA) correlations was found to be about 2% for each of the KLL transition.

TABLE 1. Calculated probabilities of double Auger processes (P_{DAP}) upon decay of K vacancy in Ne

Transition	Final state*	Br. ratio	Correlation state	Type	P_{DAP},%
KL_1L_1	$2s^0 2p^6 \ \varepsilon s^{Aug}$	0.103	$2s^1 2p^4 \ \{n,\varepsilon\} s \varepsilon s^{Aug}$	CC	1.36
			$2s^1 2p^4 \ \{n,\varepsilon\} d \varepsilon s^{Aug}$	CC	5.44
			$2s^0 2p^5 \ \{n,\varepsilon\} p \varepsilon s^{Aug}$	CC	1.36
			$2s^0 2p^5 \ \varepsilon_1 l_1 \varepsilon_2 l_2$	CA	3.95
KL_1L_{23}	$2s^1 2p^5 \ \varepsilon p^{Aug}$	0.308	$2s^2 2p^3 \ \{n,\varepsilon\} s \varepsilon p^{Aug}$	CC	0.81
			$2s^2 2p^3 \ \{n,\varepsilon\} d \varepsilon p^{Aug}$	CC	2.73
			$2s^0 2p^5 \ \{n,\varepsilon\} s \varepsilon p^{Aug}$	CC	0.10
			$2s^1 2p^4 \ \{n,\varepsilon\} p \varepsilon p^{Aug}$	CC	1.23
			$2s^0 2p^5 \ \varepsilon_1 l_1 \varepsilon_2 l_2$	CA	0.36
			$2s^1 2p^4 \ \varepsilon_1 l \varepsilon_2 l$	CA	2.08
$KL_{23}L_{23}$	$2s^2 2p^4 \ \varepsilon d^{Aug}$	0.545	$2s^0 2p^5 \ \{n,\varepsilon\} p \varepsilon d^{Aug}$	CC	0.03
			$2s^1 2p^4 \ \{n,\varepsilon\} s \varepsilon d^{Aug}$	CC	0.15
			$2s^2 2p^3 \ \{n,\varepsilon\} p \varepsilon d^{Aug}$	CC	0.93
			$2s^1 2p^4 \ \varepsilon_1 l_1 \varepsilon_2 l_2$	CA	0.59
			$2s^2 2p^3 \ \varepsilon_1 l_1 \varepsilon_2 l_2$	CA	1.40
	$2s^2 2p^4 \ \varepsilon s^{Aug}$	0.044	$2s^0 2p^5 \ \{n,\varepsilon\} p \varepsilon s^{Aug}$	CC	0.03
			$2s^1 2p^4 \ \{n,\varepsilon\} s \varepsilon s^{Aug}$	CC	0.15
			$2s^2 2p^3 \ \{n,\varepsilon\} p \varepsilon s^{Aug}$	CC	0.93
			$2s^1 2p^4 \ \varepsilon_1 l \varepsilon_2 l$	CA	0.24
			$2s^2 2p^3 \ \varepsilon_1 l_1 \varepsilon_2 l_2$	CA	3.18

Total weighted probability of DAP	5.39
Experiment	5.97 [28]; 5.8 [33]

* $1s^2$ is omitted in the notations of configurations

In this work we calculate the contribution to DAP from both core-core (CC) and core-Auger electron (CA) correlations. Both radial and angular CA correlations are included (the excitations to all possible channels with l up to 3 are considered). The CC correlations are included within the configuration interaction approach [28] while the CA correlations are described within the frame of the perturbation theory.

It should be noted that the two-electron CC correlations which are present both in initial and final state of an Auger transition (for example, $2p2p\text{-}\{n_1,\varepsilon_1\}l_1\{n_2,\varepsilon_2\}l_2$) cancel themselves out in the transition amplitudes between many-configuration states, and do not lead to any multiple processes. This follows from the fact that both configurations mixing and Auger transition amplitudes are calculated with the same operator of electrostatic interaction [48]. Inclusion of such correlations *only* in initial or *only* in final state may lead to severe mistakes. It follows then that only the CC correlations absent in initial state should be considered.

Calculated DAP probabilities are presented in Table 1. Listed in the last column of Table 1 DAP probabilities refer to specific diagram Auger transitions. Note that different Auger transitions contribute differently to DAP. Thus DAP probability upon KL_1L_1 transition is the largest, 12.1%; KL_1L_{23} and $KL_{23}L_{23}$ transitions end up in two electron ejection with the probabilities of 7.1% and 3.2%, respectively.

The contributions from $2s^0 2p^5$, $2s^1 2p^4$, and $2s^2 2p^3$ core final states are 0.71%, 2.14%, and 2.54%. These results are in contradiction to those obtained by Amusia *et al.* [34] who calculated the probability of DAP with $2s^0 2p^5$ to be 1.5%, while their estimated combined contribution from $2s^1 2p^4$ and $2s^2 2p^3$ was 2.5%.

Our calculated total DAP probability resulting from core-core and core-Auger electron correlations upon decay of the K vacancy in Ne is close to the experiment (see bottom lines of Table 1). One can conclude, therefore, that core–core and core–Auger electron correlations are the principal mechanisms of double Auger processes.

REFERENCES

1. Krause M.O., Vestal M. L., Johnston W.H., and Carlson T.A. *Phys. Rev.* **133**, A385-A390 (1964).
2. Carlson T.A., and Krause M.O. *Phys. Rev.* **137**, A1655-A1662 (1965).
3. Krause M.O., and Carlson T.A. *Phys. Rev.* **149**, 52-58 (1966).
4. Carlson T.A., Hunt W.E., and Krause M.O *Phys. Rev.* **151**, 41-47 (1966).
5. Krause M.O., and Carlson T.A., *Phys. Rev.* **158**, 18-24 (1967).
6. Zimmermann P. *Comments At. Mol. Phys.* 23, 45-53 (1989).
7 Ueda K. Shigemasa E. Sato Y., Yagishita A. Ukai M., Maezawa H., Hayaishi T. and Sasaki T. *J.Phys. B: At. Mol. Opt. Phys.* **24**, 605-613 (1991).
8. Saito N., and Suzuki I.H. *Int. J. Mass Spectrom. Ion. Proc.* **115**, 157-172 (1992).
9. Saito N., Suzuki I.H. *J. Phys. B: At. Mol. Opt. Phys.* **25**, 1785-1793 (1992).
10. Tawara H., Hayaishi T., Koizumi T., Matsuo T., Shima K., Tonuma T., and Yagishita A. *J. Phys. B: At. Mol. Opt. Phys.* **25**, 1467-1473 (1992).
11. Doppelfeld J., Anders N., Esser B., von Busch F., Scherer H., and Zinz S. *J. Phys. B: At. Mol. Opt. Phys.* **26**, 445-456 (1993).
12. Lindle D.W., Manner W.L., Steinbeck L., Villalobos E., Levin J.C., and Sellin I.A. *J. Electron Spectrosc. Relat. Phenom.* **67**, 373-385 (1994).
13. Tamenori Y., Okada K., Nagaoka S., Ibuku T., Tanimoto S., Shimizu Y., Fujii A., Haga Y., Yoshida H., Ohashi H., and Suzuki I.H. . *J. Phys. B: At. Mol. Opt. Phys.* 2002 (in press)
14. Cooper J.W., Southworth S.H., MacDonald M.A. and LeBrun T. *Phys. Rev. A.* **50**, 405-411 (1994).
15. Southworth S. H., MacDonald M. A., LeBrun T. and Deslattes R. D. *Nucl. Instr. Meth. Phys. Res.A.* **347**, 499-503 (1994).
16. von Busch F., Doppelfeld J., Günther C., and Hartmann E. *J. Phys. B.: At. Mol. Opt. Phys.* **27**, 2151-2160 (1994).
17. Omar G, and Hahn Y. *Z. Phys. D: Atoms, Mol. and Clust.* **25**, 41-46 (1992).
18. Mirakhmedov M.N., and Parilis E.S. *J. Phys. B: At. Mol. Opt. Phys.* **21**, 795-804 (1988)
19. Kochur A.G., and Sukhorukov V.L *J.Electron Spectrosc. and Relat. Phenom.* **76**, 325-328 (1995).
20. Kochur A.G., and Sukhorukov V.L. *J.Phys.B At.Mol.Opt.Phys.* **29**, 3587-3598 (1996).
21. Verkhovtseva E.T., and Pogrebnjak P.S. *J.Phys.B: At.Mol.Phys.* **13**, 3535-3543 (1980).
22. Verkhovtseva E.T., Gnatchenko E.V., Pogrebnjak P.S., and Tkachenko A.A. *J. Phys. B: At. Mol. Phys.* **19**, 2089-2108 (1986).
23. Brühl S. Fluoreszenzspektroskopie an atomarem Xenon imvakuumultravioletten Spektrlbereich. Dissertation. Hamburg,1999.
24. Mitkina Ye.B. *X-ray emission spectra and charge distribution of ions upon cascading decays of inner-shell vacancies.* Thesis cand. phys.-math. scis. Rostov State university of Transport Communication, Rostov-na-Donu:, 2001
25. Kochur A.G., Mitkina Ye.B., and Sukhorukov V.L. *J.Phys.B:At.Mol.Opt.Phys.* **31**, 5293-5300 (1998).
26. Kochur A.G., Sukhorukov V.L., and Mitkina Ye.B.. *J.Phys.B: At.Mol.Opt.Phys.* **33**, 2949-2953 (2000).

27. KochurA.G., Petrini D., and E.P. da Silva. *A&A*, 2002 (to be published).
28. Kanngießer B., Jainz M., Bruenken S., Benten W., Gerth Ch., Godehusen K., Tiedtke K., van Kampen P., Tutay A., Zimmermann P., Demekhin V.F., and Kochur A.G. *Phys. Rev. A*, **62**, 014702 (2000).
29. Luhmann T., Gerth Ch., Martins M., Richter M., and Zimmermann P. *Phys. Rev. Lett.* **76**, 4320 (1996).
30. Gerth Ch., Kochur A. G., Groen M., Luhmann T., Richter M., and Zimmermann P. *Phys. Rev. A*, **57**, 3523-3533 (1998).
31. Brünken S., Gerth Ch., Kanngießer B., Luhmann T., Richter M., and Zimmermann P. *Phys. Rev. A*, **65**, 042708 (2002).
32. Carlson T.A., and Krause M.O. *Phys. Rev. Lett.* **14**, 390-392 (1965).
33. Saito N., and Suzuki I.H.. *Phys. Script.* **49**, 80-85 (1994).
34. Amusia M.Ya., Lee I.S., and Kilin V.A.. *Phys.Rev.A.* **45**,. 4576-4587 (1992).
35. Mukoyama T. *J. Phys. Soc. Japan*. **55**,.3054-3058 (1986)
36. Mukoyama T., Tonuma T., Yagishita A., Shibata H., Matsuo T., Shima K., and Tawara H. *J. Phys. B: At. Mol. Opt. Phys.* **20**, 4453-4460 (1987)
37. Kochur A.G., Dudenko A.I., Sukhorukov V.L., and Petrov I.D. *J. Phys. B. At. Mol. Opt. Phys.* .**27**, 1709-1721 (1994).
38. Kochur A.G., Sukhorukov V.L., Dudenko A.I., Demekhin Ph.V. *J. Phys. B. At. Mol .Opt. Phys.* **28**, 387-402 (1995).
39. Sukhorukov V.L., Dudenko A.I., Vasilieva M.Ye, and Dementiev A.P. *Bull. Acad. Sci. USSR.* **55**, 2472-2477 (1991).
40. Sachenko V. P. and Demekhin V. F. *J. Expr. Theor. Phys.* **49** 765-769 (1965).
41. Kučas S., and Karazija R. *Phys. Scr.* **47**, 754-764 (1993).
42. Omar G, and Hahn Y. *Phys. Rev. A.* **43**, 4695-701 (1991).
43. Kanter E.P., Dunford R.W., Krässig B., and Southworth S.H. *Phys. Rev. Lett.* **83**, 508-511 (1999).
44. Kochur A. *J. Synchrotron Rad.* **8**, 218–219 (2001).
45. Kochur A. *J. Electron Spectrosc. Relat. Phenom.* **114–116**, 81–84 (2001)
46. Kochur A.G., Sukhorukov V.L., and Petrov I.D J.Phys.B. At. Mol. Opt. Phys. **29**, 4565-4572 (1996).
47. V.F. Demekhin, and N.V. Demekhina. *Studied in Russia (electronic journal)* **91**,1258-1270 (2000), http://zhurnal.ape.relarn.ru/articles/2000/091.pdf
48. A.G.Kochur *The processes of decay of vacancies in deep electron shells*. Thesis doct. phys.-math. scis. Rostov State University of Transport Communication, Rostov-na-Donu, 1997.

Resonant Auger for the detection of quadrupolar transitions

J. Danger[**†*], P. Le Fèvre[†], H. Magnan[†**], D. Chandesris[†], J. Jupille[‡], S. Bourgeois[§], T. Eickhoff[¶] and W. Drube[||]

*IPCMS, CNRS-Université Louis Pasteur, 67037 Strasbourg, FRANCE.
[†]LURE, CNRS-Université Paris Sud, Bât. 209d, BP 34, 91898 Orsay, FRANCE.
**SPCSI, Commissariat à l'Energie Atomique, 91191 Gif sur Yvette, FRANCE.
[‡]GPS, CNRS-Université Paris VI et Paris VII, Tour 23, 2 place Jussieu, 75251, Paris, FRANCE.
[§]LRRS, CNRS-Université de Bourgogne, BP 47870, 21078 Dijon, FRANCE.
[¶]Institut für Experimentalphysik, Universität Hamburg, 22761 Hamburg, GERMANY.
[||]HASYLAB at DESY, 22603 Hamburg, GERMANY.

Abstract. Quadrupolar transitions can play an important role in X-ray absorption spectroscopy, especially when it is used for magnetic measurements, like in X-ray Magnetic Circular Dichroism or Resonant Magnetic Scattering. We show here that resonantly excited Ti $KL_{2,3}L_{2,3}$ Auger spectra of TiO_2 (110) carry a clear signature of quadrupolar transitions from the $1s$ to localized e_g and t_{2g} d-like states. The quadrupolar nature of the observed additional spectator lines are clearly demonstrated by their angular dependence, and their intensity is used to locate and quantify the quadrupolar transitions in the absorption spectrum.

Since it obeys strict selection rules, X-ray absorption (XAS) is an element and orbital-selective probe of both crystallographic and electronic structure of materials. Beyond the dominant features of the absorption spectra arising from excitations which can be interpreted within the dipolar approximation, the absorption pre-edge and near-edge features commonly involve second order quadrupolar transitions (QT) [1]. Although these are of weak intensity, their corresponding terms in the interaction Hamiltonian often contain valuable informations. For instance, they can be predominant in X-ray Magnetic Circular Dichroism measurements at the $L_{2,3}$ edges of rare earths [2, 3] or in Resonant X-ray Magnetic Scattering, both at the $L_{2,3}$ edges of rare earths or near pre-K-edge features of $3d$ compounds [4, 5]. Associated to transitions to p and d orbitals [6, 7], these prepeaks involve both dipolar and quadrupolar transitions and are also important in magnetic measurements [5]. A picture of the occurrence of QT is therefore of prime interest to achieve a description of the electronic structure of these compounds and to fully account for their physical properties. A common way to discriminate dipolar and quadrupolar transitions is based on the interpretation of the angular dependence of the XAS cross section, since the electric dipolar ($\varepsilon.r$) and quadrupolar ($\varepsilon.r\ k.r$) terms of the interaction hamiltonian are sensitive either to the polarization direction ε of the X-rays or to both ε and the direction of the photon wave vector k [8, 9]. However, this method suffers from severe drawbacks. It can only be applied to single crystals and, to insure that the weak quadrupolar contributions can be distinguished from the more intense dipolar components, its use is restricted to the study of pre-edge features

CP652, X-Ray and Inner-Shell Processes: 19th International Conference on X-Ray and Inner-Shell Processes
edited by A. Bianconi, A. Marcelli, and N. L. Saini
© 2003 American Institute of Physics 0-7354-0111-X/03/$20.00

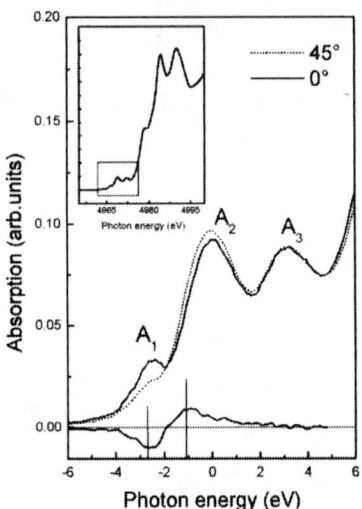

FIGURE 1. A_1, A_2 and A_3 prepeaks of the Ti K-absorption edge in rutile. The X-rays wave vector is perpendicular to the c axis and either parallel to (0°) or at 45° (45°) from the (110)-surface normal. The bottom curve is the difference between the two absorption spectra. Insert : absorption threshold at 0°.

well separate from the main absorption threshold. Moreover, it does not discriminate the different contributions. Resonant Auger spectroscopy permits to overcome those limitations. It consists in observing an Auger decay while scanning the photon energy through an absorption edge. Due to the presence of a core photoexcited electron in a localized empty orbital, additional decay lines, so-called spectator lines [10], appear at constant binding energies and at higher kinetic energies than the classical decay. In this work, the capability of the method to resolve quadrupolar and dipolar contributions is illustrated by solving the long debated question of the pre-K-edge of titanium in rutile TiO_2 [11, 12] and by discriminating the various components its features in good agreement with theoretical predictions [12, 13].

The experiments were performed at the BW2 wiggler beamline of HASYLAB, equipped with Si(220) crystals monochromator. Data were collected on a TiO_2(110) rutile single-crystal, prepared according to standard procedures in order to obtain the stoichiometric composition [14], and the angular dependence was studied by rotating the sample around a vertical axis. The electron collection direction is fixed in the horizontal plane at 45° from the incoming X-ray beam. The Ti K-edge XAS spectrum (see Fig. 1) exhibits a series of three prepeaks called A_1, A_2 and A_3. In the following, the photon energy is referenced relative to the maximum of A_2 whose absolute energy position was measured to be 4968.9 eV. A_1 and A_3 are thus peaking at -2.75 eV and +3.0 eV relative photon energy, respectively. Total electron yield spectra were recorded for two different X-rays incidence angles with respect to the (110) surface (Fig. 1). The wave vector of the X-rays is parallel to the (110)-surface normal (0°, ε at 45° from the a

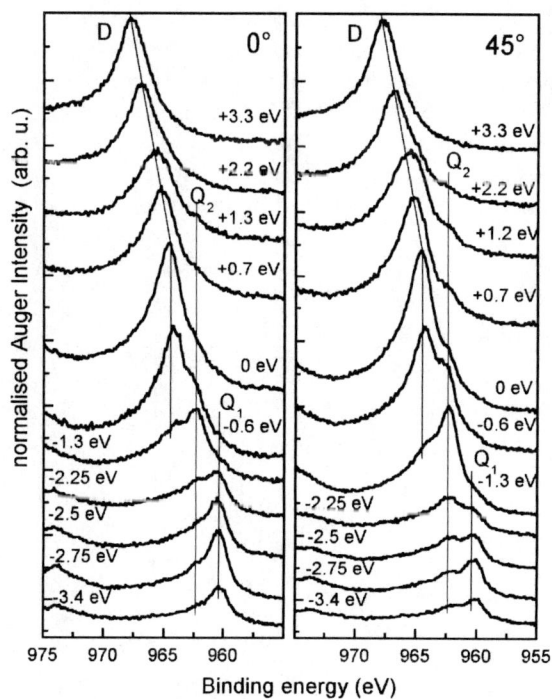

FIGURE 2. Ti KL_2L_3 ($^1D_2$) Auger line for the indicated different photon energies recorded in the 0° (left part) and the 45° (right part) geometries.

axis), or at 45° from it (45°, ε parallel to the a axis). At a K-edge, the dipolar transitions probability is not expected to vary [1], while, on the contrary, quadrupolar contributions are expected to have the largest possible intensity variation between 0° and 45°, since the d symmetry of the probed orbitals confers to their intensity a $\cos(4\phi)$ dependence (where ϕ denotes the rotation of ε and k in the (100) plane). All the angular dependences are therefore due to intensity variations of QT. In rutile, the first coordination shell of Ti atoms is composed of 6 oxygen atoms which form a distorted octahedron [11]. Using the electric quadrupolar hamiltonian formula, one can easily demonstrate that the maximum intensity at 0° and the minimum at 45° of the A_1 prepeak are indicative of QT towards t_{2g}-like orbitals. The reversed behavior of the A_2 prepeak shows that its quadrupolar component is mostly e_g-like, while the absence of any angular dependence of A_3 is assigned to a purely dipolar origin [13]. The data meet the previously published results [11, 12]. However, these did not allow an experimental decomposition of the pre-edge into its various quadrupolar and dipolar contributions. In this goal, an analysis of the Ti KLL resonant Auger lines was undertaken in the pre-edge region of the Ti K-edge. Chosen spectra collected in both 45° and 0° geometries are shown in Fig. 2. In the following, we will concentrate on the main $^1D_2$ of the normal Auger spectrum (shown on top of Fig. 3) which is the more intense. The series of resonant $^1D_2$ Auger spectra collected at 0° and 45° exhibit three components, labeled Q_1, Q_2 and D (Fig. 2) whose intensity strongly varies with the photon energy. Q_1 and Q_2 appear both always at constant binding energy (959.1 and 961.1 eV, respectively), while D has a different behavior. It first appears at a constant binding energy of 963.2 eV (Raman-Auger behavior), until it reaches the position of the $^1D_2$ Auger peak at a constant kinetic energy of 4006.3 eV. The Auger spectra were assumed to be a simple sum of theses three components using Lorentzian lineshapes and an integral background. Three examples are shown on Fig. 3 for spectra all recorded at -2.25 eV, but for different incidence angles of the X-rays.

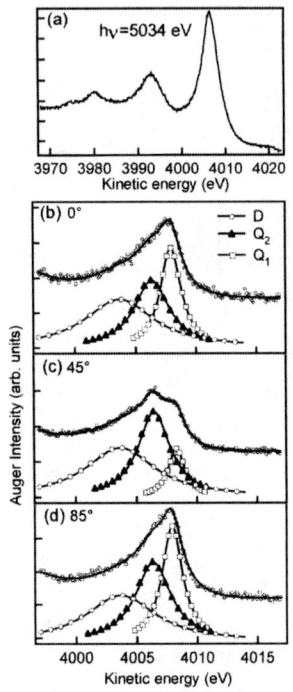

FIGURE 3. (a) Ti $KL_{2,3}L_{2,3}$ Auger spectrum in TiO_2. The main KL_2L_3 ($^1D_2$) Auger line recorded for a relative photon energy of -2.25 eV with the photon wave vector at 0° (b), 45° (c) and 85° (d) from the (110) surface, deconvoluted into the D (open circles), Q_1 and Q_2 (closed up triangles) components

The results are presented in Fig. 4. Q_1 has a maximum intensity on the maximum of the A_1 prepeak (-2.75 eV) and follows the same angular dependence. Considering both resonant behavior and angular dependence, Q_1 is assigned to an Auger decay following the quadrupolar excitation of a $1s$ electron in the t_{2g}-like orbitals. Moreover, since it appears at constant binding energy, it can be identified as a spectator line. The intensity of Q_2 shows a maximum just below A_2 (-1.0 eV) roughly where the difference between the two XAS spectra recorded at 0° and 45° is maximum (Fig. 1). Unlike Q_1, it increases from 0 to 45°, and can therefore be attributed to a spectator line due an additional valence electron promoted in the e_g-like orbitals via a QT. Last, the D peak is attributed to decay following a dipolar transition of the $1s$ electron towards the p empty states of continuum, since it finally stabilizes at the position of the classical Auger line. Consistently, its intensity does not show any significant angular variation (bottom of Fig. 4). Assuming equal Auger decay rates for the different intermediate states, this intensity analysis of the different Auger components gives an estimate for the relative contributions of quadrupolar and dipolar transitions in the prepeaks. Thus, A_1 has a purely quadrupolar origin, A_3 is purely dipolar; a quadrupolar contribution to A_2 is detected on its low photon energy side. All these results are in agreement with the calculations presented in ref. 12. The quadrupolar origin of Q_1 and Q_2 is also confirmed by the complete angular dependence we recorded at a fixed photon energy of -2.25 eV, between the A_1 and A_2 prepeaks. Three selected spectra are presented in Fig. 3. Q_1 is more intense than Q_2 at $\phi=0^o$, while a reversed behavior is observed at $\phi=45^o$. At $\phi=85^o$, the spectrum recovers its $\phi=0^o$ shape. This angular dependence, i.e. the symmetry of the orbitals involved in the resonances, provides direct evidence for the quadrupolar nature of Q_1 and Q_2.

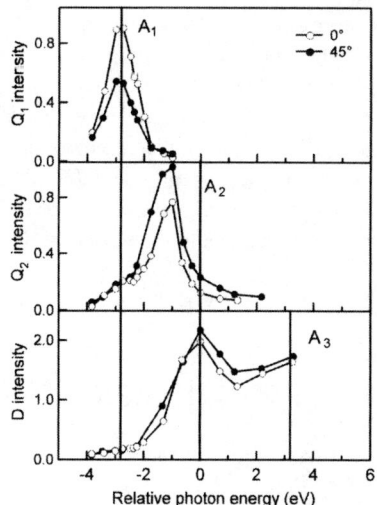

FIGURE 4. Intensities of the three peaks Q_1 (top), Q_2 (middle) and D (bottom) of the resonant Ti KL_2L_3 ($^1D_2$) Auger structure in the 0° (open circles) and 45° (closed circles) geometries. The vertical lines indicate the position of the A_1, A_2 and A_3 absorption prepeaks.

The intensity variation of Q_1 (respectively Q_2) with the X-ray incidence angle mimics the angular dependence of A_1 (A_2) in the absorption measurements and possesses the $\cos(4\phi)$ dependence expected from the electric quadrupolar hamiltonian. The clear determination of the e_g and t_{2g} quadrupolar components also allows a precise measurement of crystal field splitting between these two levels, given by the photon energy difference between the resonance positions of Q_1 and Q_2 (1.75 eV). Since the coulomb interaction between the $3d$ levels and the $1s$ core hole are weak, the $10D_q$ value can be directly extracted from the experimental results, while it is more complicated at the $L_{2,3}$ edges [15].

In the region of the Ti K-edge in TiO_2, the resonant behavior of the KLL Auger lines has been used to unravel the different contributions to the absorption spectra. Raman Auger lines following dipolar excitation towards p states, and spectator lines following QT towards e_g and t_{2g} states, split by fine crystal field effects, are clearly discriminated. These decays are unambiguously identified by analyzing their resonant energy and their angular dependence. An estimate of the t_{2g}-e_g splitting is directly obtained. Therefore, resonant Auger spectra offer direct and separate fingerprints for the different transitions occurring in the studied absorption edge, either to the continuum, or to bound states, like, e.g., quadrupolar transitions to d states at K-edges. As an experimental method to bring QT to the fore, resonant Auger is certainly more efficient than angular dependent XAS, which only allows to vary the relative weights of dipole and quadrupole contributions in the unselective mixture of the absorption signal. These two contributions are clearly separated in the Auger spectra, so that their energy position and intensity allow an identification of the QT not only in resolved absorption prepeaks. Next, angular XAS can only be performed on single crystals, whereas resonant Auger measurements can be carried out on powders or amorphous materials. In this case, the angular analysis would be pointless (left and right sides of Fig. 2 would be the same), but same Raman and spectator lines will be observed since they are purely due to the local electronic structure. Angular XAS relies on the orientation of the electronic orbitals with respect to the axis of the infinite and periodic crystallographic structure, whereas resonant Auger measurements is a real selective and local probe of the near Fermi level empty orbitals, provided they are localized enough. Last, the sensitivity of electron spectroscopy to the first atomic planes also opens new prospects to evidence differences of the electronic structure near the surface.

This work was supported by the E.U. program for access to large facilities.

REFERENCES

1. Brouder, C., *J. Phys.: Condens. Matter*, **2**, 701 (1990).
2. Baudelet, F., Giorgetti, C., Pizzini, S., Brouder, C., Dartyge, E., Fontaine, A., Kappler, J. P., and Krill, G., *J. Electron Spectrosc. Relat. Phenom.*, **62**, 153 (1993).
3. Matsuyama, H., Fukui, K., Okada, K., Harada, I., and Kotani, A., *J. Electron Spectrosc. Relat. Phenom.*, **92**, 31 (1998).
4. Dumesnil, K., Dufour, C., Stunault, A., and Mangin, P., *J. Phys.: Condens. Matter*, **12**, 3091 (2000).
5. Neubeck, W., Vettier, C., Lee, K.-B., and de Bergevin, F., *J. Phys.: Condens. Matter*, **12**, 3091 (2000).
6. Arrio, M.-A., Rossano, S., Brouder, C., Galoisy, L., and Calas, G., *Europhys. Lett.*, **51**, 454 (2000).
7. Westre, T., Kennepohl, P., DeWitt, J. G., Hedman, B., Hodgson, K. O., and Solomon, E. I., *J. Am. Chem. Soc.*, **119**, 6297 (1997).
8. Dräger, G., Frahm, R., Materlik, G., and Brümmer, O., *Phys. Stat. Sol. (b)*, **146**, 287 (1988).
9. Hahn, J. E., Scott, R. A., Hodgson, K. O., Doniach, S., Desjardins, S. R., and Solomon, E. I., *Chem. Phys. Lett.*, **88**, 595 (1982).
10. Brown, G. S., Chen, M. H., Crasseman, B., and Ice, G. E., *Phys. Rev. Lett.*, **45**, 1937 (1980).
11. Poumellec, B., Cortes, R., Tourillon, G., and Berthon, J., *Phys. Stat. Sol. (b)*, **164**, 319 (1991).
12. Uozumi, T., Okada, K., Kotani, A., Durmeyer, O., Kappler, J. P., Beaurepaire, E., and Parlebas, J. C., *Europhys. Lett.*, **18**, 85 (1992).
13. Joly, Y., Cabaret, D., Renevier, H., and Natoli, C. R., *Phys. Rev. Lett.*, **82**, 2398 (1999).
14. Pétigny, S., Sba, H. M., Domenichini, B., Lesniewska, E., Steinbrunn, A., and Bourgeois, S., *Surf. Sci.*, **410**, 250 (1998).
15. de Groot, F. M. F., Fuggle, J. C., Thole, B. T., and Sawatzky, G. A., *Phys. Rev. B*, **41**, 928 (1990).

The Two-Photon Decay of 1s2s $^1S_0$ States in Heavy He-Like Atomic Systems

P.H. Mokler [1], R.W. Dunford [2] and E.P. Kanter [2]

[1] GSI-Darmstadt, D-64291 Darmstadt, Germany
[2] ARGONNE Nat. Lab., Argonne,IL 60439, USA
(P.Mokler@gsi.de)

Abstract. In He-like systems the decay of the 1s2s $^1S_0$ excited state to the $1s^2$ $^1S_0$ ground state is not allowed. This excited state can only decay to the ground state via the emission of two photons. The spectral shape of the emitted continuum is determined by the complete structure of the atomic system as all bound and continuum P states contribute to the 2E1 decay. For very heavy atomic systems the ^3P states also have to be included and the normalized spectral shape changes with atomic number according to the relative strengths of both, the electron-electron interaction and of the relativistic effects. A brief survey on the variation of the spectral shape of the two-photon continuum with atomic number is given and compared to experiments ranging from He-like Ni to He-like Au with special emphasis on the heavy relativistic system. The data compare well with fully relativistic calculations.

INTRODUCTION

Traditionally atomic structure is probed by radiative transitions giving access to binding energies of the involved states. Atomic binding energies are calculated using matrix elements where the Hamiltonian is multiplied from both sides by the same wave function. Here, variational methods which minimize the binding energies provide accurate results. Lifetime measurements are more sensitive to the exact shape of the density distributions of the electrons as the overlap of the wavefunctions weighted by the interaction determines the transition rates. This is correspondingly true for line shapes, in particular for line widths. For inner-shell transitions in heavy atomic systems the transition rates are pretty large and hence, for fast allowed transitions, lifetimes or line widths are difficult to measure. There, only for metastable states or for "forbidden transitions" can lifetimes or linewidths be determined experimentally giving direct access to the structure of the wave functions.

In He-like systems a direct transition from the "metastable" 1s2s $^1S_0$ state to the $1s^2$ $^1S_0$ ground state is absolutely forbidden by selection rules. This state can only decay by the emission of two photons to the ground state, i.e. by the 2E1 decay. To some extent, the same is true for the excited 2s $^2S_{1/2}$ state in H-like ions; however,

CP652, *X-Ray and Inner-Shell Processes: 19th International Conference on X-Ray and Inner-Shell Processes*
edited by A. Bianconi, A. Marcelli, and N. L. Saini
© 2003 American Institute of Physics 0-7354-0111-X/03/$20.00

there a direct M1 decay to the $1s\,^2S_{1/2}$ ground state competes with the 2E1 decay. Two-photon decay was treated theoretically by Göppert-Mayer [1] more than 70 years ago. The 2E1 decay rate is determined by a summation over all two-photon transitions possible via all intermediate P states of the atomic system, where the sum energy of the two photons corresponds to the total binding energy difference between the initial and final state. The summation includes all relevant bound and continuum states. The differential transition probability $W_{2\gamma}$ has the form [1]:

$$dW_{2\gamma}/\,d\omega_1 = \{\omega_1\cdot\omega_2/\,(2\pi\cdot c)^2\}\cdot|M_{2\gamma}|^2\cdot d\Omega_1\cdot d\Omega_2$$

where ω_j is the energy and $d\Omega_j$ the solid angle for the jth photon and the transition energy ω_0 satisfies energy conservation $\omega_0 = \omega_1 + \omega_2$. The second order matrix element $M_{2\gamma}$ is given by the expression:

$$M_{2\gamma} = \varepsilon_1\cdot\varepsilon_2\ \Sigma\ \{<1\,^1S_0\ \|\ R_{E1}(\omega_2)\ \|\ n><n\ \|\ R_{E1}(\omega_1)\ \|\ 2\,^1S_0>/\ (\ E_n - E_{2\cdot S_0} + \omega_1\ )$$
$$+ <1\,^1S_0\ \|\ R_{E1}(\omega_1)\ \|\ n><n\ \|\ R_{E1}(\omega_2)\ \|\ 2\,^1S_0>/\ (\ E_n - E_{2\cdot S_0} + \omega_2\ )\ \}$$

Here, ε_j is the polarization vector for the jth photon and $R_{E1}(\omega_j)$ is the electric dipole operator. It has to be mentioned that the summation runs over all intermediate states n independent of whether they are empty or occupied [2].

Integrating finally over all possible photon energies yields the total transition rates that are sensitive to the structure of the complete atomic system. Measurements of lifetimes for the $1s2s\,^1S_0$ and the $2s\,^2S_{1/2}$ levels in He- and H-like ions, respectively, and their Z dependences give therefore important information on the total structure of heavy few-electron ions; for an overview see e.g. Refs. [3, 4]. The 2E1 transition rates increase strongly with Z – in first approximation with the sixth power of the atomic number, Z^6 – due to the increase in total transition energy with Z. For our cases of interest, heavy He-like ions, the possible different groundstate transitions and their rate dependences are indicated in the level diagram shown in Fig. 1 for a heavy L-shell excited ion. An overview of the rate dependences is given e.g. in Ref. [5] and in the literature cited there.

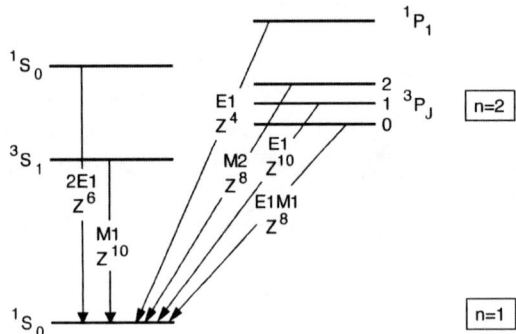

FIGURE 1. Level diagram for heavy He-like ions, possible ground state transitions and their rate dependences are shown, cf. [5].

THE TWO-PHOTON CONTINUUM

In contrast to lifetime measurements, a determination of the complete two-photon decay spectrum will provide much more detailed information on the complete structure of He-like atomic systems. Here, the shape of a measured two-photon continuum has to be compared with the energy differential transition probability, $dW_{2\gamma}/d\omega_1$, given above. As the two photons are indistinguishable, the spectrum is symmetric in each photon, i.e. it is mirror symmetric around the mid point at half the total transition energy, $\omega_0/2$. For convenience in comparing spectra from different atomic systems, the total transition energy is best normalized to "1" with the fractional energies of the two photons $f_j = \omega_j/\omega_0$ and $f_1 + f_2 = 1$. In this representation the differential transition probability for two-photon decay can be approximated by a product of a simple "phase space factor", $f_1 \cdot (1 - f_1)$, and a "structure factor", c.f. Ref. [6]:

$$dW_{2\gamma}/d\omega_1 \propto f_1 \cdot (1 - f_1) \bullet |M_{2\gamma}|^2$$

The phase space factor has the shape of a simple parabola and the structure factor is only determined by the square of the energy differential matrix element $M_{2\gamma}$.

The structure factor contains all the information on the complete atomic system and varies considerably with atomic species [7-9]. Just considering the widths (FWHM) of the spectra, the structure factor normally broadens the distributions in a way that depends on the system. For H-like systems the FWHM of the continuum starts at a value of 0.82 (in relative units of ω_0) for H and decreases with increasing atomic number due to relativistic effects. For H-like U the FWHM is about 0.71, which is also the value for a pure parabola, cf. Ref. [8]. For He the FWHM starts at 0.75, a narrower width compared to H. This narrowing is caused by the electron-electron interaction. Due to the relative decrease of this interaction with atomic number Z compared to that of the central potential the width for He-like ions widens first towards the width of H-like systems before around the mid-Z region where the relativistic effects turn the tendency back, cf. Ref. [9]. Finally the width for He-like U approaches that of H-like U – an indication that H- and He-like atomic structures approach each other for the heaviest atomic species. However, the widths are only an integral measure of the structure, more details can be gained by studying the detailed shape of the photon continuum.

A recent survey on the two-photon decay in heavy He-like systems is given in Ref. [10] with special emphasis on the change of the spectral shape with atomic number Z. There, the two-photon decay in singly K-shell ionized heavy atomic systems is also included. These quasi He-like atomic systems were investigated in detail by Ilakovac and his group [11 – 13]. Here, we concentrate only on heavy true He-like ions. In Fig. 2 the shape of the two-photon distribution and its change with Z is presented for He-like ions. The structure factors for He, Ni^{26+} and Au^{77+} are given along with the form of the parabola according to calculations by Derevianko and Johnson [9]. The structure factors for He and Au^{77+} are pretty similar whereas that for Ni^{26+} deviates considerably broadening the parabolic shape of the spectrum.

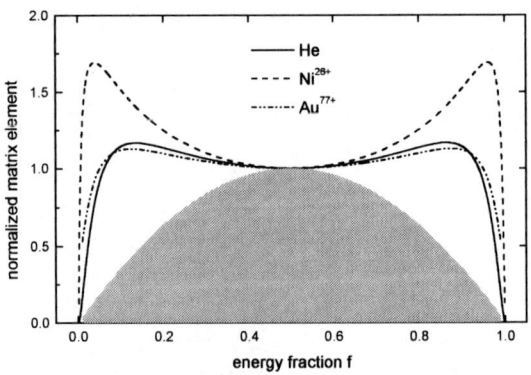

FIGURE 2. The shape factors for the 2E1 decay continuum in He-like ions. The structural factor for He, Ni^{26+} and Au^{77+} are given as normalized matrix elements (squared) along with the parabolic shape of the phase-space factor (rim of the shaded area) as a function of the fractional photon energy, acc. to [9].

EXPERIMENTAL TECHNIQUES

In order to produce He-like heavy ions the acceleration-stripping technique is normally used, where high velocities are needed for heavy ions. To efficiently produce He-like species of Ni and Au typical energies around 10 and 100 MeV/u are needed (compare the binding energies for the K electrons of about 9 and 80 keV), respectively. Then, the He-like fast ions are magnetically selected before they are excited in a second "exciter foil" with a certain probability to the $1s2s$ $^1S_0$ state. The lifetime of this state is 154 ps in Ni^{26+} [14] and only 0.32 ps in Au^{77+}[15]. The radiative decay of these excited states is detected behind the foil by at least two solid-state x-ray detectors (Si(Li) and Ge(i) detectors, respectively) looking face-to-face perpendicular to the downstream ion beam. The two photons of the decay are preferentially emitted 180^0 apart and are detected in coincidence.

Due to the short lifetimes of the Au ions these ions decay close to the exciter foil despite their high velocities. Moreover, in this relativistic velocity regime the Lorentz transformation of the photon emission angle from the fast ion system into the laboratory system leads to a forward tilt of the "90^0" observation angle [16, 17]. The experimental arrangement for the Au^{77+} case is given on the right side of Fig. 3. In order to be able to compensate for the Doppler effect of the emitted photons at different observation angles one of the Ge(i) detectors is granularly divided into stripes. For a precise determination of the true two-photon spectrum, emission angles, detector efficiencies, photon transmission losses and electronic coincidence efficiencies have to be determined with high accuracy. Fig. 3 gives a contour plot for the Au experiment of coincident photons measured with both the detectors. The ridge along the diagonal line describes the wanted two-photon continuum convoluted with all the relevant efficiencies.

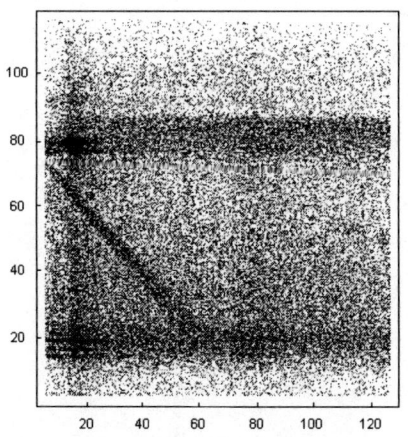 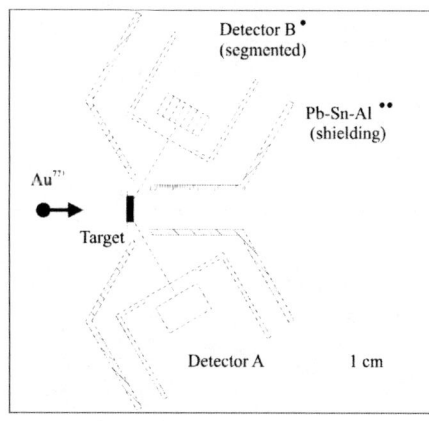

FIGURE 3. Contour plot for the coincident emission of two photons for Au ions (106 MeV/u Au^{77+} excited in Al foil; detector A vs. detector B, photon energies in keV). The experimental arrangement is given at the right side. Along the diagonal ridge in the contour plot the sum energy of both photons is constant and corresponds to the total transition energy of the 2E1 decay [16, 17].

RESULTS

Due to the complicated efficiency response of the detection system, the measured coincident two-photon spectra cannot be compared directly to theory. Instead of unfolding the measured spectra with all the uncertainties involved, the theoretical spectral distributions are normally convoluted with all the relevant efficiencies and then compared directly with the experimental ones. This convolution is done best by Monte Carlo simulations starting from the theoretical spectral distribution and then including all the experimental correction factors, cf. Ref. [4]. This method was successfully introduced for the case of the two-photon continuum in He-like Kr^{34+} ions [4]. A clear change in the relative spectral distribution for the two-photon decay compared to the one for atomic He was established. Unfortunately, the systematic errors are still somewhat large. A further improvement in accuracy was achieved for Ni ions by comparing the two-photon decay for H- and He-like species, Ni^{25+} and Ni^{24+}, respectively [6]. Assuming that the spectral distribution for H-like ions can be treated correctly all the experimental efficiencies can be extracted with high accuracy; and then, the spectral distribution for the He-like case can be compared to theory yielding the most precise determination for the two-photon continuum in heavy He-like ions done up to now. However, at this Z range the relativistic effects on the spectrum are still too small to be uniquely detected. Relativistic [9] and non-relativistic [18] calculations still deliver here quite similar distributions.

For the heaviest investigated system, He-like Au^{77+} ions, relativistic effects are strong. Hence, a comparison of the experimental spectrum with the simulation is

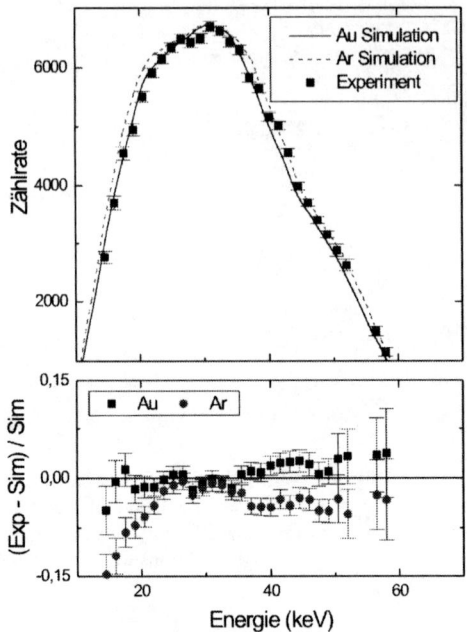

FIGURE 4. The x-ray continuum for the two-photon emission from He-like Au^{77+} ions (top part) [17]. The measured data points are shown along with Monte Carlo simulations one for fully relativistic Au^{77+} ions and one for non-relativistic Ar^{16+} ions (dashed line). At the bottom the residues of the experimental data points compared to the simulations are given.

displayed at the top of Fig. 4 [17]. Within the error band for the simulation and the experimental error bars, good agreement results between measurement and fully relativistic theory. For comparison, a corresponding simulation for the non-relativistic case of Ar^{16+} ions is also given (dashed line). At the bottom the residues of the experimental data with respect to the results of the simulations are given. It is evident that the non-relativistic approach is at variance with the experimental findings, only the fully relativistic approach is in agreement.

Taking out the trivial phase space factor from the continuous spectrum one obtains the energy dependent square of the matrix element, see Fig. 5. For comparison the normalized square of the theoretical matrix elements are shown for He-like Ni and Au ions along with the experimental data for Au^{77+}. The difference in the distributions demonstrates quite clearly the relativistic influence, whereby all intermediate ^{3}P states are included in the calculations [9]. The experimental points evidently favor the fully relativistic Au distribution; a χ^2 test gives values of 3.48 and 7.35 for the Au and Ni predictions, respectively. However, it is also evident, that additional and more precise measurements are needed for a more refined comparison with theory.

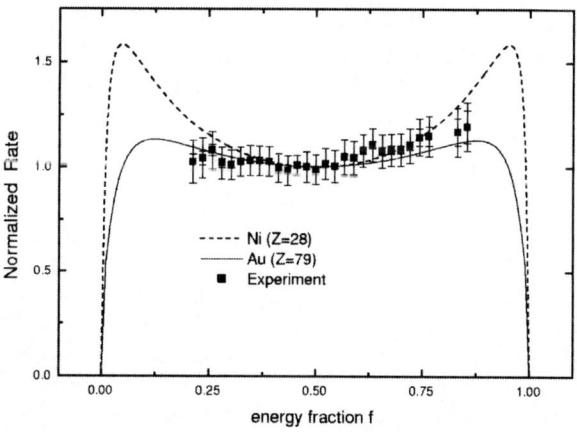

FIGURE 5. The normalized square of the energy dependent matrix element (rate) extracted from the experiment is compared to the fully relativistic theory for Au^{77+} and Ni^{26+} (full and dashed lines, respectively [9]).

OUTLOOK

Both experimentally and theoretically, important steps towards a better understanding of the complete atomic structure of He-like heavy ions were made recently by investigating the two-photon decay continuum associated with the excited $1s2s\,^1S_0$ state. Recently, non-relativistic approaches [18] valid up to the medium Z region (around 30) were complemented by fully relativistic calculations [9] including all intermediate ^3P states. Measurements in the medium Z region still compare well with both calculations as the relativistic effects are not yet dominant there [4, 6]. For a true relativistic system, Au^{77+}, the first experiment clearly favors the fully relativistic calculation. Higher accuracies in future experiments are certainly feasible and expected for the next experiments using the described beam foil excitation technique.

A recent experiment demonstrated a new method to produce the $1s2s\,^1S_0$ at moderate relativistic ion velocities. It was found that selectively ionizing the K shell of L-like ions leads with high efficiency to the $1s2s\,^1S_0$ excited state [19]. This new experimental method may give improved access to this area. Moreover, with the third generation synchrotron radiation facilities new methods are at hand to produce singly K ionized atoms of any atomic number. These singly K ionized atoms decay in a small fraction of the time via the two-photon decay branch and thus give information on the complete atomic structure of multi-electron systems which finally can be compared to the corresponding true He-like cases. In the past, most of the data on inner shell two-photon decay in singly ionized systems were obtained from nuclear unstable species decaying by nuclear K electron capture. This method was demonstrated and succesfully applied by Ilakovac and his group [11 – 13]. Further measurements, both at heavy ion accelerators and at modern synchrotron radiation facilities, will give unprecedented access to the complete atomic structure of He-like or quasi-He-like atomic systems in the strong field domain of the heaviest atomic species.

ACKNOWLEDGEMENTS

The authors highly acknowledge the collaboration of many colleagues in the experiments as well as the assistance from theory in particular we like to name S. Cheng, L.J. Curtis, C. Kozhuharov, A.E. Livingston, H.W. Schäffer, Z. Stachura, Th. Stöhlker, A. Warczak, as well as A. Derevianko, W.R. Johnson and G.W.F. Drake. The work was partially supported by the WTZ scientific-technical collaboration program of the German and Polish governments and by a NATO grant. RWD and EPK were supported by the Chemical Sciences, Geosciences, and Biosciences Division of the Office of Basic Energy Sciences, Office of Science, U. S. Department of Energy, under Contract W-31-109-Eng-38.

REFERENCES

1. M. Göppert-Mayer, Ann. Phys. (Leipzig) **9** (1931) 273
2. Y.B. Bannet, I. Freund, Phys. Rev. **A30** (1984) 299
3. R. Marrus, P.J. Mohr, Adv. At. Mol. Phys. **14** (1978) 181
4. R. Ali et al., Phys. Rev. **A55** (1997) 994
5. P.H. Mokler et al., Phys. Scripta **T51** (1994) 28
6. H.W. Schäffer et al., Phys. Rev. **A59** (1999) 245
7. G.W.F. Drake, Phys. Rev. **A34** (1986) 1182
8. W.R. Johnson, Phys. Rev. Lett. **29** (1972) 1123
9. A. Derevianko, W.R. Johnson, Phys. Rev. **A56** (1997) 1288
10. P.H. Mokler, R.W. Dunford, Fizika **A10** (2001) 105
11. K. Ilakovac et al., Phys. Rev. Lett. **56** (1986) 2469
12. K. Ilakovac et al., Phys. Rev. **A44** (1991) 7392
13. K. Ilakovac et al., Phys. Rev. **A46** (1992) 132
14. R.W. Dunford et al., Phys.Rev.Lett. **62** (1989 2809
15. W.R. Johnson et al., Adv.At.Mol.Phys. **35** (1995) 255
16. H.W. Schäffer, GSI-report Diss.99-16 (1999)
17. H.W. Schäffer et al., Phys. Lett. **A260** (1999) 489
18. G.W.F. Drake, Phys. Rev. **A34** (1986) 2871
19. Th. Stöhlker et al., GSI scientific report 2000, GSI-2001-1 (2001) 95

Nuclear Excitation involving Inner-Shell Electrons

Takeshi Mukoyama

Kansai Gaidai University, 16-1 Nakamiya-Higashinocho, Hirakata, Osaka 573-1001, Japan

Abstract. The experimental and theoretical studies on nuclear excitation involving atomic electrons are described. Several processes, such as nuclear excitation by electron transition, nuclear excitation by free-electron capture, nuclear excitation by target electron capture in ion-atom collisions, nonresonant photoexcitation, and nuclear excitation by positron annihilation, are discussed.

INTRODUCTION

Nuclear excitation takes place inside the nucleus due to the interaction of the nucleus with electromagnetic radiation or particles and is, in general, not associated with atomic electrons outside of the nucleus. However, there are several processes in which the inner-shell electrons do take part in excitation of the nucleus. In the present work, we will describe five nuclear excitation processes involving inner-shell electrons: (1) nuclear excitation by electron transition (NEET), (2) nuclear excitation by free-electron capture (NEFEC), (3) nuclear excitation by target-electron capture in ion-atom collisions (NETEC), (4) nonresonant photoexcitation (NRPE), and (5) nuclear excitation by positron annihilation (NEPA).

The Feynman diagrams for these processes are shown in Fig. 1. The bold arrows represent the nucleus, the thin arrows the electron or positron, and the wavy line the electromagnetic interaction with a photon or through the exchange of virtual photon. The dashed lines mean that the electron is bound to the nucleus. It is interesting to note that four diagrams for the NEET, the NEFEC, the NETEC, and the NEPA are similar and sometimes called the inverse process of internal conversion in nuclear decay, the *inverse internal conversion*. Their difference consists only in electron states, free or bound and electron or positron. For comparison, the diagram for the internal conversion is also shown in Fig. 1.

NUCLEAR EXCITATION BY ELECTRON TRANSITION

An inner-shell vacancy created in atoms is filled by outer-shell electrons with emission of x rays or Auger electrons. Morita proposed the third mechanism, where the energy produced in the electron transition is given to nucleus with its

CP652, *X-Ray and Inner-Shell Processes: 19th International Conference on X-Ray and Inner-Shell Processes*
edited by A. Bianconi, A. Marcelli, and N. L. Saini
© 2003 American Institute of Physics 0-7354-0111-X/03/$20.00

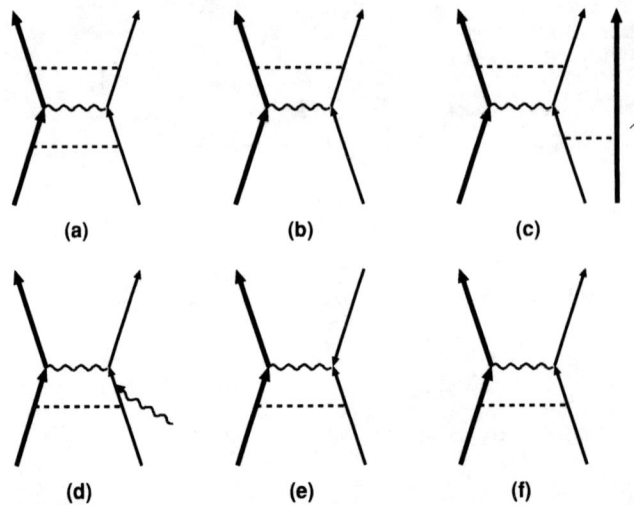

FIGURE 1. Diagrams for (a) NEET, (b) NEFEC, (c) NETEC, (d) NRPE, (e) NEPA, and (f) internal conversion.

subsequent excitation [1]. This NEET process is possible only when the atomic and nuclear transitions have the same multipolarity and almost equal transition energy. Since the first experimental study by Otozai *et al.* [2], extensive studies have been performed for various nuclides both experimentally and theoretically. The NEET probability for ^{189}Os observed by Otozai *et al.* with electron impact ionization [3] was later found due to the direct nuclear excitation by incident electrons [4]. With refinement in theoretical models and experimental measurements, both theoretical and experimental values became smaller and smaller than those expected earlier.

Since the NEET process is discussed in details elsewhere [5], only the experimental results with synchrotron radiation and the recent theoretical values are listed in Table 1. Most earlier experimental studies were made by electron bombardment or bremsstrahlung. In such cases, the possibility of other nuclear excitation processes, whose probabilities often exceed the NEET probability, cannot be excluded.

TABLE 1. Experimental and theoretical values of NEET.

Nuclide	Experimental	Ref.	Theoretical	Ref.
^{189}Os	$(5.7 \pm 1.7) \times 10^{-9}$	[6]	1.1×10^{-10}	[7]
	$< 9.0 \times 10^{-10}$	[8]	3.2×10^{-10}	[9]
	$< 1.4 \times 10^{-10}$	[10]	1.3×10^{-10}	[8]
			1.1×10^{-10}	[11]
^{179}Au	$(5.0 \pm 0.6) \times 10^{-8}$	[12]	1.3×10^{-7}	[7]
			7.2×10^{-9}	[9]
			3.6×10^{-8}	[11]

For ^{189}Os, Shinohara *et al.* [6] observed the NEET process with synchrotron radiation. However, they used a white beam, which has continuous photon spectrum. Their values is larger than the recent upper limits obtained with monochromatized photon beams [8, 10].

Only reliable evidence of the NEET is the experiment by Kishimoto *et al.* for ^{197}Au [12]. They irradiated the gold target with monochromatic x-ray beams produced at SPring-8 in a several-bunch operation mode and measured the internal conversion electrons from the isomeric state with a silicon avalanche photodiode. To discriminate the NEET signals from strong background due to atomic processes, they observed the time spectra and accepted signals only in the time region between 4.5 and 20 nsec within the period of 42 sec after the incident radiation. With this new technique they could succeed to observe the NEET process.

NUCLEAR EXCITATION BY FREE ELECTRON CAPTURE

When a vacancy is created in an atom and there exist free electrons near to the atom, the atom captures one of free electrons with emission of radiation. This process is well known as *radiative recombination*. If the excess energy released is equal to the nuclear excitation energy, it is used to excite the nucleus and the nuclear excitation takes place by free electron capture. When the free electron has the kinetic energy E and the binding energy of the atomic orbital into which the electron is capture is B, the excited energy of the nuclear level is given by

$$W = E + B . \tag{1}$$

Since the NEFEC is a resonance process, the density of free electrons with energy E should be large in the vicinity of the nucleus.

Goldanskii and Namiot [13, 14] considered that the above condition can be realized by the pulsed laser irradiation. Assume that the natural width of the nuclear level to be excited is Γ. The probability of nuclear excitation is expressed as

$$P = n_e \sigma v \tau , \tag{2}$$

where n_e is the density of free electrons with energy E within Γ, σ is the cross section for the nuclear excitation by these electrons, v is the velocity of electrons, and τ is the duration of laser plasma. When the density of free electrons is n, n_e is nearly equal to $n\Gamma/W$. We assume that the cross section is approximated as $\sigma \sim (\hbar/mv)^2$, where m is the electron mass. Then Eq. (2) is written by

$$P \approx n \frac{\Gamma}{W} \frac{\hbar^2}{m(2mE)1/2} \tau . \tag{3}$$

For the case of ^{235m}U with $W = 73$ eV and $\Gamma \approx 3 \times 10^{-19}$ eV (half life of 26 min), the probability can be given by

$$P \approx 3 \times 10^{-29} n_e (1/\text{cm}^3) \tau(\text{sec}) . \tag{4}$$

If we assume that the output energy of the laser beam is about $10J$ and $W \approx B$, we obtain $P \approx 3 \times 10^{-14}$ and this means that the number of ^{235m}U nuclei produced in one laser pulse is ≈ 1000.

A similar experiment proposed above was performed by Izawa and Yamanaka [15]. They irradiated a metallic foil of natural uranium by a CO_2 laser with energy of $1J$ in 100 nsec at 10.6 μm. After 100 laser pulses on the target with a repetition rate of 0.5 sec^{-1}, the induced activity was measured with a channel electron multiplier. When the time-independent component, which originates from α decay, is subtracted, the life time was determined to be 25.7 $\pm$ 0.4 min and in good agreement with that of ^{235m}U. However, Izawa and Yamanaka attributed this isomer production to the NEET process. On the other hand, Goldanskii and Namiot [16] pointed out that in the experimental condition of Izawa and Yamanaka the probability for the NEFEC is about 1000 times larger than that for the NEET and the induced activity observed by them is due to the NEFEC.

Arutyunyan et al. [17] repeated the similar experiments with a pulsed CO_2 laser (5 J, 200 nsec) and a ceramic target of 6%-enriched uranium, but no isomers were observed. Then they performed the experiments for plasma produced by a relativistic electron beam. Targets of 100%-enriched ^{235}U were irradiated with the high-current accelerator with electron energy of $\sim$ 500 keV, beam current of $\sim$ 150 kA and pulse duration of $\sim$ 30 nsec. They found the induced activity with half life of 24.8 $\pm$ 6.7 min. However, their cross section is approximately two or three orders of magnitude smaller than the value of Izawa and Yamanaka.

NUCLEAR EXCITATION BY TARGET ELECTRON CAPTURE

In the charge transfer process during fast ion-atom collisions, an electron in the target atom is captured into one of bound states in the projectile. It is well known that in order to satisfy the conservation of energy and momentum, the excess energy produced in capture is emitted as radiation (radiative electron capture, REC), or is used to excite a projectile electron into higher energy states (resonant transfer excitation, RTE) and to ionize another target electron (transfer ionization, TI). Cue et al. proposed a new mechanism in which the projectile nucleus is excited simultaneously with the electron capture [18]. This is a competitive process to the RTE and has the resonance character, which is possible only at a definite energy of the projectile. It should be noted, however, that the RTE is considered as the inverse process of the Auger decay in atoms, while the NETEC corresponds to the inverse internal conversion.

Since the target electrons have a momentum distribution in atoms, the resonance shape of the NETEC is characterized by their Compton profile. Taking into account this fact and considering the analogy between the NETEC and the RTE, Cue et al. obtained an order-of-magnitude estimate for the NETEC cross section, σ_N, for

TABLE 2. Cross sections of some nuclides for NETEC [18]

Shell	Nuclide	Excited state (keV)	Excitation energy (MeV/u)	σ_N (mb)
K	[165]Ho	94.7	54.2	3
	[173]Yb	78.6	12.6	20
	[185]Re	125	76.5	3
	[187]Re	134	92.7	2
	[195]Pt	98.8	14.7	7
$L_{1,2}$	[183]W	46.5	47.1	1
	[236]U	45.2	20.0	10
	[239]Pu	57.3	38.7	4

several nuclides in terms of the RTE cross section, σ_A:

$$\sigma_N = g \left(\frac{E_A}{E_N}\right)^{3/2} \left(\frac{\Gamma_N}{\Gamma_A}\right) \sigma_A \,, \tag{5}$$

where g includes the factor due to atomic vacancy and electronic and nuclear spin weighting, E_A and E_N are the resonance energy corresponding to the RTE and the NETEC, and Γ_N is the internal conversion rate for the nuclear transition, and Γ_A is the Auger decay rate for atoms.

The calculated results for bare projectiles are shown in Table 2. These cross sections are smaller than those for other atomic processes. However, the characteristic time scale of the atomic processes are much shorter than the nuclear lifetime and it is possible to discriminate the NETEC experimentally by the use of time dependence.

On the other hand, the background due to the nuclear event, such as Coulomb excitation of the projectile nucleus by the target nucleus, is difficult to eliminate. Cue *et al.* proposed to use a single crystal target in an axial channelling geometry. This technique allows ions to pass through the region of high electron density, but decreases the probabilities for collisions with small impact parameters and is expected to reduce the contribution from the Coulomb excitation. Kimball *et al.* calculated directly without atomic data the NETEC cross sections for ions channelling along the Si $< 110 >$ axis [19]. The cross sections thus obtained are in good agreement with the values in Table 2.

NONRESONANT PHOTOEXCITATION

Nuclear photoexcitation process was explained by two mechanisms: resonant absorption and Compton scattering of a photon by the nucleus. When the existence of atomic electron is taken into consideration, there is an additional process in which atomic electrons play an important role. In ordinary photoelectric effect, a photoelectron carries off the whole energy available, but it is also possible that a part of its energy is used to excite the nucleus. It should be noted that most experimental studies for photoexcitation were performed by using photons from radioactive sources or bremsstrahlung radiation from accelerators. In such cases, the contributions from the nonresonant process may be appreciable.

This process was considered by two groups independently. Batkin called it the Compton excitation of nuclear levels [20] and calculated the cross section for excitation of low-lying nuclear states by synchrotron radiation [21]. On the other hand, the same process was called the inelastic photoelectric effect by Ljubičić *et al.* [22].

The experimental studies on the nonresonant photoexcitation was made by Ljubičić *et al.* for the 1078-keV level of ^{115}In using ^{60}Co γ rays [22]. The nuclear excitation was confirmed by observing 336-keV γ rays from the isomeric state, which was produced through de-excitation processes, i.e. the so-called (γ,γ') reaction (Fig. 2). When the nonresonant process is taken into consideration, the excitation probability can be written by

$$P = \phi_R(E_R)\sigma_R + \phi_{NR}\sigma_{NR}, \tag{6}$$

where ϕ_R is the photon flux at the resonance energy E_R, ϕ_{NR} represents the photon flux for nonresonant process integrated with respect to energy, and σ_R and σ_{NR} are the excitation cross sections for resonance and total nonresonant process, respectively.

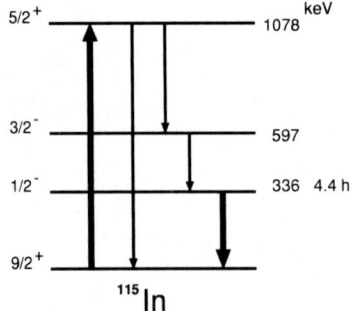

FIGURE 2. Level scheme of ^{115}In.

The experiments were performed with 680-Ci ^{60}Co source. They inserted lead absorbers with various thicknesses between the ^{60}Co source and the indium target during irradiation. These absorbers distort the photon spectrum in the target, which was estimated beforehand using a much weaker source in the same geometry. The observed induced activity as a function of absorber thickness was fitted to Eq. (6) to determine σ_R and σ_{NR}. They found that the photoactivation of ^{115}In contains a dominant contributions from the nonresonant process. The total level width of the 1078-keV level thus obtained is smaller than other experimental values by photoexcitation [23]. They also measured the resonant and nonresonant cross sections for various nuclides, such as ^{87}Sr, ^{111}Cd, ^{113}In, and ^{115}In [24–28].

The similar measurement was made by Bikit *et al.*. They found negligibly small contribution from the nonrelativistic process for ^{111}Cd [29], but confirmed the necessity of introducing the nonresonant photoexcitation process in ^{115}In [30]. Lee *et al.* obtained the evidence of the nonresonant contribution for ^{111}Cd [31].

On the other hand, Yoshihara *et al.* [32] reinvestigated the (γ,γ') process for ^{115}In by ^{60}Co γ rays and showed with the hot-atom chemistry technique that the nonresonant contribution is less than 3.3% of the resonance effect. Collins *et al.* used the single pulses of intense bremsstrahlung to excite ^{115}In [33] and ^{111}Cd

[34]. By measuring the induced activity as a function of the end-point energy of electrons producing bremsstrahlung, they obtained the integrated excitation cross sections for the (γ, γ') reactions. For both nuclides, they found no evidence of any importance of nonresonant channels. They also made comment [35] on the results of Krčmar et al. for ^{111}Cd [27]. Using the Monte Carlo simulation, they pointed out that the small resonant cross section of Krčmar et al. is due to neglection of Compton scattering from environments and no need to introduce the nonresonant contribution. In reply to this comment, Krčmar et al. stated that the Monte Carlo calculations seriously overestimated the intensity of environmental Compton scattering [36].

Theoretically Ljubičić et al. [22] estimated the cross sections for the nonresonant process for ^{115}In. Pisk et al. calculated the dependence of the total cross sections on the photon energy as well as the nuclear excitation energy [37]. However, the theoretical cross sections are much smaller than their experimental results. Later Ljubičić applied the two indistinguishable quantum oscillator (IQO) model for the nonresonant nuclear excitation process and obtained good agreement with the available experimental data [9]. Durkarev studied the nonresonant nuclear excitation process by photons and obtained the relation between the nonresonant and resonant cross sections [38].

NUCLEAR EXCITATION BY POSITRON ANNIHILATION

When a positron annihilates with an electron, at least two quanta are emitted from the conservation law of momentum. However, in the case where the electron is strongly bounded to the nucleus, the nucleus can absorb the excess momentum and several special annihilation modes are possible [39]. In 1951, Present and Chen [40] proposed a new mode of annihilation, in which the energy liberated in annihilation process is transferred to the nucleus with its subsequent excitation. In this way, the annihilation takes place without emission of any radiation and this process is called the nuclear excitation by positron annihilation. The NEPA is a resonance process and occurs only for positrons whose kinetic energy corresponds to the nuclear excited level.

The first experimental evidence of the NEPA was established by Mukoyama and Shimizu [41]. They irradiated the indium target with positrons from a ^{22}Na radioactive source and observed internal conversion electrons from the isomeric state, in the manner similar to the (γ, γ') reaction. The effective cross section for excitation is expressed as

$$\sigma_{\text{eff}} = n\sigma\Gamma_{\text{iso}}/\Gamma, \tag{7}$$

where Γ and Γ_{iso} are the total width of the excited level and its partial width for the de-excitation to the isomeric state, respectively, σ is the resonance excitation cross section, and n is the ratio of the number of positrons in the target foil within the resonance width to the total number of incident positrons. The experimental NEPA cross sections for the 1078-keV level was in order of $\sim 10^{-24}$ cm^2. This value was

TABLE 3. Cross sections for the 1078-keV level of ^{115}In.

Positron source	Method	Cross section (cm^2)	Ref.
^{22}Na	e^-	$\sim 10^{-24}$	[41]
^{22}Na	γ	$(4.6\pm3.2)\times10^{-24}$	[42]
^{64}Cu	γ	$(8.7\pm4.2)\times10^{-24}$	[42]
^{64}Cu	γ	$(3.9\pm1.4)\times10^{-24}$	[43]
^{64}Cu	γ	$(4.8\pm2.1)\times10^{-24}$	[46]

later confirmed by measuring γ rays for different experimental conditions, as shown in Table 3. Similar experiments were performed by us for ^{111}Cd [44] and ^{176}Lu [45]. All the experimental data are summarized in two review articles [47, 48]. These experimental values are much larger than the theoretical predictions [43, 49].

Raghavan and Mills [50] pointed out that the large discrepancy between theory and experiment can be explained by taking into account the contributions from the nonresonant process, which is quite similar to the NRPE in photoactivation. Batkin and Churakova also made similar calculations [51]. However, Ljubičić et al. showed that the calculations of Raghavan and Mills are too simple and overestimate the nonresonant contribution by many order of magnitude [52]. They tried to include the radiative nuclear excitation process by positron annihilation, but found that this process cannot explain the discrepancy [53].

Recently Ljubičić and Logan analyzed the NEPA process using the IQO model [54]. The total resonance width in Eq. (7) was chosen as the sum of the natural line width of the nuclear level and the width of the K-shell vacancy in the atom. From the σ_{eff} value of Watanabe et al. [43], they obtained $\sigma = (1.4 \pm 0.4) \times 10^{-26}$ cm^2 for the 1078-keV level of ^{115}In. This value is in good agreement with the theoretical value of Kaliman et al. [55], 2.06×10^{-26} cm^2. Kaliman and Orlić also calculated the NEPA cross sections for a screened potential [56] and using the Coulomb wave functions [57]. In both cases, the calculated values for ^{115}In agree with the experimental value modified above.

Finally it is interesting to note that other nuclear reaction processes by positron annihilation with inner-shell electrons have been considered. Present and Chen [40] already pointed out the possibility of neutron emission and nuclear fission in annihilation process. Borozents et al. measured the probability of neutron emission for ^{9}Be by detecting x and γ rays from ^{152}Eu used as a neutron detector [58]. By the use of ^{45}Ti and ^{11}C positron sources, they obtained the value in order of $\sim 10^{-30}$ cm^2, which is by three order of magnitude larger than the theoretical estimate [40]. On the other hand, Grechukhin and Soldatov calculated the probability for excitation of a spontaneously fissile isomer by positron annihilation with K-shell electrons [59].

CONCLUSION

Nuclear excitation processes involving inner-shell electrons are described. They are very interesting to understand the interplay between the atomic and nuclear physics. Except for the NETEC, other processes have been observed experimentally, but in most cases there still remains considerable discrepancy between theory and experiment. It is hoped that more elaborate experimental studies as well as more rigorous theoretical calculations for these processes be performed to elucidate this discrepancy in near future.

ACKNOWLEDGMENTS

The author would like to express his thanks to Prof. A. Ljubičić at Zagreb and Prof. S. Kishimoto at KEK, Tsukuba, for sending him preprints and reprints.

REFERENCES

1. M. Morita, Prog. Theor. Phys. **49**, 1574 (1973).
2. K. Otozai, R. Arakawa, and M. Morita, Prog. Theor. Phys. **50**, 1771 (1973).
3. K. Otozai, R. Arakawa, and T. Saito, Nucl. Phys. **A297**, 97 (9178).
4. T. Sato, N. Katsushima, and H. Ohtsubo, Prog. Theor. Phys. **89**, 103 (1993).
5. D. S. Gemmell, this proceedings.
6. A. Shinohara, T. Saito, M. Shoji, A. Yokogama, H. Baba, M. Ando, and K. Taniguchi, Nucl. Phys. **A472**, 15 (1987).
7. E. V. Tkalya, Nucl. Phys. **A539**, 209 (1992).
8. I. Ahmad, R. W. Dunford, H. Esbensen, D. S. Gemmell, E. P. Kanter, U. Rütt, and S. H. Southworth, Phys. Rev. C **61**, 051304(R) (2000).
9. A. Ljubičić, in *Proceedings of the Internat. Meeting on Frontiers of Physics, 1998, Kuala Lumpur*, ed. by S. P. Chia and D. A. Bradley, World Scientific, Singapore, 1998, p. 133.
10. K. Aoki, K. Hosono, K. Tanimoto, M. Terasawa, H. Yamaoka, M. Tosaki, Y. Ito, A. M. Vlaicu, K. Taniguchi, and J. Tsuji, Phys. Rev. C **64**, 044609 (2001).
11. M. R. Harston, Nucl. Phys. **A690**, 447 (2001).
12. S. Kishimoto, Y. Yoda, M. Seto, Y. Kobayashi, S. Kitao, R. Haruki, T. Kawauchi, K. Fukutani, and T. Okano, Phys. Rev. Lett. **83**, 1831 (2000).
13. V. I. Gol'danskiĭ and V. A. Namiot, Pis'ma Zh. Eksp. Teor. Fiz. **23**, 495 (1976) [JETP Lett. **23**, 451 (1976)].
14. V. I. Goldanskii and V. A. Namiot, Phys. Lett. **62B**, 393 (1976).
15. Y. Izawa and C. Yamanaka, Phys. Lett. **88B**, 59 (1979).
16. V. I. Goldanskii and V. A. Namiot, Yad. Fiz. **33**, 319 (1981) [Sov. J. Nucl. Phys. **33**, 169 (1981)].
17. R. V. Arutyunyan, L. A. Bol'shov, V. D. Vikharev, S. A. Dorshakov, V. A. Kornilo, A. A. Krivolapov, V. P. Smirnov, and E. V. Tkalya, Yad. Fiz. **53**, 36 (1991) [Sov. J. Nucl. Phys. **53**, 23 (1991)].
18. N. Cue, J.-C. Poizat, and J. Remillieux, Europhys. Lett. **8**, 19 (1989); N. Cue, Nucl. Instr. and Meth. **B40/41**, 25 (1989).
19. J. C. Kimball, D. Bittel, and N. Cue, Phys. Lett. A **152**, 367 (1991).
20. I. S. Batkin, Yad. Fiz. **29**, 903 (1979) [Sov. J. Nucl. Phys. **29**, 464 (1979)].
21. I. S. Batkin and M. I. Berkman, Yad. Fiz. **32**, 972 (1980) [Sov. J. Nucl. Phys. **32**, 502 (1980)].
22. A. Ljubičić and K. Pisk, Phys. Rev. C **23**, 2238 (1981).
23. Y. Watanabe and T. Mukoyama, Nucl. Sci. and Eng. **80**, 92 (1982).

24. M. Krčmar, A. Ljubičić, K. Pisk, B. A. Logan, and M. Vrtar, Phys. Rev. C **25**, 2097 (1982).
25. M. Krčmar, A. Ljubičić, B. A. Logan, and M. Bistrović, Phys. Rev. C **33**, 293 (1986).
26. M. Krčmar, A. Ljubičić, K. Pisk, B. A. Logan, and M. Bistrović, FIZIKA **18**, 171 (1986).
27. M. Krčmar, S. Kaučić, T. Tustonić, A. Ljubičić, B. A. Logan, and M. Bistrović, Phys. Rev. C **41**, 771 (1990).
28. D. A. Bradley, I. A. Jalil, M. Krčmar, and A. Ljubičić, J. Radioanal. Nucl. Chem. **244**, 475 (2000).
29. I. Bikit, J. Slivka, I. V. Aničin, L. Marinkov, A. Rudć, and W. D. Hamilton, Phys. Rev. C **35**, 1943 (1987).
30. I. Bikit, J. Slivka, I. V. Aničin, L. Marinkov, and A. Rudć, Prorodno-Matematički Fakultet – Univerzitet u Novom Sadu, Zbornik radova, knjiga **17**, 55 (1987).
31. C.-B. Lee, D. A. Bradley, I. A. Jalil, Y. M. Amin, M. J. Maah, and K. Z. M. Dahlan, Rad. Phys. and Chem. **61**, 367 (2001).
32. K. Yoshihara, Zs. Németh, L. Lakosi, I. Pavlicsek, and Á. Veres, Phys. Rev. C **33**, 728 (1986).
33. C. B. Collins, J. A. Anderson, Y. Paiss, C. D. Eberhard, R. J. Peterson, and W. L. Hodge, Phys. Rev. C **38**, 1852 (1988).
34. J. A. Anderson, M. J. Byrd, and C. B. Collins, Phys. Rev. C **38**, 2838 (1988).
35. P. von Neumann-Cosel, A. Richter, J. J. Carroll, and C. B. Collins, Phys. Rev. C **44**, 554 (1991).
36. M. Krčmar, A. Ljubičić, B. A. Logan, and M. Bistrović, Phys. Rev. C **47**, 906 (1993).
37. K. Pisk, M. Krčmar, A. Ljubičić, and B. A. Logan, Phys. Rev. C **25**, 2226 (1982).
38. E. G. Drukarev, in *X-Ray and Inner-Shell Process, 18th Internat. Conf.*, ed. by R. W. Dunford *et al.*, AIP Conference Proceedings 506, American Institute of Physics, New York, 2000, pp. 496–500.
39. D. Berényi, in *Proc. Intern. Conf. on Inner-Shell Ionization Phenomena and Future Applications, Atlanta, Georgia, 1972*, ed. by R. W. Fink *et al.*, USAEC, Oak Ridge, Tenn., 1973, p. 2175.
40. R. D. Present and S. C. Chen, Phys. Rev. **83**, 238 (1951); **85**, 447 (1952).
41. T. Mukoyama and S. Shimizu, Phys. Rev. C **5**, 95 (1972).
42. Y. Watanabe, Master Thesis, Dept. of Nuclear Engineering, Kyoto University, 1977, unpublished.
43. Y. Watanabe, T. Mukoyama, and S. Shimizu, Phys. Rev. C **19**, 32 (1979).
44. Y. Watanabe, T. Mukoyama, and S. Shimizu, Phys. Rev. C **21**, 1753 (1980).
45. Y. Watanabe, T. Mukoyama, and R. Katano, Phys. Rev. C **23**, 695 (1981).
46. I. N. Vishnevskiĭ, V. A. Zheltonozhskiĭ, V. P. Svyato, and V. V. Trishin, Pis'ma Zh. Eksp. Teor. Fiz. **30**, 394 (1979) [JETP Lett. **30**, 366 (1979)].
47. T. Mukoyama, ATOMKI Közlemények **23**, 89 (1981).
48. I. N. Vishnevskiĭ, V. A. Zheltonozhskiĭ, and V. M. Kolomiets, Fiz. Elem. At. Yadra **19**, 237 (1988) [Sov. J. Pat. Nucl. **19**, 101 (1988)].
49. D. P. Grechukhin and A. A. Soldatov, Zh. Eksp. Teor. Fiz. **74**, 13 (1978) [Sov. Phys. JETP **47**, 6 (1978)].
50. R. S. Raghavan and A. P. Mills, Jr., Phys. Rev. C **24**, 1814 (1981).
51. I. S. Batkin and T. A. Churakova, Yad. Fiz. **35**, 282 (1982) [Sov. J. Nucl. Phys. **35**, 161 (1982)].
52. A. Ljubičić, M. Krčmar, K. Pisk, and B. A. Logan, Phys. Rev. C **30**, 209 (1984).
53. K. Pisk, M. Krčmar, A. Ljubičić, and B. A. Logan, Phys. Rev. C **32**, 83 (1985).
54. A. Ljubičić and B. A. Logan, Phys. Lett. **325B**, 297 (1994).
55. Z. Kaliman, K. Pisk, and B. A. Logan, Phys. Rev. C **35**, 1661 (1987).
56. Z. Kaliman and N. Orlić, Rad. Phys. and Chem. **61**, 355 (2001).
57. N. Orlić and Z. Kaliman, Rad. Phys. and Chem. **61**, 387 (2001).
58. G. P. Borozenets, I. N. Vishnevskiĭ, and V. A. Zheltonozhskiĭ, Yad. Fiz. **46**, 1320 (1987) [Sov. J. Nucl.Phys. **46**, 774 (1987)].
59. D. P. Grechukhin and A. A. Soldatov, Yad. Fiz. **29**, 296 (1979) [Sov. J. Nucl. Phys. **29**, 146 (1979)].

X-Ray Spectroscopy On Neon-Like Heavy Ions

Nobuyuki Nakamura*, Daiji Kato* and Shunsuke Ohtani*†

*Cold Trapped Ions Project, ICORP, JST, Kawaguchi, Saitama 332-0012, Japan
†The University of Electro-Communications, Chofu, Tokyo 182-8585, Japan

Abstract. The atomic structure and electron-impact-excitation processes of neonlike heavy ions have been systematically investigated through X-ray spectroscopy with the Tokyo electron beam ion trap. Several $n = 3$-2 transitions for $Z = 50$ to 56 were observed with a flat crystal spectrometer. From the observed spectra, wavelengths and relative line intensities were obtained as functions of Z and electron energy. Through the wavelength measurements, strong configuration interaction among $n = 3$ excited levels were investigated, while the electron-impact-excitation processes were investigated through the line intensity measurements. Theoretical calculation has also been performed to reproduce the experimental spectra. The theoretical results were found to agree well with the experiment both for wavelength and for line intensity. Near future plan with a new spectrometer which is recently developed for hard X-rays (> 15 keV) are also presented.

INTRODUCTION

An electron beam ion trap (EBIT) [1] is a devise which is well suited for spectroscopic studies of highly charged ions. We have constructed a high-energy EBIT (called Tokyo EBIT) [2, 3, 4] and started physics experiments with it at late 1997. To date, spectroscopic studies at the Tokyo EBIT have been performed mainly for two wavelength range; one is the visible (and near-UV) range and another is the X-ray range. For the visible range, M1 transitions between the fine-structure levels of the $(3d^4)^5 D_J (J = 2, 3)$ ground terms in titaniumlike ions have been systematically observed [5] because it shows interesting behavior that the wavelength stays in the visible (and near-UV) range over a wide range of Z. For the X-ray range, neonlike ions have been mainly investigated. Since the neonlike ion has a closed shell structure, its abundance in hot plasmas is high for a wide range of plasma parameters. Neonlike ions can therefore be widely used for many kinds of application, such as X-ray lasers and plasma diagnostics [6]. For these applications, systematic studies of transition wavelengths, oscillator strengths and collision strengths in neonlike ions are strongly needed. To date many theoretical [7, 8, 9, 10] and experimental [11, 12, 13] studies have been done. In the present paper, systematic studies on neonlike ions with the Tokyo EBIT are summarized.

CP652, X-Ray and Inner-Shell Processes: 19th International Conference on X-Ray and Inner-Shell Processes
edited by A. Bianconi, A. Marcelli, and N. L. Saini
© 2003 American Institute of Physics 0-7354-0111-X/03/$20.00

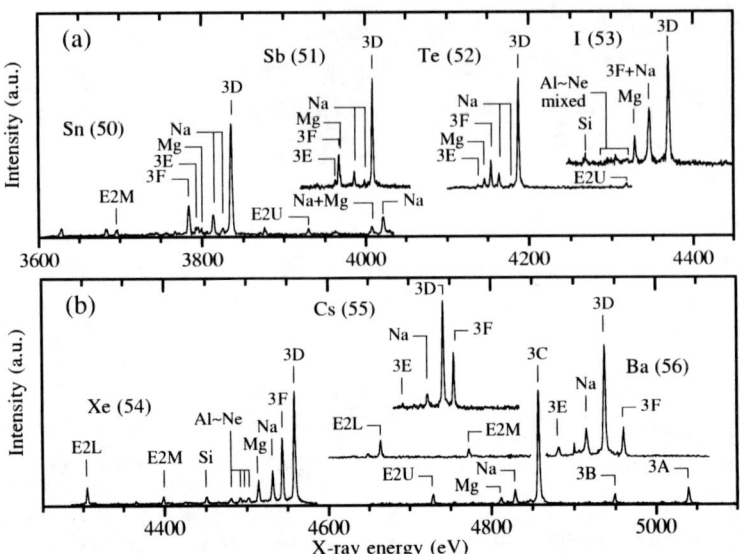

FIGURE 1. X-ray spectra for n=3 to 2 transitions in highly charged ions with (a) $Z = 50\text{-}53$ and (b) $Z = 54\text{-}56$. The final state is the ground state for all the transitions. The notations, "Si", "Al", "Mg", "Na" and "Ne", denote the X-ray transition in the given isoelectronic sequences for the corresponding lines ("Si" denotes the siliconlike ion for example). The other notations [11, 17] denote transitions in neonlike ions: 3A to 3F are electric dipole (E1) transitions from the upper states $(2s^{-1}3p_{3/2})_{J=1}$, $(2s^{-1}3p_{1/2})_{J=1}$, $(2p_{1/2}^{-1}3d_{3/2})_{J=1}$, $(2p_{3/2}^{-1}3d_{5/2})_{J=1}$, $(2p_{3/2}^{-1}3d_{3/2})_{J=1}$, and $(2p_{1/2}^{-1}3s)_{J=1}$; E2L, E2M and E2U are electric quadrupole (E2) transitions from the upper states $(2p_{3/2}^{-1}3p_{1/2})_{J=2}$, $(2p_{3/2}^{-1}3p_{3/2})_{J=2}$ and $(2p_{1/2}^{-1}3p_{3/2})_{J=2}$.

INVESTIGATION OF STRONG CONFIGURATION INTERACTION

In the atomic system of an isoelectronic sequence, as the atomic number Z increases the proper coupling scheme of angular momentum changes from LS to jj. In the course of the variation of the coupling scheme, the order of energy levels can change at a certain value of Z. In such a Z region, the energy levels of the corresponding states become degenerate, so that the wave functions of those states can mix strongly. For the neonlike isoelectronic sequence, strong configuration mixing among the three levels $(2p_{3/2}^{-1}3d_{5/2})_{J=1}$, $(2p_{3/2}^{-1}3d_{3/2})_{J=1}$, and $(2p_{1/2}^{-1}3s)_{J=1}$ is found at $Z \sim 50$. We have studied this configuration interaction by measuring wavelengths for the transitions from these excited levels to the ground state [14, 15].

Figure 1 shows X-ray spectra for $n = 3$ to 2 transitions in highly charged ions with $Z = 50\text{-}56$ obtained with an flat crystal spectrometer [16]. In an EBIT, an electron beam emitted from a cathode is accelerated towards an ion trap, and compressed by an axial magnetic field produced by superconducting magnets. The ion trap consists of three successive drift tubes (DTs) where positive ions can be

TABLE 1. Elements studied in the present study. E_e and I_e represent the electron energy and current at which the spectrum was obtained.

Element (Z)	Injection method	E_e (keV)	I_e (mA)
Sn (50)	MEVVA*	5.8	110
Sb (51)	Compound gas Sb(CH$_3$)$_3$	6.2	155
Te (52)	Compound gas Te(CH$_3$)$_2$	6.5	110
I (53)	I$_2$ gas	6.8	100
Xe (54)	Xe gas	7.0	80
Cs (55)	Cs ion source[†]	7.4	65
Ba (56)	evaporation from the cathode	7.7	160

* Metal vapor vacuum arc ion source [18]

[†] HeatWave product No.1141

trapped axially by applying a positive bias to the two outer DTs. The electron beam successively ionize trapped ions, and highly charged ions are produced and trapped in the middle DT. Table 1 lists the methods of injecting the source elements into the EBIT, together with the electron energies and currents at which the spectra were observed. The electron energies were so selected that the abundance of the neonlike charge state became dominant. It is noted that the electron energy in the table represents the potential difference between the cathode and the middle DT. The actual electron energy (electron - ion interaction energy) is thus considered to be slightly lower than the value listed in the table due to the space charge of the electron beam. Since the radiation source in the EBIT is a line source whose width is about 60 μm, it is possible to use wavelength dispersive spectrometers without an entrance slit. The present spectrometer consisted of a flat crystal and a position sensitive proportional counter (PSPC) [14, 16] with a backgammon-type cathode. Two types of the crystal were used according to objective wavelengths; one was LiF(200) with an area of 100×50 mm^2 and another was Si(111) with an area of 120×70 mm^2. In the present observations, the crystal was placed at 620 - 980 mm away from the center of the trap and the PSPC at 220 - 740 mm away from the crystal. The spectrometer was operated in vacuo ($\sim 10^{-7}$ torr) to avoid absorption by air. A beryllium foil with a thickness of 50 μm was used to separate the vacuum of the EBIT ($\sim 10^{-9}$ torr) from that of the spectrometer.

Figure 2 shows the Z dependence of wavelength for the transitions from $(2p_{3/2}^{-1}3d_{5/2})_{J=1}$, $(2p_{3/2}^{-1}3d_{3/2})_{J=1}$, and $(2p_{1/2}^{-1}3s)_{J=1}$ to the ground state in neonlike ions. In this Z region, the upper levels of these three lines have a similar energy, so that the configuration interaction among them is strong. To see the degree of the interaction, theoretical transition energies were calculated both with and without taking the configuration interaction into account. The experimental results are in good agreement with the theory with the configuration interaction. The strong interaction between $(2p_{3/2}^{-1}3d_{5/2})_{J=1}$ and $(2p_{1/2}^{-1}3s)_{J=1}$ is found as an avoided crossing between $Z = 54$ and 55. On the other hand, the near crossing between $(2p_{3/2}^{-1}3d_{3/2})_{J=1}$ and $(2p_{1/2}^{-1}3s)_{J=1}$ at $Z = 51$ suggests weak interaction between them. This difference between these two crossings is clearly

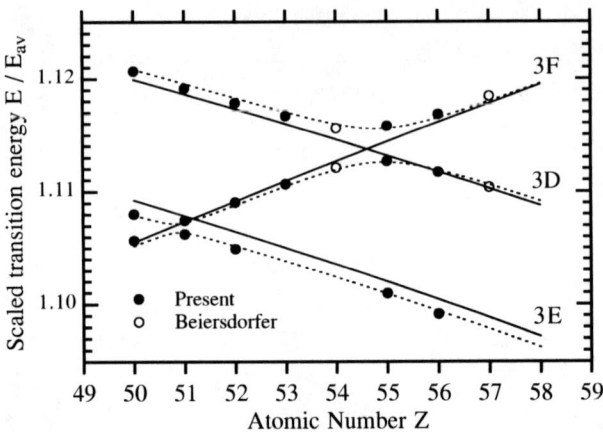

FIGURE 2. Wavelengths of the $3D$ $((2p_{3/2}^{-1}3d_{5/2})_{J=1} \rightarrow 2p^6)$, $3E$ $((2p_{3/2}^{-1}3d_{3/2})_{J=1} \rightarrow 2p^6)$ and $3F$ $((2p_{1/2}^{-1}3s)_{J=1} \rightarrow 2p^6)$ lines in neonlike ions. The vertical axis is the transition energy scaled by the configuration averaged energy. Dotted and solid lines represent the theoretical transition energies calculated with and without taking the configuration interaction, respectively.

explained in terms of the jK-coupling scheme [15]. We found that the three atomic states $(2p_{3/2}^{-1}3d_{5/2})_{J=1}$, $(2p_{3/2}^{-1}3d_{3/2})_{J=1}$ and $(2p_{1/2}^{-1}3s)_{J=1}$ can be represented as $(2p^{-1}3d)\frac{3}{2}\left[\frac{1}{2}\right]_{1-}$, $(2p^{-1}3d)\frac{3}{2}\left[\frac{3}{2}\right]_{1-}$ and $(2p^{-1}3s)\frac{1}{2}\left[\frac{1}{2}\right]_{1-}$ in the $(2l^{-1}3l')j[K]_{J\Pi}$ notation, and that K is approximately good quantum number in this Z-region. According to ordinal quantum theories, states with different K weakly interact through small perturbation, while states with the same K strongly interact when K is good quantum number. The near crossing between the $(2p_{3/2}^{-1}3d_{3/2})_{J=1}$ and $(2p_{1/2}^{-1}3s)_{J=1}$ is therefore attributed to the weak configuration interaction between the states with different K , while the large avoided crossing between $(2p_{3/2}^{-1}3d_{5/2})_{J=1}$ and $(2p_{1/2}^{-1}3s)_{J=1}$ is due to the strong configuration interaction between the states with the same K.

ELECTRON-IMPACT-EXCITATION PROCESSES

Electron-impact-excitation processes of highly charged ions are one of important atomic processes which produce high-energy line emissions in high-temperature plasmas. However, direct measurements of the excitation cross sections are still challenging for the highly charged ions. The crossed beam technique is inefficient for very highly charged ions mainly due to low target density achievable at present. The merged beam technique can not be applied for high-energy collisions. In an EBIT, a nearly mono-energetic electron-beam excites trapped highly charged ions followed by X-ray line emissions. Thus, the electron-impact-excitation cross sections at discrete energies can be measured by observing intensities of the X-

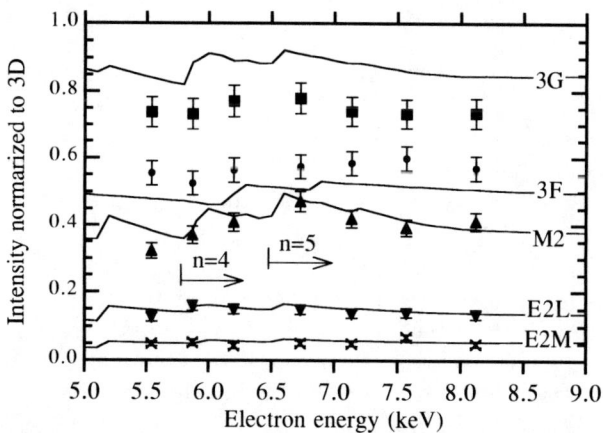

FIGURE 3. Energy dependence of the X-ray line intensities for neonlike Xe^{44+}. Notations are the same as those in Fig 1 except for 3G which is E1 transition from the upper state $(2p_{3/2}^{-1}3s)_{J=1}$ and M2 which is magnetic quadrupole transition from $(2p_{3/2}^{-1}3s)_{J=2}$. All the intensities are normalized to that of 3D.

ray lines excited in an EBIT. The measurements of the excitation cross sections for several highly charged ions have been made by the X-ray spectroscopy with EBITs [1, 19, 20, 21].

In this chapter, we present some measurements of the line intensities in the X-ray spectra of highly charged neonlike ions, and analysis by using a collisional-radiative model for the X-ray line emissions excited in the EBIT. All atomic data in the model were calculated using the HULLAC code [22]. The line intensities of neonlike ions are analyzed as a function of the electron-beam energy. Ionization balance and energy transport in the EBIT are discussed, and satellite line intensities of sodiumlike or lower charge states are calculated and compared with the measurements.

Figure 3 shows variation of X-ray line intensities for $n = 3$-2 transitions in neonlike Xe^{44+} as a function of the electron-beam energy. A simple collisional-radiative (CR) model was developed to explain the measurements. Typical electron density in the EBIT may be as low as 10^{12} cm^{-3}. Thus, most population is essentially in the ground level; excited levels are depleted via the spontaneous decay to lower levels immediately after they are excited by electron-impact. Beside the direct electron-impact excitation, the radiative recombination of fluorinelike Xe^{45+} and the L-shell ionization of sodiumlike Xe^{43+} can, in principle, participate in the populations of the n=3 excited levels of neonlike Xe^{44+}. All the measurements in Fig. 1 are, however, at the electron-beam energies below both the ionization threshold of the neonlike ions (7.6 keV) and the L-shell ionization threshold of the sodiumlike ions (> 7.5 keV), except for the measurements at the highest two energies of 7.6 keV and 8.1 keV. As the first approximation, we, therefore, assume a single-charge-state CR model for the neonlike ions. In the low electron density, a set of rate equations for population densities of excited levels is written as,

$$\frac{dn_i}{dt} = n_e v_e \sigma^{col}_{g \to i}(E_e) n_g + \sum_{E_j > E_i} A^{rad}_{j \to i} n_j - A^{rad}_i n_i, \tag{1}$$

where n_e and v_e are the electron density and velocity at a given electron energy E_e, respectively, A^{rad}'s are the radiative decay probabilities, and $\sigma^{col}(E_e)$ is the collisional cross section. The electron density in the EBIT was estimated on the faith of the Herrmann's magneto-hydrodynamics theory [23].

Equation (1) describes an excitation mechanism to create the level i in the EBIT; the first term is the direct electron-impact excitation from the ground level, the second term indirect excitations through spontaneous radiative cascades from upper levels, and the third term the line emissions by the spontaneous decay from the level i to the lower levels. Since the excitation cross sections of ions have finite values at the threshold energies due to the long-rage Coulomb filed of the target ions, the indirect excitation rates show discontinuous increase as the electron-beam energy passes the threshold energies of the upper levels. Apparent bumps seen in the calculated intensities of Fig. 3 can be ascribed to the discontinuous increase of the indirect excitation rates. The calculated intensities agree with all the measurements, though the model used is very simple. Some discrepancies are found for the 3G and 3F lines. In the present calculations, the angular distribution of the line intensities are not taken into account, while the emissions were measured at the right angle to the electron-beam axis. For detailed comparison, we may need to take account for alignments in the populations of the excited levels created by collisions with the unidirectional electron-beam of the EBIT. At low electron-beam energies as 5.54 keV, the satellite lines of the sodiumlike ion are as prominent as lines of the neonlike ion. From the observation for intensities of the satellite lines, the abundance of sodiumlike ion is inferred to be much larger than that estimated from so-called ionization-equilibrium in which the ion abundance is determined only by balance between the ionization and recombination rates. In the EBIT, however, a certain amount of the trapped ions is leaking out (ion escape) over the electrostatic potential well via diffusion of the ions caused by ion-ion Coulomb collisions. Thus, plasmas in the EBIT are not of the ionization-equilibrium; the ion escape has influence on the ionization balance. Since the escape rate depends on temperature of the trapped ions, the energy transport in the EBIT also must be taken into account. We developed a model to simulate ionization and temperature balance in the EBIT. A set of rate equations for ion densities and temperatures may be written as,

$$\begin{aligned}
\frac{dn_q}{dt} &= n_e R^{ion}_{q-1 \to q} n_{q-1} + n_e R^{rec}_{q+1 \to q} n_{q+1} - (n_e R^{ion}_{q \to q+1} + n_e R^{rec}_{q \to q-1} + n_q R^{esc}_q) n_q, \\
\frac{dkT_q}{dt} &= n_e R^{Coul}_{e-q} \delta E + \sum_{q' \neq q} n_{q'} R^{Coul}_{q'-q} (kT_{q'} - kT_q) - n_q R^{esc}_q (qeV + kT_q)
\end{aligned} \tag{2}$$

where R^{ion}, R^{rec}, R^{esc} and R^{Coul} are rate coefficients of electron-impact ionization, recombination, the ion escape, and two-body Coulomb-collision, respectively,

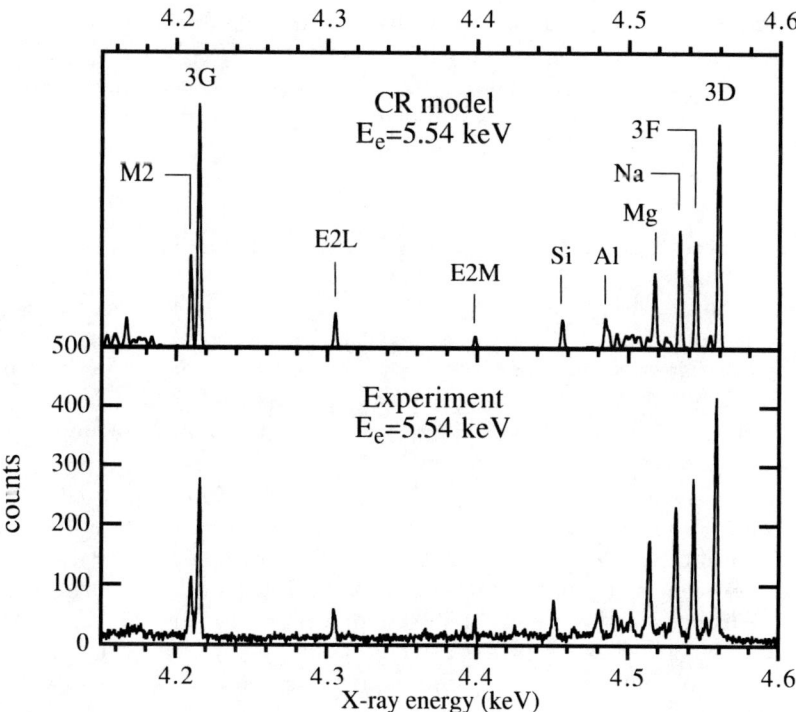

FIGURE 4. X-ray spectrum of the highly charged xenon ions. Notations are the same as in Figs. 1 and 3.

and δE represents effective energy transfer from an electron to the ions via the Coulomb-collision. The right-hand side of the first set of the equations includes nothing but the ionization-equilibrium, except for the last term, i.e. the ion escape term. We used an expression for the ion escape rate coefficient which has been derived within the Fokker-Planck approximation [24]. The dielectronic-recombination was omitted in the present model. The second set of the equations describes the energy transport in the EBIT; the first term corresponds to electron-beam heating of the trapped ions, the second term energy exchange of the ions between different charge states, and the third term evaporative self-cooling of the ions due to the ion escape. In the present calculations, neutral Xe gas is assumed to be always injected into the trap so that the neutral density is maintained at $n_0 - 10^5$ cm^{-3}

Figure 4 shows the preliminary result of the calculation together with the experimental spectrum obtained at an electron beam energy of 5.54 keV for highly charged xenon ions. The calculated spectrum was convoluted with the Gaussian distribution function having a width of 2.1 eV (full-width at half-maximum). The width is the root sum square of an instrumental width of 2 eV and a Doppler width of 0.6 eV which is obtained using the ion temperature of 786 eV in the stationary state. The line intensities were calculated with the CR model of Eq. (1)

separately for each charge state. Then, the spectra of different charge states were superimposed, weighting the line intensities with their respective ion densities in the stationary state. The ion temperature and density for each charge state were determined by solving Eq. (2) numerically. The calculation predicts accurately strong relative intensities of the satellite lines.

SPECTROSCOPY OF HARD X-RAYS

In general, the transition energy of a specific transition in an isoelectronic sequence increases in proportion to Z^2. Thus, a spectrometer which is useful for hard X-rays is needed in order to study neonlike ions with higher Z. For example, the transition energy of $n = 3$-2 transitions in neonlike ions reaches near 20 keV for $Z = 92$.

For spectroscopic studies of such high-Z highly charged ions, we have developed a bent crystal spectrometer. It consists of a crystal bent in the Johann geometry and a image sensor for hard X-rays. The crystal is Ge(400) with a useful area of 75×16 mm^2, which is mounted in a crystal bender so as to make the radius of curvature variable. The uniformity of the curvature is measured with an accurate laser displacement meter to be less than 1 μm for ± 20 mm region when the crystal was bent with a curvature radius of 5800 mm. The image sensor is HAMAMATSU V5102UCsI, which consists of a CsI scintillator and an image intensifier. The effective area of the sensor is 17.5 mm in diameter and the thickness of the scintillator is 150 μm. The image intensifier has three micro channel plates to have the sufficient gain. The quantum efficiency of the sensor is mainly dominated by the absorption in the scintillator ,which is about $\sim 70\%$ for 20 keV. The crystal and the sensor are so settled on goniometers that the position against the EBIT source can be adjusted precisely. The size of the goniometer stage is 1.2×1.2 m^2.

Figure 5 shows spectrum of the characteristic lines of Ag obtained in order to examine the spectrometer. In this measurement, the crystal curvature was correspondent to the Rowland circle of 2900 mm. The characteristic lines were observed *in situ* through interaction of a Ag wire inserted into the EBIT with an electron beam whose current was less than 1μA. This X-ray source can be used as a reference for the absolute wavelength measurements. A peak width of ~ 20 eV (which corresponds to $E/\Delta E \sim 1100$) FWHM was obtained as shown in the figure. The spectrometer is still under development; the resolution is expected to be improved furthermore.

SUMMARY

We have systematically investigated X-ray transitions in neonlike heavy ions using the Tokyo EBIT. Up to now a spectrometer whose useful range is about 3 to 8 keV has been used, where the subjects of investigation were ions with medium Z ($Z \sim 50$-60). To investigate higher-Z ions, we have developed a bent crystal spectrometer with an image sensor which is useful for high energy X-rays. The

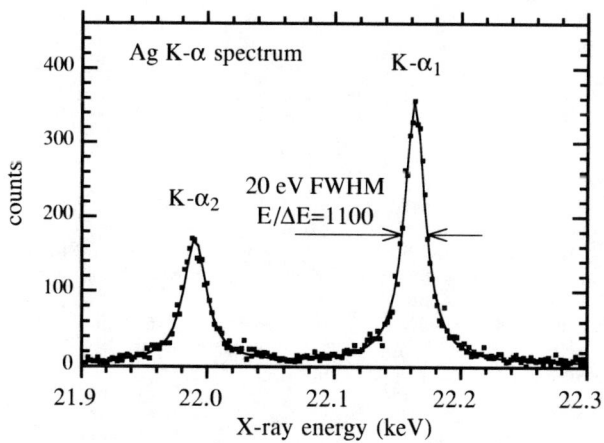

FIGURE 5. Spectrum of Ag $K\alpha$ obtained with the Johann spectrometer. The line width of $K\alpha_1$ is 20 eV FWHM, which contains the natural width of about 8 eV.

useful range of the new spectrometer is about 15 to 25 keV; neonlike ions with much higher Z ($Z \sim 90$) thus become the subject of investigation in the near future. In addition to observation of L shell transitions in neonlike ions, K shell transitions in few electron ions, such as hydrogen-, helium- and lithiumlike ions with medium Z, can also be possible subjects of investigation. One of the research subjects planned for few electron system is $1s$ Lamb shift measurement through the inter-comparison method [25] between Lyman-α of hydrogenlike In and Lyman-β of hydrogenlike Rh. The relativistic and quantum electrodynamics effects will be more clearly investigated by using such high-Z ions.

REFERENCES

1. Marrs, R. E., Levine, M. A., Knapp, D. A., and Henderson, J. R., *Phys. Rev. Lett.*, **60**, 1715 (1988).
2. Currell, F. J., Asada, J., Ishii, K., Minoh, A., Motohashi, K., Nakamura, N., Nishizawa, K., Ohtani, S., Okazaki, K., Sakurai, M., Shiraishi, H., Tsurubuchi, S., and Watanabe, H., *J. Phys. Soc. Jpn.*, **65**, 3186 (1996).
3. Watanabe, H., Asada, J., Currell, F. J., Fukami, T., Hirayama, T., Motohashi, K., Nakamura, N., Nojikawa, E., Ohtani, S., Okazaki, K., Sakurai, M., Shimizu, H., Tada, N., and Tsurubuchi, S., *J. Phys. Soc. Jpn.*, **66**, 3795 (1997).
4. Nakamura, N., Asada, J., Currell, F. J., Fukami, T., Hirayama, T., Kato, D., Motohashi, K., Nojikawa, E., Ohtani, S., Okazaki, K., Sakurai, M., Shimizu, H., Tada, N., Tsurubuchi, S., and Watanabe, H., *Rev. Sci. Instrum.*, **69**, 694 (1998).
5. Watanabe, H., Crosby, D., Currell, F. J., Fukami, T., Kato, D., Ohtani, S., Silver, J. D., and Yamada, C., *Phys. Rev. A*, **63**, 042513 (2001).
6. Beyer, H. F., Kluge, H. J., and Shevelko, V. P., *X-ray Radiation of Highly Charged Ions*, vol. 2, Springer, Berlin, 1997.
7. Quinet, P., Gorlia, T., and Biémont, E., *Phys. Scr.*, **44**, 164 (1991).
8. Kagawa, T., Honda, Y., and Kiyokawa, S., *Phys. Rev. A*, **44**, 7092 (1991).
9. Avgoustoglou, E., and Liu, Z. W., *Phys. Rev. A*, **54**, 1351 (1996).

10. Safronova, U. I., and Wyart, J. F., *Phys. Scr.*, **46**, 134 (1992).
11. Beiersdorfer, P., von Goeler, S., Hinnov, M., Bell, R., Bernabei, S., Felt, J., Hill, K. W., Hulse, R., Stevens, J., Suckewer, S., Timberlake, J., Wouters, A., Chen, M. H., Scofield, J. H., Dietrich, D. D., Gerassimenko, M., Silver, E., Walling, R. S., and Hagelstein, P. L., *Phys. Rev. A*, **37**, 4153 (1988).
12. Aglitskii, E. V., Ivanova, E. P., Panin, S. A., Safronova, U. I., and Ulityn, S. I., *Phys. Scr.*, **40**, 601 (1989).
13. Rice, J. E., Fournier, K. B., Goetz, J. A., Marmar, E. S., and Terry, J. L., *J. Phys. B*, **33**, 5435 (2000).
14. Nakamura, N., Kato, D., and Ohtani, S., *Phys. Rev. A*, **61**, 052510 (2000).
15. Kato, D., Nakamura, N., Ohtani, S., and Sasaki, A., *Phys. Scr.*, **T92**, 126 (2001).
16. Nakamura, N., *Rev. Sci. Instrum.*, **71**, 4065 (2000).
17. Loulergue, M., and Nussbaumer, H., *Astron. Astrophys.*, **45**, 125 (1975).
18. Nakamura, N., Kinugawa, T., Shimizu, H., Watanabe, H., Ito, S., Ohtnai, S., Yamada, C., Okazaki, K., and Sakurai, M., *Rev. Sci. Instrum.*, **71**, 684 (2000).
19. Chantrenne, S., Beiersdorfer, P., Cauble, R., and Schneider, M. B., *Phys. Rev. Lett.*, **69**, 265 (1992).
20. Wong, K. L., Beiersdorfer, P., Reed, K. J., and Vogel, D. A., *Phys. Rev. A*, **51**, 1214 (1995).
21. Nakamura, N., Kato, D., Miura, N., and Ohtani, S., *J. Phys. Soc. Jpn.*, **69**, 3228 (2000).
22. Bar-Shalom, A., Klapisch, M., Goldstein, W. H., and Oreg, J., *"The HULLAC package computer set of codes for atomic structure and processes in plasmas"* (unpublished).
23. Hermann, G., *J. Appl. Phys.*, **29**, 127 (1958).
24. Khudik, V. N., *Nucl. Fusion*, **37**, 189 (1997).
25. Nakamura, N., Kato, D., Nakahara, T., and Ohtani, S., *J. Chin. Chem. Soc.*, **48**, 535 (2001).

PCI Effects on Coincidence Spectra Associated with the Emission of Two Auger Electrons

S.Sheinerman†, P.Lablanquie¶, F.Penent‡, R.I.Hall‡, M.Ahmad‡,
Y.Hikosaka§ and K.Ito§

†St.Petersburg State Maritime Technical University, 198262 St. Petersburg, Russia
¶LURE, Centre Universitaire Paris–Sud, 91898 Orsay, France
‡DIAM, Universite P. & M. Curie, 75252 Paris Cedex 05, France
§IMSS, Photon Factory, Oho 1-1, Tsukuba 305–0801, Japan

Abstract. Experimental investigation of the threshold electron / fast electron coincidences allows one to select lines which are associated with two Auger electron emission. Such an investigation carried out for near threshold photoionization of Xe $4d$ shell reveals a considerable distortion of the lineshapes due to Post Collision Interaction (PCI). Analysis of the PCI influence on the Auger lineshapes allows us to clarify dynamics of the two Auger electron ejection. Our study shows that both double Auger decay (DA) and cascade Auger decay (CA) could contribute to the dynamics of the decay process.

INTRODUCTION

Photoionization of an innershell produces abundant threshold electrons. These threshold electrons are mainly of two sorts: either photoelectrons ejected from the selected innershell, or secondary ones released upon Auger decay of the hole. The nature of the secondary electrons is not fully understood. In order to clarify this point an efficient method of characterization by observing coincidences between these threshold electrons and the associated electrons has been proposed recently [1,2]. The purpose of this work is to gain insight into the dynamics of Auger decay by means of such threshold electron / fast electron coincidence measurements. To do this we analyse the effect of post-collision interaction (PCI) on the lineshapes

CP652, *X-Ray and Inner-Shell Processes: 19th International Conference on X-Ray and Inner-Shell Processes*
edited by A. Bianconi, A. Marcelli, and N. L. Saini
© 2003 American Institute of Physics 0-7354-0111-X/03/$20.00

of the coincidence spectra. It is well established that the threshold photoelectron peak itself is distorted by the PCI effect. This effect can be understood as being due to the interaction of the charged particles in the intermediate and final states. In the case of the usual Auger decay, the interaction is between the photoelectron, the Auger electron and the ion [3,4].

In our study we are faced with the case where the Auger decay results in the release of two Auger electrons. PCI in Auger decays with two emitted electrons is less well understood. Nevertheless, existing PCI theories allow us to analyse the coincidence spectra in the case where several Auger electrons are ejected.

In this work we focus on $4d_{3/2}$ and $4d_{5/2}$ ionization of Xenon. The conditions of experiment and spectra were presented in [2]. Excess energies E_{ex} of 7.5, 1 and 0.5 eV above the threshold have been selected for the analysis. Our previous analysis done within the eikonal approach revealed that the dominant process is decay of $Xe^{+}(4d^{-1})$ to Xe^{3+} through cascade emission of a threshold Auger electron followed by a fast Auger electron [2]. The role of the double Auger emission appeared to be negligible. In this paper we present a more refined analysis which goes beyond the eikonal approach. This analysis demonstrates that both double Auger (DA) and cascade Auger (CA) processes could contribute to the experimental spectrum.

DYNAMICS OF TWO AUGER ELECTRON EMISSION

We consider the Auger decay processes of the Xe^{+*} ion leading to the ejection of two electrons, one of which has zero kinetic energy. There are two such possibilities. Firstly, double Auger decay (DA) where two electrons are ejected simultaneously:

$$\gamma + Xe \rightarrow e_{ph} + Xe^{+*}(4d_{3/2,5/2}^{-1}) \rightarrow$$

$$\rightarrow e_{ph} + e_{A1}(E \approx 0) + e_{A2}(E_{fast}) + Xe^{3+}((5p^{-3})\,^4S)\,. \qquad (1)$$

Secondly, cascade Auger decay where the two electrons are emitted sequentially. This process, in turn, is divided into two processes. One is cascade Auger decay (CA1) where the initial emission of a fast electron forms an intermediate state of the Xe^{2+*} ion which decays to Xe^{3+} yielding a zero energy electron:

$$\gamma + Xe \rightarrow e_{ph} + Xe^{+*}(4d_{3/2,5/2}^{-1}) \rightarrow e_{ph} + e_{A1}(E_{fast}) + Xe^{2+*} \rightarrow$$

$$\rightarrow e_{ph} + e_{A1}(E_{fast}) + e_{A2}(E \approx 0) + Xe^{3+}((5p^{-3})\,^4S)\,. \qquad (2)$$

The other is cascade Auger decay (CA2) where the initial emission of a zero energy electron forms Xe^{2+*} and its decay to Xe^{3+}, the fast electron:

$$\gamma + Xe \rightarrow e_{ph} + Xe^{+*}(4d_{3/2,5/2}^{-1}) \rightarrow e_{ph} + e_{A1}(E \approx 0) + Xe^{2+*} \rightarrow$$

$$\rightarrow e_{ph} + e_{A1}(E \approx 0) + e_{A2}(E_{fast}) + Xe^{3+}((5p^{-3})\,^4S)\,. \tag{3}$$

Here E_{fast} is the energy of the fast Auger electron which is close to the energy difference between the Xe^{+*} and $Xe^{3+}((5p^{-3})\,^4S)$ states, namely, $E_{fast} \approx 3.4$ eV for the $4d_{5/2}$ subshell and $E_{fast} \approx 5.4$ eV for the $4d_{3/2}$ subshell.

All three processes lead to a final state with four charged particles: the photo-electron, two Auger electrons and the residual ion. The Coulomb interaction of these particles in the intermediate and final states of processes (1)–(3) is regarded as PCI and can be expected to modify the lineshapes of the spectra. Of course, the DA, CA1 and CA2 processes are coherent and one should add their amplitudes in order to describe the lineshape precisely. However, the relative values of the magnitudes and phases of the DA, CA1 and CA2 amplitudes are generally unknown. Therefore, to estimate the role of each of the processes (1)–(3) we can calculate incoherent contributions of these processes to the cross section. If one of the processes dominates, the lineshape (especially near the maximum of the line) will be determined by its cross section. Accordingly, we have calculated separately the cross sections of the DA, CA1 and CA2 processes.

PCI ANALYSIS

Our earlier analysis of the PCI influence on the processes (1)-(3) was done in the framework of the eikonal approach [2]. This approach is based on the PCI models for CA and DA processes [5] and implies that the potential energy of the interacting particles is small, much less than the kinetic energy of their relative motion. However, in the case of a slow photoelectron, $E_{ph} \leq 1$ eV, the validity of the eikonal approach is questionable. Therefore the interaction of the slow photoelectron with the others electrons and the target ion has to be taken into account as accurately as possible.

Although there are no PCI models for the processes (1)-(3) apart from the eikonal ones, the conditions of our experiment, namely the very low energies of the slow Auger electrons, $E_A(slow) \simeq 0$ eV, allow us to develop another approach and reduce the model with four particle in the final state to that with three particles. This simplification is based on the fact that the interaction of the slow photoelectron with the field of the target ion and two Auger electrons is the main contribution to PCI. Even though the energy of the photoelectron is small, $E_{ph} \leq 1$ eV, its interaction with the triply charged ion is screened by the threshold Auger electron, and the photoelectron can be considered as moving in the field of the doubly charged ion. In this case, the PCI model for the DA process can be reduced to a model with three charged particles in the final state: the photoelectron, the doubly charged ion and the fast Auger electron. Consequently, the lineshape of the fast Auger electron measured in coincidence with the threshold Auger electron in the true double Auger

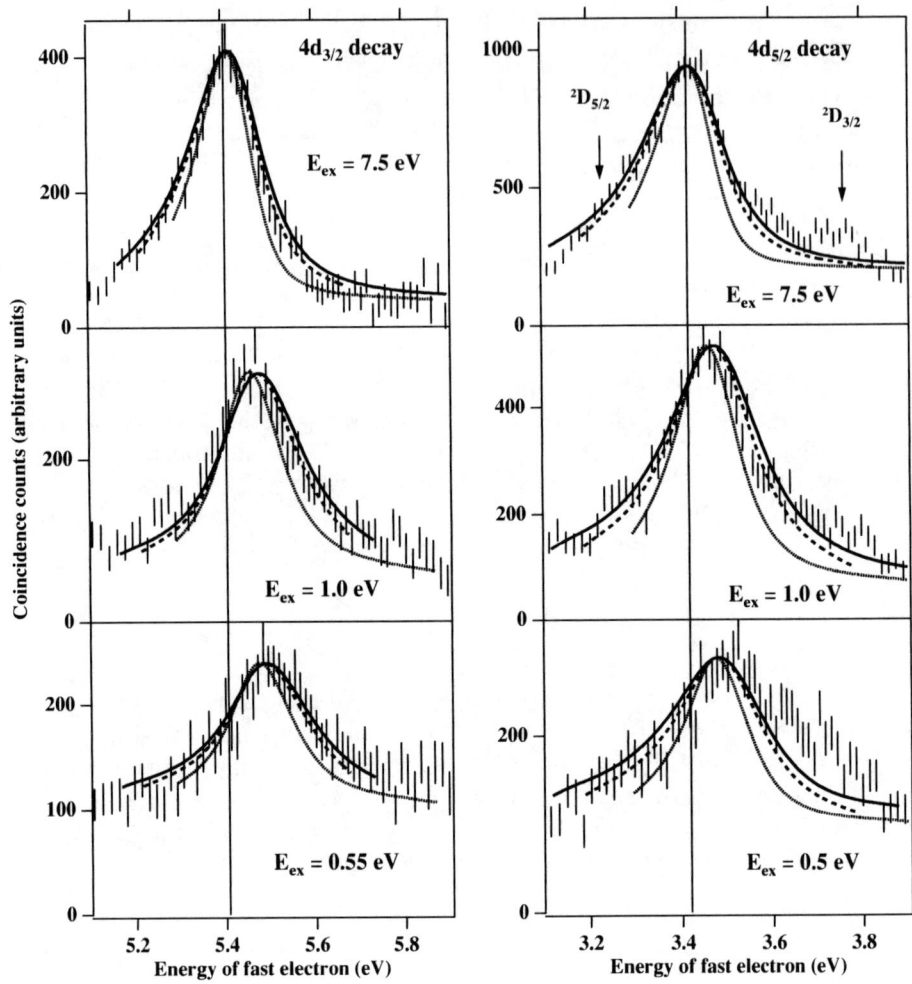

FIGURE 1. Comparison of the coincidence spectra for the threshold Auger electron with the fast Auger electron [2] with calculations for the $4d_{3/2,5/2}$ decay into the Xe^{3+} ^4S ground state for different excess energies E_{ex} above the thresholds. Solid lines are the DA or CA1 process, dotted lines are the CA2 process with the intermediate Xe^{2+} state width $\Gamma_2 = 60$ meV, and dashed lines are those for the CA2 process with $\Gamma_2 = 400$ meV.

decay following inner shell photoionization should be similar to that of the Auger electron ejected in the single Auger decay of the same vacancy with the same energy difference between the initial and final states:

$$\gamma + Xe \rightarrow e_{ph} + Xe^{+*}(4d_{3/2,5/2}^{-1}) \rightarrow e_{ph} + e_{A2}(E_{fast}) + Xe^{2+} \tag{4}$$

This simplification allows us to take into account the PCI effects within the framework of the quantum mechanical approach [6]. In this model we treat the interaction between the slow photoelectron and the ion exactly, and account for the interaction between the photoelectron and the Auger electron using eikonal approximation.

Our analysis shows that due to the zero kinetic energy of the second Auger electron the lineshape of the fast Auger electron in the CA1 process (2) is identical to that in the DA process (1) which, in turn, is reduced to process (4). Thus, the lineshapes of the DA and CA1 processes must be very similar and can be calculated using expressions of Ref.[6]. This way we have calculated the lineshapes of the fast Auger electron for the DA and CA1 processes. Convolution with the detector function was carried out in the same way as in [2]. The lineshapes obtained are shown by solid lines in figure 1 for the $4d_{3/2}$ and $4d_{5/2}$ vacancies, respectively.

The figures show a rather good agreement between the calculated and measured shapes. We may then conclude that the DA (and the indistinguishable from it CA1) processes contribute to the observed spectra. Note that only a precise analysis beyond the eikonal approach reveals the probable contribution of the DA process.

Now consider the CA2 process (3). As explained in [2], this process can be simulated by a DA process (1) with a combined width for the inner vacancy: $\Gamma_{comb} = \Gamma_1\Gamma_2/(\Gamma_1 + \Gamma_2)$, where Γ_1 is the true width of inner vacancy ($\Gamma_{3/2} = 90$ meV, $\Gamma_{5/2} = 110$ meV [7]) and Γ_2 is the unknown width of the intermediate Xe^{2+*} state. In turn this DA process can be replaced by the single Auger decay process (4) due to the low energy of the slow, threshold Auger electron. It should be noted that the lineshape of the Auger electrons ejected in this process depends on the unknown value of the width Γ_2. The value $\Gamma_2 = 60$ meV which gave a good agreement between the eikonal model and experiment [2], now gives a much less satisfactory agreement with the measured spectra for all selected energies. However, choosing a larger value of Γ_2 allows one to obtain good agreement with the experimental data as can be seen in figure 1 where the calculated CA2 lineshapes are presented for $\Gamma_2 = 60$ meV (dotted lines) and $\Gamma_2 = 400$ meV (dashed lines). Note that the larger the value of Γ_2 the closer is the CA2 lineshape to that of DA. Of course, it would be highly desirable to have some information about the value of Γ_2 from other, independent, measurements to make a final conclusion on the CA2 contribution.

In summary, a refined analysis of PCI effects shows that both the DA (and CA1 indistinguishable from it) and CA2 processes can contribute to the Auger decay.

REFERENCES

1. P.Lablanquie *et al*, Int. Workshop on Photoionization (IWP'97), (Chester), Abstracts, 96 (1997)
2. P.Lablanquie *et al*, Phys.Rev.Lett. **87** 53001 (2001)
3. M. Yu. Kuchiev and S. A. Sheinerman, Sov. Phys. - Usp. **32**, 569 (1989)
4. V. Schmidt, Rep. Prog. Phys. **55**, 1483 (1992)
5. S. Sheinerman, J.Phys. B **27**, L571 (1994); S. Sheinerman, J.Phys. B **31**, L361 (1998)
6. M.Kuchiev and S.Sheinerman, J.Phys.B **21** 2027 (1988)
7. V. Schmidt, *Electron Spectrometry of Atoms using Synchrotron Radiation*, (University Press, Cambridge) (1997)

Multiple-Auger electron ejection after inner-shell ionization and excitation

Jens Viefhaus

Fritz-Haber-Institut der Max-Planck-Gesellschaft, Faradayweg 4-6, 14195 Berlin, Germany

Abstract. Results on the Auger decay of core-ionized and core-exited Ar atoms above the Ar $2p$ threshold and at the Ar $2p_{3/2} \rightarrow 3d$ resonance leading to double and triple ionization states are presented. Using a multiple time-of-flight analyzer arrangement for electron-electron coincidences, we directly observe for the first time a double Auger continuum following core electron ionization. Our results show clear evidence for continuously distributed Auger electron intensity over a 160 eV range of kinetic energies. This double Auger decay represents roughly 10 % of the normal single Auger channels. In the case of the resonant Auger decay we also observe a two-electron continuum of the same order of magnitude as in the non-resonant case which can be explained due to the existence of excited states of the doubly charged ion in the vicinity or just above the triple ionization threshold. In the latter case these states can further decay via emission of a low kinetic energy electron, which makes it possible to study the electron emission characteristics of the triple electron process. Both double- and triple-electron emission Auger processes will make it possible to study electron correlations undisturbed by the symmetry properties of the photoabsorption process.

INTRODUCTION

Whereas the study of double electron processes in direct photoionization is a rather mature field of intense research [1], the study of triple electron processes is just at its beginning. Only very recently such processes became accessible to experimental verification and quantitative analysis. It was the triple photoionization of lithium measured by Wehlitz et al. [2], which opened the field to experimental investigation.

There were two important points in that study: the very low triple ionization probability - by three orders of magnitude lower than the double ionization rate -, and the rather high probability of double ionization with additional excitation - 50 % of the double ionization yield. A more detailed study of the latter one would require the highly differential method of electron-electron coincidence spectroscopy, instead of the more integrated ion spectroscopy method used.

Instead of studying the triple ionization processes in photoionization we concentrated on Auger electron emission which is unaffected by symmetries imposed on the system by the photoabsorption process. In particular we performed measurements on the Auger decay of core-ionized and core-exited Ar atoms at the Ar $2p_{3/2} \rightarrow 3d$ resonance [3, 4]. The corresponding Ar energy level diagram is depicted in fig. 1 together with possible decay pathways. Previous ion yield measurements [5, 6, 7] have shown that triply charged final states are contributing to nearly 10 % of the total cross section in the region around the Ar $2p$ edge. Our main aim was therefore to identify the decay routes leading to triple electron ejection by an electron-electron coincidence study.

CP652, *X-Ray and Inner-Shell Processes: 19th International Conference on X-Ray and Inner-Shell Processes*
edited by A. Bianconi, A. Marcelli, and N. L. Saini
© 2003 American Institute of Physics 0-7354-0111-X/03/$20.00

Argon energy levels

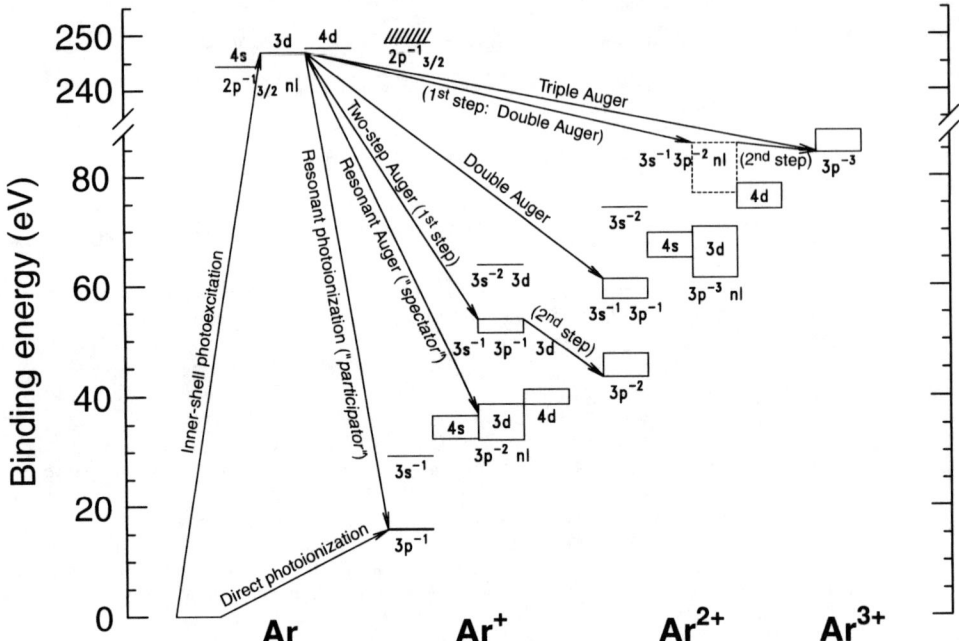

FIGURE 1. The energy level diagram of Ar together with possible decay pathways of the $2p$ inner-shell excited Ar. The Ar II levels are from ref. [8, 9], the Ar III levels from ref. [10, 11] together with estimated levels of higher excitation energy (broken line) and the Ar IV levels are adapted from ref. [12].

EXPERIMENTAL

The measurements were performed at the BW3-SX700 undulator beamline [13, 14] of the DORIS III storage ring at the Hamburger Synchrotronstrahlungslabor (HASYLAB) of DESY in Hamburg. The monochromator resolution was adjusted to give a bandpass of 100 meV. By taking total electron yield scans in the region of the Ar $2p$ inner-shell excitation it was verified that the resolving power was sufficient to isolate the different excitation states. The monochromatized synchrotron radiation was crossed by an effusive beam of Ar atoms using a gas inlet made out of a Mo needle having an inner diameter of 300 μm and a distance of about 4 mm with respect to the interaction region. The background pressure was below 5×10^{-6} hPa whereas the target pressure in the interaction region was about 100 times higher. The size of the interaction region centered at the focus of the beamline was estimated to be less than 1 mm in diameter with about 2-3 mm along the light beam direction.

The experimental setup shown in fig. 2 consists of seven separate electron time-of-flight analyzers. During installation special care was taken that all of the analyzers are perfectly mounted in the plane perpendicular to the incoming light as well as that all an-

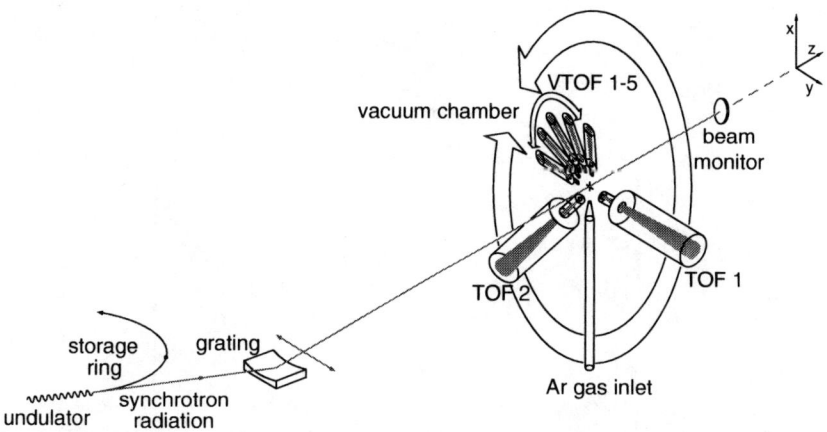

FIGURE 2. A schematic of the experimental setup used for multiple time-of-flight electron-electron coincidence spectroscopy.

alyzers point to the same interaction region, which is necessary for coincidence studies. For high transmission of low energy electrons the earth magnetic field was compensated by external coils down to less than 1 % of its original strength which was verified by a three-axis magnetic field sensor placed at the interaction region. The DORIS III storage ring was running in five-bunch mode leading to a pulse period of at least 192 ns between each light pulses. In order to be able to detect electrons down to less than 1 eV kinetic energy the necessary acceleration potential was applied to the flight tubes of the time-of-flight analyzers. The detector signals were pre-amplified and discriminated using the constant fraction discriminator technique. The time measurements were performed by the standard time-to-analog, analog-to-digital conversion technique using the bunch marker signal derived from the high-frequency accelerator of the storage ring as stop trigger. The overall time resolution achieved was below 200 ps, which includes both the contributions of the electronics and the synchrotron light source.

The parallel stream of time-of-flight data of up to six analyzers is fed into a buffer interface which bins the events into single and coincidence time-of-flight spectra. The corresponding analyzer angles with respect to the plane of linear polarization were $\theta = -54.7°, 0°, 22.5°, 45°, 90°, 180°$. The total measurement time was slightly more than 100,000 s for each spectrum recorded in slices of 3600 s. Further processing of the data included subtraction of the random coincidences, which took advantage of the simultaneously recorded singles spectra as well as time-to-energy conversion using the Xe NOO Auger lines for calibration purposes [15, 16]. The count rate for the singles spectra was in the order of kHz whereas the total true double coincidence rate was about 10 Hz with a true-to-random ratio of greater than 1 in the regions of interest.

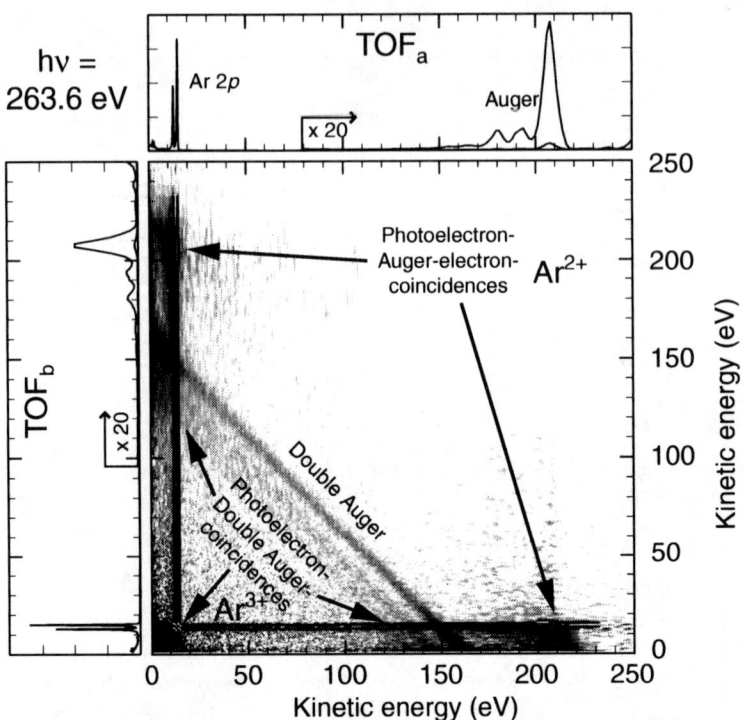

FIGURE 3. Two-dimensional electron-electron coincidence spectrum of Ar taken at $h\nu$=263.6 eV along with the two corresponding non-coincident spectra. Except for small structures that are barely visible in the kinetic energy range of 1-10 eV the main portion of the triply charged states is populated via direct double Auger decay (the peak structure in the singles spectra below 5 eV is due to the scattered synchrotron light produced by the following bunch).

RESULTS AND DISCUSSION

Above the Ar 2p threshold

The spectra plotted in fig. 3 show two non-coincident singles spectra plus a coincidence gray scale map of all 15 double coincidence combinations added together. Considering the angular range covered by the six analyzers participating in the measurement it seems reasonable to assume that the presented data is a representative of the possible decay paths regardless of the imposed angular restrictions.

The strongest features of the coincidence map are coincidences between the two lines of the Ar 2p multiplett and their corresponding principal Auger lines which are located at the edges of the map (large arrows at high excess energies, which is the sum of the kinetic energies of the emitted electrons). These are well known transitions to doubly charged final states[17]. In addition continuously distributed intensity is found forming several diagonal lines of constant excess energy which have a lower excess energy compared to the doubly charged final states mentioned before. Therefore this

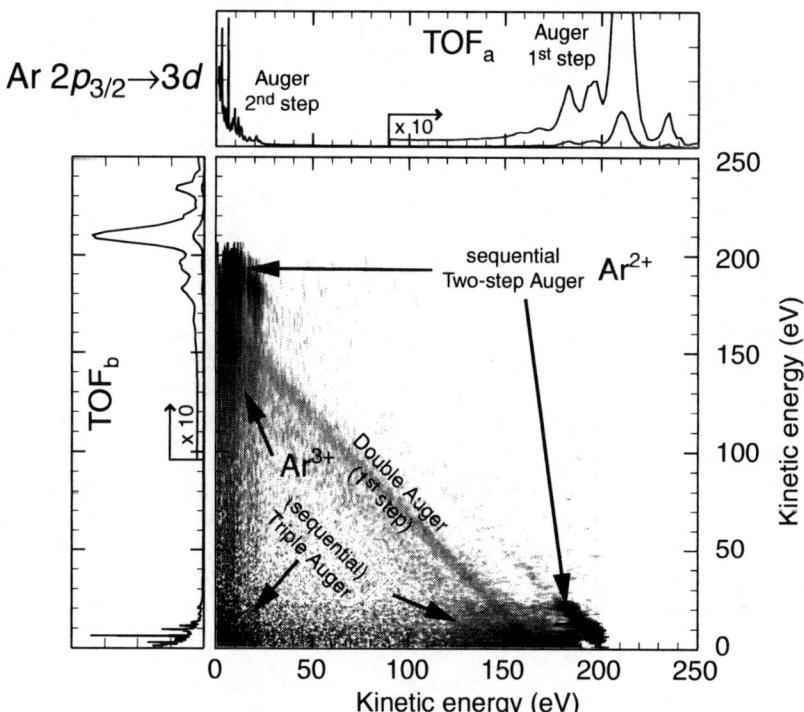

FIGURE 4. Two-dimensional electron-electron coincidence spectrum of Ar taken at the Ar $2p_{3/2} \to 3d$ excitation ($h\nu$=246.93eV) along with the two corresponding non-coincident spectra.

intensity is attributed to triply charged final states where the Ar $2p$ photoionization is followed by direct double Auger emission[18]. Consequently three electrons have to be emitted during the photoionization process. As only double coincidence events are depicted in fig. 3 two different coincidence patterns are observed. The horizontal and vertical lines located at the kinetic energies of the two photoelectron lines of the Ar $2p$ multiplett are photoelectron-Auger electron coincidences including one of the two double Auger electrons whereas the diagonal structure is a direct representation of the two simultaneously ejected double Auger electrons. The double Auger intensity shows a clear preference for asymmetric energy sharing. Comparing the intensities observed the probability for triply charged states is in the order of 10 % in agreement with photo-ion threshold-electron coincidence measurements[19].

On the Ar $2p_{3/2} \to 3d$ resonance

The compilation of double coincidence spectra shown in fig. 4 clearly shows strong discrete structures at high excess energies close to the edges of the coincidence map. Discrete structures can directly be interpreted as sequential decay to doubly charged final states. The measurements show that this two-step decay route is strongly dominant for all of the low binding energy states corroborating previous measurements[19]. The participating intermediate states coincide with those already obtained by coincidence measurements[20]. The present results can also verify the assignments of the high-kinetic energy lines of the first step as well as the low kinetic energy lines of the second step involved in these transitions.

The major new finding is again continuously distributed coincidence intensity at lower excess energy corresponding to higher binding energy of the doubly charged state. Due to the absence of clear structure as well as the fact that the intensity covers the whole range of energy sharing of the two electrons we can conclude that these states are populated via simultaneous - namely direct - double Auger decay. Doubly charged states above the triple ionization threshold will most likely decay via autoionization to the triply charged final state. As the separation of the two levels participating in this final step of the decay is very small (in the order of a few eV or even less) the corresponding kinetic energy of the emitted electron is very low. As the density of highly excited double charged states close to the triple ionization threshold must be rather high it is not possible to observe particular discrete features in the measured coincidence and/or singles spectra. Therefore we cannot fully exclude a contribution of direct triple Auger processes superimposed on the observed coincidence yield. The naive assumption that this process is very unlikely awaits still a thorough theoretical investigation.

SUMMARY

Taking advantage of the time-of-flight electron-electron coincidence method we have measured both electrons of the direct double Auger process for the first time. The results obtained after exciting or ionizing an Ar $2p$ inner-shell electron indicate that the double Auger intensity is distributed over the full range of energetically allowed excess energies with a clear preference for asymmetric energy sharing. Preliminary results obtained for the coincident angular distribution show that a study of electron correlations undisturbed by the properties of the photoabsorption process will be possible using the existing setup. Further measurements on other elements such as Ne are in progress. This will allow in particular systematic studies of the excess energy dependence of the direct double Auger process.

ACKNOWLEDGMENTS

I would like to thank Slobodan Cvejanović, Burkhard Langer and Toralf Lischke for their assistance during the measurements as well as Uwe Becker for continuous support and stimulating discussions. The help of the HASYLAB staff in particular by Ralph Dörmann und Thomas Möller is gratefully acknowledged.

REFERENCES

1. J. S. Briggs and V. Schmidt, J. Phys. B **33**, R1 (2000)
2. R. Wehlitz et al., Phys. Rev. Lett. **81**, 1813 (1998)
3. J. Mursu et al., J. Phys. B **29**, 4387 (1996)
4. B. Langer et al., J. Phys. B **30**, 4255 (1997)
5. T. Hayaishi et al., J. Phys. B **17**, 3511 (1984)
6. J. A. R. Samson et al., Phys. Rev. A **54**, 2099 (1996)
7. F. v. Busch et al., J. Phys. B **29**, 5434 (1996)
8. L. Minnhagen, J. Opt. Soc. of America **61**,1257 (1971)
9. A. Kikas, et al., J. Electron Spectrosc. Relat. Phen. **77**, 241 (1996)
10. V. Kaufman and W. Whaling, J. Res. Nat. Inst. Stand. Tech. **101**, 691 (1996)
11. L. Avaldi et al., J. Phys. B **30**, 5197 (1997)
12. National Institute of Standards and Technology (NIST) Atomic Spectra Database Energy Levels Data available online at http://physics.nist.gov/cgi-bin/AtData/levels_form
13. A. R. B. de Castro and R. Reininger, Rev. Sci. Instrum. **63**, 1317 (1992)
14. C. U. S. Larsson et al., Nucl. Instrum. Methods Phys. Res. A **337**, 603 (1994)
15. H. Aksela, S. Aksela, and H. Pulkkinen, Phys. Rev. A **30**, 865 (1984)
16. U. Becker et al., Phys. Rev. A **39**, 3902 (1989)
17. L. O. Werme, T. Bergmark, and K. Siegbahn, Phys. Scr. **8**, 149 (1973)
18. T. A. Carlson and M. O. Krause, Phys. Rev. Lett. **17**, 1079 (1966)
19. T. Hayaishi et al., J. Phys. B **21**, 3203 (1988)
20. E. v. Plate-v. Raven, Ph.D. thesis, University of Hamburg, Hamburg, 1992; E. v. Raven et al., J. Electron Spectrosc. Relat. Phen. **52**, 677 (1990)

VI. X-RAY SCATTERING

A Novel Method for Studying Thermal Motion and Point Defects in Crystals by X-Ray Resonant Diffraction

V. E. Dmitrienko[*], K. Ishida[†], A. Kirfel[**], J. Kokubun[†] and
E. N. Ovchinnikova[‡]

[*]A. V. Shubnikov Institute of Crystallography, 59 Leninski prospekt, 117333, Moscow, Russia
[†]Tokyo University of Science, Noda, Chiba 278-8510, Japan
[**]Mineralogisch-Petrologisches Institut der Universität Bonn, Poppelsdorfer Schloss, D-53115 Bonn, Germany
[‡]Physical Department, Moscow State University, Moscow, Russia

Abstract. After an introductory survey of the X-ray resonant anisotropy, we present a novel X-ray method to observe thermal-motion-induced (TMI) and point-defect-induced (PDI) distortions of electronic states of atoms. This method uses the idea that, in general, the local atomic environment becomes less symmetric owing to point defects and/or the thermal vibrations of the atoms in a crystal. As a result of this phenomenon, an additional anisotropy of the resonant scattering factors can occur and "forbidden" Bragg reflections can be excited near the absorption edges. Examples of crystals are discussed (Ge, $Y_3Fe_5O_{12}$) where TMI and PDI reflections can be found. The tensor structure factors of the both types of reflections are calculated. According to our theory, the TMI reflection structure factors are proportional to the vibration correlations, $u_{\parallel}^2$ and $u_{\perp}^2$, of neighboring atoms, and it is inferred that $u_{\parallel}^2$ provides the main contribution to the thermal-motion-induced anisotropy of X-ray resonant scattering. The TMI reflections in Ge were recently observed by Kokubun et al. (Phys. Rev. B64, 073203 (2001)), Kirfel et al. (in press), and Colella et al. (in press) in accordance with our prediction. For the 006 reflection, the intensity increases about 25 times with the temperature increasing from 30 to 735 K. Owing to their resonant character, the PDI reflections allow to separately study both impurity atoms and host atoms of different types. The considered phenomena can provide a very sensitive method for studying point defects because only the atoms that are affected by defects contribute to the PDI reflections.

INTRODUCTION

The aim of this paper is to attract researchers attention to a novel X-ray probe, sensitive to thermal motion of atoms and point defects in crystals. The near-edge X-ray absorption (XANES, EXAFS) is well known for its selective sensitivity to the environment of an absorbing atom. However, even in single crystals, it probes an environment that is averaged over the unit cell. The combination of absorption with diffraction, which is called diffraction anomalous fine structure (DAFS) or X-ray resonant diffraction, is a more sophisticated tool: in principle, it allows to determine the real part, f', and the imaginary part, f'', of the anomalous scattering corrections to the atomic scattering factor for all the atoms separately (see a review in [1]). In other words, the resonant diffraction combines the chemical and short-range sensitivity of the near-edge absorption with the site

CP652, X-Ray and Inner-Shell Processes: 19th International Conference on X-Ray and Inner-Shell Processes
edited by A. Bianconi, A. Marcelli, and N. L. Saini
© 2003 American Institute of Physics 0-7354-0111-X/03/$20.00

sensitivity of diffraction.

A very important feature of the resonant diffraction are the unusual polarization properties. For conventional diffraction, f' and f'' are polarization independent and the polarization properties are rather simple: an initial beam with linear $\sigma(\pi)$ polarization yields a diffracted beam with the same $\sigma(\pi)$ polarization (the vectors of σ and π polarizations are perpendicular and parallel to the scattering plane, respectively). The $\sigma\sigma$ and $\pi\pi$ reflection coefficients can be easily calculated for both kinematic and dynamic cases. Near absorption edges both f' and f'' become tensors (we will denote tensors by hats, $\hat{f}'$ and $\hat{f}''$), and the structure amplitudes of reflections also become tensors. This changes drastically the polarization properties of the reflections: $\sigma\pi$ and $\pi\sigma$ reflections can become allowed and sometimes dominant, and the reflection coefficient may be different for right-handed and left-handed circular polarizations, *etc.* [2, 3]. The physical reason for the tensor anisotropy of $\hat{f}'$ and $\hat{f}''$ is the anisotropy of the environment which can by rather asymmetric, even in cubic crystals, and which distorts the valence electron states of the absorbing atom.

It was recognized many years ago [4] that the anisotropy of the X-ray susceptibility allows the excitement of reflections otherwise forbidden by screw-axis and glide-plane symmetry operations. After the theoretical development [5, 6] and the first experimental observation [7, 8] of such 'forbidden' reflections in $NaBrO_3$, they were studied in many more crystals, e.g. Cu_2O [9], TiO_2, MnF_2 [10], $Ba(BrO_3)_2.H_2O$ [11], Fe_3O_4 [12, 13, 14], FeS_2 [15], $La_{0.5}Sr_{1.5}MnO_4$ [16] (orbital ordering), and others. In all these cases, the dipole anisotropy of the X-ray susceptibility arises near the absorption edges of resonant scattering atoms because the electronic states of these atoms are distorted by their asymmetric environments.

THERMAL-MOTION-INDUCED REFLECTIONS: GERMANIUM

A more intriguing situation occurs in crystals like Ge where the atoms are in positions with so high a symmetry that the dipole anisotropy is absent . In this case, 'forbidden' reflections can be excited owing to a higher-rank-tensor anisotropy. For example, the $0kl, k+l = 4n+2$ reflections in germanium can appear as a consequence of a mixed dipole-quadrupole transition [17]. However, these reflections can also be excited owing to a thermal-motion-induced (TMI) anisotropy [18, 19]. Indeed, if the resonant atom moves in arbitrary direction away from its equilibrium position, then the point symmetry of its instantaneous environment is lowered, its electronic wave functions are distorted, and hence the dipole-dipole anisotropy can apply. Contrary to the first described situation which is associated with an intensity decrease with increasing temperature, as observed for cuprite [20], the TMI anisotropy should grow (in absolute value) with rising temperature.

For germanium, it is found [21, 22] that the intensity of the 006 reflection increases significantly with temperature (see Fig. 1). This effect can be quantitatively explained if we suppose that the TMI anisotropy is proportional to the relative displacement of neighboring atoms. In this case, the structure amplitude of the TMI reflections is proportional to the vibration correlations, $u_{\parallel}^2$ and $u_{\perp}^2$, of neighboring atoms with complex

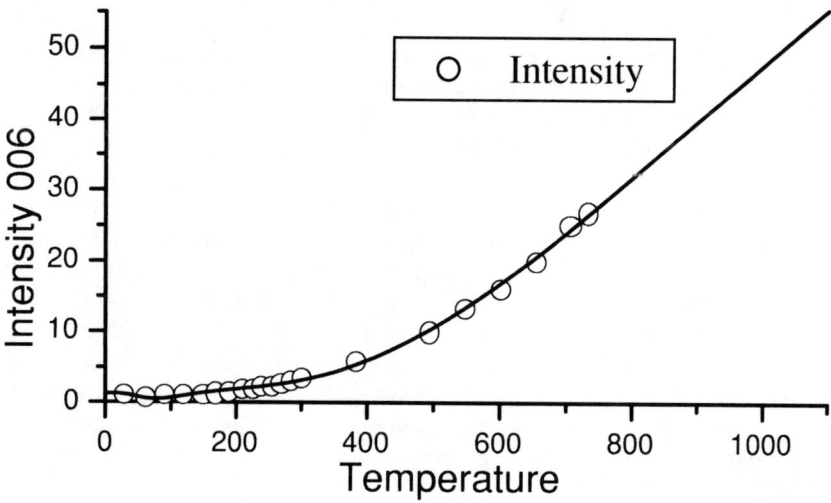

FIGURE 1. Temperature dependence of the 006 reflection intensity in Ge (points – experiment, curve – theoretical fitting extrapolated up to the Ge melting temperature).

coefficients. Correspondingly, the 006 reflection intensity can be written as $I(006, Ge) = |C + P_\perp u_\perp^2 + P_\parallel u_\parallel^2|^2$. The coefficients C, $P_\perp$, and $P_\parallel$ are considered as fitting parameters but they can as well be calculated from the microscopic theory of X-ray resonant diffraction. It was found that $u_\parallel^2$ provides the main contribution to $I(006, Ge)$. The fact that a temperature-independent contribution C is also found may be attributed to either a dipole-quadrupole scattering or to other static effects.

POINT-DEFECT-INDUCED REFLECTIONS

Ferrimagnetic iron garnets are widely used in magnetic devices owing to their useful magnetic properties, which strongly depend on the crystal composition, in particular, on different impurities which can be considered as point defects. We shall consider garnets with the composition described as $M_\delta Y_{3-\delta} Fe_5 O_{12}$, where M are rare-earth atoms, and $(Mc_2O_3)_\delta \cdot 3Y_2O_3 \cdot (Fe_2O_3)_{5-\delta}$, where Me are non-magnetic atoms. Resonant X-ray diffraction is an experimental method, which allows to study in detail the structural and electronic distortions around impurity positions with the help of "forbidden" Bragg reflections near the absorption edges of Fe, Y, M and Me. At first, let us consider the resonant X-ray diffraction in a single crystal of a pure yttrium-iron garnet (YIG, $Y_3Fe_5O_{12}$). Its symmetry is described by the space group $Ia\bar{3}d$. The magnetic ions Fe^{3+} occupy two sites: the octahedral 16(a) with $\bar{3}$ local symmetry and the tetrahedral 24(d) with $\bar{4}$ symmetry. The Y^{3+} ions occupy the position 24(c) with 222 site symmetry, and the oxygen atoms are in 96(e) positions. Near the K-edge of Fe (7.1 KeV), "forbidden" reflections can occur, which are separately induced by the 16(a) and 24(d) positions of iron. Near the absorption edge of yttrium (17.04 KeV K-edge), we can find "forbidden" reflections induced by yttrium in the 24(c) position. The 16(a) position gives contributions to the following set of forbidden reflections: $h = 2n + 1, k = 2n' + 1, l = 0$ (n and n' arbitrary integers), and $h = k = 2n + 1, l = 4n'$ (for example, (110)). The 24(d) and 24(c) sites yield "forbidden" reflections with $h = k = 4n, l = 4n' + 2$ (for example, (002)).

We also can find some reflections, which are forbidden for one of the positions but allowed for another. For example, $h = k = 2n + 1, l = 4n + 2$ reflections become allowed for the 24(d) position, but remain strictly forbidden for 16(a). The reflections $h = k = 4n + 2, l = 4n' + 2$ still remain forbidden in dipole-dipole approximation. Below we neglect the combined magnetic effects, which also can cause the appearance of additional reflections [23]. Such "forbidden" reflections, similar to the considered ones, have been studied theoretically and observed experimentally in Mössbauer nuclear diffraction [24, 25, 26, 27, 28].

The forbidden reflection (002) is described by the following structure amplitude:

$$\hat{F}(002)^{Fe} \sim 8(\hat{f}_{[010]} - \hat{f}_{[100]}) = 8(f_{xx} - f_{zz}) \begin{pmatrix} 1 & 0 & 0 \\ 0 & -1 & 0 \\ 0 & 0 & 0 \end{pmatrix}, \tag{1}$$

where $\hat{f}_{[010]}$ and $\hat{f}_{[100]}$ belong to the iron atoms on the $\bar{4}$ axes parallel [010] and [100], whereas the eight iron atoms on the $\bar{4}$ axis along [001] do not contribute to the (002) reflection.

In analogy, near the yttrium absorption edge, the "forbidden" reflection (002) is described by:

$$\hat{F}(002)^{Y} \sim 8(\hat{f}_{[010]} - \hat{f}_{[100]}) = 8 \begin{pmatrix} f_{xx} - f_{zz} & 0 & 0 \\ 0 & f_{zz} - f_{yy} & 0 \\ 0 & 0 & f_{yy} - f_{xx} \end{pmatrix}. \tag{2}$$

Similar to the 24(d) iron position, only 16 yttrium atoms contribute to this reflection.

Again, near the iron absorption edge, the reflection (110) is described by:

$$\hat{F}(110)^{Fe} \sim 4(\hat{f}_{[\bar{1}11]} - \hat{f}_{[1\bar{1}1]}) = 8f_{xz} \begin{pmatrix} 0 & 0 & -1 \\ 0 & 0 & 1 \\ -1 & 1 & 0 \end{pmatrix}, \tag{3}$$

where $\hat{f}_{[ijk]}$ denote the scattering amplitudes of the iron atoms on the three-fold axes.

Let us now suppose that Me atoms substitute the iron atoms. They can then be found in both the 16(a) and the 24(d) positions. If they do not influence each other, both the (110) and (200) reflections will appear near the absorption edges of iron and Me. In the garnets, where yttrium atoms are substituted by M atoms, the (002) reflection can be observed near the absorption edges of yttrium and M, but the (110) reflection remains truly forbidden. It was demonstrated in [19] that impurities (point defects), which cause atomic displacements, can give rise to "forbidden" PDI reflections in germanium. Similar to the technique developed in [19], we construct the scattering amplitude in the presence of such impurities as follows (neglecting the highest order term):

$$F_{ij}(H) = \sum_s \exp(iHr^s)(f_{ij}^s + f_{ijk}^s u_k + f_{ijkl}^s u_k u_l)(1 + iH_k u_l^s) \tag{4}$$

$$= F(H)_{ij}^0 + \sum_s \exp(iHr^s)(f_{ijk}^s u_k^s + i f_{ij}^s H_k u_k^s) + i f_{ijk}^s H_k u_l^s u_l^s + f_{ijkl}^s u_k^s u_l^s).$$

We see that the local symmetry of the atomic site causes the sensitivity of the resonant scattering to the atomic displacements. For example, for the 16(a) position with $\bar{3}$ symmetry, the third-rank tensor gives no contribution to the PDI reflections, opposite to the 24(d) position. The term $i f_{ij}^s H_k u_l$ does contribute neither to the (110) nor to the (002) reflection. The calculations with the help of expression (4) show also that owing to the displacements induced by impurities in the 24(d) or the 24(c) positions, the 16(a) iron sites can give a contribution to the (002) forbidden reflection near the absorption edge of iron. This is described by the term $f_{ijkl}^s u_k^s u_l^s$. Such a contribution was observed in a Mössbauer diffraction pattern [28]. Similarly owing to defects, the 24(d) iron position can give a contribution to the (110) reflection. We can also see, that impurities in the iron positions can provide displacements of yttrium and, correspondingly, excite the (110) PDI reflection near the absorption edges of yttrium. For example, if we take into account the impurities closest to the 24(d) position (first coordinadion sphere), which induce displacements of the 24(c) yttrium atoms into 48(f) positions, then the correction to the scattering amplitude will be equal to:

$$\Delta \hat{F}(110)^Y \sim \Delta f \cos 2\pi u \begin{pmatrix} 0 & 0 & 1 \\ 0 & 0 & -1 \\ 1 & -1 & 0 \end{pmatrix}, \tag{5}$$

where u denotes the yttrium displacement and Δf describes the change of the scattering amplitude. The non-diagonal terms appear in (5), because (due to the displacements) the local symmetry of the yttrium position becomes 2 instead of 222.

Thus, in summary, it follows from the above considerations that the observation of "forbidden" reflections near the absorption edges of Fe, Y, M and Me would allow to find the properties of the M and Me atoms in all crystallographic positions and to study the atomic displacements in a crystal.

ACKNOWLEDGMENTS

This work was partly supported by the INTAS grant 01-0822 and by the Bundesminister für Bildung und Forschung (BMBF contract 05 KS1PDA). V.E.D. is grateful to the organizers of the 19th International Conference on X-Ray and Inner-Shell Processes for financial support.

REFERENCES

1. Hodeau, J.-L., Favre-Nicolin, V., Bos, S., Renevier, H., Lorenzo, E., and Berar, J. F., *Chemical Reviews*, **101**, 1843-1867 (2001).
2. Belyakov, V. A., and Dmitrienko, V. E., *Soviet Physics Uspekhi*, **32**, 697–719 (1989).
3. Kirfel, A., and Petkov, A., *Zeitschrift fur Kristallographie*, **195**, 1–15 (1991).
4. Templeton, D. H., and Templeton, L. K., *Acta Crystallographica A*, **36**, 237–241 (1980).
5. Dmitrienko, V. E., *Acta Crystallographica A*, **39**, 29–37 (1983).
6. Dmitrienko, V. E., *Acta Crystallographica A*, **40**, 89–95 (1984).
7. Templeton, D. H., and Templeton, L. K., *Acta Crystallographica A*, **41**, 133–142 (1985).
8. Templeton, D. H., and Templeton, L. K., *Acta Crystallographica A*, **42**, 478–481 (1986).
9. Eichhorn, K., Kirfel, A., and Fischer, K., *Zeitschrift fur Naturforschung*, **43a**, 391–392 (1988).
10. Kirfel, A., Petkov, A., and Eichhorn, K., *Acta Crystallographica A*, **47**, 180–195 (1991).
11. Templeton, D. H., and Templeton, L. K., *Acta Crystallographica A*, **48**, 746–751 (1992).
12. Kirfel, A., Lippmann, T., and Morgenroth, W., Anisotropy of anomalous scattering. iv. dipole-quadrupole type forbidden reflections in magnetite, Fe_3O_4., Tech. rep., HASYLAB Jahresbercht (1995).
13. Hagiwara, K., Kanazawa, M., Horie, K., Kokubun, J., and Ishida, K., *Journal of the Physical Society of Japan*, **68**, 1592–1597 (1999).
14. Garcia, J., Subias, G., Proietti, M. G., Renevier, H., Joly, Y., Hodeau, J. L., Blasco, J., Sanchez, M. C., and Berar, J. F., *Physical Review Letters*, **85**, 578–581 (2000).
15. Nagano, T., Kokubun, J., Yazawa, I., , Kurasawa, T., Kuribayashi, M., Tsuji, E., Ishida, K., Sasaki, S., Mori, T., Kishimoto, S., and Murakami, Y., *Journal of the Physical Society of Japan*, **65**, 3060–3067 (1996).
16. Murakami, Y., Kawada, H., Kawata, H., Tanaka, M., Arima, T., Moritomo, Y., and Tokura, Y., *Physical Review Letters*, **80**, 1932–1935 (1998).
17. Templeton, D. H., and Templeton, L. K., *Physical Review B*, **49**, 14850–14853 (1994).
18. Dmitrienko, V. E., Ovchinnikova, E. N., and Ishida, K., *Letters to Journal of Experimental and Theoretical Physics (JETP Lett.)*, **69**, 938–942 (1999).
19. Dmitrienko, V. E., and Ovchinnikova, E. N., *Acta Crystallographica A*, **56**, 340–347 (2000).
20. Kirfel, A., and Krane, H.-G., Low temperature study of anisotropic anomalous scattering in cuprite, Cu_2O., Tech. rep., HASYLAB Jahresbercht (1999).
21. Kokubun, J., Kanazawa, M., Ishida, K., and Dmitrienko, V. E., *Physical Review B*, **64**, 073203–1–073203–4 (2001).
22. Kirfel, A., Grybos, J., and Dmitrienko, V. E., *Physical Review B*, submitted (2002).
23. Ovchinnikova, E. N., and Dmitrienko, V. E., *Acta Crystallographica A*, **53**, 388–395 (1997).
24. H.Winkler, R.Eisberg, E.Alp, Ruffer, R., E.Gerdau, A.X.Trautwein, V.Grodzicki, and A.Vera, *Z.Phys.*, **B49**, 331–341 (1983).
25. E.Gerdau, R.Ruffer, H.Winkler, W.Tolksdorf, C.P.Klages, and J.P.Hannon, *Phys.Rev.Lett.*, **54**, 835–838 (1985).
26. R.Ruffer, E.Gerdau, H.D.Ruter, W.Sturhahn, A.Hollatz, and A.Schneider, *Phys. Rev. Lett.*, **63**, 2677–2679 (1989).
27. G.Balestrino, E.Gerdau, M.Grove, R.Hollatz, E.Milani, A.Paoletti, P.Paroli, R.Ruffer, H.D.Ruter, and W.Sturhahn, *Europhys. Letters*, **7**, 329–335 (1988).
28. Labushkin, V., Ovchinnikova, E., Smirnov, E., Sarkisov, E., and Uspenskii, M. N. *Crystallography Reports*, **40**, 900–1050 (1995).

Spin Projection of Empty Partial Density of States by Resonant X-ray Scattering (RXS): Application to Materials with Different Magnetic Ordering

Günter Dräger and Pavel Machek

Fachbereich Physik, Martin-Luther-Universität Halle-Wittenberg, 06108 Halle, Germany

Abstract. We report the first experimental spin projections of empty partial density of states in antiferromagnetic NiO and CuO, paramagnetic MnO and in ferrimagnetic $Dy_3Fe_5O_{12}$ by means of resonant X-ray scattering (RXS). Resolving resonantly scattered $K\alpha_{1,2}$, $K\beta_{1,3}$, $L\alpha_1$ and Ll core line spectra into their spin-up and spin-down components the spin character of the dipole- and quadrupole-excited conduction band states can quantitatively be analyzed. Since the method employs spin conservation in the RXS process and local spin references, it needs neither circularly polarized radiation nor sample magnetization for measuring the spectra. Hence, antiferro- and paramagnetic materials can be investigated as well.

In the paper, the basic idea of the novel method, its experimental realization and the data treatment are reported including the spectra decomposition into the spin-up and spin-down components by using Principal Component Analysis (PCA). New and unambiguous results will be presented providing the opportunity to verify experimentally the results of spin-dependent (LSDA+U) calculations. So we argue the new spectroscopy complements X-ray magnetic dichroism, which is silent for antiferro- and paramagnetic materials. In fact, the novel method gives insight into the spin polarization of conduction band states in correlated materials, independently on their magnetic ordering.

INTRODUCTION

X-ray absorption spectroscopy is a powerful tool for probing empty electronic states of solid state materials. Utilizing the effect of Magnetic Circular Dichroism (MCD) the spin-polarized Partial Density of States (PDOS) of magnetized ferro- and ferrimagnetic materials can be investigated. Antiferro- and paramagnetic materials can be studied qualitatively by the local-spin-selective X-ray absorption spectroscopy /1,2/.

In this paper we report the first quantitative spin projection of empty PDOS of materials independent on their magnetic ordering. The basic idea is, to resolve the Near-Edge X-ray Absorption Fine Structure (NEXAFS) transmission spectra into their spin-up and spin-down components by means of Resonant X-ray Scattering (RXS) spectra. Since the method employs spin conservation and local spin references, it

CP652, *X-Ray and Inner-Shell Processes: 19th International Conference on X-Ray and Inner-Shell Processes*
edited by A. Bianconi, A. Marcelli, and N. L. Saini
© 2003 American Institute of Physics 0-7354-0111-X/03/$20.00

needs no circularly polarized radiation and no sample magnetization. Hence, the spectra and the PDOS of antiferromagnetic and paramagnetic materials can be resolved as well. Moreover, by angular-dependent measurements of spectral data the orbital-resolved conduction band PDOS can also be obtained.

METHODICAL

The novel technique for the quantitative resolution of K and L absorption spectra (and of the corresponding empty PDOS) into the spin-up and spin-down components has been developed after the following idea:

Supposing only spin-conserving transitions in the resonant X-ray scattering process the complementary spin parts x for the up spin $\uparrow$ and $(1-x)$ for the down spin $\downarrow$ in the K- and L-absorption spectrum should determine the intensity ratio of appropriate spin-polarized core emission components. Therefore, in the case of analyzing the $K\alpha_{1,2}$ emission, for example, we suppose the ratio $x/(1-x)$ of spin-up and spin-down states to be the same for the resonantly scattered $K\alpha_{1,2}$ lines (and the resulting $2p_{3/2,1/2}$ holes in the final state) as for the excited conduction band states (and the produced 1s holes in the intermediate states). Following this, by resolving the resonantly scattered $K\alpha_{1,2}$ emission into their spin-projected components we can determine the $x/(1-x)$ ratio. This ratio can be used to resolve the corresponding K-absorption spectrum into its spin-up and spin-down components.

For practical data analysis we describe the total absorption coefficient $\mu(E_1)$ as measured integral or angular-dependently in the transmission geometry in terms of the desired spin-up and spin-down components $\mu\uparrow(E_1)$ and $\mu\downarrow(E_1)$, respect., by

$$\mu(E_1) = \mu\uparrow(E_1) + \mu\downarrow(E_1) = x(E_1)\cdot\mu(E_1) + [1-x(E_1)]\cdot\mu(E_1). \qquad (1)$$

Here, the complementary spin-up and spin-down parts x and $(1-x)$, respect., depend also on the excitation energy E_1. To determine x and $(1-x)$ the resonantly scattered $K\alpha_{1,2}$ must be resolved into their spin-up and spin-down components $u(E_1,E_2)$ and $d(E_1,E_2)$, respect., which depend on E_1 and on the emission energy E_2. For the resolving procedure the measured intensity distribution $I(E_1,E_2)$ can be written as

$$I(E_1,E_2) = x\cdot u(E_1,E_2) + (1-x)\cdot d(E_1,E_2), \qquad (2)$$

were the experimental $I(E_1,E_2)$ are normalized with respect to the integral intensity. By this normalization the $I(E_1,E_2)$ measured at different E_1 are reduced to an „unit transition" in the concerned resonant scattering process. At the same time the dependence on the beam power and effects of self-absorption are eliminated or strongly reduced, respect.

With eq. (2) and the normalization procedure the problem of spin resolving spectral densities is well quantified. It demands the solution of nonlinear equations like eq. (2) to obtain the spin-up x and the spin-down part $(1-x)$ of the „unit transition" under consideration. In practice, one has to analyze two or more RXS emission spectra, measured at one E_1 but angular-dependently for several sample orientations or, alternatively, some spectra measured at one orientation (or integral at polycrystalline samples) and excited at several, but closely neighbouring E_1. In this case, the comparatively small E_1 dependence of $u(E_1,E_2)$ and $d(E_1,E_2)$ can be neglected. Then one can resolve the spectra into the u and d and determine the x and $(1-x)$ belonging to the considered spectra. For the $K\alpha_{1,2}$ and $K\beta_{1,3}\beta'$ RXS spectra we have solved the problem by special methods of Principal Component Analysis (PCA).

To get an overview on the spin parts x and $(1-x)$ the area normalised intensities $I(E_1,E_2)$ after eq.(2) can be plot in a two-dimensional (E_1,E_2) – contour line diagram. By analysis of such diagrams semi-quantitative information on x and $(1-x)$ and on the spin-dependent absorption thresholds can be obtained /3/.

Finally, with the x and $(1-x)$ determined with one of the two methods mentioned above the spin-resolved metal K-absorption spectra can be derived by the relations $\mu\uparrow(E_1) = x(E_1) \cdot \mu(E_1)$ and $\mu\downarrow(E_1) = [1-x(E_1)] \cdot \mu(E_1)$ as defined in eq. (1).

EXPERIMENTAL

At the beamlines X21 (NSLS/BNL) and BW1 (HASYLAB/DESY) we have measured well resolved metal K and L core line spectra of several compounds with different magnetic ordering (antiferromagnetic NiO and CuO, paramagnetic MnO, ferrimagnetic Fe_2O_3 and $Dy_3Fe_5O_{12}$ a.o.m.). To obtain the RXS spectra the incoming linearly polarized and highly monochromatized radiation was tuned over an energy range of some ten eV from below to above the metal K- and L-absorption edges. Measuring, as in most cases, at several selected orientations of the single-crystal scattering samples with respect to the linear polarization vector **e** and the wave vector **k** of the incoming radiation allowed to differentiate between dipole ($1s \rightarrow np$) and quadrupole ($1s \rightarrow 3d$) excitation of the RXS spectra and to determine the orbital character of the empty d-states.

The respective K- and L-NEXAFS spectra has been measured in a transmission geometry at the A1 and E4 stations at HASYLAB/DESY.

RESULTS AND DISCUSSION

In this paper we will exemplary present and discuss results on the spin-projected empty Ni p- and d-like PDOS in antiferromagnetic NiO. These are obtained from the Ni K-NEXAFS and the resonantly scattered Ni $K\alpha_{1,2}$ emission, both spectra measured angular-dependently at single-crystal NiO samples. By special geometries and orientations of the samples used for the transmission NEXAFS and in the scattering experiment well defined transitions are selected for the absorption and excitation process, respect. /4/. Here, in the so-called „(100) excited" and „(110) excited" RXS, for example, besides the isotropic $1s \rightarrow np$ dipole excitations, only $1s \rightarrow 3d(e_g)$ and

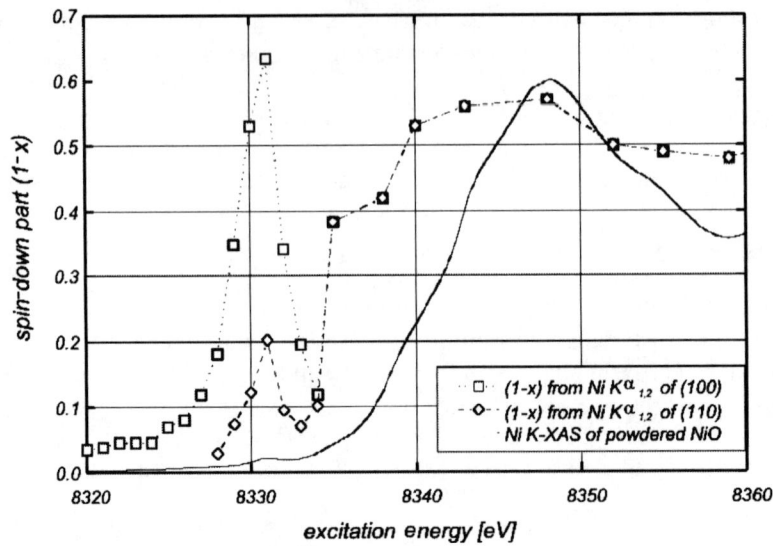

FIGURE 1. The spin-down parts *(1-x)* quantitatively deduced from the Ni Kα$_{1,2}$ RXS spectra, and the Ni K-NEXAFS of powdered NiO for an overview.

1s → 3d(t$_{2g}$) quadrupole transitions, respect., take part in the absorption and excitation process and can be observed separately.

Resolving the Ni Kα$_{1,2}$ RXS spectra measured for the two orientations (100) and (110) of the NiO sample we have obtained the spin-up and spin-down components $u(E_1, E_2)$ and $d(E_1, E_2)$, respect., by the PCA method /3/. Moreover, PCA provides the desired spin parts x and (1-x) as the relative weights of the "principal components" u and d in the unresolved spectra.

From the spin-down parts (1-x) in Fig.1 obtained for the two sample orientations (100) and (110) follows that the orbital character of the empty 3d states (at about 8331 eV in the Ni K-NEXAFS) is predominantly e$_g$-like, but only little t$_{2g}$-like, and that both empty 3d states are spin-down.

With eqn.(1) and the spin-down parts *(1-x)* for the (100) and (110) orientation, as shown in Fig.1, the Ni K-NEXAFS measured for the corresponding orientations has been resolved into the spin-up and spin-down components $\mu \uparrow(E_1)$ and $\mu \downarrow(E_1)$, respectively (Figs. 2a,b).

The p-like empty PDOS shows a pronounced spin-polarization similar to that in Ref.1. The most interesting result firstly now observed is the existence and the nearly absence of the small pre-edge peak at about 8331 eV in the (100) and (110) excited total NEXAFS spectra, respect., showing clearly the existence of the dominant 3d(e$_g$)-like conduction band states (Figs. 2a,b). The comparatively strong appearance of the peak in the spin-down component $\mu \downarrow(E_1)$ provides the information on the spin-down character of these states. Both results on the orbital and spin symmetry 3d(e$_g\downarrow$) agree very well with the results of spin-polarized (LSDA+U) band structure and cluster calculations /4,5/, which here are confirmed for the first time in a direct experimental way.

Analyzing the Cu K-NEXAFS of antiferromagnetic CuO by means of the Cu $K\alpha_{1,2}$ RXS spectra /6/ the big hump in the Cu K-edge obviously can be explained by strongly spin-polarized empty Cu p-like states in the edge region.

The spin analysis of the Mn K-NEXAFS of paramagnetic MnO by means of resonantly scattered $K\beta_{1,3}\beta'$ spectra provides two well resolved pre-edge absorption peaks split by about 0.8 eV. They could be identified as $3d(e_g\downarrow)$ and $3d(t_{2g})$ states /7/, both in a very good agreement with the theory.

Finally, it can be reported, that the spin-polarization analysis of the Dy L_3-

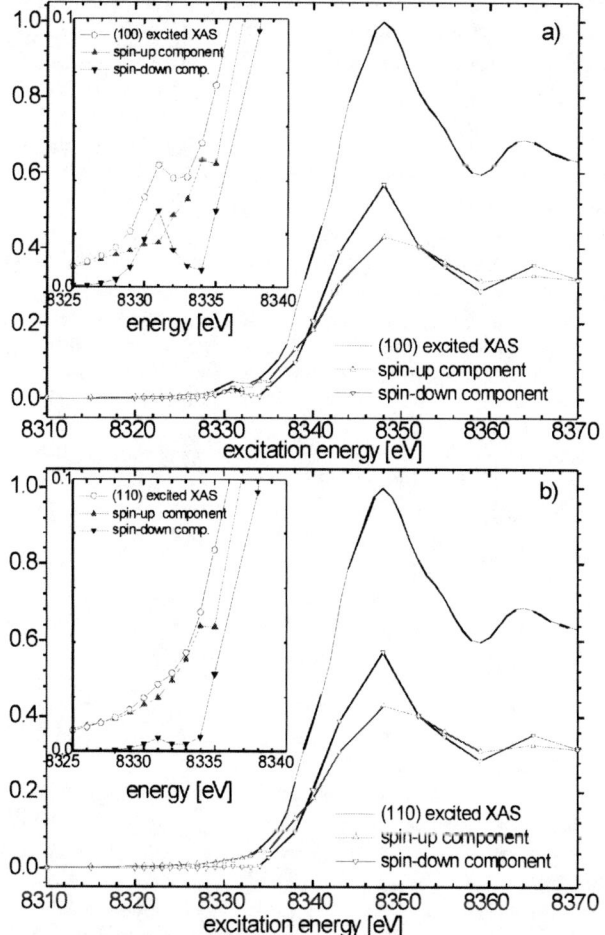

FIGURE 2. The Ni K-NEXAFS of single-crystal NiO measured for the (100) and (110) orientations where a) the $1s \rightarrow 3d$ (e_g) and b) the $1s \rightarrow 3d$ (t_{2g}) quadrupole transitions, respectively, are excited separately. The spin-up and spin-down projected NEXAFS components are shown by the signs Δ and ∇ only at those excitation energies where they have been resolved by means of the Ni $K_{1,2}$ RXS spectra.

NEXAFS of ferrimagnetic $Dy_3Fe_5O_{12}$ by means of the Dy $L\alpha_1$ RXS spectra confirms the existence of $p \rightarrow f$ quadrupole transitions and enables to determine the spin character of the empty f-states.

CONCLUSIONS

In conclusion, a novel experimental method of spin-resolving NEXAFS by means of resonant X-ray scattering has been developed. Because the spin-resolved spectral densities reflect the spin-projected PDOS it provides the first opportunity for proving results of spin-polarized (LSDA+U) band structure and cluster calculations in a direct experimental way. The method needs neither circularly polarized radiation nor sample magnetization and, therefore, it works independently on the magnetic ordering of the sample material.

REFERENCES

1. Hämäläinen, K., Kao, C.C., Hastings, J.B., Siddons, D.P., Berman, L.E., Stojanoff, V. and Cramer, S.P., *Phys. Rev. B* **46**, 14274 (1992).
2. Wang, X., de Groot, F.M.F., and Cramer, S.P., *Phys. Rev. B* **56**, 4553 (1997).
3. Because of the lack of space the spectra treatment by means of PCA and alternatively by analysis of contour line diagrams constructed from the normalized RXS spectra will be described in a detailed publication on the basis of the NiO spectra exemplary considered here: G.Dräger, in preparation
4. Vedrinskii, R.V., Kraizman, V.L., Novakovich, A.A., Elyafi, Sh.M., Bocharov, S., Kirchner, Th., and Dräger, G., *phys. stat. sol. (b)* **226**, 203 (2001).
5. Terakura, K., Oguchi, T., Williams, A.R.., and Kübler, J., *Phys. Rev. B* **30**, 4734 (1984); Hugel, J., and Kamal, M., *Sol. Stat. Comm.* **100**, 457 (1996); Massida, S., Continenza, A., Pasternak, M., and Baldareschi, A., *Phys. Rev. B* **55**, 13494 (1997).
6. Dräger, G., Kirchner, Th., Bocharov, S., and Kao, C.C., *J. Synchrotron Rad.* **8**, 398 (2001).
7. Dräger, G., Kirchner, Th., Bocharov, S., and Kao, C.C., *Appl. Phys. A* **73**, 687 (2001).

Magnetic Circular Dichroism of Resonant Inelastic X-ray Scattering in Magnetic Materials

Toshiaki Iwazumi

Photon Factory, Institute of Materials Structure Science,1-1 Oho, Tsukuba, Ibaraki 305-0801, Japan

Abstract. Magnetic circular dichroism (MCD) measurement of resonant inelastic X-ray scattering (RIXS) is a direct experimental technique used to obtain the final spin state of excited elements in a ferro- or ferri-magnetic system. In the lanthanoid elements we can observe a large MCD-RIXS signal in the radiative transitions to the 2p levels, because the 2p core excitation always exhibits the MCD expected from the transition probability determined by the Clebsch-Gordon coefficient. In this article, we discuss our recent results of the MCD-RIXS by the electric quadrupole excitation in the lanthanoid systems after the brief review of the origin of MCD-RIXS.

INTRODUCTION

During the past decade, x-ray emission spectroscopy (XES) using monochromatic incident photons obtained by synchrotron radiation has been under intensive investigation both experimentally and theoretically [1-29]. XES is a second-order optical process, whose intermediate state is the same as the final state of first-order optical processes, x-ray absorption spectroscopy (XAS) and x-ray photoemission spectroscopy. If we take into account the XAS-type excitation of a deep core electron to the absorption threshold and the radiative decay from a shallow core or a valence band to the excited deep core, we have the resonant XES or the resonant inelastic x-ray scattering (RIXS). Almost all of the recent experimental and theoretical developments mentioned above have been carried out for the RIXS. Using RIXS, we can obtain the joint density of states between the conduction and valence bands [1, 2], as well as information concerning the charge-transfer excitation [3, 4] and the electric quadrupole (E2) excitation at the pre-edge of the inner-shell absorption [5-7]. The magnetic circular dichroism (MCD) of the RIXS (referred to as MCD-RIXS) has also been researched [8-30]. The MCD-RIXS is defined by the difference of RIXS for incident photons with positive and negative helicities, where the helicity of the emitted photon is not analyzed, and is a direct experimental technique to determine the final spin state of the excited elements in a ferro- or ferri-magnetic system.

CP652, *X-Ray and Inner-Shell Processes: 19ᵗʰ International Conference on X-Ray and Inner-Shell Processes*
edited by A. Bianconi, A. Marcelli, and N. L. Saini
© 2003 American Institute of Physics 0-7354-0111-X/03/$20.00

In this article we report recent experimental results of the MCD-RIXS by the E2 excitation for Sm $3d_{5/2,3/2}$-$2p_{3/2}$ radiative decay in Sm-Co amorphous alloy. Before going into the main subject of this article, we briefly review a qualitative interpretation for the MCD in the L emissions of the lanthanoid elements by the electric dipole (E1) excitation above the L absorption edges in Section 2. Experimental details and the results are presented in Sections 3 and 4. Comparison of the experimental results with recent theoretical ones is also reported.

MCD-RIXS OF LANTHANOID ELEMENTS

Let us consider, for example, the MCD of the $3d_{5/2,3/2}$-$2p_{3/2}$ radiative decay following the $2p_{3/2}$ absorption process of the lanthanoid elements. To simplify the explanation, the MCD-RIXS is considered as a two-step process, starting with an excitation from the $2p_{3/2}$ level to the conduction band by the circularly polarized X-rays, after which the $2p_{3/2}$ hole decays into the $3d_{5/2,3/2}$ core hole while a photon is emitted.

It has been well known that the asymmetry of the core holes is induced when the core electron is excited by the circularly polarized X-rays [31]. The E1 transition probability from the $2p_{3/2}$ level to the 5d or higher d-symmetry level by the positive (negative) helicity incident light is represented by the Clebsch-Gordon coefficient, and is shown schematically in Fig. 1. Because the $2p_{3/2}$ sublevels are all occupied in the initial state and the conduction sublevels are all unoccupied in the lanthanoid elements, the spin asymmetry of the $2p_{3/2}$ holes made by the circularly polarized X-rays is always 25%.

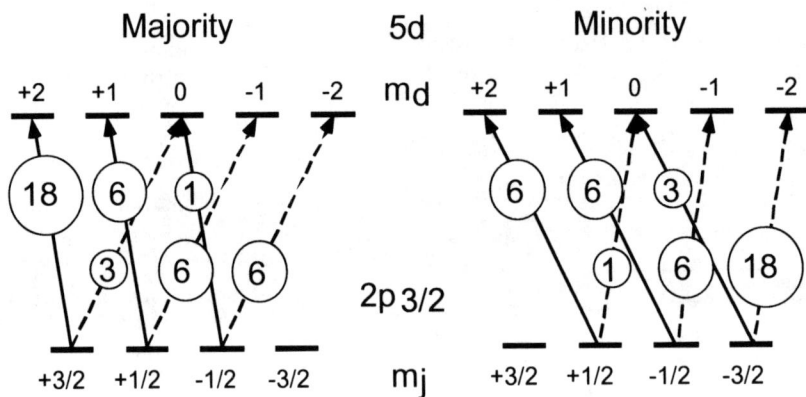

FIGURE 1. The 5d and $2p_{3/2}$ orbits specified by the spin and magnetic quantum numbers are shown. The number in the circle denotes the integer ratio of the excitation from the $2p_{3/2}$ level to the 5d or higher d-symmetry level by the positive helicity incident light (solid arrows) and the negative one (broken arrows).

On the other hand, it has been also well known that the spectral shape of the radiative decay from the shallow to the deep core levels of the transition-metal or lanthanoid elements is closely related to the number of unpaired valence electrons [32]. This relation can be understood in terms of multiplet splitting of the levels caused by spin-orbit and electrostatic exchanges of the inner electrons with the unpaired outer electrons. The final state multiplets can be divided into two groups; one is a parallel orientation of the resultant electron spin between the shallow and the valence levels, and another is an antiparallel. The emission energy caused by a parallel spin orientation is higher than that by an antiparallel [33-35].

As mentioned in the introduction, the MCD-RIXS is defined by the difference of RIXS for incident photons with positive and negative helicities, where the helicity of the emitted photon is not analyzed. Assuming the conservation of the spin direction in decay process, the asymmetry of the core holes excited by the circularly polarized X-rays introduces the asymmetry of the emission intensity caused by the two orientations of the resultant electron spin. Therefore, we can always observe a large MCD-RIXS signal in the radiative transitions to the 2p levels of the lanthanoid elements, not the same as the MCD of the XAS (referred to as MCD-XAS), even if there is no interaction between the magnetic 4f valence states and the 5d or higher conduction bands. An example of the MCD-RIXS is shown in Fig. 2. The positive (negative) MCD-RIXS is explained by a multiplet family of the majority (minority) spin character.

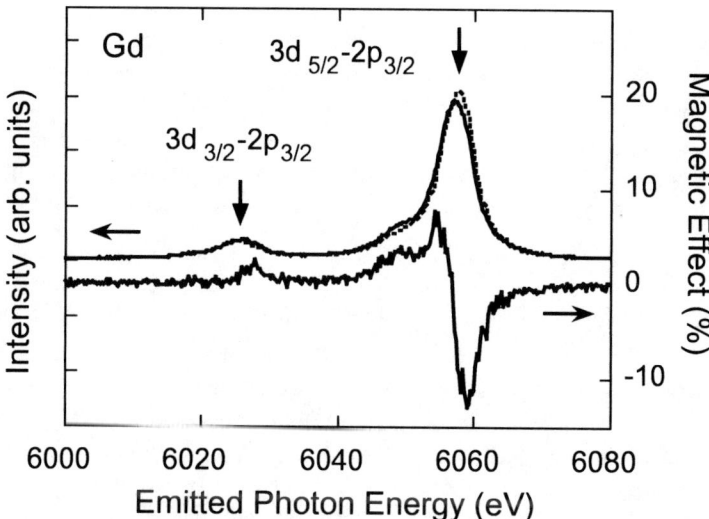

FIGURE 2. Example of the RIXS (top) and MCD-RIXS (bottom) spectra of Gd $3d_{5/2,3/2}$-$2p_{3/2}$ radiative decay in the Gd-Co amorphous when the incident photon energy was enough higher than the Gd $2p_{3/2}$ absorption edge [29]. The solid and broken curves in the top figure are the RIXS spectra after the excitations by the positive and negative helicities, respectively.

EXPERIMENTAL DETAILS

The sample used in this experiment was an amorphous Sm-Co thin film of 2 μm thickness sputtered on a polyimide film. The composition of Sm-Co was 21 at. % Sm and 79 at. % Co, measured by the inductively coupled plasma method. The MCD-XAS and MCD-RIXS experiments were performed at the elliptical multipole wiggler beamline (28B) of the Photon Factory, Institute of Materials Structure Science [36]. This beamline is equipped with focusing optics using a double-crystal monochromator between the two bent mirrors, and provides a 2.3 mm horizontal and 0.3 mm vertical focused beam. With a Si(111) monochromator, the flux of the incident beam is estimated to be $\sim 10^{11}$ photons/s, and the degree of circular polarization of this beam is estimated to -0.51 in the energy range around the Sm $2p_{3/2}$ absorption edge. The sample film was mounted at the focus point and between the pole pieces of an electro-magnet for the MCD-XAS measurement and a rotatable permanent-magnet for the MCD-RIXS so that the magnetization of the sample could be reversed periodically. All measurements were done in the room temperature.

The MCD-XAS spectrum at the Sm $2p_{3/2}$ absorption edge was measured with the transmission geometry using two ion chambers. The applied magnetic field of 0.6 T was reversed every 4 seconds to obtain the MCD effect. We define the positive (negative) direction of a magnetic field when the **B** vector is directed parallel (antiparallel) to the x-ray wave vector. The MCD-XAS was defined as subtracting the absorption coefficients detected on the positive direction from that on the negative one.

The MCD-RIXS spectra were measured as follows: the angle between the incident X-rays and a sample plane was chosen to be 36.9°, which provided an 80% MCD effect compared with that at 0°. The scattered radiation was analyzed at a 54.7° scattering angle (the so-called the magic angle [14]) in the vertical plane by a cylindrically bent InSb(444) crystal for the energy range around the Sm $3d_{5/2,3/2}$-$2p_{3/2}$ emission lines. The analyzed X-rays were detected by a position-sensitive proportional counter (PSPC) with the charge-division method. The resistive anode of the PSPC was a carbon fiber 7 μm in diameter, and its resistance was 4.2 kΩ per cm. The entrance window was made of a 1-mm thick beryllium plate, and the sensitive area of the window was 10 x 100 mm^2. The distance between the anode and the entrance window was 10 mm. The pressure of the flow gas (Ar + 10% CH_4) was controlled to be 7 atm. Both the sample and the detector were arranged on the center axis of the analyzing crystal cylinder, so that we could obtain sagital-focused and meridional-energy dispersive x-rays. The total energy resolutions of the present apparatus are 1.2 eV around the Sm $3d_{5/2,3/2}$-$2p_{3/2}$ emission lines. The MCD-RIXS was defined by subtracting the emission X-rays detected on the positive direction from that on the negative one, and normalized by the peak intensity of the Sm $3d_{5/2}$-$2p_{3/2}$ emission line.

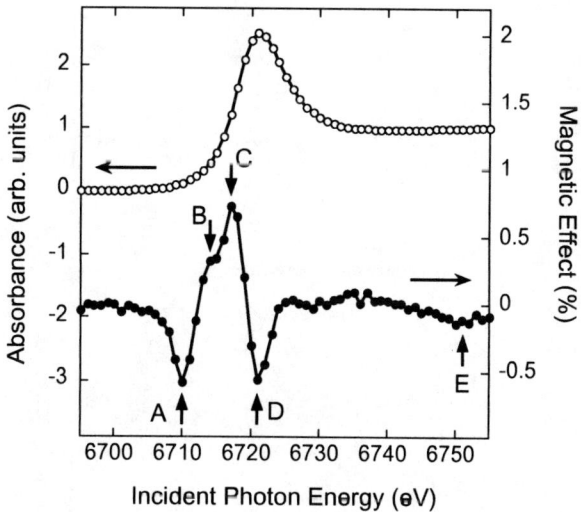

FIGURE 3. XAS (open circles) and MCD-XAS (filled circles) at the Sm $2p_{3/2}$ absorption edge in Sm-Co amorphous at room temperature. The MCD-RIXS were measured at the energies labelled by A~E with vertical arrows.

RESULTS AND DISCUSSIONS

The XAS and MCD-XAS at the Sm $2p_{3/2}$ absorption edge in the Sm-Co amorphous are shown in Fig. 3. Peaks A and B on the MCD-XAS spectrum in Fig. 3, which have the opposite signs and locate in the pre-edge region, are attributed to correspond the E2 transition to the 4f states. The energy labeled A~E in Fig. 3 are chosen as the energy at which MCD-RIXS spectra are measured, where the energy labeled E can be regarded as a high-energy off-resonant region.

The RIXS and MCD-RIXS spectra of the Sm $3d_{5/2,3/2}$-$2p_{3/2}$ emission lines are shown in Fig. 4 as a function of the emitted photon energies (ω_2) for a number of the incident photon energies (ω_1) labeled A~E in Fig. 3. In the RIXS spectra for ω_1 = 6710 and 6714 eV, the excess emission lines labeled Q_1 and Q_2 are observed, which can be explained as the $3d_{5/2,3/2}$-$2p_{3/2}$ E1 transition after the $2p_{3/2}$-4f E2 excitation [5, 6]. The energy difference of the ω_2 between the $3d_{5/2}$-$2p_{3/2}$ main peak and the Q_1 (Q_2) peak is about 9.0 (3.6) eV. These relations are the same as those reported by Bartolomé et al. [7].

All MCD-RIXS spectra above ω_1 = 6717 eV display a derivative-like behavior similar to that of the Gd as shown in Fig. 2 except its polarity. This opposite tendency is caused by the difference of the occupied orbit in the magnetized Hund ground state between the less-than-half and the more-than-half filled 4f shells [18, 20]. In the MCD-RIXS spectra for ω_1 = 6710 and 6714 eV, the MCD features associating with the Q_1 and Q_2 peaks are observed with the negative and positive signs, respectively.

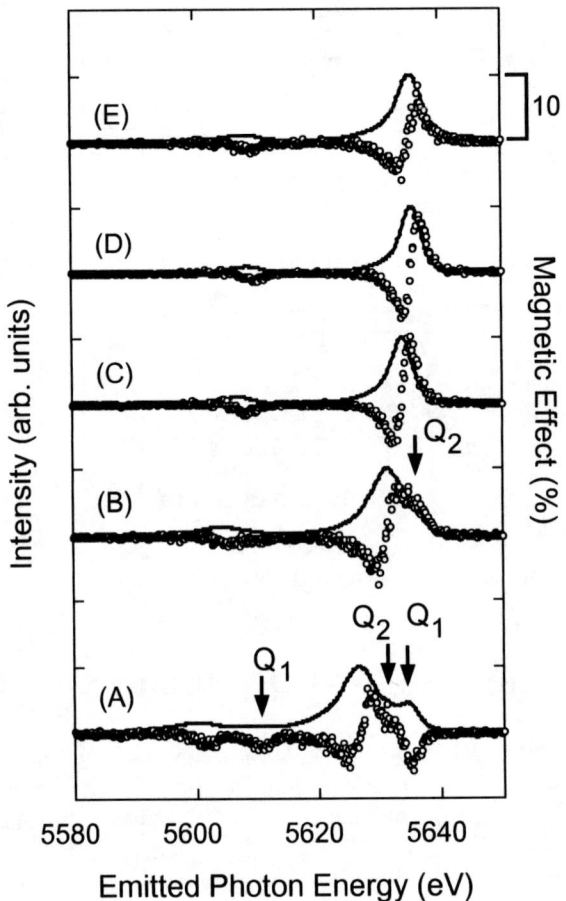

FIGURE 4. Observed RIXS (solid lines) and MCD-RIXS (open circles) spectra for the Sm $3d_{5/2,3/2}$-$2p_{3/2}$ emission lines. The incident photon energy of each spectrum labeled A~E is consistent with consistent with those in Fig. 3.

More importantly, these features are not the derivative-like behavior similar to that of the $3d_{5/2}$-$2p_{3/2}$ main peak but the single-peak-like behavior. This observation is a good indication that the Q_1 (Q_2) peak is brought about the contribution of only the majority (minority) spin of 4f subshell. When ω_1 = 6710 eV, the MCD of the Q_1 peak is larger than that of the Q_2, so the ω_2 integration of the MCD-RIXS spectrum becomes negative. When ω_1 = 6714 eV, the ω_2 integration is positive. These signs are consistent with those of the MCD-XAS at the corresponding ω_1. As shown in the bottom two spectra of Fig. 4, the E2 contributions are confirmed to be the dominant origin of the corresponding MCD-XAS for ω_1 labeled A and B, because the E1 parts

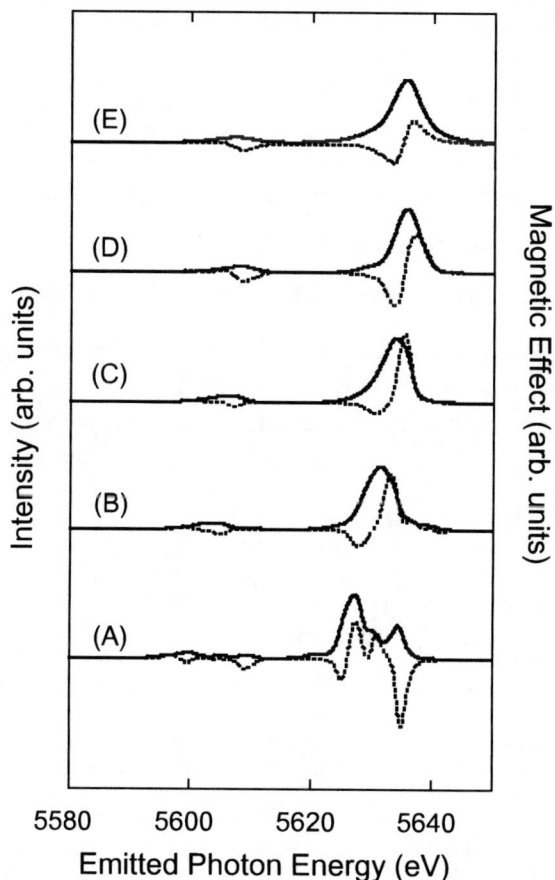

FIGURE 5. Calculated RIXS (solid lines) and MCD-RIXS (open circles) spectra for the Sm $3d_{5/2,3/2}$-$2p_{3/2}$ emission lines. The incident photon energy of each spectrum labeled A~E is consistent with consistent with those in Fig. 3.

of the MCD-RIXS spectra have the dispersive shapes of which integration seem to be almost zero. Bartolomé et al. have assumed the Q_1 (Q_2) peak as originated in the E2 excitation to the majority (minority) 4f subshell by comparing the ω_1 dependence of the Q_1 or Q_2 peak to the MCD-XAS [7]. From the MCD-RIXS measurement, we can obtain more direct information about the spin direction of the excited electron.

Theoretical calculation based on the formula of the coherent second order optical process has been carried out for all the experimental MCD-RIXS spectra [37], and these results are shown in Fig. 5. In the calculation, the multiplet coupling effect and the enhancement of the 2p-5d dipole matrix elements proposed for the MCD-XAS [38] were taken into account. As shown in Figs. 4 and 5, the calculated MCD-RIXS spectra have reproduced all the experimental results very successfully.

In summary, we have studied the RIXS and MCD-RIXS for the Sm $3d_{5/2,3/2}$-$2p_{3/2}$ emission lines in the ferromagnetic Sm-Co amorphous alloy. From the MCD-RIXS measurements, the information about the contributed spin direction of two kinds of the E1 transition after the E2 excitation is directly determined. The theoretical calculation based on the formula of the coherent second order optical process shows the excellent agreements with all the measured MCD-RIXS spectra.

ACKNOWLEDGMENTS

The author would like to thank Dr. T. Nakamura, Dr. H. Shoji, Professor S. Nanao, Dr. K. Fukui, Dr. H. Ogasawara, Professor A. Kotani, Professor I. Harada, Dr. R. Katano, and Professor Y. Isozumi for collaborations and discussions.

REFERENCES

1. Carlisle, J. A., Shirley, E. L., Hudson, E. A., Terminello, L. J., Callcott, T. A., Jia, J. J., Ederer, D. L., Perera, R. C. C., and Himpsel, F. J., Phys. Rev. Lett. 74, 1234 (1995).
2. Minami, T., and Nasu, K., Phys. Rev. B57, 12084 (1998).
3. Kao, C.-C., Caliebe, W. A. L., Hastings, J. B., and Gillet, J.-M., Phys. Rev. B54, 16361 (1996).
4. Shoji, H., Kobayashi, K., Iwazumi, T., Katano, R., Isozumi, Y., Kishimoto, S., and Nanao, S., Jpn. J. Appl. Phys. Sup 38-1, 592 (1999).
5. Udagawa, Y., Hayashi, H., Tohji, K., and Mizushima, T., J. Phys. Soc. Jpn., 63, 1713 (1994).
6. Krisch, M. H., Kao, C.-C., Sette, F., Caliebe, W. A., Hämäläinen, K., and Hastings, J. B., Phys. Rev. Lett. 74, 4931 (1995).
7. Bartolomé, F., Tonnerre, J. M., Sève, L., Chaboy, J., García, L. M., Krisch, M., and Kao, C.-C., Phys. Rev. Lett. 79, 3775 (1997).
8. Strange, P., Durham, P. J., and Gyorffy, B. L., Phys. Rev. Lett. 67, 3590 (1991).
9. Hague, C. F., Mariot, J.-M., Strange, P., Durham, P. J., and Gyorffy, B. L., Phys. Rev. B48, 3560 (1993).
10. Duda, L.-C., Stöhr, J., Mancini, D.C., Nilsson, A., Wassdahl, N., Nordgren, J., and Samant, M. G., Phys. Rev. B50, 16758 (1994).
11. Hague, C. F., Mariot, J.-M., Guo, G. Y., Hricovini, K., and Krill, G., Phys. Rev. B51, 1370 (1995).
12. Krisch, M. H., Sette, F., Bergmann, U., Masciovecchio, C., Verbeni, R., Goulon, J., Caliebe, W., and Kao, C.-C., Phys. Rev. B54, R12673 (1996).
13. Caliebe, W. A., Kao, C.-C., Berman, L. E., Hastings, J. B., Krisch, M. H., Sette, F., and Hämäläinen, K., J. Appl. Phys. 79, 6509 (1996).
14. de Groot, F. M. F., Nakazawa, M., Kotani, A., Krisch, M. H., and Sette, F. , Phys. Rev. B56, 7285 (1997).
15. Eisebitt, S., Lüning, J., Rubensson, J.-E., Schmitz, D., Blügel, S., and Eberhardt, W., Solid State Commun. 104, 173 (1997).
16. Braicovich, L., Dallera, C., Ghiringhelli, G., Brookes, N. B., and Goedkoop, J. B., Phys. Rev. B55, R14729 (1997).
17. Iwazumi, T., Kobayashi, K., Kishimoto, S., Nakamura, T., Nanao, S., Ohsawa, D., Katano, R., and Isozumi, Y., Phys. Rev. B56, R14267 (1997).
18. Iwazumi, T., Kobayashi, K., Kishimoto, S., Nakamura, T., Nanao, S., Ohsawa, D., Katano, R., Isozumi, Y., and Maruyama, H., J. Electron Spectrosc. and Relat. Phenom., 92, 257 (1998).
19. Nakamura, T., Nanao, S., Iwazumi, T., Kobayashi, K., Kishimoto, S., Ohsawa, D., Katano, R., and Isozumi, Y., J. Electron Spectrosc. and Relat. Phenom., 92, 261 (1998).

20. Jo, T., and Tanaka, A., J. Phys. Soc. Jpn., **67**, 1457 (1998).
21. Braicovich, L., Dallera, C., Ghiringhelli, G., Brookes, N. B., and Goedkoop, J. B., Solid State Commun. **105**, 263 (1998).
22. Braicovich, L., van der Laan, G., Ghiringhelli, G., Tagliaferri, A., van Veenendaal, M. A., Brookes, N. B., Chervinskii, M. M., Dallera, C., De Michelis, B., and Dürr, H. A., Phys. Rev. Lett. **82**, 1566 (1999).
23. Iwazumi, T., Kobayashi, K., Tominaga, T., Nakamura, T., Nanao, S., Kishimoto, S., Ohsawa, D, Katano, R., and Isozumi, Y., J. Synchrotron Rad., **6**, 685 (1999).
24. Jo, T., and Parlevas, J.-C., J. Phys. Soc. Jpn. **68**, 1392 (1999).
25. Iwazumi, T., Nakamura, T., Shoji, H., Kobayashi, K., Kishimoto, S., Katano, R., Isozumi, Y., and Nanao, S., J. Phys. Chem. Solid **61**, 453-456 (2000).
26. Wittkop, C., Schülke, W., and de Groot, F. M. F., Phys. Rev. B**61**, 7176 (2000).
27. Nakamura, T., Shoji, H., Iwazumi, T., Nanao, S., Kishimoto, S., Katano, R., and Isozumi, Y., Phys. Rev. B**62**, 5301 (2000).
28. Nakamura, T., Kawamura, N., Iwazumi, T., Maruyama, H., Urata, A., Shoji, H., Nanao,S., Kishimoto, S., Katano, R., and Isozumi, Y., J. Synchrotron Rad. **8**, 428 (2001).
29. Fukui, K., Ogasawara, H., Kotani, A., Iwazumi, T., Shoji, H., and Nakamura, T., J. Phys. Soc. Jpn. **70**, 1230 (2001).
30. Fukui, K, Ogasawara, H., Kotani, A., Iwazumi, T., Shoji, H., and Nakamura, T., J. Phys. Soc. Jpn. **70**, 3457 (2001).
31. See, for instance, Ebert, H., and Schütz, G., (eds.), *Spin-Orbit-Influenced Spectroscopies of Magnetic Solids*, Springer-Verlag, Berlin, 1996.
32. See, for instance, Meisel, A., Leonhardt, G. and Szargan, R., *X-Ray Spectra and Chemical Binding*, Springer-Verlag, Berlin, 1989.
33. de Groot, F. M. F., Fontaine A., Kao, C.-C., and Krisch, M., J. Phys.: Condens. Matter **6**, 6875 (1994).
34. Peng, G., de Groot, F. M. F., Hämäläinen, K., Moore, J. A., Wang, X., Grush, M. M., Hastings, J. B., Siddons, D. P., Armstrong, W. H., Mullins, O. C., and Cramer, S. P., J. Am. Chem. Soc. **116**, 2914 (1994).
35. Wang, X., de Groot, F. M. F., and Cramer, S. P., Phys. Rev. B**56**, 4553 (1997).
36. Iwazumi, T., Koyama, A., and Sakurai, Y., Rev. Sci. Instrum. **66**, 1691 (1995).
37. Nakamura, T., Shoji, H., Hirai, E., Nanao, S., Fukui, K., Ogasawara, H., Kotani, A., Iwazumi, T., Harada, I., Katano, R., and Isozumi, Y., submitted to Phys. Rev. B.
38. Matsuyama, H., Harada, I., and Kotani, A., J. Phys. Soc. Jpn. 66, 337 (1997).

Theory of Resonant Inelastic X-ray Scattering in f and d Electron Systems

Akio Kotani

Institute for Solid State Physics, University of Tokyo,
5-1-5 Kashiwanoha, Kashiwa, Chiba 277-8581, Japan
and
The Institute of Physical and Chemical Research (RIKEN),
1-1-1 Kouto, Mikazuki, Sayo, Hyogo 679-5143, Japan

Abstract. Two examples are presented on theory of angle-dependence in resonant inelastic X-ray scattering (RIXS) for f and d electron systems. Incident and scattering angle-dependence in RIXS is calculated for electric quadrupole excitation in MnO. In addition to numerical calculations of RIXS, an angle-dependent factor is obtained analytically by a group theoretical method. For ferromagnetic materials, magnetic circular dichroism (MCD-RIXS) occurs in RIXS. It is shown that when the incident X-ray direction is parallel to the magnetization, the MCD-RIXS is given by the diagonal term of the second order quantum process, while for the incident X-ray direction perpendicular to the magnetization the MCD-RIXS is given by the 100% interference term. The calculated results, in both examples, are compared with experimental ones.

INTRODUCTION

Resonant inelastic X-ray scattering (RIXS) has recently been a subject of remarkable progress due to the advent of high-brilliance synchrotron radiation sources [1]. In RIXS, a core electron is excited by incident X-ray to the absorption threshold and this excited state decays by emitting X-ray. The polarization-dependence and the angle-dependence of RIXS provide us with important information on the symmetry of electronic states [2–4]. These dependences should be different depending on whether the relevant optical transition is electric dipole or quadrupole one, although the effect of the quadrupole transition on RIXS has not been studied widely. Furthermore, in ferromagnetic systems, the magnetic circular dichroism occurs in RIXS, as an interesting aspect of the polarization-dependence.

In this paper we report very recent theoretical development in the polarization- and angle-dependences of RIXS. As two examples, we present the angle-dependence in RIXS by quadrupole excitation for MnO in Sec. 2 [5] and the magnetic circular dichroism (MCD) in RIXS in Sec. 3 [7]. Comparison of the theoretical results with recent experimental ones is also given.

CP652, *X-Ray and Inner-Shell Processes: 19th International Conference on X-Ray and Inner-Shell Processes*
edited by A. Bianconi, A. Marcelli, and N. L. Saini
© 2003 American Institute of Physics 0-7354-0111-X/03/$20.00

ANGLE-DEPENDENCE IN RIXS BY ELECTRIC QUADRUPOLE EXCITATION

We consider a RIXS process for MnO, where the Mn $1s$-$3d$ electric quadrupole (EQ) excitation at the pre-edge of Mn K absorption is followed by the Mn $3p$-$1s$ electric dipole (ED) transition of X-ray emission [5]. For the electronic states of MnO, we use an MnO_6 cluster model with cubic symmetry O_h. The geometrical alignment is shown in Figure 1. We take the coordinate axes x, y and z parallel to the Mn-O cubic axes of the cluster. The direction of the incident X-ray is given by the polar coordinate (θ_i, ϕ_i), and the direction of the emitted X-ray is described by the scattering angle θ, where the scattering plane is assumed to include the z axis. The polarization of the incident photon is taken to be perpendicular to the scattering plane (so that in the z' direction) and the polarization of the emitted photon is not analyzed (so that the y'' and z'' polarizations are detected), because the experimental observation of the emitted photon polarization is very difficult.

The intensity of the pre-edge absorption is very weak, but it is known that the EQ excitation is split into T_{2g} and E_g peaks due to the crystal field splitting of the Mn $3d$ level. In the following, we calculate the dependence of RIXS spectra on the

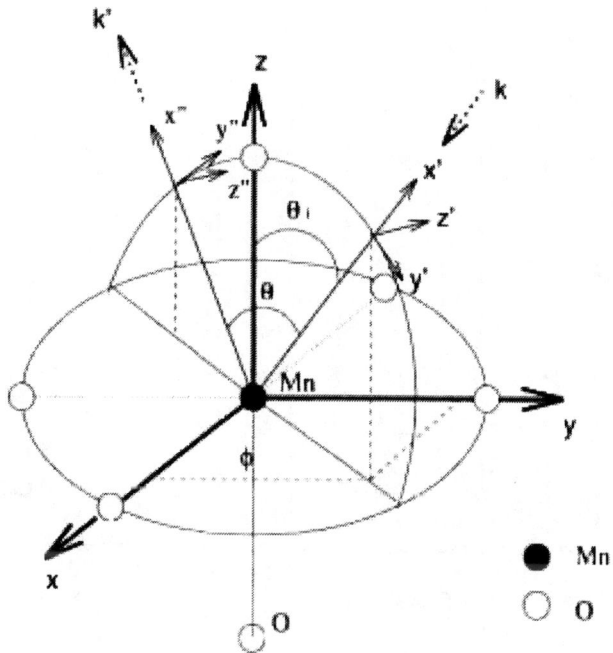

FIGURE 1. Geometrical arrangement of incident and emitted photons and the MnO_6 cluster.

angles, θ_i, ϕ_i and θ, when the incident photon energy is tuned at the T_{2g} and E_g excitation energies. The RIXS spectrum is expressed as

$$F(\Omega, \omega) = \sum_{T'} \sum_f \left| \sum_m \frac{< f|T'|m >< m|T|g >}{E_g + \Omega - E_m + i\gamma} \right|^2 \frac{\gamma'/\pi}{(E_g + \Omega - E_f - \omega)^2 + \gamma'^2}, \quad (1)$$

where Ω and ω are, respectively, the incident and emitted photon energies, $|g >$, $|m >$, and $|f >$ are initial, intermediate, and final states of the material system, respectively, E_g, E_m, and E_f are their energies, and γ and γ' represent the spectral broadening due to the core-hole lifetime in the intermediate and final states, respectively. T is the EQ excitation operator, and T' is the ED de-excitation operator, where the summation for T' in Eq.(1) is taken over the two polarization directions y'' and z'' of the emitted photon.

Numerical calculations are performed with the MnO_6 cluster model taking into account the intra-atomic multiplet coupling and the interatomic charge transfer interaction. The parameter values of the cluster model are essentially the same as those used by Taguchi et al. [6]. The calculated results are compared with recent experimental data.

As an example, the experimental (a) and theoretical (b) results of θ_i dependence of RIXS are shown in Figure 2, where ϕ_i and θ are fixed to 45° and 90°, respectively

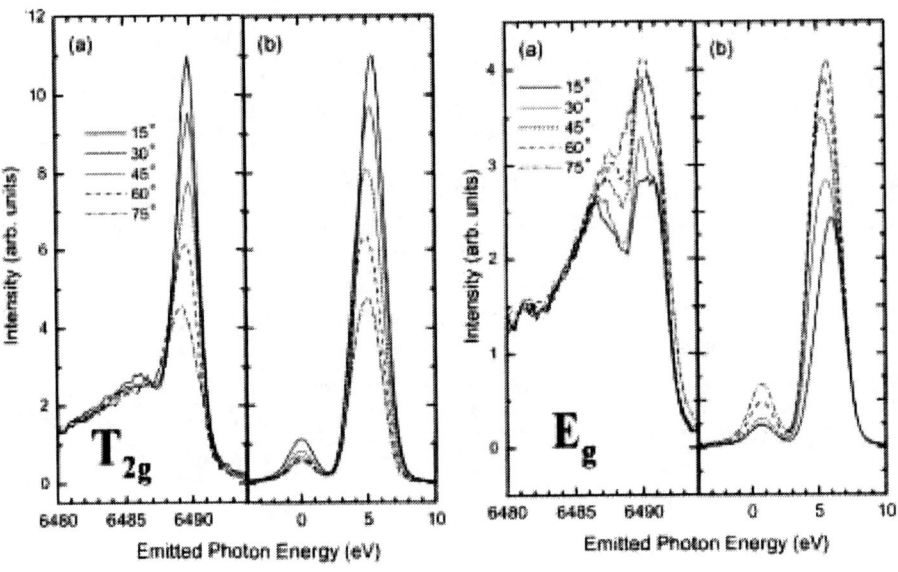

FIGURE 2. The θ_i dependence of RIXS for MnO at the T_{2g} excitation (left) and the E_g excitation (right). Experimental and theoretical results are shown in (a) and (b), respectively.

TABLE 1. Behavior of angle-dependent factor $\beta_{\Gamma_f}(\Gamma_m)$ at T_{2g} ans E_g excitations for each final state Γ_f

$\beta_{\Gamma_f}(\Gamma_m)$	θ_i dependence	θ dependence	ϕ_i dependence
$\beta_{A_{2u}}(T_{2g})$	$\frac{1}{9}\cos^2\theta_i$	$\frac{1}{9}\cos^2 70°$	$\frac{1}{18}(1+\cos^2 2\phi_i)$
$\beta_{E_u}(T_{2g})$	$\frac{1}{18}\cos^2\theta_i(1+3\sin^2\theta_i)$	$\frac{1}{18}\cos^2 70°(1+3\cos^2(\theta-70°))$	$\frac{1}{72}(4+\sin^2 2\phi_i)$
$\beta_{T_{1u}}(T_{2g})$	$\frac{1}{6}\cos^2\theta_i(1+\cos^2\theta_i)$	$\frac{1}{6}\cos^2 70°(1+\sin^2(\theta-70°))$	$\frac{1}{24}(1+5\cos^2 2\phi_i)$
$\beta_{T_{2u}}(T_{2g})$	$\frac{1}{6}\cos^2\theta_i$	$\frac{1}{6}\cos^2 70°$	$\frac{1}{24}(1+\cos^2 2\phi_i)$
$\beta_{T_{1u}}(E_g)$	$\frac{1}{4}\sin^2\theta_i(1+\sin^2\theta_i)$	$\frac{1}{4}\sin^2 70°(1+\cos^2(\theta-70°))$	$\frac{3}{16}\sin^2 2\phi_i$
$\beta_{T_{2u}}(E_g)$	$\frac{1}{12}\sin^2\theta_i(2+3\sin^2\theta_i)$	$\frac{1}{12}\sin^2 70°(2+3\sin^2(\theta-70°))$	$\frac{7}{48}\sin^2 2\phi_i$

[5]. The figures on the left and right sides are the results for the T_{2g} excitation and the E_g excitation, respectively. It is seen that with increase in θ_i the RIXS intensity with the T_{2g} excitation decreases, while that with the E_g excitation increases, in both theoretical and experimental results. The tendency of the shift in the main RIXS peak position with θ_i in the theoretical result also agrees with that in the experimental one. The main difference between the theoretical and experimental results is the existence and absence of the background in the experimental and theoretical results, respectively, but this background originates from the Mn $1s$-$4p$ ED transition, which is disregarded in the calculation.

Next we calculate analytic expressions for θ_i, ϕ_i and θ dependence of RIXS spectra. We describe only the outline of the calculation here. We expand T and T' with respect to the irreducible tensor operators, where the irreducible representations of T are T_{2g} and E_g, and that of T' is T_{1u}. The expansion coefficients are expressed as functions of θ_i, ϕ_i and θ. Then, these T and T' are inserted in Eq.(1), and we calculate the transition matrix elements between $|g>$ and $|m>$ and those between $|m>$ and $|f>$. It is to be noted that the ground state of MnO is the orbital singlet state A_{1g} (with $S = 5/2$), the intermediate states are T_{2g} and E_g, and the final states are A_{2u}, E_u, T_{1u} and T_{2u} via the T_{2g} intermediate state, and T_{1u} and T_{2u} via the E_g intermediate state. In calculating the matrix elements, we use the Wigner-Eckart theorem to perform explicitly the angular integration, which enables us to obtain analytic expressions for θ_i, ϕ_i and θ dependence of RIXS spectra.

The result is given by

$$F(\Omega,\omega) = \sum_{\Gamma_f}\sum_{\Gamma_m} \frac{1}{3\{\Gamma_f\}}|f_{\Gamma_f}(\Gamma_m)|^2 \beta_{\Gamma_f}(\Gamma_m)\frac{\gamma'/\pi}{(E_g+\Omega-E_f-\omega)^2+\gamma'^2}. \qquad (2)$$

Here $\beta_{\Gamma_f}(\Gamma_m)$ is the angle-dependent factor for the intermediate state denoted by the irreducible representation Γ_m and the final state Γ_f, $|f_{\Gamma_f}(\Gamma_m)|^2$ is the angle-independent factor which includes the reduced matrix elements coming from the Wigner-Eckart theorem, and $\{\Gamma_f\}$ is the dimension of the irreducible representation Γ_f. The expression of $\beta_{\Gamma_f}(\Gamma_m)$ is listed in Table 1. Here, in each of the θ_i, θ and ϕ_i dependences of $\beta_{\Gamma_f}(\Gamma_m)$, the other angles are fixed at two of the angles $\theta_i = 70°$, $\theta = 90°$ and $\phi = 45°$.

Let us consider, for example, the θ_i dependence of RIXS. According to Table 1, the factor $\beta_{\Gamma_f}(T_{2g})$ is proportional to $\cos^2 \theta_i$ for all the final states, so that the RIXS intensity for $\Gamma_m = T_{2g}$ decreases when θ_i increases from 0 to 90°. On the other hand, the RIXS intensity for $\Gamma_m = E_g$ increases when θ_i increases from 0 to 90°, because $\beta_{\Gamma_f}(E_g)$ is proportional to $\sin^2 \theta_i$ for all the final states. This is quite consistent with the results shown in Figure 2.

The angle-dependence of RIXS can be a powerful tool to distinguish the ED and EQ excitations in the pre-edge region, because the angle-dependence should be quite different between them. In the system without inversion symmetry or with inversion symmetry but having large hybridization, it is not clear which of the ED and EQ excitations contributes more dominantly to the pre-edge structure. In order to distinguish them, the angle-dependence of X-ray absorption has so far been used. However, the angle-dependence of RIXS gives more precise information than the X-ray absorption, including the information by the X-ray absorption.

MAGNETIC CIRCULAR DICHROISM IN RIXS

Let us consider the situation shown in Figure 3. The magnetization of a ferromagnetic thin-film sample is parallel to the sample surface and the angle between the incident X-ray (emitted X-ray) and the magnetization is θ_1 (θ_2). The MCD in RIXS (referred to as MCD-RIXS) is defined by the difference of RIXS for incident photons with + and - helicities, where the helicity of the emitted photon is not analyzed.

Before calculating MCD-RIXS, we decompose the expression (1) of RIXS spectrum as follows:

$$F(\Omega, \omega) = \sum_{T'} \sum_f \sum_m \left| \frac{< f|T'|m ><m|T|g >}{E_g + \Omega - E_m + i\gamma} \right|^2 \frac{\gamma'/\pi}{(E_g + \Omega - E_f - \omega)^2 + i\gamma'^2}$$

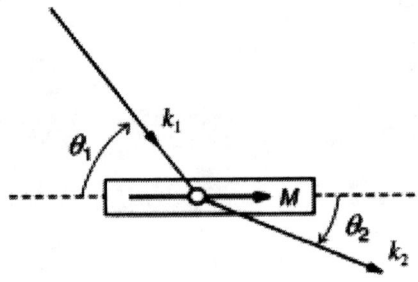

FIGURE 3. Geometrical alignment of MCD-RIXS.

$$+ \sum_{T'} \sum_{f} \sum_{m,m'} \left\{ \frac{< f|T'|m >< m|T|g > (< f|T'|m' >< m'|T|g >)^*}{(E_g + \Omega - E_m + i\gamma)(E_g + \Omega - E_{m'} - i\gamma)} + \text{c.c.} \right\}$$

$$\times \frac{\gamma'/\pi}{(E_g + \Omega - E_f - \omega)^2 + \gamma'^2}, \tag{3}$$

where the first and the second terms of Eq.(3) are the diagonal and cross terms in the expansion of $|\dots|^2$ in Eq.(1). Therefore, the second term represents the quantum mechanical interference effect in RIXS, and is called the interference term.

Until very recently, the role of the interference process in RIXS was not well understood, but very recently it has been revealed theoretically that the interference process plays an essential role in magnetic circular dichroism (MCD) of RIXS in a special geometry for ferromagnetic materials, and this finding has been confirmed by experimental observations [7].

With the atomic model, we have calculated the dependence of MCD-RIXS on θ_1 and θ_2 by assuming the ED transition both for excitation and de-excitation processes. By group theoretical consideration, we obtain the analytic expression within atomic model. The result of MCD-RIXS, ΔF, is given by

$$\Delta F = \sum_{f} \left\{ -\frac{1}{2} \cos\theta_1 \left[(1 + \cos^2\theta_2) \left(|f_{+1,+1}|^2 + |f_{-1,+1}|^2 - |f_{-1,-1}|^2 - |f_{+1,-1}|^2 \right) \right. \right.$$

$$+ 2\sin^2\theta_2 \left(|f_{0,+1}|^2 - |f_{0,-1}|^2 \right) \Big]$$

$$- \frac{1}{4} \sin\theta_1 \sin 2\theta_2 \left[(f_{-1,0}^* f_{0,-1} + f_{0,-1}^* f_{-1,0}) - (f_{+1,0}^* f_{0,+1} + f_{0,+1}^* f_{+1,0}) \right.$$

$$\left. \left. + (f_{-1,+1}^* f_{0,0} + f_{0,0}^* f_{-1,+1}) - (f_{+1,-1}^* f_{0,0} + f_{0,0}^* f_{+1,-1}) \right] \right\}$$

$$\times \frac{\gamma'/\pi}{(E_g + \Omega - E_f - \omega)^2 + \gamma'^2}, \tag{4}$$

where f_{q_2,q_1} is an angle-independent function defined by

$$f_{q_2,q_1} \equiv \sum_{m} \frac{< f|rC_{q_2}^{(1)}|m >< m|rC_{q_1}^{(1)}|g >}{E_g + \Omega - E_m + i\gamma} \tag{5}$$

with the spherical tensor operator of the ED transition $rC_q^{(1)}$.

This result indicates remarkable facts: The intensity of the diagonal term is proportional to $\cos\theta_1$, while that of the interference term to $\sin\theta_1$. Therefore, when the incident photon direction is parallel to the magnetization (denoted by the longitudinal geometry), we have the 100% diagonal term contribution to MCD-RIXS, while for the incident X-ray perpendicular to the magnetization (denoted by the transverse geometry) we have the 100% interference term contribution. This result is surprising, because the interference effect in RIXS is often considered to be negligibly small. Another remarkable point of the calculated results is that the θ_2 dependence of MCD-RIXS is given simply by $\sin 2\theta_2$.

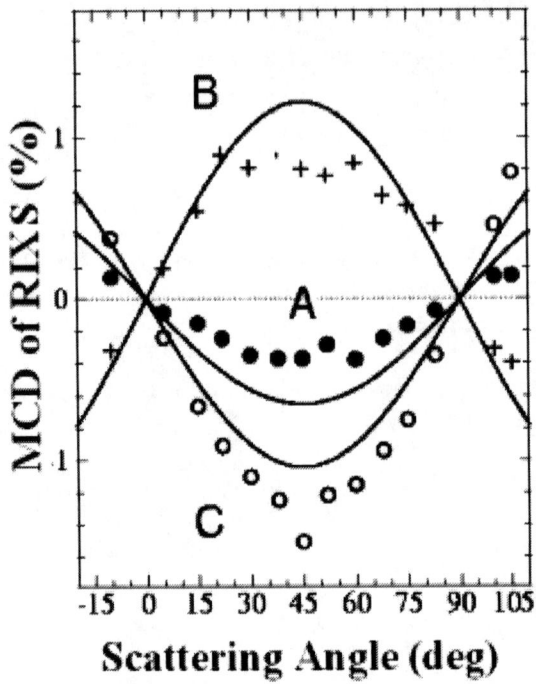

FIGURE 4. The θ_2 dependence of MCD-RIXS in the transverse geometry. Experimental results for three different emitted photon energies A (closed circle), B (cross) and C (open circle) for Gd-Co amorphous alloy is compared with the calculated ones (solid curves).

Experimental observations of MCD-RIXS for longitudinal and transverse geometries have been made by Iwazumi *et al.* for the Gd $2p_{3/2}$-$5d$ excitation and the $3d$-$2p_{3/2}$ de-excitation of a Gd-Co amorphous alloy sample [7]. The observed results are in good agreement with the calculated ones both in the spectral shape and the relative intensity of the spectra in the longitudinal and transverse geometries. Furthermore, the θ_2 dependence of MCD-RIXS in the transverse geometry has also been measured and compared with the calculated result. The result is shown in Figure 4, where A(●), B(+) and C(○) are intensities of an MCD-RIXS peak by the $3d_{3/2}$-$2p_{3/2}$ transition, and of two MCD-RIXS peaks by the $3d_{5/2}$-$2p_{3/2}$ transition, respectively, and they are displayed as a function of the angle θ_2 (actually, as a function of the scattering angle $\theta_2+90°$). The theoretical result is shown with the solid curves, and is found to be in reasonable agreement with the experimental result. This is the evidence that almost pure interference contribution has been observed in MCD-RIXS experiments in the transverse geometry.

Here we should mention that our expression of Eq.(4) is quite general (within the atomic model) and applicable to any atomic species and any core-level transitions. So far, experimental observations, as shown in Figure 4, have been made only for Gd, but it is desirable to confirm the theoretical angle-dependence for other magnetic elements and other core-level transitions. Another interesting challenge of experimental observations will be to detect MCD-RIXS by EQ excitation. As a pre-edge structure in the Mn K threshold of MnO, in the case of rare earth systems, a $2p$-$4f$ EQ excitation coexists with the $2p$-$5d$ ED excitation. Of cource, the intensity of the EQ excitation is much smaller than that of the ED excitation, the measuremennts of MCD-RIXS by the EQ excitation will be extremely difficult. However, with super brilliant synchrotron radiation souces such measurements might be possible in the future. Fukui *et al.* [7] have also given a theoretical expression of angle-dependence of MCD-RIXS by EQ excitation and ED de-excitation, where the expression is somewhat different from that of Eq.(4).

Finally, we would like to point out that a similar angle-dependence will be observed in MCD of resonant X-ray two-photon absorption spectroscopy (RXTAS) instead of RIXS. In MCD-RXTAS, two photons are incident on a ferromagnetic sample and resonantly absorbed, and then the dependence of MCD-RXTAS on the incident angles of the two photons is studied. RXTAS is also represented by a coherent second-order optical formula somewhat similar to Eq.(1), where the X-ray emission transition in Eq.(1) is replaced by an X-ray absorption transition of the second incident photon. One of the advantages of MCD-RXTAS over the MCD-RIXS is that the circular polarizations of both incident photons can be controlled, whereas in MCD-RIXS the polarization of the emitted photon is very difficult to be observed experimentally. Very recently, the angle-dependences of MCD-RXTAS have also been calculated theoretically by Fukui and Kotani [8], and it would be desirable to observe them in future experiments.

ACKNOWLEDGMENTS

The author would like to thank Dr. M. Taguchi, Dr. K. Fukui, Dr. H. Ogasawara, Dr. M. Matsubara, Dr. T. Uozumi, Professor K. Okada, Dr. T. Iwazumi, Dr. H. Shoji, Dr. T. Nakamura, Dr. E. Hirai, Professor S. Nanao, Dr. Y. Isozumi and Professor S. Shin for collaborations and discussions.

REFERENCES

1. Kotani, A., and Shin, S., *Rev. Mod. Phys.* **73**, 203 (2001).
2. Taguchi, M., Parlebas, J. C., Uozumi, T., Kotani, A., and Kao, C.-C., *Phys. Rev. B* **61**, 2553 (2000).
3. Matsubara, M., Uozumi, T., Kotani, A., Harada, Y., and Shin, S., *J. Phys. Soc. Jpn.* **69**, 1558 (2000).
4. Nakazawa, M., Ogasawara, H., and Kotani, A., *J. Phys. Soc. Jpn.* **69**, 4071 (2000).

5. Shoji, H., Hirai, E., Nanao, S., Taguchi, M., Iwazumi, T., Kotani, A., and Isozumi, Y., preprint.

6. Taguchi, M., Uozumi, T., and Kotani, A.,*J. Phys. Soc. Jpn.*, **66**, 247 (1997).

7. Fukui, K., Ogasawara, H., Kotani, A., Iwazumi. T., Shoji, H., and Nakamura, T., *J. Phys. Soc. Jpn.* **70**, 3457 (2001).

8. Fukui, K., and Kotani, A., in preparation.

VII. X-RAY APPLICATIONS
TO SOLIDS AND SURFACES

EXAFS studies of local thermal expansion

S. a Beccara*, G. Dalba*, P. Fornasini*, R. Grisenti*, A. Sanson*,
F. Rocca[†], J. Purans** and D. Diop[‡]

*INFM and Dipartimento di Fisica, Università di Trento, I-38050 Povo (Trento), Italy
[†]IFN, Istituto di Fotonica e nanotecnologie del CNR, I-38050 Povo (Trento), Italy
**Institute of Solid State Physics, Latvia University, Riga, Latvia
[‡]Physics Department, University of Dakar, Senegal

Abstract. Original information on local thermal expansion can be obtained through a cumulant analysis of EXAFS. The difference between first and third EXAFS cumulants, and the comparison with Bragg diffraction results, can help in disentangling the contributions to thermal expansion of potential anharmonicity and geometrical effects. In germanium, the perpendicular Mean Square Relative Displacement has been obtained from EXAFS. In Ag_2O, whose framework structure exhibits negative thermal expansion, a positive expansion of the Ag–O bond has been measured and the deformation of the Ag_4O structural units monitored.

INTRODUCTION

Thermal expansion in crystals is basically due to the anharmonicity of the crystal potential. Geometrical effects, independent of anharmonicity, can also be relevant, for example in accounting for the difference between average interatomic distance and distance between average atomic positions [1]. As a further example, the Negative Thermal Expansion (NTE) observed in some framework structures, like ZrW_2O_8, is often attributed to a network folding induced by low-frequency Rigid Unit Modes (RUM) [2, 3]. EXAFS oscillations are directly related to inter-atomic distance, so that temperature-dependent EXAFS can give peculiar information on local thermal expansion, in particular contributing to disentangle anharmonicity and geometrical effects.

Thermal vibrations spread atomic positions into three-dimensional distributions, and EXAFS spectra result from one-dimensional distributions of distances. In case of moderate disorder, a careful analysis of EXAFS allows to recover the basic properties of the distance distributions, parametrised in terms of lowest-order cumulants [4, 5]. The first cumulant directly measures the average distance; its comparison with the distance between average positions, measured by Bragg diffraction, gives original information on the correlation of vibrational motion. The third cumulant measures the asymmetry of the distribution, and is directly connected to the anharmonicity of the one-dimensional effective pair potential. The difference between thermal expansions measured by the first and the third cumulant in several systems has been explained in terms of a rigid shift of the effective pair potential [6, 7, 8]. A complete understanding of the relations between EXAFS cumulants and structural and dynamical properties of crystals, as well as of the relation between effective pair potential and crystal potential, is however still lacking.

This paper is a contribution to the advancement in these topics. A short account will be

CP652, *X-Ray and Inner-Shell Processes: 19th International Conference on X-Ray and Inner-Shell Processes*
edited by A. Bianconi, A. Marcelli, and N. L. Saini
© 2003 American Institute of Physics 0-7354-0111-X/03/$20.00

given of cumulants and their use for thermal expansion studies, updating and extending previous treatments [1, 5]. The still rather unexplored possibilities offered by EXAFS in this field will be illustrated by two experimental examples. In the case of the first shell of germanium, the Mean Square Relative Displacement (MSRD) perpendicular to the bond direction was for the first time obtained from EXAFS spectra [7]. In the case of Ag_2O, which is characterised by a NTE of the lattice parameter from 10 to 450 K, the average Ag-O nearest-neighbours distance expands upon heating, while the Ag-Ag next-nearest-neighbours distance contracts. The comparison between 1st and 3rd cumulant allows to monitor the distortion of the basic Ag_4O tetrahedral units, giving original insight on the local behaviour of a NTE solid [9].

THEORY

Let us consider the relations connecting the leading cumulants C_i^* of the one-dimensional distribution of distances $\rho(r,T)$ to the three-dimensional structural and dynamical properties of crystals. It is convenient to introduce the inter-atomic distance R_0 for an ideal classical state of absolute rest, and to decompose the instantaneous relative displacement $\Delta \vec{u}$ due to thermal motion into its projections $\Delta u_\parallel$ and $\Delta u_\perp$ along the bond direction and in the perpendicular plane, respectively. The instantaneous bond distance is then, to first approximation [1],

$$r = \simeq R_0 + \Delta u_\parallel + \Delta u_\perp^2 / 2R_0. \tag{1}$$

The distribution $\rho(r,T)$ can be connected to an effective pair potential V_e. In the classical approximation

$$\rho(r,T) = \exp[-\beta V_e(r)]\left\{\int \exp[-\beta V_e(r)]\, dr\right\}^{-1}. \tag{2}$$

In quasi-harmonic approximation, the parallel relative displacement in Eq. (1) can be decomposed as $\Delta u_\parallel = a + (\Delta u_\parallel)_h$, where a is the thermal expansion due to the anharmonicity of the effective potential V_e, while $(\Delta u_\parallel)_h$ is a purely harmonic contribution.

The first EXAFS cumulant is the average distance, $C_1^* = \langle r \rangle$. We want now to clarify its relation with both the anharmonic thermal expansion a and the crystallographic distance R_c measured by Bragg diffraction. Let us consider two limiting cases. For a purely *translational* relative motion, the three-dimensional harmonic distribution of relative distances is an ellipsoid, so that $\langle \Delta u_\parallel \rangle_h = 0$, and

$$C_1^* \simeq R_0 + a + \langle \Delta u_\perp^2 \rangle / 2R_0; \qquad R_c = R_0 + a. \tag{3}$$

For a purely *librational* relative motion, the ellipsoid is curved along a spherical surface, so that $\langle \Delta u_\parallel \rangle_h \simeq -\langle \Delta u_\perp^2 \rangle / 2R_0$, and

$$C_1^* = R_0 + a; \qquad R_c \simeq R_0 + a - \langle \Delta u_\perp^2 \rangle / 2R_0. \tag{4}$$

Both a and $\langle \Delta u_\perp^2 \rangle$ have finite values at zero kelvin, due to the zero point motion. Relative translational motion exhaustively describes the behaviour of crystals with simple, highly symmetrical structures. Relative librational motion characterizes the dynamics of rigid units in molecular crystals and framework structures [10]. In both cases of translation and libration, the average distance measured by EXAFS is larger than the distance between average positions measured by Bragg diffraction, due to the effect of perpendicular MSRD:

$$C_1^* \simeq R_c + \langle \Delta u_\perp^2 \rangle / 2R_0. \tag{5}$$

The 2nd cumulant C_2^* measures the width of the distribution $\rho(r, T)$ and depends on the 2nd order force constant k_0 of the effective potential V_e; to a good approximation, $C_2^* \simeq \langle \Delta u_\parallel^2 \rangle$. The third cumulant measures the asymmetry of $\rho(r, T)$, and depends on the 3rd order force constant k_3 of V_e. Frenkel and Rehr [11] have shown, on the basis of a perturbative quantum approach in quasi-harmonic approximation, that the quantity a appearing in Eqs. (3) and (4) can be expressed as

$$u = -3k_3 C_2^* / k_0, \tag{6}$$

where k_0 and k_3 can be obtained by fitting suitable models to the temperature dependence of the 2nd and 3rd EXAFS cumulants.

EXAFS is sensitive to an effective distribution of distances [4]. The difference between cumulants C_i^* and C_i of the real and effective distributions, respectively, is always relevant for the first one,

$$C_1^* \simeq C_1 + (2C_2/C_1)(1 + C_1/\lambda) \tag{7}$$

where λ is the photoelectron mean free path [12]. For higher order cumulants the difference is generally not considered, although in some cases it could be not negligible [5]. Actually, only a finite number of polynomial coefficients $\tilde{C}_i$ can be derived from EXAFS analysis. The problem of the accuracy by which the polynomial coefficients $\tilde{C}_n$ approximate the cumulants C_n has been addressed elsewhere [5, 13]. We only remember here that accurate relative values of interatomic distances, as required by thermal expansion studies, can be obtained by the ratio method, when multiple scattering effects are negligible.

GERMANIUM

In germanium only translational relative motion is expected between nearest-neighbours, and Eq. (3) should hold. The results of the 1st-shell EXAFS analysis [7] are summarised in Fig. 1. The crystallographic thermal expansion δR_c [14] is represented by the continuous line. The temperature dependence of the quantity a, calculated from experimental EXAFS data according to Eq. (6), is shown as crossed squares. The value of a at zero kelvin, amounting to about 0.003 Å, has been here subtracted. The good agreement between δa and the crystallographic thermal expansion δR_c suggests that the anharmonicity of the effective potential V_e reflects the anharmonicity of the

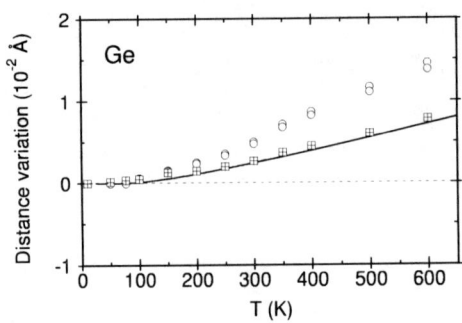

FIGURE 1. Temperature dependence of the first EXAFS cumulants C_1^* for the 1st-shell of germanium (open circles): upper and lower values correspond to a mean free path λ of 6 or 12 Å, respectively. The crossed squares refer to the anharmonicity parameter a, defined in the text and determined from the third EXAFS cumulants.

crystal potential. The temperature dependence of the first EXAFS cumulant C_1^* is shown as open circles. Slightly different values are obtained by assuming different λ values in Eq. (5). The difference between the EXAFS δC_1^* values and the crystallographic δR_c values is due, according to Eq. (3), to the perpendicular MSRD. This difference was exploited in Ref. [7] to calculate $\langle \Delta u_\perp^2 \rangle$ as a function of temperature. The values of $\langle \Delta u_\perp^2 \rangle$ so obtained have been recently reproduced by D. Strauch *et al.* through ab initio dynamical calculations [15]. The ratio $\gamma = \langle \Delta u_\perp^2 \rangle / \langle \Delta u_\parallel^2 \rangle$, which is about 6 at high temperatures, is in good agreement also with the calculations made for silicon using an adiabatic bond charge model [16]. The results obtained for germanium are consistent with the following phenomenological picture. According to Eq. (3), the thermal expansion δC_1^* of the average distance is the sum of two effects: one, measured by δa, depends on the asymmetric shape, assumed temperature independent, of the potential V_e; the other depends on the increase of the perpendicular MSRD, and is connected with a positive rigid shift of V_e with temperature. This picture has been confirmed by simulations: the relative motion perpendicular to the bond induces a shift of the effective potential, without significantly affecting its shape [5].

SILVER OXIDE

Silver oxide Ag_2O shares with Cu_2O the *cuprite* structure. Each Ag atom is linearly coordinated to two O atoms, and each O atom is tetrahedrally coordinated to four Ag atoms. The structure can also be described as a framework of two interpenetrating networks of corner-sharing Ag_4O tetrahedra [17]. NTE of the lattice parameter extending over large temperature ranges is frequently observed in framework structures [2], and has been measured in Ag_2O from 10 to 450 K [18]. The thermal expansion of framework structures results from the competition between a positive contribution of the potential anharmonicity and a negative contribution which is often attributed to a geometrical effect of RUMs [3]. To test RUM theories, it is important to directly measure the thermal

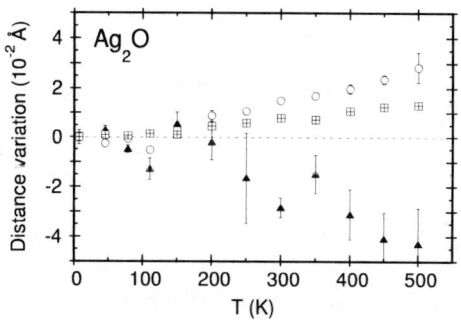

FIGURE 2. Temperature dependence of the first EXAFS cumulants C_1^* for the 1st-shell Ag–O (open circles) and 2nd-shell Ag–Ag (triangles). The crossed squares refer to the 1st-shell anharmonicity parameter a, defined in the text and determined from the third EXAFS cumulants.

expansion of nearest-neighbour bonds and the degree of rigidity of corner-sharing basic units. To this aim, EXAFS measurements have been performed at the K edge of silver in Ag_2O, in the temperature range from 7 to 500 K, at the Gilda beamline (ESRF, Grenoble). The main results are summarized in Fig. 2. The next-nearest-neighbours Ag-Ag distance decreases with temperature, in qualitative agreement with the lattice parameter NTE. The nearest-neigbours Ag-O distance on the contrary increases. This behaviour is consistent with the positive thermal expansion of the rather rigid nearest-neighbour bond with oxygen found in other framework structures affected by lattice parameter NTE [19].

Let us now focalise our attention on the first-shell Ag-O distance. A direct comparison of the average distance δC_1^* with the distance between average atomic positions δR_c cannot be done in this case, since no experimental data from diffraction are as yet available. The presence of two very different atomic species and the rather complex framework structure prevents from simply scaling the lattice parameter thermal expansion to the first-shell distance, as was done for germanium. It is then impossible to obtain $\langle \Delta u_\perp^2 \rangle$ by inversion of Eq. (5). Anyway, further information can be obtained by calculating the quantity a from the experimental 2nd and 3rd cumulants, according to Eq. (6). The resulting δa values are shown in Fig. 2 as crossed squares. The contribution of zero point vibrations, here subtracted, amounted to about 0.01 Å. According to Eq. (4), for a purely librational motion, as would be expected for perfectly rigid tetrahedral Ag_4O units, $\delta C_1^* = \delta a$. The discrepancy between δC_1^* and δa indicates that the nearest-neighbour relative motion is not purely librational, but contains a significant translational contribution. By properly inverting Eq. (3), one can recover the values $\langle \Delta u_\perp^2 \rangle_{tr}$, where "tr" indicates the translational contribution to the total perpendicular MSRD. The ratio $\gamma = \langle \Delta u_\perp^2 \rangle_{tr} / \langle \Delta u_\parallel^2 \rangle$ is about 9 at high temperatures. This high value indicates that the Ag-O bond, which is rather stiff with respect to stretching, is much looser with respect to bending. This result is consistent with the high values found for the 2nd-shell parallel MSRD, and with diffraction mesaurements on Cu_2O [20]. The low-frequency modes responsible for the bond bending monitored by EXAFS cannot be identified as RUMs; the translational character of the perpendicular MSRD is instead a clear indication of a distortion of the Ag_4O tetrahedra.

CONCLUSIONS

Temperature dependent EXAFS measurements allow to directly measure the thermal expansion of nearest-neighbours bonds. The comparison between first and third cumulants and with Bragg diffraction results gives further information on the geometrical contribution to thermal expansion. These properties can be exploited to gain a deeper insight on local dynamical properties, in particular on correlation of vibrational motion. In the case of NTE framework structures, one can also monitor the degree of distortion of the basic structural units.

A generalisation of the results presented in this paper is however far from trivial, and requires further experimental work on different systems. As a matter of fact, other more subtle phenomena can influence EXAFS. For example, evidence has been found in AgI and CuBr of a negative shift of the effective pair potential at increasing temperature [6, 21]; the shift has been tentatively attributed to cluster-induced local distortions [8]. Besides, an accurate treatment of coordination shells beyond the first one is still difficult.

ACKNOWLEDGMENTS

One of the authors (D.D.) is grateful to ICTP (Trieste) for support within the program "Training and Research in Italian Laboratories".

REFERENCES

1. Fornasini, P., *J. Phys.: Condens. Matter*, **13**, 7859–7872 (2001).
2. Evans, J. S. O., *J. Chem. Soc. Dalton Trans.*, **19**, 3317–3326 (1999).
3. Heine, V., Welche, P. R. L., and Dove, M. T., *J. Am. Ceram. Soc.*, **82**, 1793–1802 (1999).
4. Bunker, G., *Nucl. Instrum. Methods Phys. Res.*, **207**, 437–444 (1983).
5. Fornasini, P., Monti, F., and Sanson, A., *J. Synchrotron Radiat.*, **8**, 1214–1220 (2001).
6. Dalba, G., Fornasini, P., and Rocca, F., *Phys. Rev. B*, **52**, 149–157 (1995).
7. Dalba, G., Fornasini, P., Grisenti, R., and Purans, J., *Phys. Rev. Lett.*, **82**, 4240–4243 (1999).
8. Ishii, T., *J. Phys. Soc. Jpn.*, **70**, 159–166 (2001).
9. a Beccara, S., Dalba, G., Fornasini, P., Grisenti, R., Sanson, A., and Rocca, F., *Phys. Rev. Lett.*, **89**, 25503 (2002).
10. Willis, B. T. M., and Pryor, A. W., *Thermal Vibrations in Crystallography*, Cambridge University Press, 1975.
11. Frenkel, A. I., and Rehr, J. J., *Phys. Rev. B*, **48**, 585–588 (1993).
12. Freund, J., Ingalls, R., and Crozier, E. D., *Phys. Rev. B*, **39**, 12537–12547 (1989).
13. Dalba, G., Fornasini, P., and Rocca, F., *Phys. Rev. B*, **47**, 8502–8514 (1993).
14. Touloukian, Y. S., Kirby, R. K., Taylor, R. E., and Desai, P. D., *Thermophysical Properties of matter*, vol. 13, Plenum, New York, 1977.
15. Birner, G., Strauch, D., and Pavone, P. (2001), private commun.
16. Nielsen, O. H., and Weber, W., *J. Phys. C: Solid St. Phys.*, **13**, 2449–2460 (1980).
17. Zuo, J., Kim, M., O'Keeffe, M., and Spence, J., *Nature (London)*, **401**, 49–52 (1999).
18. Artioli, G., and Dapiaggi, M. (2002), private commun.
19. Tucker, M. G., Dove, M. T., and Keen, D. A., *J. Phys. Condens. Matter*, **12**, L425–431 (2000).
20. Lippman, T., and Schneider, J. R., *J. Appl. Crystallogr.*, **33**, 156–167 (2000).
21. Kamishima, O., Ishii, T., Maeda, H., and Hashino, S., *Solid St. Commun.*, **103**, 141–144 (1997).

X-ray absorption spectroscopy study of nanotubes

S. Bellucci[1*], S. Botti[2], A. Marcelli[1], K. Ibrahim and Z.Y. Wu[3,1,**]

[1]Laboratori Nazionali di Frascati, Istituto Nazionale di Fisica Nucleare,
Via Enrico Fermi 40, I-00044 Frascati, Italy
[2]ENEA, Divisione Fisica Applicata, P.O. Box 65, 00044 Frascati, Italy
[3]Beijing Synchrotron Radiation Facility, Institute of High Energy Physics,
Chinese Academy of Sciences, P.O Box 918, 100039 Beijing, P.R. China
*) Email: bellucci@lnf.infn.it
**) Email: wuzy@lhep.ac.cn

Abstract. X-ray absorption spectroscopy at the C K edge has been performed on carbon nanotubes to demonstrate the capability of the technique to investigate the electronic structure of these systems. We compare and discuss the experimental results achieved on both graphite and nanotubes trying also to determine the relationship between multiple scattering structures and different topology and surface structure of the samples.

INTRODUCTION

Nanomaterials are a new class of systems widely studied, characterized by a large surface to volume ratio and consequently a large surface energy, which affects the physical and chemical properties with respect to bulk counterparts.
Among nanomaterials in these last years attention has been focused on nanotubes: long tubular structures with diameter of the tube ranging from a few nanometers to few tens of nanometers. The length of the structures could be hundreds of nanometers or even several micrometers. These latter can be actually considered one dimensional structures, but various other forms of nanotubes also exist like nanoropes or nanowires, nanocoils etc. The exceptional physical properties of nanotubes make them unique candidates to fabricate the strongest lightweight fibers, as well as the smallest metallic wires known. This research gained a lot of attention and importance in the recent past mainly because of its potential applications in diverse fields as catalysis, energy storage, gas storage tanks, electron emitters, gas and chemical sensors, fast acting switches, molecular electronics, etc.
Although nano-systems and in particular nanotubes can be made of different materials like carbon, boron carbide, tungsten sulphide, molybdenum sulphide, the most widely studied are carbon nanotubes [1]. This is because carbon in itself has wide ranging

CP652, X-Ray and Inner-Shell Processes: 19th International Conference on X-Ray and Inner-Shell Processes
edited by A. Bianconi, A. Marcelli, and N. L. Saini
© 2003 American Institute of Physics 0-7354-0111-X/03/$20.00

applications and is biocompatible. In particular, carbon nanotubes stick out in the field of nanostructures, owing to their exceptional mechanical, capillarity, electronic transport and superconducting properties [2-4]. They are made purely of carbon atoms on a single sheet of graphite wrapped around a cylindrical axis. The same material with slight variations in the geometrical arrangement yields different electronic properties, ranging from metallic to semiconducting behavior. The very same geometry of the nanotube determines also its mechanical properties. Moreover, nanotube channeling is just emerging as a particle beam instrument, replacing the well-established technique of crystal channeling: provided nanotubes can efficiently channel and deflect particle beams, they offer an interesting opportunity to make clean beams of potentially very small size, down to 1 square nanometer if needed [5-11].

Within the realm of nanotubes there are two main sets, i.e. single walled and multiple walled nanotubes, e.g. systems where many tubes of smaller diameter are present inside bigger nanotubes. Because of their very small size, they can be observed only by a powerful electron microscope, e.g., a Scanning Tunneling Microscope or a Transmission Electron Microscope.

Nanotubes are predominantly prepared by cathodic or plasma arc deposition methods and other chemical methods. Among the different methods to fabricate carbon nanotubes there is a technique using as solid precursor laser-synthesized nanoparticles, without requiring metal catalyst [12]. By following this procedure it is possible to obtain a dense array of single walled nanotubes (SWNTs) with 1.1 ± 0.3 nm average size. They are strictly interlaced in bundles, maintaining a constant rope diameter 60 ± 30 nm, over the whole length.

To investigate and to understand the properties of such materials, which of course are affected by the different manufacture procedures, we considered a non destructive local structural probe, which is also sensitive to the local electronic properties: the x-ray absorption spectroscopy.

X-ray absorption spectroscopy (XAS) is a powerful local technique of analysis based on the excitation of electronic transition from an inner level to outer unoccupied states. The photoabsorption process yields information on the local electronic structure and coordination environment around the absorbing atom. XAS techniques such as EXAFS (Extended x-ray absorption fine structure) and XANES (X-ray absorption near edge structure) have been used extensively to investigate the bulk local structure but also the surface local structure, as well as bonding formation of a variety of solid-state materials [13-15].

In this manuscript we present data on nanotubes obtained by solid precursors deposited by impaction onto a Si(100) substrate, for film formation. After deposition, the substrate is resistively heated up to 950° C. For the two samples presented here the deposition time was 30 minutes (sample A) and 60 minutes (sample B), and for both well aligned carbon nanotubes, oriented perpendicularly to the substrate surface and homogeneously distributed, over an area of 50 mm x 50 mm were detected. This method does not use any kind of solvents in the nanotube manufacture.

From SEM images, we estimated a bundle diameter of 60 ± 30 nm, whereas a length of about 3 ± 1 μm can be only roughly evaluated because the bundle extremities are not easy to be detected by electron microscope. By increasing the deposition time up to

60 min, fibers with a larger diameter, laid down on the substrate with long noodle shapes are observed. Most of the carbon nanowires are smoothly curved with some short straight sections while some of them possess many kinks and bends. Due to the high nano-fiber density and curved disordered structure, we could not identify the wire ends.

The experimental set-up for the investigation of absorption spectra at the C K-edge of nanotubes is described in details in ref. 16. Samples were loaded in an UHV chamber and maintained in a background pressure of $\sim 8 \times 10^{-10}$ torr, reaching a pressure up to $\sim 1 \times 10^{-9}$ torr during the data acquisition. The spectra were recorded using the total electron yield (TEY) mode, a surface-sensitive detection method with a typical probing depth of a few nm. The photon energies used to record the C K-edge absorption spectrum range from 275 to 320 eV, and the experimental resolution at these photon energies was about 0.3 eV. In molecules the C 1s core-hole lifetime is around 0.1 eV and goes towards 0.2 eV or more in a bulk system such as graphite. If we look at the FWHM of the first peak in the XANES spectra in Fig. 1, it appears significantly larger than 0.3 eV, so that a resolution of ~ 0.3 eV is consistent with the observed features in solid systems such graphite and nanotubes are. All spectra have been normalized using the standard procedure of the XANES spectroscopy at low energy. Calibration for energy was made using graphite. Spectra were recorded at steps of 0.1 eV. The recorded spectra were fitted with a linear function to account for the base line, and normalized to one at high energy some 40/50 eV above the threshold. Spectra of nanotubes have been compared with graphite, a well known carbon based layered material. As a consequence, the XAS data have to exhibit an angular behaviour. We performed all experiments using synchrotron radiation with a fixed geometry and the

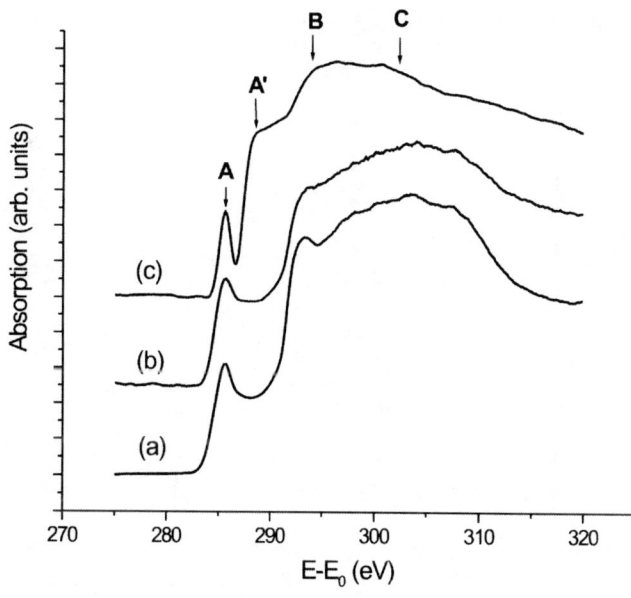

FIGURE 1. Total electron yield spectra at the C K edge of a graphite sample (a), vs. two different nanotube samples: A (curve c) and B (curve b).

light was hitting the surface sample at an angle of about 30° from the normal to the surface of the sample. Experiments performed in this condition, e.g., almost at the magic angle, would not be affected by angular dependence.

In Fig.1 we compare the C K-edge XANES of sample A and B and the spectrum of graphite. This latter spectrum exhibits a reasonable agreement with previous published data [17-20]. If we look at the C K edge of graphite we may characterize this spectrum as composed by three main features, i.e., an almost atomic contribution at about 286 eV, a peak at about 293 eV and a wide multiple scattering (MS) resonance centered at about 305 eV, labeled respectively, A, B and C. The resonance at about 286 eV is attributed to a transition from C 1 s core level to π states of locally unsaturated carbon bonds. Comparing the curves in Fig. 1 we have to notice the presence of an additional spectral feature (A') in the sample A in the region of 287-289 eV between the A and B peaks. This feature A' may be recognized as a σ^* resonance characteristic of C-H bonds in amorphous carbon [20-23] and may be attributed to C-H bonding present in sample A, assigned to a certain amount of amorphous carbon, i.e. the pristine particles. Actually, previous studies demonstrated that laser-synthesized particles retain hydrogen from synthesis process, due to the incomplete hydrocarbon decomposition [12].

The presence of surface C-H σ bonds may be verified qualitatively comparing XANES experiments recorded after different treating conditions. Figure 2 shows the C K-edge XANES spectra of sample A without treatment (c), after heating at 350° C for 4 hrs in vacuum (b) and after a further Ar bombardment at 150 V and 20 mA for 1 hr followed by a gentle baking for 1 hr (a). While the above treatments do not affect the A feature, large changes may be observed in the region of peak A' associated to C-H bonds. As expected, the C-H σ^* intensity decreases going from (c) to (a) showing that the surface is gradually cleaned by the presence of hydrogen. This effect is particularly

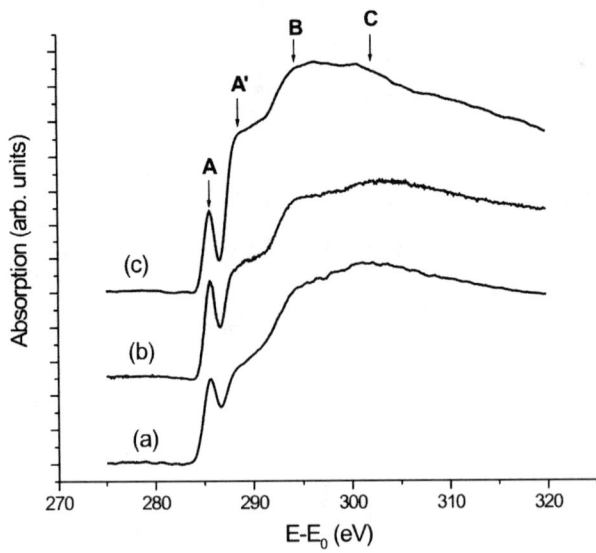

FIGURE 2. Comparison of the C K-edge XANES spectra of sample A measured after different treating processes: (c) without any treatment, (b) after heating at 350° C for 4 hrs and (a) a further Ar bombardment at 150 V and 20 mA for 1 hr and an additional baking of 1 hr.

strong in the first phase of the treatment of the sample, i.e., an heating process at moderate temperature for several hours in vacuum.

If we consider now the C K-edge XANES spectrum of sample B (shown in Fig. 1) and that of graphite, the intensity of the peak A does not change significantly compared to graphite, indicating that the structure just above the Fermi level is mainly determined by a molecular interaction between the central atom and its nearest neighbor coordination shell (short-range effects). It is worth noticing that XANES of sample B does not exhibit a complex pre-edge structure, which addresses also a reasonable ordering of the carbon matrix. However, a difference between the spectra of sample B and graphite can be recognized looking at the 289-320 eV region: the transition $1s \rightarrow \sigma^*$-like states is less intense and fairly broadened in the sample B of carbon nanotube.

In the XAS spectroscopy, the cross section of an absorbing atom, and in particular in the XANES region depends on the scattering of the generated photoelectron wave from neighboring atoms, thereby including information about short-range and medium-range structures around the absorber. [24] In addition to structural effects, important effects may be observed at threshold, where depending by the edge, by the symmetry and by the system, a more or less structured pre-edge may be observed. Important electronic contributions may be observed in this region at low energy edges (e.g., carbon or oxygen K edge) in particular in correlated systems like oxides or superconductors. [25] Looking at the C K edge spectra, in the language of molecular orbital theory the pre-edge feature A is generally attributed to transitions from 1s to π^* antibonding states. After the pre-edge, the threshold region may be understood and reproduced by using the multiple scattering (MS) calculations. Within the MS framework XANES simulations may show that peak B can be reproduced by taking into account the higher coordination neighbors. [26] The final state allowed by this core electron transition, is not a simple atomic or molecular state, and medium- to long-range effects may influence the threshold region. [27-29] It is evident that, in the spectra of carbon nanotube, the intensity of peak B decreases. In the framework of the MS theory this behavior implies that the "medium-range" structure around C atoms deviates from bulk. Actually, more C atoms are exposed on the surface where an anisotropic environment exists, and where a C atom bonds to both inner and outer C atoms simultaneously. Then, the differences in the C K-edge XANES spectra should arise from the outer-lying atomic shells, indicating that the long-range effects have to play an important role in determining the near-edge structures.

The nanotube behavior implies that the "long- or medium-range" atomic order deviates from graphite one (e.g. disorder and/or distortion effects). Actually, in a nanotube more C atoms are surface ones, experiencing an anisotropic geometry, where a C atom is bonded to inner and outer atoms simultaneously. The anisotropic environment of the surface C atoms can be certainly one of the causes of the distortion of the original trigonal structure. This mechanism, in turn, increases the number of distinct photoelectron scattering paths in a nanotube, yielding a broadening of the XANES spectrum: the B feature disappears and C broadens. These effects may be due to a random phase decoherence among each set of closely similar multiple scattering paths and the variation in the threshold E_0 in the absorption coefficient [30]. Such effects could be also ascribed to the fact that the tubes in the bundles are not infinitely long nor perfectly straight. However, also the compositional disorder, i.e. the presence of residual amorphous carbon, leads to the smearing of near-edge features.

CONCLUSIONS

In summary, we investigated carbon nanotubes synthesized by nanoparticles without catalyst. Different kinds of tubular nanostructures can be formed by controlling the deposition time. When the carbon nanoparticles are deposited onto the silicon substrate, the high temperature causes the transformation of amorphous hydrogenated particles in nanotubes and leaves some pristine particles (sample A). The achieved results are also consistent with the expected chemical environment of a carbon nanotube. Moreover, in a qualitative way, the data at the C K edge in both graphite and nanotubes show that structural and bonding configurations can be recognized looking at the characteristic pre-edge features but also that significant differences in the multiple-scattering region exist addressing the presence of important differences in topology and degree of order of these materials.

ACKNOWLEDGEMENTS

This work was partially supported by INFN - Gruppo V, as NANO experiment. Z.Y. Wu acknowledges the financial support of the *100-Talent Research Program* of the Chinese Academy of Sciences and of the *Outstanding Youth Fund* (10125523) and Key Important Nano-Research Project (90206032) of the National Natural Science Foundation of China.

REFERENCES

1. S. Iijima, Nature **354**, 56 (1991) *Appl. Phys. Lett.* 80, 2973 (2002); R. Saito, G. Dresselhaus, M. S. Dresselhaus, *Physical Properties of Carbon Nanotubes* (Imperial College Press, London, 1998).
2. M. Bockrath, D.H. Cobden, J. Lu, A.G. Rinzler, R.E. Smalley, L. Balents and P.L. McEuen., *Nature* **397**, 598 (1999); Z. Yao, H.W.J. Postma, L. Balents and C. Dekker, *Nature* **402**, 273 (1999).
3. See e.g. S. Bellucci and J. Gonzalez, *Eur. Phys. J. B* 18, 3 (2000); ibid. *Phys. Rev. B* **64**, 201106 (2001) (Rapid Comm.).
4. A. Yu. Kasumov, et al., *Science* 284,1508 (1999); M. Kociak, et al., *Phys. Rev. Lett.* **86**, 2416 (2001).
5. V.V. Klimov and V.S. Letokhov, *Phys. Lett. A* **222**, 424 (1996).
6. L.G. Gevorgian, K.A. Ispirian and R.K. Ispirian. *JETP Lett.* **66**, 322 (1997).
7. N.K. Zhevago and V.I. Glebov, *Phys. Lett. A* **250**, 360 (1998).
8. V.M. Biryukov and S. Bellucci, *Phys. Lett. B* **542**, 111 (2002), [ArXiv physics/0205023].
9. S. Bellucci, V.M. Biryukov, Yu.A. Chesnokov, V. Guidi and W. Scandale., *Nucl. Instr. and Meth. B (in press)*; [ArXiv:physics/0208081].

10. S. Bellucci and S.B. Dabagov, On X-Ray Channeling in mu- and n-capillaries, *Phys. Lett. B (submitted)*, http://arxiv.org/abs/physics/0209011

11. S. Bellucci, V.M. Biryukov, Yu.A. Chesnokov, V. Guidi and W. Scandale, Making Micro- and Nano-Beams by channeling in Micro- and Nano- Structures, *Phys. Rev. Lett. (submitted)*; [ArXiv: physics/0209057].

12. S. Botti, R. Ciardi, M.L. Terranova, S. Piccirillo, V. Sessa, M. Rossi and M. Vittori-Antisari, *Appl. Phys. Lett.* **80**, 1441 (2002); S. Botti, R. Ciardi, M.L. Terranova, S. Piccirillo, V. Sessa and M. Rossi, *Chem. Phys. Lett.* **355**, 395 (2002).

13. K. S. Hamad, R. Roth, J. Rockenberger, T. van Buuren, and A. P. Alivisatos, *Phys. Rev. Lett.* **83**, 3474 (1999).

14. J.J. Rehr and R.C. Albers, *Rev. Mod. Phys.* **72**, 621 (2000)

15. Z.Y. Wu, J. Zhang, K. Ibrahim, D. C. Xian, G. Li, Y. Tao, T.D. Hu, S. Bellucci, A. Marcelli, Q. H. Zhang, L. Gao and Z. Z. Chen, *Appl. Phys. Lett.* **80**, 2973 (2002).

16. F.Q. Liu, K. Ibrahim, H.J. Qian, Y. Yang, X. P. Tao, J. F. Jia and Y. H. Dong, *J. Elect. Spectrosc. & Rel. Phenom.* **80**, 409 (1996).

17. X.D. Weng, P. Rez and H. Ma, *Phys. Rev. B* **40**, 4175 (1989).

18. P.A. Bruhwiler, A.J. Maxwell, C. Puglia, A. Nilsson, S. Andersson and N. Martensson, *Phys. Rev. Lett.* **74**, 614 (1995).

19. J. Bruley, D.B. Williams, J.J. Cuomo and D.P. Pappas, *J. Microscopy* **180**, 22 (1995).

20. Y.H. Tang, P. Zhang, P.S. Kim, T.K. Sham, Y.F. Hu, X.H. Sun, N.B. Wong, M.K. Fung, Y.F. Zheng, C.S. Lee and S.T. Lee, *Appl. Phys. Lett.* **79**, 3773 (2001).

21. J. Stöhr, *NEXAFS Spectroscopy* (Springer, New York, 1992).

22. I. Ishii and A.P. Hitchcock, J. Electron Spectrosc. Relat. Phenom. **46**, 55 (1988).

23. D. Wesner, S. Krummacher, R. Carr, T.K. Sham, M. Strongin, W. Eberhart, S.L. Weng, G. Williams, M. Howells, F. Kampas, S. Heald and F.W. Smith, *Phys. Rev. B* **28**, 2152 (1983).

24. M. Benfatto, C.R. Natoli, A. Bianconi, J. Garcia, A. Marcelli, M. Fanfoni and I. Davoli, *Phys. Rev. B* **34**, 5774 (1986)

25. I. Davoli, M. Tomellini, A. Marcelli, A. Bianconi and M. Fanfoni, *Phys. Rev. B* **33**, 2979 (1986).

26. Z.Y. Wu, unpublished data.

27. A.V. Soldatov T. S. Ivanchenko, S. Della Longa, A. Kotani, Y. Iwamoto and A. Bianconi, *Phys. Rev. B* **50**, 5074, (1994).

28. Z.Y. Wu, G. Ouvrard, P. Moreau and C.R. Natoli, *Phys. Rev. B* **55**, 9508 (1997)

29. Z.Y. Wu, S. Gota, F. Jollet, M. Pollak, M. Gautier–Soyer and C.R. Natoli, *Phys. Rev. B* **55**, 2570 (1997)

30. F. Farges, G.E. Brown and J.J. Rehr, Phys. Rev. *B* **56**, 1809 (1997); F. Farges, G.E. Brown, A. Navrotsky, H. Gan and J.J. Rehr, *Geochim. Cosmochim. Acta* **60**, 3023 (1996).

The MXAN procedure: a new method of modeling the XANES spectra to obtain structural quantitative information

M. Benfatto[a], S. Della Longa[b] and P. D'Angelo[c]

[a]*Laboratori Nazionali di Frascati dell'INFN - C.P.13 - 00044 Frascati, ITALY*
[b]*Universita' dell'Aquila - via Vetoio, loc. Coppito II - 67100 l'Aquila, ITALY*
[c]*Dipartimento di Chimica, Universita' di Roma " La Sapienza", Piazzale A. Moro 5, 00185 Rome, ITALY*

Abstract. In this paper we present a new method, named MXAN, able to fit the XANES energy range (from the edge to about 200 eV) of experimental X-ray absorption data to obtain geometrical information. This method is based on the comparison between the experimental spectrum and several theoretical calculations generated by changing the relevant geometrical parameters of the site around the absorbing atom. The theoretical spectra are derived in the framework of full multiple scattering approach. Our procedure is able to recover the right information on the symmetry and atomic distances and the solution is found to be independent on the starting conditions.

INTRODUCTION

X-ray absorption spectroscopy (XAS) is one of the most powerful methods to obtain structural and electronic information on the local environment of an excited atom. During the past ten years, much theoretical and computational effort has been made that has led to the development of ab-initio methods for XAS calculations in arbitrary systems [1]. Over 50 eV above the rising edge an important approximation can be made which leads to the reduction of the many-body process in that of a photoelectron scattering in an effective potential. This allows a quantitative analysis of the extended x-ray-absorption fine-structure (EXAFS) region of the XAS spectra. The EXAFS technique has found many applications ranging from chemistry to molecular biology, including liquid and solid-state problems in condensed-matter physics.

The situation is different concerning the low-energy part of the x-ray absorption cross section extending up to around 50-100 eV above the threshold, the so called XANES (X-ray absorption near-edge structure) region. This part is very sensitive to the geometrical details of the absorbing site (overall symmetry, distances and bond angles), so that, in principle, an almost complete recovery of the geometrical structure

CP652, *X-Ray and Inner-Shell Processes: 19th International Conference on X-Ray and Inner-Shell Processes*
edited by A. Bianconi, A. Marcelli, and N. L. Saini
© 2003 American Institute of Physics 0-7354-0111-X/03/$20.00

within 6-7 Å from the absorbing site can be achieved from the experimental data with atomic resolution. However, the quantitative analysis of the full XAS spectrum, including the edge, is a complex many-body problem that requires an adequate treatment and the need for heavy time-consuming algorithms to calculate the absorbing cross section including all multiple-scattering (MS) contributions of any order, the so-called full MS approach. Due to these difficulties, the analysis of the pre-edge and first part of the XAS spectra (up to 50-100 eV) has been exploited so far only on a qualitative grounds, by comparison with model compounds or as an aid for EXAFS studies or more advanced investigations, such as the ones based on the analysis of contributions related to correlation functions of orders higher than two [2].

Few attempts have been made to quantify the theoretical sensitivity of the low-energy part of the spectrum to the structural parameters and a few examples of quantitative comparisons between experimental data and "ab-initio" calculations can be found in the literature and almost related to known structural compounds. This was due mainly to the lack of a fitting procedure based on the full MS approach which allows the exact calculation of the photo-absorption cross section from the edge, avoiding any "a-priori" selection of the relevant MS paths.

Recently some of us have proposed [3] and applied to several systems [4] a new method to perform a quantitative analysis of the XANES energy range, i.e. from the edge up to 200 eV. The method, called MXAN, is based on the comparison between experimental data and many theoretical calculations performed by varying selected structural parameters starting from a putative structure, i.e. from a well defined initial geometrical configuration around the absorber. The calculation of XANES spectra related to the hundreds of different geometrical configurations needed to obtain the best fit of the experimental data is done in a reasonable time and the optimization in the space of parameters is achieved by the minimization of the square residual function in the parameter space. The calculations are performed in the energy space without involving any Fourier transform algorithm; polarized spectra can be easily analyzed because the calculations are performed by the full MS approach.

In this paper we present the results obtained for some transition metals in aqueous solution, these systems are ideal test cases, with the aim to further validate the MXAN method and define its potentiality and limitation.

The MXAN procedure

The MXAN procedure uses the set of programs developed by the Frascati theory group [1,5]; in particular VGEN, a generator of muffin-tin potentials, and the CONTINUUM code for the full multiple scattering cross section calculation. The optimization in the space of the parameters is achieved using the MINUIT routines of the CERN library; a single best fit procedure takes typically 8 hours on a UNIX scalar α-VAX machine for a calculation involving 6 fitting parameters in a cluster of 35 atoms. The MINUIT routines minimize the square residual function

$$S^2 = n \frac{\sum\limits_{i=1}^{m} w_i \left[\left(y_i^{th} - y_i^{exp} \right) \varepsilon_i^{-1} \right]^2}{\sum\limits_{i=1}^{m} w_i} \qquad (1)$$

where n is the number of independent parameters, m the number of data points, y^{th}_i and y^{exp}_i the theoretical and experimental values of absorption, ε_i the individual errors in the experimental data set, and w_i is a statistical weight. For w_i=constant=1, the square residual function S^2 becomes the statistical χ^2 function. The application of the software package to several test cases shows that the best-fit solution is independent from the minimization strategy and to the starting conditions.

The MXAN method is based on the muffin-tin approximation for the shape of the potential and the use of the concept of complex optical potential, based on the local density approximation of the self-energy of the excited photoelectron. The effects of the non-MT corrections on the XANES calculation are still not well understood, nevertheless evidence exists that their influence, if present, is confined within the first 20-30 eV from the edge [3,6,7]. The application of the MXAN procedure to several test cases confirms this indication because a weak influence of the non-MT corrections to the structural determination is found [8]. It is not surprising as it is possible to demonstrate that the absorption cross section can be written as the sum of the MT calculated cross section plus other terms containing all the non-MT corrections [3]. These corrections depend on the system and go to zero as the energy increases. Their influence is strongly reduced making the fits in the energy range from the edge up to 200 eV. In this way the geometrical arrangements restrains the numerical results of the fitting procedure.

The real part of self-energy is calculated either by the X-α approximation or by using the Hedin-Lundqvist (HL) potential [1]. To avoid the over damping at low energies of the complex part of the HL potential in the case of covalent molecular systems, the MXAN method can alternatively account for all the inelastic processes by a convolution with a broadening lorentzian function having an energy dependent width of the form $\Gamma(E) = \Gamma_c + \Gamma_{mfp}(E)$. The constant part Γ_c includes the core hole lifetime and the experimental resolution, while the energy dependent term represents all the intrinsic and extrinsic inelastic processes. The $\Gamma_{mfp}(E)$ function is zero below an onset energy E_s (which in extended systems corresponds to the plasmon excitation energy) and begins to increase from a value A_s following the universal functional form of the mean free path in solids. Both the onset energy E_s and the jump A_s are introduced in the $\Gamma_{mfp}(E)$ function via an arctangent functional form to avoid discontinuities and to simulate the electron-hole pair excitations. Their numerical values are derived at each step of computation (i.e. for each geometrical configuration) on the basis of a Monte Carlo fit similarly to the procedure used in optimisation by simulated annealing [9]. This type of approach can be justified on the basis of a multi-channel multiple scattering theory [10]. In the sudden limit, the net absorption is given by a sum over all the possible excited states of the (N-1)-electron system [1,10]. By assuming that the channels coming from the excitation of the N-1 electrons are near in energy, the total absorption is given by a convolution of the one-particle spectrum, calculated with the full-relaxed potential, with a spectral function $A(\omega)$ representing the weight of the other excited states. Hence the total XAS cross-section can be written as

$$\mu = \sum_n \mu_n \xrightarrow[\Delta E \,\rangle 0]{} \int \mu(\omega - \omega')A(\omega')d\omega' \qquad (2)$$

where the "ansatz" is made that the spectral function $A(\omega)$ is well approximated by a Lorentzian function with the energy dependent width $\Gamma(E)$ previously defined. Obviously, when contributions from one or more of these excited states become relevant, they must be considered explicitly in the calculations.

It is also possible to demonstrate that this convolution procedure is equivalent to a calculation performed with a potential containing an appropriate complex part derived by reducing a multi-channels process to a single-channel one [10]. Therefore the $\Gamma(E)$ function is characterized by parameters which have a clear physical meaning and they are not free to assume any value, but are varied in a well defined interval.

The MXAN method introduces as a total four non-structural parameters, whose influence seems limited to an increase of a few percentage of the error value in the structural parameters determination.

Metal ions in solution

We show the MXAN analysis of the K-edges of Ni^{2+} and Fe^{2+} ions in aqueous solution. Both ions are often used as test cases due to the well-defined formal valency of the ionic species and the very simple geometry around the absorber. The data at the K-edges have been recorded in transmission mode using Mylar cell at beam station 7.1 of the Daresbury Laboratory. A double-crystal Si(111) monochromator was used and the storage ring was operated at 2 GeV with an average current of about 150 mA. The samples have been prepared to obtain a water solution of 5 mM and 50 mM of Ni^{2+} and Fe^{2+} ions respectively. The pH has been controlled in order to have the hydrated species. The background contribution from previous edges has been fitted with a linear function and subtracted from the raw data. The XANES spectrum of Zn^{2+} at the Zn K edge has been recorded in transmission mode using the EMBL spectrometer at DESY.

In Fig.1 we report the comparison between the experimental data and the calculations related to the best-fit structures for both ions. These best fits correspond to an octahedral symmetry with an oxygen-metal distance of 2.03±0.03 Å and 2.08+0.02 Å for Ni^{2+} and Fe^{2+} solutions respectively. Hydrogen atoms are included in the calculations.

The agreement between the experimental data and the best fit theoretical curves are good in the whole energy range, small discrepancies remaining in the intensity of the resonance at 15 eV, essentially due to the muffin-tin approximation for the shape of the potential. We note that the inclusion of the hydrogen atoms in the calculation does change the metal-oxygen distance determination, going from the value of 2.00 Å

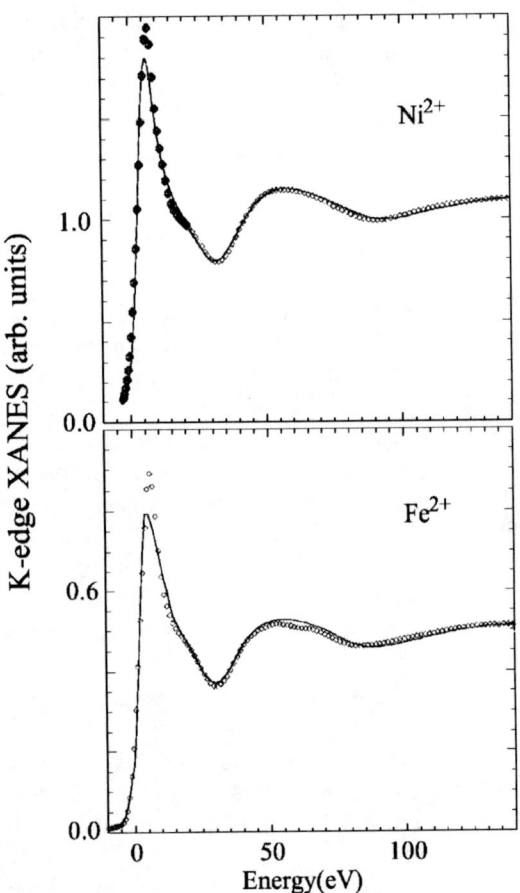

FIGURE 1. Upper frame: Experimental (circles) and best-fit calculation (solid line) of the Ni K-edge XANES spectrum of Ni^{2+} aqua ion. Lower frame: Experimental (circles) and best fit calculation (solid line) of the Fe K-edge XANES spectrum of Fe^{2+} aqua ion.

previously calculated without hydrogens [3] to 2.03 Å. This structural determination is in good agreement with the one obtained by the GNXAS analysis using the EXAFS energy region [8,11].

In Fig.2 we show the comparison between the experimental data related to the Ni^{2+} and the best-fit calculation obtained by using the complex HL potential. In this case the best fit structure corresponds to an octahedron with a oxygen-metal distance of 2.04±0.02 Å. It is noticeable that we obtain the right geometry although the presence of relevant discrepancies between the experimental data and the best-fit calculation in the first 40 eV due to the peculiar behavior of the complex part of HL potential. This is a clear demonstration of the importance of using the phenomenological approach for the inelastic losses calculations to have a full access to the structural information contained in the low energy part of the XANES spectrum. This is particularly

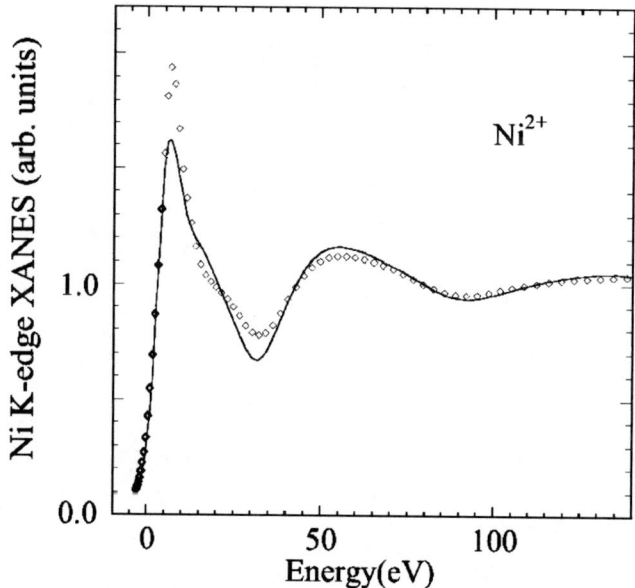

FIGURE 2. Comparison between the experimental (circles) and best-fit calculation (solid line) of the Ni K-edge XANES spectrum of Ni^{2+} aqua ion. In this case we have used the complex HL potential to account for the inelastic losses of the photoelectron.

important in the study of biological systems for which most of the experimental data present a low S/N ratio and a limited k-range data for the analysis [12].

In order to gain a deeper insight into the effect of the potential details on both the quality of the fits and the accuracy of the structural determination, we report in Fig. 3 the best fit results for the Zn^{2+} ion performed with a different MT radius respect to best choice (1.35 Å) [8] and with the X-α potential. In the upper panel we show the comparison between the experimental data and theoretical best-fit calculations performed with a different Zn MT radius (1.20 Å). From the minimization procedure the Zn-O distance is now 2.05 ± 0.05 Å face to the value of 2.06 ± 0.02 obtained with the MT radius of 1.35 Å. The error function R_{sq} [13] increases to twice the value obtained in the previous cases. The corresponding GNXAS analysis [11] gives a Zn-O distance of 2.078 ± 0.002.

In the lower panel of Fig. 3 the best-fit analysis of the calculations performed with the X-α potential is reported. In this case [8] a shorter Zn-O distance value has been obtained (2.04 ± 0.04 Å) while the agreement between the theoretical and experimental spectra is of the same quality of the one reported in Ref. 8, Fig. 1. This finding demonstrates that XANES is essentially dominated by the geometrical arrangement of atoms around the photoabsorber, while it is less sensitive to the potential details.

367

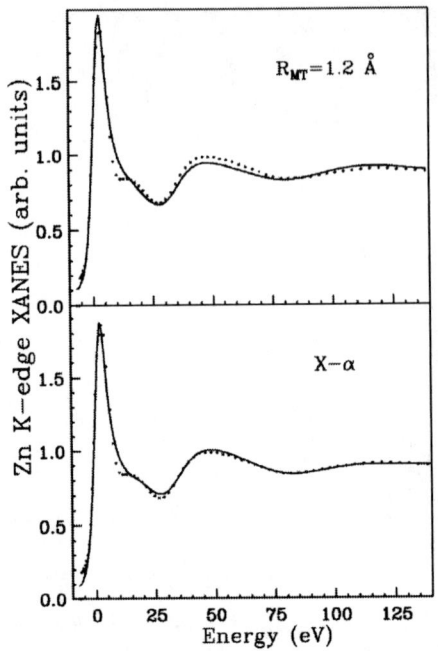

FIGURE 3. Comparison between the XANES experimental spectrum (dotted lines) of the Zn^{2+} ion and best-fit theoretical calculation (solid lines) calculated with different choice of the Zn MT radius (upper panel) and with the X-α potential (lower panel).

CONCLUSIONS

In conclusion, we have presented a new method for a quantitative analysis of XANES spectra based on a fitting procedure in the framework of full multiple scattering calculations. In this way we achieve a full access to the whole XANES energy range allowing a complete structural determinations of the local geometry around the absorber. The MXAN method has been proven to provide reliable structural results confirming the validity of the application of the full MS scheme in the framework of the MT approximation. The results of the present investigation represent a step forward in understanding the role that the XANES technique can play in providing quantitative structural information on chemical systems.

REFERENCES

1. T.A. Tyson, K.O. Hodgson, C.R. Natoli and M. Benfatto, *Phys. Rev B* **46**, 5997 (1992); J.J. Rehr and R.C. Albers, *Rev. Mod. Phys.* **72**, 621 (2000).
2. A. Filipponi and A. Di Cicco, *Phys. Rev B* **52**, 15135 (1995).
3. M. Benfatto and S. Della Longa, *J. Synchrotron Rad.* **8**, 1087 (2001).

4. S. Della Longa, A. Arcovito, M. Girasole, J.L. Hazemann and M. Benfatto, *Phys. Rev. Lett.* **87**, 155501 (2001).
5. C.R. Natoli and M. Benfatto, *J. Phys. (France) Colloq.* **47**, C8:11-23 (1986).
6. D. Cabaret, Y. Joly, H. Renevier and C.R. Natoli, J. *Synchrotron Rad.* **6**, 258-260 (1999).
7. Y.Joly, *Phys. Rev. B* **63**, 125120 (2001).
8. P. D'Angelo, M. Benfatto, S. Della Longa, and N.V. Pavel, *Phys. Rev B* **66**, 064209 (2002).
9. S. Kirkpatrick, C.D. Gelatt Jr, and M.P. Vecchi, *Science* **220**, 671-680 (1983).
10. C.R. Natoli, M. Benfatto, C. Brouder, M.F. Ruiz Lopez and D.L. Foulis, *Phys. Rev.B* **42**, 1944 (1990); C.R. Natoli, M. Benfatto, S. Della Longa and K. Hatada, *J. Synchrotron Rad.*, to be published (2002).
11. P. Dangelo, V. Barone, G. Chillemi, N. Sanna, W. Meyer-Klaucke and N.V. Pavel, *J.Am.Chem.Soc.* **124**, 1958 (2002).
12. S.S. Hasnain and K.O. Hodgson, *J. Synchrotron Rad.* **6**, 852-864 (1999).
13. M. Benfatto, P. D' Angelo, S. Della Longa and N.V. Pavel, *Phys. Rev. B* **65**, 174205 (2002)

Atomic form factors and photoelectric absorption cross-sections near absorption edges in the soft X-ray region

C. T. Chantler

School of Physics, University of Melbourne, Vic. 3010, Australia

Abstract. Reliable knowledge of the complex X-ray form factor [Re(f) and Im(f)] and the photo-electric attenuation coefficient (σ_{PE}) is required for crystallography, medical diagnosis, radiation safety and XAFS studies. Key discrepancies in earlier theoretical work are due to the smoothing of edge structure, the use of non-relativistic wave functions, and the lack of appropriate convergence of wave functions. These discrepancies lead to significant corrections for most comprehensive (i.e. all-Z) tabulations. This work has led to a major comprehensive database tabulation [Chantler, C. T. (2000). J. Phys. Chem. Ref. Data, 29, 597-1048] which serves as a sequel and companion to earlier relativistic Dirac-Fock computations [Chantler, C. T. (1995). J. Phys. Chem. Ref. Data, 24, 71-643]. The paper finds that earlier work needs improvement in the near-edge region for soft X-ray energies, and derives new theoretical results of substantially higher accuracy in near-edge soft X-ray regions. Fine grids near edges are tabulated demonstrating the current comparison with alternate theory and with available experimental data. The best experimental data and the observed experimental structure as a function of energy are strong indicators of the validity of the current approach. New developments in experimental measurement hold great promise in making critical comparisons with theory in the near future. This work forms the latest component of the FFAST NIST database [http://physics.nist.gov/PhysRefData/FFast02/Text/cover.html].

INTRODUCTION

Tables for form factors and anomalous dispersion are of general use in the UV, x-ray and γ-ray communities. Much of the recent theoretical basis for these was contributed by Cromer, Mann and Liberman[1] while much of the experimental data was synthesised by Henke et al.[2] The generality of these works has entailed numerous simplifications compared to detailed relativistic S-matrix calculations. Detailed S-matrix results do not appear to give convenient tabular application for the range of Z and energy of general interest, while the tables have limited validity across extended regimes.

Earlier relativistic Dirac-Fock computations[3] addressed the primary interactions of X-rays with isolated atoms from Z = 1 (hydrogen) to Z = 92 (uranium) and computed them within a self-consistent Dirac-Hartree-Fock framework. This has general application across the range of energy from 1-10 eV to 400-1000 keV, with limitations as the low- and high-energy extremes are approached. Tabulations are provided for the f_1 and f_2 components of the form factors, together with the photoelectric attenuation coefficient for the atom, μ, as functions of energy and wavelength. This work has lead to significant quantitative improvement above 30 keV to 60 keV energies, near absorption edges, and at 0.03 keV to 3 keV energies. Recent experimental syntheses are often complementary

CP652, *X-Ray and Inner-Shell Processes: 19th International Conference on X-Ray and Inner-Shell Processes*
edited by A. Bianconi, A. Marcelli, and N. L. Saini
© 2003 American Institute of Physics 0-7354-0111-X/03/$20.00

to this sort of approach.

Discrepancies between currently used theoretical approaches [4, 5, 6, 7, 3] of 200 % exist for numerous elements from 1 keV to 3 keV X-ray energies. This level of inconsistency may be surprising to some users who have conventionally viewed log-log plots covering decades in energy and attenuation coefficient, but these discrepancies have been present in the literature for decades.

A major comprehensive database tabulation [8] addresses these key discrepancies and derives new theoretical results of substantially higher accuracy in near-edge soft X-ray regions (0.1 keV to 10 keV). The grid size and spacing of the reported tabulation is given with synchrotron users in mind, where fine grids near edges are necessary and continuous energy scans are possible. All energies above 0.1 keV, and all elements to Z=92, were investigated in this computation. Estimates for the expected accuracy of the pair of publications[3, 8] is given across the full range of Z and energy, including near-edge limitations of wavefunction convergence and near-edge structure itself. An obvious point, for XAFS and MAD users, is the absence of near-edge structure, and perhaps a deviation of the edge onset from a particular solid-state system. This is an advantage, which serves to separate the oscillatory near-edge structure and discrete lines or near-edge zeroes from a reference baseline for the given element.

This paper summarises new results in areas of critical recent discussion, and some preparation towards a proper resolution of theoretical and experimental flaws.

MAJOR DISCREPANCIES

Compilations of experimental data for form factors are widespread, particularly for common elements such as silicon, copper, silver and gold over the central X-ray energies[5]. These are particularly useful in evaluating the reliability of a particular measurement, or the difficulty of an experiment in a given energy regime. However, the range of the imaginary coefficient in such compilations often varies by 10% to 30% [Figure 1]. This implies in general that claimed experimental accuracies of 1%-3% are not reliable.

Strictly, one or two of the results in such compilations could indeed reach the claimed accuracy, but the remainder must then be in error by up to 10 σ. The effect of a 10% error is similar to a 10% error in the thickness of the sample, or a 10% error in the exponent of the probability of photoabsorption through a sample. This variation seems almost independent of the year of the experiment, or the specification for high or low energy measurements.

A second general source for an experimental best-fit line is given by the Centre for X-Ray Optics, Lawrence Berkeley Laboratory[2, 6, 9]. These references present experimental-theoretical syntheses for the complex form factor in the softer X-ray regime. As a weighted evaluation of experimental data, they are extremely useful. They are also extremely valuable in the very soft X-ray regime where multi-electron interactions occur which are not addressed by general IPA (independent particle approximation) theories. However, no variation or error bar is associated with this single fit, and in soft X-ray regimes, near-edge regimes and other areas the result may be in sharp discrepancy with theory and expected results, or with the best available data. The deviations lie at the

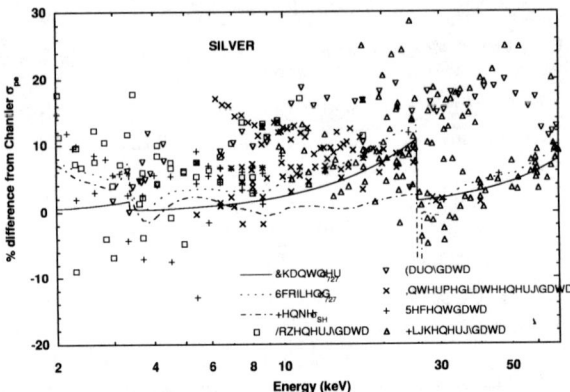

FIGURE 1. Attenuation in silver relative to Chantler (1995) σ_{PE}. Scattering indicated by Chantler σ_{TOT}. Silver shows an array of data [7] quoted at 1% - 3%, but with variation of 28%, or 16% for more recent measurements. Theory shows variations of 7%.

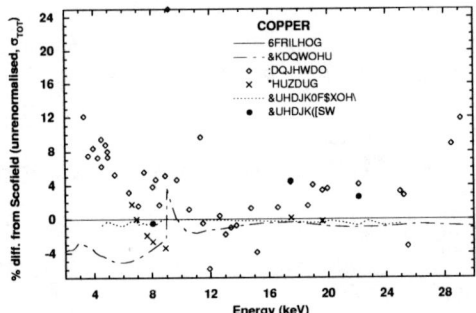

FIGURE 2. Major discrepancies in the form factor of copper. Scattering is minor. Quoted experimental uncertainty given by symbol size [7, 10, 11, 17]. Measurements of f'' or σ_{PE} for copper have quoted 1% accuracy, but discrepancies reach 25% near edges and 8% away from edges. This has been stressed recently. Theories disagree by 6%, and experiment and theory disagree by 12% away from edges.

same 10% - 30% level as the variation of less critical compilations.

RELIABILITY CRITERIA FOR EXPERIMENTAL RESULTS: WHICH SHOULD ONE CONSULT?

Turning exclusively to measurements claiming high accuracy, i.e. 1%, for copper, we narrow the spectrum of measurements only slightly [10, 11] [Figure 2]. Rather than referencing papers claiming high accuracy, one can require that optimal or prefered experimental techniques are followed in papers used for comparison or reference. This would include

(i) accurate determination of thickness, composition and purity of samples

(ii) accurate energy calibration and monochromatic sources

(iii) accurate orientation and alignment techniques

(iv) control of harmonic contamination and source divergence

(v) other criteria including statistical precision and detection linearity[12, 13, 14].

In the case of silicon, these criteria were carefully evaluated as part of an international effort to address such variations. The result was a very accurate and consistent set of measurements for silicon, including contributions by Creagh, Barnea, Gerward, Kerr del Grande and others [12, 15]. These same principles were applied, in the experimental references cited, and specifically for the copper examples given in Fig. 2; and yet the variation and discrepancies remain. Hence, a number of systematic error sources are not accounted for. For experimentalists these (unknown) error sources represent an intriguing limitation to X-ray investigations of all types, and hence an issue of fundamental importance.

UNCERTAINTIES NEAR SOFT X-RAY LII, LIII, MIV, MV EDGES

The greatest discrepancies between these theories occur near edges, with deviations by factors of 5 or more between predictions. The cause of near-edge error in theoretical computations is often inadequate interpolation, extrapolation or integration methods. These introduce oscillations or discontinuities into the data [16]. The cause of near-edge error in experimental compilations is often due to neglect of the edge region or smoothing through edge structure [6]. The cause of near-edge error in specific experiments is often due to the dramatic variation of form factor with energy, requiring both accurate absolute intensity measurement and also precision energy calibration [13].

Assuming that these issues have been correctly addressed, theory will disagree with experiment near edges by large factors due to XAFS and related structure. This can reach a 200% discrepancy between IPA theory and a solid-state experiment [17]. Even if the experiment is performed on a monatomic gas, there may be pressure-dependent structure and other strong oscillatory behaviour near edges. Some of this structure (shape resonances and Cooper minima) may be qualitatively predicted by some theoretical approaches, but often the experimental result will show significant quantitative discrepancy [18].

The largest discrepancies between Chantler (1995) and the Scofield theory are not due to any of these causes. Chantler (1995) claims uncertainties of up to a factor of two in soft X-ray near-edge regions. Saloman, Hubbell and Scofield (1988) refers to 10% - 20% discrepancies from experimental data in the medium-Z regime, which may be taken as an uncertainty estimate. In most elements and regions, the near-edge variation falls within these error bars. Such experimental data is not sufficiently precise to distinguish between these two theories, or even to observe edge structure.

In the region 1 keV – 2 keV for particular edges in medium or high-Z elements, large discrepancies are observed between these two theoretical treatments. This is illustrated in Fig. 3 for Zn, Z=30. This is not due to XAFS or any such near edge oscillation.

Relative to appropriate high-energy theory, which would yield well-defined edges and smooth behaviour for each orbital on a log-log plot, the results of Scofield, Chantler

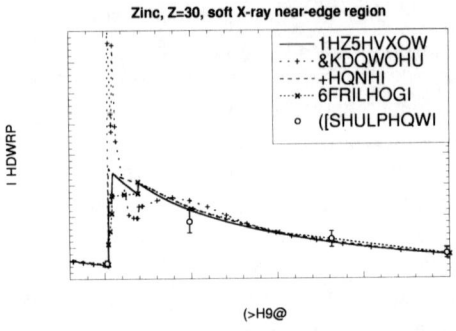

Zinc, Z=30, soft X-ray near-edge region

	1HZ5HVXOW
---·+···	&KDQWOHU
------	+HQNHI
···*···	6FRILHOGI
○	([SHULPHQWI

I HDWRP

(>H9@

FIGURE 3. Attenuation in Zn (Z=30), represented by Im (f). Experimental data from [7]. Refs [3, 2, 4] agree with available experiment, but all have large error, as indicated by the new result [8].

and Henke are all in error. This error arises from an accumulation of minor errors in inner shell electronic wavefunctions. Particularly for near-edge energies, these errors accumulate, which is a strong reason for the low accuracy claimed by theory in this region. The K and LI shell are accurately computed, and the form factors for these sub-shells are accurate; but the errors for LII and LIII are amplified, and also fall in increasingly difficult soft X-ray energies. Hence the wavefunction solution for the orbital radial electron density, which leads to the computation of the near-edge form factor, becomes unreliable and increasingly inaccurate.

Within the convergence criteria for the DHF wavefunctions, this may be more or less difficult to address, depending upon the exchange potential and method used. We have been able to retain the original formalism and to require a better and more uniform convergence in these regions.

When the wavefunctions are thereby improved and this issue is addressed, we obtain the 'New Result' [Figure 3]. Appropriate high-energy theory would expect a behaviour very closely following this. This then obtains the theoretically expected IPA edge structure. The precision of these results is dramatically improved. The accuracy is still limited as discussed above. We would claim no better than 20% - 30% accuracy in this region, even though experiment may agree to better than 10%.

The largest effects are represented by Zn Z=30 and Pm Z=61. These represent 1.5 σ errors for the 1995 tabulation, where σ is estimated as 50%, as stated above. In these cases the Scofield result yields 160% and 220% errors near the edge (or 4-5 σ errors); conversely, [3] yielded maximum 68% and 87% errors respectively at the same locations. We believe that the cause of the Scofield discrepancies lies in the same problem regarding the electron distribution. This will be affected by the formalism used to derive wavefunctions. The Chantler (1995) errors tended to be extended over slightly larger energy ranges (i.e. 40-50% versus 20-30% above the edge).

Usually the experimental data is inadequate to make a critical comparison of theory. However, the general trend is given by Kr Z=36 [Fig. 4] [19]. The predicted new structure matches up well with such optimum experiments, as opposed to alternative theoretical structures.

Recent work by our research group has reinvestigated copper explicitly, and yielded

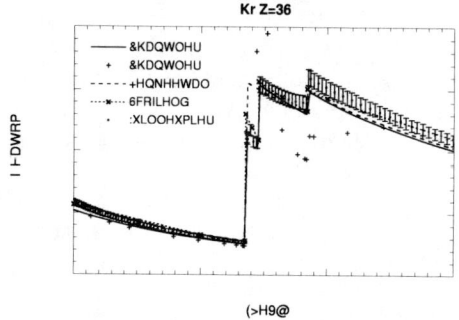

FIGURE 4. Plot illustrating the consistency of near-edge structure between experiment [19] and [8], as opposed to earlier theory. Im (f) = f_2. Experimental values include contributions from scattering. $[\mu/\rho]$ (in cm2/g) = f_2 (e/atom) $\times 5.02152 \times 10^5$.

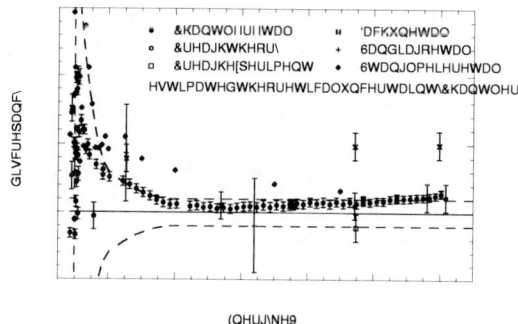

FIGURE 5. Comparison between [14] and earlier measurements [20, 10, 21, 17] for copper. Data are compared to theory [3, 8], $\{\% = \frac{[\mu/\rho]-[\mu/\rho]_{theory}}{[\mu/\rho]_{theory}}\}$, with theoretical uncertainty given by the region between dashed lines, which increases to 20% near the K-edge. The comparison of mass absorption coefficient is identical to that of Im(f). An alternate theory [17] agrees with the reference theory, and agreement of the experimental data with current theory is good.

dramatically reduced error bars [Fig. 5]. This result with 0.3% uncertainty is able to verify theory in a conclusive manner, in this energy regime, for the first time.

CURRENT EXPERIMENTAL ISSUES

Figures 4,5 show a potential significant absolute experimental offset at the 1 σ level. In fact, there is a common tendency towards experimental results lying higher than theoretical predictions. In some cases this is partly due to an inadequate control of scattering, divergence and alignment. Misalignment or significant divergence will both lead to measured attenuation coefficients lying higher than correct values, while the effect of uncertainty in scattering depends on the model assumed for the given sample.

This remains an issue for experimental comparisons. This sort of discrepancy is strong motivation for high accuracy experiments to address these issues.

Experimental and theoretical difficulties remain in central X-ray energies, but also at VUV energies where the IPA assumption fails, and at very high energies. In both regimes experimental measurement difficulties arise, and at high energies it is very difficult to isolate problems of the photoeffect computation from those of the computation of scattering contributions. The other difficulty at high energies relates to exactly what type of sample (gas, foil, crystal or other) is under investigation, because this will change the dominant scattering contributions by orders of magnitude.

CONCLUSION

Several generic difficulties with theoretical determinations of the atomic form factor in the X-ray region have been resolved. Selected experimental data sets suggest the accuracy of [8] compared to most alternatives. Key discrepancies are often due to the smoothing of edge structure, the use of non-relativistic wavefunctions, and the lack of appropriate convergence of wavefunctions.

In general experimental data are not sufficiently accurate to assess theory at the level required. However, the best experimental data and the observed experimental structure as a function of energy are strong indicators of the validity of the current approach. New developments in experimental measurement hold great promise in making critical comparisons with theory in the near future.

ACKNOWLEDGMENTS

We would like to thank several colleagues for helpful discussions, including J. Hubbell, J.-L. Staudenmann, Z. Barnea and D. C. Creagh, and the experimental collaboration.

REFERENCES

1. Cromer, D. T., Liberman, D., J. Chem. Phys. 53 (1970) 1891-1898; Cromer, D. T., Mann, J. B., Acta Cryst. A24 (1968) 321-324; Cromer, D. T., Liberman, D. A., Acta Cryst. A37 (1981) 267-268.
2. Henke, B.L., Davis, J.C., Gullikon, E.C., Perera, R.C.C., (1988) LBL-26259 UC-411, 376 pages.
3. Chantler, C.T., (1995) J. Phys. Chem. Ref. Data 24 71-643
4. Scofield J.H. (1973), LLNL Report UCRI-51326
5. Saloman, E.B., Hubbell, J.H., (1986) NBSIR 86-3431
6. Henke B. L., Gullikson E. C., Davis J. C. (1993), At.Dat.Nucl.Dat.Tables 54 181-342
7. Saloman E.B., Hubbell J.H., Scofield J.H. (1988), At.Dat.Nucl.Dat.Tables 38 1-197
8. Chantler, C. T. (2000). J. Phys. Chem. Ref. Data, 29, 597-1048
9. Cullen D.E., Hubbell J.H., Kissel L. (1997) EPDL97: The evaluated photon data library Lawrence Livermore National Library Report UCRL-50400 Vol 6 Rev 5.
10. Wang, D., Ding, X., Wang, X., Yang, H., Zhou, H., Shen, X., Zhu, G. (1992) NIM B71, 241.
11. Gerward, L. (1989) J.Phys. B22, 1963.
12. Creagh D.C. and Hubbell J.H. (1987) Acta Cryst. A43, 102-112.
13. Chantler CT, Barnea Z, Tran CQ, Tiller J, Paterson D. Optical & Quantum Elec. 1999;31:495-505.

14. Chantler CT, Tran CQ, Paterson D, Cookson DJ & Barnea Z. *Phys. Lett. A* 2001;**286**:338-346.
15. Mika J.F., Martin L.J., Barnea Z. (1985) J.Phys. C 18, 5215-5223.
16. Chantler C.T. (1994) pp61-78 in *Resonant Anomalous X-ray Scattering Theory and Applications*, G. Materlik, C.J. Sparks, K. Fischer (Eds) (Elsevier).
17. Creagh D.C. and McAuley W. (1995) section 4.2.6 in *International Tables for Crystallography, Vol. C*, A.J.C. Wilson, Ed. (Kluwer Academic).
18. Zhou B., Kissel L., Pratt R.H. (1992) Phys Rev A45 2983
19. Wuilleumier F. (1972) Phys. Rev. A6, 2067-2077 and references therein.
20. F. Stanglmeier, B. Lengeler, W. Weber, H. Göbel, and M. Schuster, Acta Cryst. **A48**, 626 (1992).
21. T.K.U. Sandiago and R. Gowda, Pramana **48**, 1077 (1997).

Organization Around Cations in Oxide Glasses Using X-Ray Absorption Spectroscopy

Laurent Cormier, Laurence Galoisy, Georges Calas

Laboratoire de Minéralogie-Cristallographie de Paris, Universités Paris 6 et 7, Institut de Physique du Globe de Paris, UMR CNRS7590
4 place Jussieu, 75005 Paris, France

Abstract. X-ray absorption spectroscopy (XANES and EXAFS) has been used to determine the environment of cations (Ni, Zn, Zr, Fe and Mo) in oxide glasses and their redox state. Direct quantitative structural information can be extracted which indicates that cations are often present in unusually low coordination number compared to crystals. Medium range environment can be assessed with second and further neighbors. This yields to define structural models on the connectivity between cation polyhedra and the network structure.

1. INTRODUCTION

Oxide glasses are important materials used in a wide range of applications (structural, optical, waste materials ...). The amorphous nature and the complex chemical composition do not allow the construction of a unique structural model, as in crystalline compounds. The glass structure may be defined by the coexistence of a polymeric network (silicate, borate, etc. units) and of cations which may act either as modifying elements which break the connectivity of the polymeric network, or as charge-compensating cations which balance local charge deficits due to the substitution of silicon by lower charged elements such as aluminium or boron [1-3]. The polymeric network may be investigated with vibrational spectroscopy or nuclear magnetic resonance [4]. The cationic organization is more difficult to investigate due to the diversity of their local surroundings, which may give rise to drastic effects on the physico-chemical properties of the glasses. Structural information can be gained by using chemically selective structural methods such as X-ray Absorption Spectroscopy (XAS) [5,6]. XAS allows the characterization of local cation surrounding and sometimes reveals the presence of a well-defined medium-range order (MRO).

Synchrotron radiation possesses high brilliance and white beam character that make it well-suited for XAS investigations of amorphous solids. Synchrotron-radiation XAS is now recognized as one of the most powerful element-specific probes which can be used to determine the local structure around a given element in a crystalline or non-crystalline material and more specifically around minor/trace elements in silicate melts/glasses [5,6]. XAS spectra consist of two regions that contain different kinds of information. In the vicinity of the X-ray absorption edge, X-ray Absorption Near Edge Structure (XANES) provides information on the average site geometry and oxidation

CP652, *X-Ray and Inner-Shell Processes: 19th International Conference on X-Ray and Inner-Shell Processes*
edited by A. Bianconi, A. Marcelli, and N. L. Saini
© 2003 American Institute of Physics 0-7354-0111-X/03/$20.00

state of the element investigated. At higher energy, Extended X-ray Absorption Fine Structure (EXAFS) provides structural parameters, such as average interatomic distances (including a Debye-Waller parameter that accounts for radial disorder) and the nature and number of neighbors. By taking advantage of the complementary information brought by XANES and EXAFS, it is possible to distinguish among the structural environment of the various glass-forming elements.

This paper provides a brief review of recent XAS studies obtained on the structural environments of cations in oxide glasses. XAS data have provided an extensive data set on oxidation states, cation coordination and MRO for a number of transition elements in oxide glasses.

2. THE LOCAL ORGANIZATION AROUND CATION

2.1. Unusual Coordination Sites

Cations often occupy unusual sites in oxides glasses and a general observation is a trend towards lower coordination numbers in glasses than in crystals. Furthermore, cations which are encountered in a strong association with the polymeric framework, such as network forming cations Zn and Fe^{3+}, have a different role than the modifier cations, which do not show fixed relationships with the polymeric framework. The former are generally in a four-fold coordinated site connected to the network, while the latter correspond to higher coordinated sites, with the frequent occurrence of 5-fold coordinated transition metal ions. If site geometry does not vary much as a function of the glass composition, including cation content, their relative site occupancies are composition dependent. These coordination states have different structural significances.

In several silicate and alumino-silicate glasses, XAS data have shown that Ni occurs in four- and five- coordination, [4]Ni and [5]Ni [7]. K-edge XANES spectra of transition elements show a pre-edge feature (A) for which the intensity is enhanced in non-centrosymmetric sites (3d-np mixing) and T_d symmetry (electric dipole allowed transitions). This intensity increases by a factor four from [6]Ni to [5]Ni and [4]Ni, with a mixture of [4]Ni and [5]Ni species in silicate glasses (Figure 1). There is a regular dependence of the intensity of both the pre-edge and shape resonance (C) on the [4]Ni content of silicate glasses (Figure 2). This sensitivity of XANES for determining cation coordination states in glasses is in agreement with studies on Fe- and Ti-bearing silicate glasses [8,9].

The modulus of the Fourier Transform (FT) of the k^3-weighted EXAFS functions (k is the photoelectron momentum) correspond to a Pair Distribution Function (PDF) relative to the absorbing atom, uncorrected for phase-shifts of the photoelectron wave. The first prominent peak of the PDF corresponds to the Ni first coordination shell. The EXAFS parameters can be derived by fitting Fourier back-transforms of this peak. Ni-O distances and Ni coordination number determined by EXAFS are consistent with a mixture of [4]Ni and [5]Ni, in agreement with XANES data. A small Debye-Waller term indicates a low radial disorder. The absence of [6]Ni in the silicate glasses investigated in [7] is a major conclusion derived from XAS. It demonstrates that the local structure of a glass may differ from that of the corresponding crystal [10]. This unusual 5-coordination may be more common in the glassy structure than in the crystalline state, as in oxide glasses containing Fe^{2+} [11], Cu^{2+} and Zn^{2+} [12,13] and Ti^{4+} [9,14].

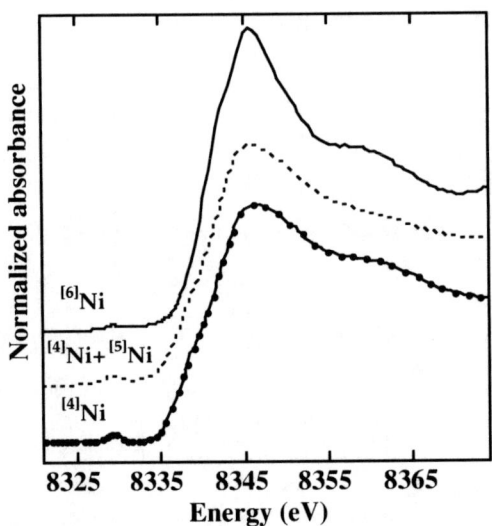

FIGURE 1. Normalized Ni K-edge XANES spectra of LiB_9O_{14} glass containing 2Wt% NiO (top), $Na_2NiSi_3O_8$ glass (middle) and $K_2NiSi_3O_8$ glass, with their respective main coordination for Ni site.

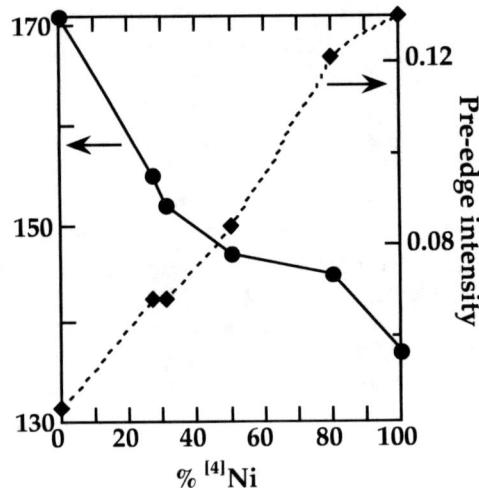

FIGURE 2. Variation of the XANES parameters as a function of the relative proportion of [4]Ni estimated from crystal field spectra in silicate glasses [7]. The solid line corresponds to the shape resonance intensity of the XANES spectra (left axis) and the dashed line corresponds to the pre-edge intensity (right axis).

These modifier ions are reactive intermediate species which indicate associative/dissociative reactions [15]. For instance, [5]Si and [5]Al are associated to atomic motions at the origin of the viscous flow and chemical diffusion which characterize the molten state [16]. [5]Ni has been extensively studied in glasses by a combination of spectroscopic techniques [7]. By contrast to [4]Ni, [5]Ni species do not show a well-defined geometrical relationship with the silicate framework and neutron diffraction indicates the presence of other cations (Ca, Ni) as second neighbors [17]. [5]Ni may be formed either from [4]Ni by an associative reaction involving the bonding of a [4]Ni species to a fifth oxygen, e.g., a bridging oxygen, or from unstable [6]Ni species by a dissociative reaction implying the loss of a sixth ligand because of the oxygen motion accompanying the viscous flow. Indeed, the frustration of ligand spatial correlation is likely to be at the origin of the small coordination numbers of most cations in glasses and melts, one of the major finding of EXAFS and neutron diffraction studies of glasses [18,5]. A dissociative reaction, together with an easier cation-oxygen bond breaking ability, may explain why, by contrast to [5]Si species which only affect a minority of the Si atoms, [5]Ni is the predominant species for Ni in most silicate glasses.

2.2. Oxidation States and Cation Coordination Numbers

High resolution XANES spectra of iron show the effects of the coordination numbers on the quantification of redox values in synthetic and volcanic glasses. Volcanic glasses show split pre-edge features of the Fe K-edge XANES spectra, arising from a bimodal distribution between the relative contributions of ferric and ferrous iron (Figure 3). Pre-edge spectra have been reproduced with about 90% confidence using a linear combination between a Fe^{2+} augite glass and $[4]Fe^{3+}$ and $[6]Fe^{3+}$ references [19]. Ferrous iron is mostly 5-fold coordinated and minority 4-fold coordinated while ferric iron occurs in 4- and 6-fold coordinated sites.

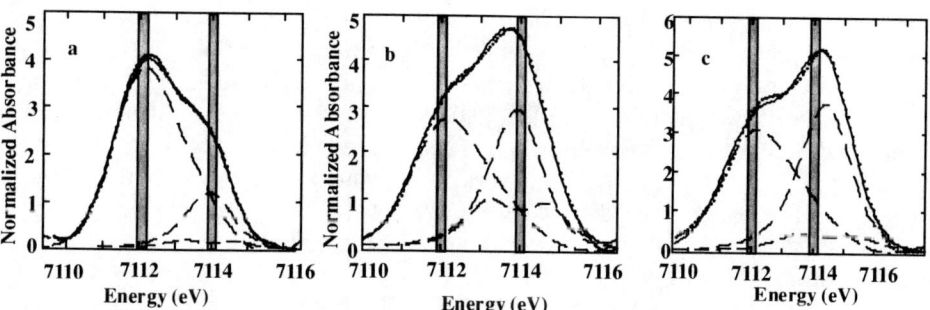

FIGURE 3. Fe pre-edge extracted from the K-edge XANES spectra of various volcanic glasses: (a) Erta'Ale basaltic glass, (b) oxidized basalt glass and (c) Boina pantelleritic glass. The experimental spectrum (dotted line) is compared with the fit (plain line) obtained using the Augite glass reference spectrum (model for Fe^{2+}) and two pseudo-Voigt components (model for $[4]Fe^{3+}$ and $[6]Fe^{3+}$) represented as dashed lines. The shaded zones indicate the position of the contribution of Fe^{2+} and Fe^{3+}, at 7112 eV and 7114 eV, respectively [reprinted from Ref. 19, © 2001, with permission from Elsevier Science].

An important set of information may be gained on the sites occupied by ferrous and ferric iron in volcanic glasses by analyzing the pre-edge in XANES spectra. The importance of [6]Fe in basaltic glasses and the prevalence of [4]Fe in silicic pantelleritic glasses are a result of the structural mechanisms which drive the redox equilibria in silicate melts. The increase of the proportion of tetrahedral Fe^{3+}, accompanied by more covalent Fe^{3+}-O bonds, is consistent with the chemical dependence of redox equilibria in silicate systems, in which the alkali rich melts correspond to more oxidizing compositions.

3. EVIDENCE OF MEDIUM RANGE ORDERING AROUND CATIONS

3.1. Presence Of Second Neighbors And Bond Valence Models

A major structural problem concerning the cation environment in oxide glasses is the link between cationic sites and the polymeric network. In most glasses, there is a significant contribution of second neighbors due to the network former cations which occur beyond the first oxygen shell. The second neighbors determined in the EXAFS signal give access to the type of linkages between polyhedra through the determination of the interatomic distances between the probed atom and its second neighbors, as well as their nature.

In alkali silicate glasses, low content of ZnO (<5 %) usually increases the mechanical properties or the chemical durability of the glass. By contrast, Zn has a nucleating role in alkaline earth silicate and aluminosilicate glasses. The driving force for Zn being a glass stabilizer or a nucleating element is not clearly understood. Zn has been found in a tetrahedral site in various alkali silicate glasses, with similar Zn-O distances (1.96±0.02Å) [20]. The mean interatomic Zn-Si distances are close to 3.2 Å, which gives a mean Zn-O-(Si, Al, Mg) angle of 130° and indicates corner-sharing ZnO_4 and SiO_4 tetrahedra. Zr is a highly charged transition element which has been investigated in silicate, borosilicate and aluminosilicate glasses [21,22]. Zr K-edge EXAFS spectra indicate that this element is in an octahedral site with a mean interatomic distance d(Zr-O) = 2.084± 0.01Å [23]. Despite a coordination number which is different from that of Zn in similar glass compositions, the connection of the ZrO_6 octahedra with the silicate network is shown by the presence of Si second neighbors. The mean Zr-Si distance varies between 3.37 Å and 3.71 Å among the glasses. This variation indicates that Zr is a good structural probe which is sensitive to local modifications of the glass structure. The Zr-Si value is compatible with ZrO_6 octahedra linked by corners to SiO_4 tetrahedra. In contrast to Zn and Zr, Mo in the oxidized form (Mo^{6+}) is one of the only transition elements which does not show the contribution to the EXAFS signal of Si second neighbors and hence no direct connection with the silicate network. Mo K-edge EXAFS indicates that Mo belongs to molybdate groups in a tetrahedral site and no second shell can be detected around Mo [24]. This geometry appears independent of glass composition.

The relationship between cation and the network may be rationalized using the Pauling rules which indicate how the electrical neutrality of the glass components is locally ensured. We have used these rules to propose simple models for the local structure of silicate glasses. The basic Pauling rules dealing with the coordination polyhedron of the cations and with the valence of the surrounding anions can be extended using the bond-valence bond-length correlation approach [25]. The stability of the structural environment around transition elements can be derived using this model by an evaluation of any significant overbonding or underbonding around a cation or an anion [6,7].

The apparent valences for the first oxygen neighbors (v_{OX}) around Mo in tetrahedral site is of 1.41-1.44 v.u.. This value precludes any kind of bonding between the molybdate groups and SiO_4 tetrahedra because of a oversaturation of the oxygen that would link Mo and Si. The charge compensating can only be ensured by alkali or alkaline-earth elements. This approach shows that Mo is not connected to the silicate network and that a tetrahedral coordination may not be a systematic indication of a network-forming position [24].

For $^{[4]}Zn$, the model derived from EXAFS data takes into account the v_{OX} value derived with Si as second neighbors (Figure 4) [20]. The residual charge on the oxygen linking Zn to Si needs to be compensated by alkalis or alkaline-earths cations (Na^+, Ca^{2+}...). The first and second EXAFS-derived distance are in agreement with a position of network former for $^{[4]}Zn$, with the charge compensation ensured by low field strength cations such as Na^+ or K^+ [7], like $^{[4]}Ni$ in silicate glasses. ZnO_4 tetrahedra tend to bring together the polymerized domains of the glass, thus reinforcing the glass stability. In glasses containing high filed strength cations such as Mg^+, Zn is present in an octahedral site which favors the nucleating role for Zn [7].

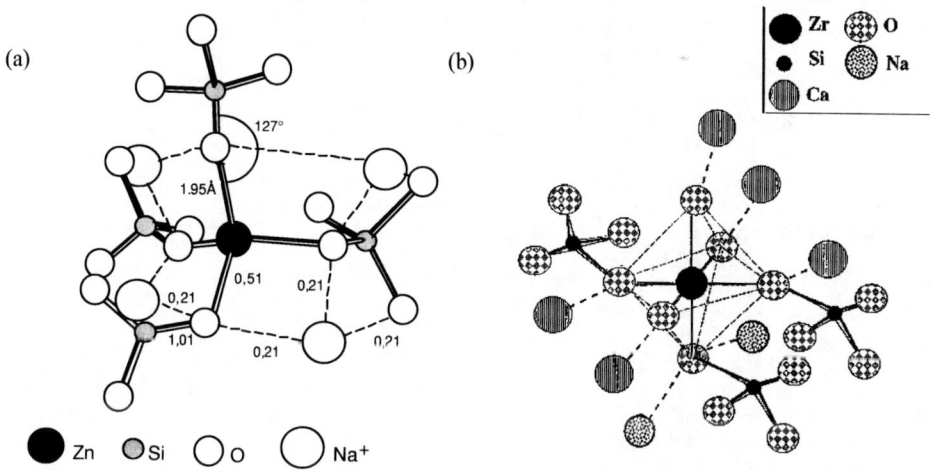

FIGURE 4. Schematic representation of the connectivity of the silicate network with (a) the ZnO_4 tetrahedra and (b) the ZrO_6 octahedra. The number of charge compensating atoms (Na+, K+ or Ca2+) is based on the application of the Pauling rules.

As the value of v_{OX} (1.8 u.v.) for [6]Zr is close to the valence of oxygen, the only possibility is to compensate the charge of the oxygen of the Zr-O-Si bonding by alkali or alkaline-earth cations (Figure 4) [23]. This explains the high solubility of zircon in alkali-bearing glasses, as well as the nucleation in the case where charge compensation is no more ensured [21].

3.2. Ni-Ordered Domains In Borate Glasses

Alkali borate glasses have received much attention as several properties present an unusual behavior as a function of the alkali content, e.g., boron coordination, density, ionic conductivity, coordination of cations such as Ni which gives these glasses peculiar colorations. These cations, present in small concentrations, may be used as local probes of the glass structure. The medium range ordering around Ni in low-alkali borate glasses (10% Li_2O, Na_2O or K_2O) has been investigated using Ni-K edge EXAFS [26]. In these glasses, Ni is present in a regular octahedral symmetry, a coordination state which is not observed in other oxide glasses. The Fourier Transforms (Figure 5) present striking structural features extending up to 6 Å, while EXAFS information is usually limited by the large topological disorder to the first and sometimes second neighboring shell in most oxide glasses.

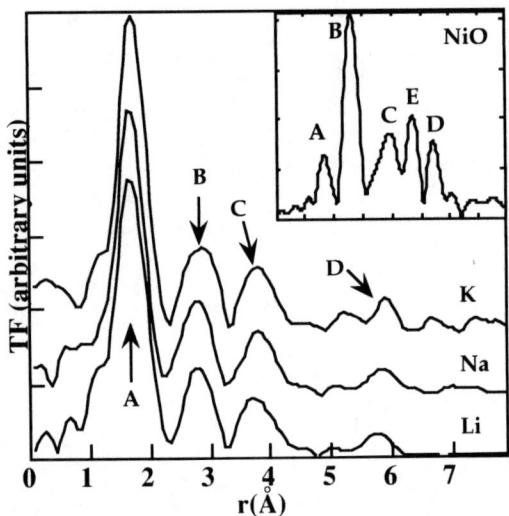

FIGURE 5. Fourier Transforms of the Ni K-edge EXAFS signals for K-, Na- and Li-containing borate glasses (top to bottom). Peak A corresponds to the first coordination shell due to the [6]Ni site. Peaks B, C and D include Multiple Scattering effects due to Ni-Ni correlations. The inset is the Fourier Transfom of the EXAFS signal for c-NiO.

At such large distance, single scattering events are unlikely and Multiple Scattering (MS) effects have to be taken in account. The spectra shown in Figure 5 present important similarities with that of NiO [27], which suggests a similar Ni surrounding. A MS analysis of these spectra was done using the *ab initio* code FEFF 6 [28]. This allowed the determination of Ni-Ni correlations at about 3, 4.1 and 6 Å, the latter being due to a focusing effect between three collinear Ni atoms with a Ni-Ni separation of 3 Å. The first contribution at 3 Å indicates a non-random distribution of Ni in borate glasses, despite the high dilution level. The distances at 4.1 and 6 Å are characteristic distances for corner and edge-sharing [6]Ni octahedra, respectively. These results indicate that Ni is present in three-dimensional domains, which mimics the oxygen close-packed structure of NiO. However, important differences in the peak amplitudes, the absence of features at 5 Å in the EXAFS spectra of borate glasses and distinct optical absorption spectra between c-NiO and borate glasses, indicates that the Ni-containing ordered domains shown by EXAFS in the investigated glasses do not correspond to unreacted NiO crystallites. These structural similarities explain the low NiO activity in alkali borate glasses [29], since the increasing size of the Ni-enriched domains rapidly lead to the formation of nucleation centers. The occurrence of this ordered domains may result from the ordered framework structure encountered in borate glasses, due to the presence of large, rigid superstructural units at low alkali content. The oxygen atoms bounded to [6]Ni sites require charge compensation which can be ensured by the four-coordinated boron atoms which are present in some borate units. This may account also for the differences in the Ni site geometry between borate and silicate glasses. Indeed, the regular octahedral geometry observed in low-alkali borate glasses needs a more ordered surrounding than the more flexible 5-coordinated sites observed in silicate glasses. Such environment is also found for other transition elements such as Co and Cu [30].

4. CONCLUSIONS

X-ray absorption spectroscopy (XANES and EXAFS) is a powerful tool for investigating the environment of specific cations in multicomponent glasses. The data obtained by EXAFS and XANES have pointed out the importance of original cationic coordination sites, for instance Ni in four and five-fold coordination state in silicate glasses or in six-fold coordination state in low alkali borate glasses. These different coordination states indicate distinct connections with the silicate glass structure, which can be evidence by the presence of second neighbors in EXAFS data. Information on the site geometry and oxidation state can also be extracted from the pre-edge region. Indeed, high-resolution XANES spectra of iron in minerals give a basis for determining the sites occupied by ferrous and ferric ions in synthetic and volcanic glasses, in which strong structural differences are observed as a function of oxidation state and glass composition. Few examples exist on the possibility of extracting medium range information from XAS, such as the ordered domains found by Ni-K edge EXAFS in low alkali borate glasses. XAS may also be used to better understand the local structure of various chemical elements in complex materials such as waste glasses. We used the XAS method to show the peculiarity of the sites occupied by some structural probe of waste glasses such as Zn, Mo or Zr. Quantitative information (coordination number, interatomic distances) can be extracted from XAS data and led to develop plausible models of medium range structure. XAS measurements is also well suited to investigate *in situ* structural modifications at high pressure or high temperature. Another major development is expected on the determination of the structure of the glass surfaces, using electron detection, in order to observe chemical and structural modifications at the glass surface.

ACKNOWLEDGMENTS

We are grateful to V. Briois, S. Belin and the staff of LURE for experimental assistance on the D44 beamline at the DCI-LURE synchrotron (Orsay, France). Special thanks are due to P.H. Gaskell and G.E. Brown Jr. for fruitful discussions. This is IPGP contribution N°1848.

REFERENCES

1. Gaskell P.H., "Models for the Structure of Amorphous Solids," in Materials Science and Technology, edited by R. W. Cahn, P. Haasen, E. J. Craner, VCH, Weinheim, 1991, Vol. 9, pp. 175-278.
2. Greaves G.N., Miner. Mag. 64, 441-446 (2000).
3. Galoisy L., Cormier L., Rossano S., Ramos A., Calas G., Gaskell P.H., Le Grand M., Miner. Mag. 64, 409-424 (2000).
4. "Structure, Dynamics and Properties of Silicate Melts", in Reviews in Mineralogy, edited by J. F. Stebbins D. B. Dingwell P. F. McMillan, Mineralogical Society of America, Washington, 1995, vol. 32.
5. Brown G.E. Jr., Calas G., Waychunas G.A., Petiau J., Rev. Miner. 18, 431-512 (1988).
6. Brown G.E. Jr., Farges F., Calas, G., Rev. Miner. 32, 317-410 (1995).
7. Galoisy L., Calas G., Geochim. Cosmochim. Acta 57, 3613-3626 (1993); Geochim. Cosmochim. Acta 57, 3627-3633 (1993)
8. Calas G., Petiau J., Bull. Minér. 106, 33-55 (1983).
9. Farges F., Brown G.E. Jr., Navrotsky A., Gan H., Rehr J.J., Geochim. Cosmochim. Acta 60, 3039-3053 (1996).
10. Galoisy L., Calas G., Am. Miner. 76, 1777-1780 (1991).
11. Brown G.E. Jr., Jackson W.E., Waychunas G.A., EOS-Trans. Amer. Geophys. Union 73, 356 (1992).
12. Matsubarata E., Waseda Y., Ashizuka M., Ishida E., J. Non-Cryst. Solids 103, 117 (1988).
13. Musinu A., Piccaluga G., Pinna G., Marducci D., Pizzini S., J. Non-Cryst. Solids 111, 221 (1988).
14. Cormier L., Gaskell P.H., Calas G., Soper A.K., Phys. Rev. B 58, 11322-11330 (1998).
15. Basolo F., Pearson R. G., Mechanism of Inorganic Reactions, J. Wiley & Sons (1967).
16. Farnan I., Stebbins J.F., Science 265, 1206-1208 (1994).
17. Cormier L., Calas G., Gaskell P.H., Chem. Geol. 174, 349-363 (2001).
18. Calas G., Brown G.E. Jr., Waychunas G.A., Petiau J., Phys. Chem. Miner. 15, 19-29 (1987).
19. Galoisy L., Calas G., Arrio M.-A., Chem. Geol. 174, 307-319 (2001).
20. Le Grand M., Ramos A.Y., Calas G., Galoisy L., Ghaleb D., Pacaud F., J. Mater. Res. 15, 2015-2019 (2000).
21. Dumas T., Ramos A;, Gandais M. Petiau J., J. Mater. Sci. Lett. 4, 129-132 (1985).
22. Farges F., Calas G., Am. Miner. 76, 60-73 (1991).
23. Galoisy L., Pellegrin E., Arrio M.-A., Ildefonse P., Calas G., J. Am. Ceram. Soc. 82, 2219-2224 (1999).
24. Le Grand M., Thesis Université Paris VII (France) (unpublished).
25. Brese N.E., O'Keefe M., Acta Cryst. B 47, 192-197 (1991).
26. Cormier L., Galoisy L., Calas G., Europhys. Lett. 45, 572-578 (1999).
27. Pickering I.J., George G.N., Lewandowski J.T., Jacobson A.J., J. Am. Chem. Soc. 115, 4137-4144 (1993).

28. Rehr J.J., Mustre de Leon J., Zabinski S.I., Albers R.C., J. Amer. Chem. Soc. 113, 5135-5140 (1991).
29. Paul A., J. Mater. Sci. 10, 422-426 (1975).
30. Galoisy L., Cormier L., Calas G., Briois V., J. Non-Cryst. Solids 293-295, 105-111 (2001).

ReflEXAFS technique: a powerful tool for structural study in new materials

Ivan Davoli[a], Hoang Ngoc Thanh[a]* and Francesco d'Acapito[b]

[a]INFM-Dipartimento di Fisica, Università di Roma "Tor Vergata", Via della Ricerca Scientifica 1, Roma. I-00133 Italy
[b]INFM-O.G.G., 6 rue Jules Horowitz, Grenoble. F-38043 France

Abstract. We report the use of X-rays absorption technique, detected in total reflection mode, to obtain structural information on new materials. After a brief description of the ReflEXAFS technique, we present the results obtained in the study of two very peculiar solid-state problem: a) the effect of the Sb as surfactant in the Si/Ge multilayers and b) the understanding of the very early stage of the spinel formation. We show that the use of a Sb film is not enough to completely stop the interdiffusion process and the quality of the interface is quantify in terms of the interdiffusion of Ge in Si. The second case deals with the structural study of the very earl stage of the $NiAl_2O_4$ spinel formation; this solid-state reaction requires high temperature and long time of exposition in O_2 atmosphere. The progresses of the reaction have been followed by several ReflEXAFS measurement, taken after each thermal treatment.

INTRODUCTION

The X-ray absorption spectroscopy is not a surface technique by itself. It may become sensitive to the first few layers when the incoming beam shines the sample surface with grazing angle. Indeed in the range of the X-ray energy, the refractive index of most materials is less than 1, and for an incident angle of few milliradiants, the beam experiences total reflection. In such geometry only the evanescent wave interact with the sample and the thickness probed is on the nanometer scale [1, 2]. Since Spring 2000 a dedicated station for ReflEXAFS measurement is part of the GILDA (General-purpose Italian beam Line for Diffraction and Absorption) at ESRF (European Synchrotron Radiation Facility). The aim of this work is to report on this new opportunity and to present the result reported in two interesting material science cases.

In the X-ray energy region the n refraction index has the form:

$$n = 1 - \delta - i\beta$$

where δ and β are of the order of $10^{-5} - 10^{-6}$ [1] and from the Snell's laws we find that if the incident X-ray impinge on the sample surface with a grazing angle of few

CP652, X-Ray and Inner-Shell Processes: 19th International Conference on X-Ray and Inner-Shell Processes
edited by A. Bianconi, A. Marcelli, and N. L. Saini
© 2003 American Institute of Physics 0-7354-0111-X/03/$20.00

milliradiants is totally reflected. The critical angle, ϕ_c, weakly depends on the beam energy and on the electronic density of the material:

$$\phi_c = \sqrt{2\delta} \cong 0.5 \div 5 mrad .$$

For an incident angle, smaller than the critical angle, the thickness z probed by the evanescent wave is given by the distance where the intensity of the incident beam is reduced to $1/e$ of the incident intensity that is:

$$z_{1/e} = \frac{\lambda}{4\pi\phi_c} \cong 20 \div 50 A$$

In Fig. 1, the surface sensitivity gained by the grazing angle geometry, is shown and is compared with the thickness probed by a standard X-ray transmission experiment. Barchewitz et al. have reported for the first time the measure of the reflectivity vs. the energy [3]. Since then several methods have been developed in order to extract the $\chi(k)$ from reflectivity spectra [4]. Here is used a method that results to be very simple providing to work at an angle $\phi < \phi_0$ and at an energy far from the edge [5]. Under such conditions the absorption coefficient vs. the reflectivity spectra is related by

$$\mu \approx \frac{1 - R(E)}{1 + R(E)}$$

where R(E) is a measure of absolute reflectivity.

Another way to collect the absorption measurement from a surface slab is collecting the fluorescence of a selected atom induced by the evanescent wave in the first few Angstrom behind the mirror plane. With the condition to have, in the thickness probed, a very diluted atomic species the intensity of the fluorescence is directly proportional to the absorption coefficient μ.

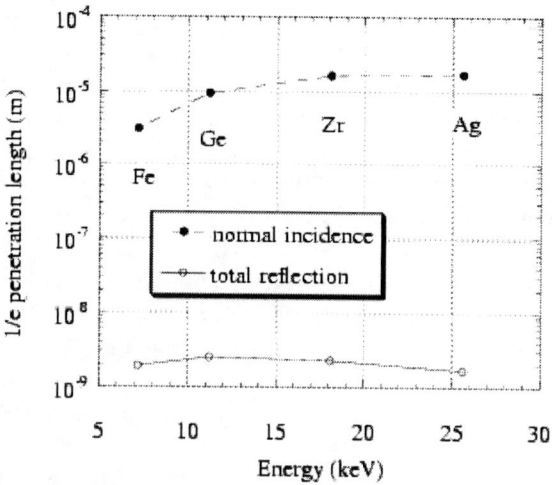

FIGURE 1. Comparison of the sample thickness probed by the standard transmission mode (on top) and the total reflection mode (on bottom).

The fluorescence intensity is given by

$$I_f = I_0 |T|^2 \frac{\mu_c}{\phi} z_c \varepsilon \frac{\Omega}{4\pi}$$

I_0 the intensity of the incoming beam

μ_c absorption coefficient of the sample

z_c the thickness probed for $z_c << z_{1/e}$

ε the fluorescence yield of the selected atom

Ω the solid angle collected by the detector

T measures the ratio, per unit of surface, of the intensity of the refracted and the incoming beam.

A special attention is deserved for the factor T. Indeed its value became

T = 4 when $\phi = \phi_c$,

T = 1 when $\phi >> \phi_c$, and

T $\to$ 0 for $\phi << \phi_c$

while out of a grazing incident geometry the transmission worth T = 1, and ϕ (typically 45°) should be replaced by $sin\phi$ [6]. In total reflection mode the ϕ is of order $\phi = 0.2$ and its $sin\phi$ may easily be hundred times smaller, meaning to increase the fluorescence intensity by a factor 100. Furthermore in very grazing incident geometry the study of a film deposited on a single crystal prevent the refracted beam to reach the crystalline substrate, avoiding in this way the disturbing effect of the reflected Bragg's peaks.

The ReflEXAFS measurements were done at the GILDA CRG [7] of the European Synchrotron Radiation Facility in Grenoble. The beam (40μm x 2000μm) is obtained collimating the synchrotron radiation by a couple of slit placed 2 meter before the sample. The very grazing incident geometry (0.17 degree in the case of Ge K-edge) makes the sample been shined on an area of 15mm x 3mm. In the measurements performed by collecting fluorescence spectra we paying attention, in order to ensure the detection linearity, to have the count rate does not exceed the 20 kcps (kilo-count per second). The [110] direction of the Si(100) substrate surface was oriented parallel to the polarization vector $\underline{\varepsilon}$ of the radiation beam.

The experimental cases

The first studied case is related to the tailoring of a new material for an optoelectronic device based on Si and utilize the well-developed Si technology, instead of the expensive III-V semiconductors. This is why in the last years a lots of research group have been involved in the growing and in the characterization of the Si/Ge multilayers. In such field one of the more debated topics is the way to form a sharp interface. Indeed Si and Ge have a natural tendency to diffuse each other compromising the effect of the super lattice mechanism. It has been found that an

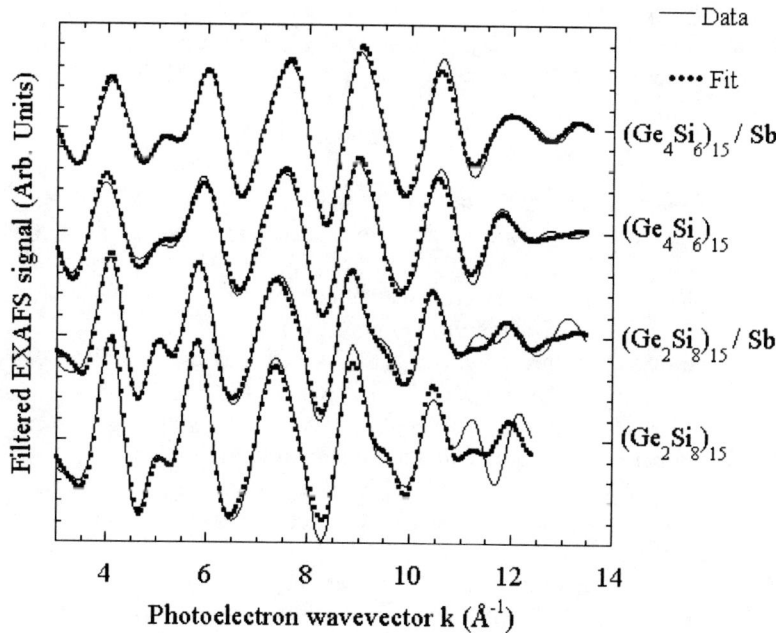

FIGURE 2. Fourier filtered EXAFS spectra in the range 1.5 – 3.6 Å.

antimony surfactant film deposited on the Si substrate before the alternative deposition of the Ge and Si film may reduce the diffusion process. We report on the study carried out on two different sets of multilayers samples prepared at the University of Camerino. The samples were grown by MBE (Molecular Beam Epitaxy) on a two Si inches substrate. In each set there were multilayers with two different chemical composition: Ge_2Si_8 and Ge_4Si_6. The two sets of samples were different by the presence of a monolayer of Sb surfactant. All the samples where capped by 22 monolayers of Si. The Sb was deposited on a hot substrate which temperature was about 450 °C. The subsequent annealing at 600 °C gets rid of the Sb excess. The correctness of the procedure has been checked, step-by-step, through RHEED patterns [8]. In Fig. 2 are reported the Fourier filtered data from where we obtained the results of Tab.1. From the data analysis, made by FEFF-8, we find that the role of the substrate temperature and the amount of Sb is crucial in the improvement of the interface quality. In particular in the system $(Ge_2Si_8)_{15}$ the Sb reduces the diffusion of the Si passing from 60% (case without Sb) to 30% (case with Sb). No effect is observed in the system $(Ge_4Si_6)_{15}$. Detailed analyses of this experiment are reported in ref. 9.

The second system investigated concern heterogeneous solid-state reaction. This is a process involving (at least) two solid phases to give rise to a solid product. They are important both from the point of view of technology (preparation by solid state synthesis of the ceramic materials), and in mineral metamorphism. The prototypical and most extensively studied example is the reaction between nickel and aluminium oxides to give the spinel [10]:

$$NiO + Al_2O_3 \longrightarrow NiAl_2O_4$$

This kind of reaction is usually studied by means of the so called "diffusion couple" method [11], in which two oxides are put in contact and heated at high temperature for a given amount of time. The theory, after the initial stage, is well known and assumes that the local chemical equilibrium is attained between each reagent phase and the product phase. At the contrary almost nothing is known in the very early stage when the local chemical equilibrium is not jet obtained and the chemical kinetics is not driven by long-range diffusion. The ReflEXAFS technique has been used as an experimental tool to investigate the very first stage of the solid-state kinetic processes. The sample preparation was made in a three-step process:

1) on a crystalline Al_2O_3 substrate, kept at room temperature, was evaporated 150Å of Ni monitored by a quartz balance and than the thickness was checked by X-ray reflectivity;

2) the film was exposed to an oxygen atmosphere of 400 mbar and annealed at 700 °C for about 30 min. The stoichiometry of the oxidized nickel was checked by EXAFS and compared with the standard NiO spectrum detected in transmission mode;

3) the solid-state reaction from the two oxides to the spinel structure was carried out by sequential annealing followed by ReflEXAFS measurement cycles. The temperature of the annealing was 1000 °C for 120 min in a 60 mbar of O_2 atmosphere. The treatment stop when the ReflEXAFS data were comparable with the EXAFS of the standard $NiAl_2O_4$ spinel.

From the XANES spectra as well from the fast Fourier transform is possible to observe

TABLE 1. Quantitative results of the EXAFS analysis on the first coordination shell. Error were calculated from the square roots of the diagonal elements of the correlation matrix

Samples	1st shell N	1st shell R (Å)	First shell I (±0.1)
Ge crystal	4.0±0.1Ge	2.458 ± 0.002	-
$(Ge_4 Si_6)_{15}$ Sb	2.1±0.2 Si 1.9±0.2Ge	2.43±0.02 2.45±0.02	-0.4
$(Ge_4Si_6)_{15}$	2.2±0.2 Si 1.8±0.2Ge	2.44±0.02 2.46±0.02	-0.4
$(Ge_2Si_8)_{15}$ Sb	2.5±0.3 Si 1.5±0.2 Ge	2.40±0.02 2.46±0.02	-0.3
$(Ge_2Si_8)_{15}$	3.2±0.2 Si 0.8±0.2Ge	2.40±0.02 2.47±0.02	-0.6
$Ge_{0.01}Si_{0.99}$	4.0±0.2 Si	2.387±0.004	-

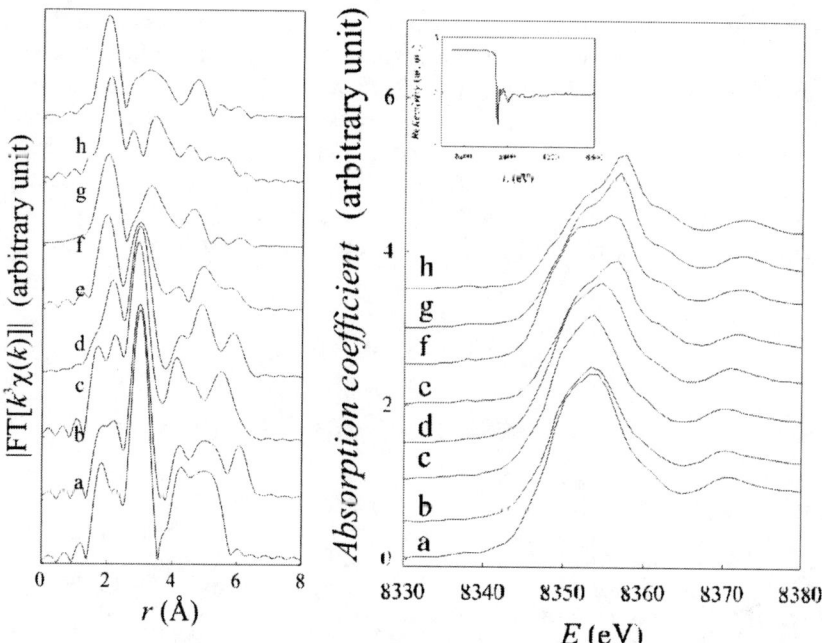

FIGURE 3. Left panel: XANES spectra after different treatment; a) NiO as grown, b) after 2h at 930 °C, c) after 4h at 930 °C; d) after 2h at 1000 °C; e) after 4h at 1000 °C; f) after 7h at 1000 °C; g) after 9h at 1000 °C; h) NiAl$_2$O$_4$. The inset shows the raw data reflectivity spectrum of g for reference. Right panel: Fourier Transform in the k = 1.5 – 11 A^{-1} range with k^3 weight. Letters have the same meaning of left panel.

the reaction advancement degree. With the increasing of the reaction the spectra starting from resembling the pure NiO become more and more similar to the spectrum of the NiAl$_2$O$_4$ (with the exception of spectrum f). Although a quantitative analysis is still in progress, we may qualitatively say that the decreasing of the Ni-O correlation seen in Fig.3b may be attributed to the formation of a disordered NiO phase, probably with some amount of Al dissolved. Also the reduction of the Ni-Ni peak (about 3Å) can be interpreted in this way.

CONCLUSION

The ReflEXAFS technique combine the atomic species selectivity, typical of the X-ray absorption spectra with the surface sensitivity obtained adopting the very grazing incident angle geometry. This combination makes the technique very interesting in the field of new material science. Indeed most of the new materials studied nowadays have their own domain on the nanoscale, exactly in the thickness range probed by ReflEXAFS. Furthermore this technique could study buried interfaces and does not need U.H.V. (Ultra High Vacuum) environment.

The cases reported are typical examples of two scientific topics faced by Physicist and Chemist and has been reported to show the interdisciplinary character of the X-ray absorption study. In the first case ReflEXAFS gave a quantitative estimation of the Sb surfactant role in the sharp interface formation, and have given the quantitative percentage of the interface quality. In the second case the technique has been exploit to monitor the kinetic process of the very early stage of the spinel solid solution. Although the detailed analysis in not jet complete, it is possible to say that the early stage of the reaction proceed by the substitution of Al atoms in the Ni sites.

REFERENCES

1. R.W. James, *The Optical Principle of the Diffraction of X-rays*, OX Bow Press. 1948 Woodbridge, Connecticut, USA.
2. L.G. Parratt, *Phys. Rev.* **95** (1954) 359
3. R. Barchewitz, M. Cremonese-Visigato, G. Onori, *J. Phys.* **C** 11 (1978) 4439
4. G. Martens and P. Rabe, *J. Phys. C* **14** (1981) 1523; S.M. Heald, H. Chen and J.M. Tranquada, *Phys. Rev. B* **38** (1988) 1016; B. Poumellec, R. Cortes, F. Lagnel and G. Tourillion, *Physica B* **158** (1989) 282; P. Borthen and H.H. Strehblow, *J. Phys.: Condens. Matter* **7** (1995) 3779
5. S. Pizzini, PhD Thesys University of Starthclyde, 1990
6. A. Naudon and D.J. Thiaudiere, *J. Appl. Crist.* **30** (1997) 822
7. S. Pascarelli, F. Boscherini, F. D'Acapito, C. Meneghini, S. Mobilio, *J. Synch.Rad.* **3** (1996) 147
8. R. Gunnella, P. Castrucci, N. Pinto, I. Davoli, D. Sebilleau, M. De Crescenzi, Phys. Rev. B **54** (1996) 8882
9. F. D'Acapito, P. Castrucci, N. Pinto, R. Gunnella, M. De Crescenzi and I. Davoli, *Surf. Sci.* (2002) in press
10. P.G. Kotula, C.B.Carter, *J. Am. Ceram. Soc.,* **81** [11] (1998) 2869
11. H. Schmalzried, *Solid State Reactions* Verlag Chemie ed. 1981

Developing XAFS Method Designed for Characterization of Materials Containing Nanostructures (Ge/Si Systems)

S.B. Erenburg[1], N.V. Bausk[1], L.N. Mazalov[1], A.I. Nikiforov[2], A.I. Yakimov[2]

[1]Institute of Inorganic Chemistry SB RAS, Lavrentiev Ave. 3,Novosibirsk, 630090, Russia,
[2]Institute of Semiconductors Physics SB RAS, Lavrentiev Ave. 13, Novosibirsk, 630090, Russia

Abstract. Here we report the use of EXAFS (Extended X-ray Absorption Fine Structure) and XANES (X-ray Absorption Near Edge Structure) spectroscopy method to study the spatial and electronic structure of Ge/Si heterostructures. The pyramid-like Ge islands deposited on Si(001) substrate using molecular beam epitaxy at 300°C reveal quantum dots (QDs) properties. Measurements of X-ray absorption fine structure at germanium K edge (GeK XAFS measurements) have been performed using total electron yield and fluorescent detection mode. It is revealed that the pure Ge nanoclusters are covered by 1,2-monolayer film with 50% Si atom impurity caused by interface diffusion at 500°C. The Ge QDs are characterized by interatomic Ge-Ge distances of 2.41Å which is 0.04 Å less than in bulk Ge. The influence of effective thickness of the germanium film, Ge nanocluster sizes and Ge, Si deposition temperature on the QDs microstructure parameters is revealed. The first attempt to extract from X-ray absorption spectra (XANES spectra) an information about the energy structure of the free states of the quantum dot was made.

INTRODUCTION

Development of technique allowing to determine spatial and electronic structure parameters on the surface of materials will provide the controllable fabrication of structures containing nanoclusters with discrete electronic spectra. The fabrication of such systems will allow achieving the success in the traditional trend toward miniaturization and in element engineering for quantum computer.

Along with the traditional surface-sensitive methods, such as Auger spectroscopy, low-energy electron diffraction, attempts are made to use the traditional physical methods which earlier were considered to be unsuitable for solving such tasks. Here we report the use of EXAFS (Extended X-ray Absorption Fine Structure) and XANES (X-ray Absorption Near Edge Structure) spectroscopy method to study the spatial and electronic structure of Ge/Si heterostructures. Measurements of X-ray absorption fine structure at germanium K edge (GeK XAFS measurements) have been performed using surface sensitive total electron yield and fluorescent detection mode.

CP652, *X-Ray and Inner-Shell Processes: 19th International Conference on X-Ray and Inner-Shell Processes*
edited by A. Bianconi, A. Marcelli, and N. L. Saini
© 2003 American Institute of Physics 0-7354-0111-X/03/$20.00

Following formulation basic ideas of the modern theory EXAFS[1,2] this method finds application for structural studies of substances and materials which have no long-range order and is unique in that it allows one to obtain direct structural information for such systems which is unobtainable by the traditional methods of X-ray structural analysis and electron-diffraction. It proved to be very efficient for studying liquid, liquid-crystalline, amorphous and poorly crystallized systems.

The phenomenon of self-organization in the process of heteroepitaxial semiconductor system growth allows fabrication of dense, extended, well-ordered structures containing islands of uniform shape and size. Such structures are especially important for high-technology applications [3,4]. In semiconductor nanostructures with spontaneously ordered inclusions of a narrower bandgap material within a broader bandgap matrix the limiting case of the dimensional quantization is realized within a certain interval of microinclusion sizes with the formation of so-called quantum dots (QDs) characterized by discrete electronic spectra.

In a heteroepitaxial system with a mismatch between the lattice constants, the initial growth may occur by layers. Formation of a thicker layers leads to a tendency toward elastic strain relaxation and elastic energy decrease by disturbing the two-dimensional growth and forming isolated islands, the so-called Stranski-Krastanov growth.

In a series of comparatively recent studies it was shown by different experimental methods (capacitance spectroscopy [5], hopping transport [6], admittance spectroscopy [7], optical spectroscopy [8]) that at a certain thickness of the epitaxial Ge film on Si(001) in the electronic spectra of the heterostructures there appear features associated with the zero-dimensional density of the states. These features are due to the dimensional quantization of the hole spectrum in the Ge islands appearing during disturbed two-dimensional growth.

The spectrum of states in the self-organizing nanoclusters may be largely influenced by the elastic deformation at the boundaries arising from a mismatch of the lattice parameters of the nanocluster and substrate. In the Ge/Si system the lattice mismatch amounts to 4.2%. This will cause changes in the local structure: local distortions of the symmetry, changes in the valence angles and interatomic distances. Such structural changes may change the energy spectrum by a magnitude of the order of 0.1 eV [9], which is comparable with the dimensional quantization energy in a QD.

The local structural changes in the thin layers and nanoclusters are not detectable by the traditional X-ray structural analysis or electron diffraction, because such systems have no long-range ordering. EXAFS (extended X-ray absorption fine structure) and XANES (X-ray absorption near-edge structure) spectroscopy provides a unique possibility for solving such problems [10]. These methods allow determination of parameters of the local environment of atoms and electronic parameters of nanoclusters. Thus in [11], using XANES spectroscopy, a broadening of the forbidden gap with decreasing size of the CVD grown germanium nanoclusters in silicon was found. In [12, 13], using EXAFS spectroscopy, the local environment of Ge adatoms on Si(001), the structure of thin Ge layers on Si(001) and a superlattice with strained $(Ge_4/Si_4)_5$ layers were studied.

The growth of self-organizing Ge islands and later of the blocking Si layer is evidently accompanied by the temperature-dependent surface diffusion of Si and Ge atoms, affecting the composition of the wetting layer composition and the composition and width of the intermediate layer at the Si-Ge interface. These characteristics of the transitional layers should have a substantial influence on the features in the QD energy spectrum. This local structural information is obtainable from EXAFS data. Thus a high degree of interlayer mixing of Ge and Si in strained $(Ge_4/Si_4)_5$ layers was found in [13].

Therefore, energy spectrum calculation for QD, interpretation peculiarities of their experimental energy spectra as well as design of elements with given electronic properties must take into account variation of the QD local structure.

In this work, a surface-sensitive EXAFS measurement based on total electron yield and fluorescent detection was used to study peculiarities of Ge/Si structures.

SECTION 1. EXPERIMENTAL

Two structures prepared by molecular beam epitaxy (MBE) on two halves of a (001) Si substrate were studied experimentally. Both structures contained layers of Ge separated by 10 nm thick blocking layers of Si and differed from each other only in the thickness of their Ge layers. In one structure each Ge layer had a thickness of 4 monolayers, in the other structure this thickness was equal to 6, 8, 10 monolayers. It was suggested that in the structure of the first type a pseudomorphous film of Ge would be formed, and in the second type structure disturbances of the two-dimensional growth after thickness of 4 monolayers will occur with the formation of pyramidal Ge nanoclusters. 10 monolayer clusters have lateral dimensions of ~15 nm, a height of ~1.5 and the separations between the islands of ~5 nm. The wetting layer thickness and the size of the Ge nanoclusters on Si grown under disturbed growth conditions were determined by electron diffraction, high-resolution electron microscopy and scanning tunneling electron microscopy [6]. The Ge film growth temperature was equal to 300°C, the growth rate was 0.035 nm/s, the deposition temperature of the blocking layer was 500°C and 300°C. For use as reference compounds, thick films (300 nm) of solid solutions of a different composition (30% Ge, 50% Ge and 75% Ge) were prepared by MBE under the same conditions.

The germanium K edge (GeK) EXAFS and XANES spectra in the Ge/Si (100) heterostructures were measured using the synchrotron radiation (SR) of the VEPP-3 storage ring at the Budker Institute of Nuclear Physics, Novosibirsk. The X-ray energy was defined by a double crystal monochromator with a channel-cut Si (111) single crystal. The spectra were recorded using the surface-sensitive EXAFS technique based on the measurements of the flux of electrons of the whole energy spectrum (total electron yield technique) and of the flux of fluorescent X-rays (fluorescent detection mode). For anisotropy study of the Ge atom environment in heterostructures, the GeK spectra have been measured at the parallel ($\parallel$) and perpendicular ($\perp$) Si(001) orientations relative to the electric field vector E of the linearly-polarized SR beam.

The obtained data were processed using the EXCURV92 software package [14]. In the data processing, the phase and amplitude characteristics were calculated in the X_α - DW approximation, using the procedures of the package [14].

For the analysis of the Ge local environment the Fourier-filtered data were fitted using the k and k^2 weighing procedure in the photoelectron wave vector interval from 2.5 Å^{-1} to 13 Å^{-1}. The error in determining interatomic distances with the fitting procedure was ±0.01 Å. The amplitude-damping factor S_0^2 was determined by data fitting for massive Ge and was equal to 0.8.

SECTION 2. RESULTS AND DISCUSSION

Indeed, while the amplitude and phase characteristics of $k\chi(k)$ for the spectra of the Ge_xSi_{1-x} films with different stoichiometries are significantly different, comparison of $k\chi(k)$ of the EXAFS spectra of a film of the $Ge_{0.50}Si_{0.50}$ solid solution and the 4-monolayer film show their complete similarity both in the amplitude and phase. These facts does not mean that the four monolayers that grew are a uniform solid solution; but this fact establishes the presence of an appreciable exchange of atoms between the Ge and Si phases, which decreases the elastic strain in the system during the deposition of the blocking Si layer (500°C). Thus the monolayers forming the so-called pseudomorphous Ge films embedded in a Si matrix contain up to 50% of Si atoms. It is essential that our results (see below) show that only the atoms in the immediate vicinity to the boundary are really participating in the exchange. On processing our data the fitting was performed using the experimental data Fourier-filtered in the interval 1.5 Å < R < 2.6 Å. The values of the Debye-Waller factor ($\sigma^2 =$ 0.0034 Å^2) and of the energy Eo = 10.6 eV were determined as a result of fitting for pure Ge with interatomic distances R(Ge-Ge) = 2.450 Å and the coordination number n = 4 and were fixed in all other cases.

Simplest models No 2, 3, 4 included one averaged variant of the environment for the Ge atoms with the distances R_1(Ge-Ge), R_2(Ge-Si) and the coordination numbers of Ge as to germanium atoms (n_1(Ge)) and to silicon atoms (n_2(Si)) (Table I). To a certain degree, such models are correct for the 4-monolayer films. The models should include at least two types of Ge atoms for heterostructures with quantum dots: the atoms at the boundary and those within the bulk of Ge nanoclusters. For the samples with QDs the models included two types of Ge atoms: 1) Ge atoms at the boundary in the 2-monolayer thick transition layer – such atoms account for about one half of all the Ge atoms, 2) the atoms inside the quantum dots accounting for the other half of the Ge atoms. The thickness of the transition layer was estimated using the conclusion of this work that the pseudomorphous 4-monolayer Ge films embedded in the Si matrix and having the thickness of two transition layers contain about 50% of Si atoms. This assumption turned out to be true, since with the use for the quantum dots of the same transition layer parameters (of the interatomic distances R_1(Ge-Ge), R_2(Ge-Si) and the coordination numbers n_1(Ge) in germanium and n_2(Si) in silicon) as in the 4-

TABLE 1. Parameters obtained by the fitting procedure for pseudomorphous Ge films and structures with pyramid-like Ge islands on Si(001). R_1(Ge-Ge)-, R_2(Ge-Si) – interatomic distances; n_1(Ge)-, n_2(Si) – the coordination numbers of Ge as to germanium atoms and to silicon atoms; The quality of the fit is determines by the index F [14].

No	sample	R_1(Ge-Ge)	R_2(Ge-Si)	n_1(Ge)	n_2(Si)	F
1	c-Gc	2.45 Å		4		0.8
2	Film ($E\parallel$)	2.41 Å	2.37 Å	1.70	2.35	0.6
3	QD ($E\parallel$)	2.40 Å	2.38 Å	3.16	1.67	1.0
3a	QD ($E\parallel$)	2.41 Å	2.37 Å	1.70	2.34	1.7
		2.40 Å	2.38 Å	4.14	0.54	
4	Film ($E\perp$)	2.41 Å	2.36 Å	1.78	2.00	3.9
5a	QD ($E\perp$)	2.41 Å	2.36 Å	1.78	2.00	1.1
		2.41 Å	2.38Å	4.5	0.10	

monolayer films, the Ge atoms in the quantum dots have practically no silicon atoms in the first sphere of their environment.

Table I shows that the interatomic distances R_1(Ge-Gc) decrease in the thin films and structures with QDs by 0.04 Å as compared with pure Ge and are equal to 2.41 Å. The R_2(Ge-Si) distances decrease by 0.03 Å as compared with the sum of the covalent radii of Ge and Si and are equal to 2.37 Å. Further refinement does not appear correct at this stage because of the assumptions and suggestion made.

Fig. 1 illustrates the monotonic growth of the fraction of Ge atoms inside purely germanium islands with entirely Ge atoms in their environment as the effective

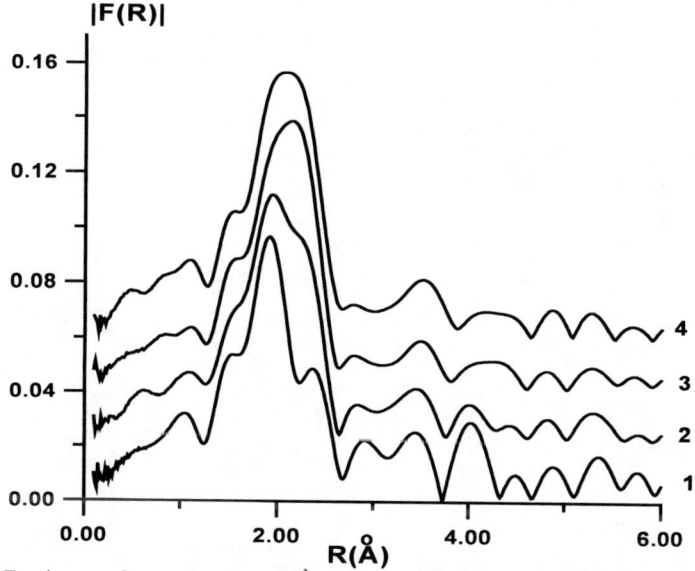

FIGURE 1. Fourier transform magnitude of $k^3\chi(k)$ GeK EXAFS data at $E\parallel$Si(001): for pseudomorphous 4-monolayer (4ML) 2D- films on Si(001) – *curve 1*, for Ge nanoclusters on pseudomorphous 4- monolayer films on Si(001) with effective thickness equal to 6ML – *curve 2*, equal to 8ML – *curve 3*, equal to 10ML – *curve 4*.

TABLE 2. Comparison of parameters obtained by the fitting procedure for pseudomorphous Ge films and structures with pyramid-like Ge islands on Si(001) with blocking Si layer deposited at 500°C and 300°C. R_1(Ge-Ge)-, R_2(Ge-Si) – interatomic distances; n_1(Ge)-, n_2(Si) – the coordination numbers of Ge as to germanium atoms and to silicon atoms; T - blocking Si layer deposition temperature. The quality of the fit is determines by the index F. [$E \parallel$ Si(001)]

T°C	sample	R_1(Ge-Ge)	R_2(Ge-Si)	n_1(Ge)	n_2(Si)	F
500°C	Film	2.41 Å	2.37 Å	1.7	2.3	1.0
300°C	Film	2.42 Å	2.37 Å	2.7	1.3	2.2
500°C	QD	2.40 Å	2.36 Å	2.6	1.4	0.5
500°C	QD	2.41 Å	2.37 Å	1.7	2.3	1.4
		2.41 Å	2.37 Å	3.7	0.3	
300°C	QD	2.42 Å	2.37 Å	2.7	1.3	2.9
		2.42 Å	2.38Å	4.0	0.0	

thickness of the germanium film increases from 4 to 10 monolayers, i.e. the monotonic size evolution of germanium nanoclusters as a function of film thickness in this series. It was of interest to study analogous systems prepared using slightly different temperatures for the film growth and the deposition of the blocking silicon layer.At the first point, structures were prepared using the same temperature for the film growth (300°C) but a lower (300°C) deposition temperature of the blocking Si layer. Table 2 displays influence of blocking Si layer deposition temperature on local structure parameters (n_1(Ge), n_2(Si)). As is seen from Table 2 with the change in the temperature conditions there is no appreciable change in interatomic distances but the partial coordination numbers of germanium with respect to Ge - (n_1(Ge) and Si - (n_1(Ge) are seriously altered. I.e. such decrease in the temperature leads to a substantial decrease of the diffusion between the phases Ge/Si and the formation of sharper phase boundaries.

As is seen from Fig. 2 the amplitude characteristics of the EXAFS spectra for such

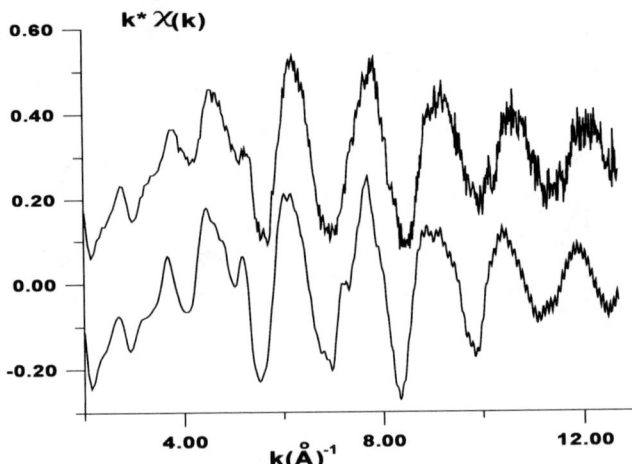

FIGURE 2. Ge K k-weighted normalized oscillating part of the X-ray absorption coefficient measured on pure Ge film (1000 Å) - *bottom curve* and on Ge nanoclusters (~15nm×~1.5nm) on Si(001) with blocking Si layer deposited at 300°C - *top curve*.

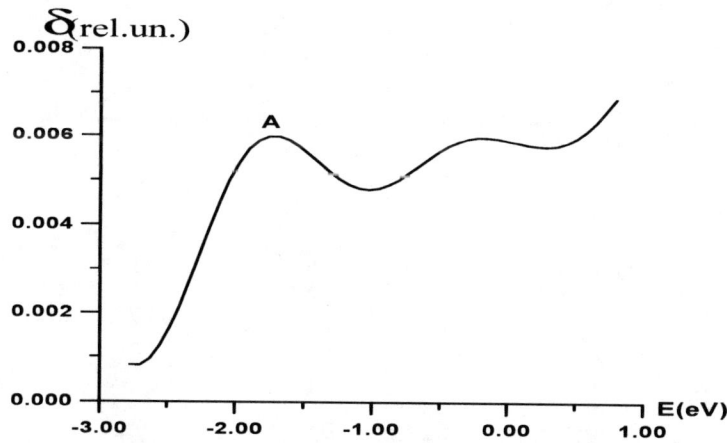

FIGURE 3. Relative difference absorption intensity of 8 ML Ge samples with and without boron dopants

films do not differ from those of the massive Ge and there is only a small phase shift. This effect may be explained by the scattering potentials of the environment for the absorbing atoms being similar not only in the first sphere of its environment. Therefore in this case we have structural similarity not only for the first coordination sphere of Ge but also an ordering in the farther coordination spheres of Ge in thin films analogous to that in the massive Ge.

On the other hand, the phase shift at large energies for the nanoclasters relative to the massive germanium is apparently due to a decrease in the Ge-Ge interatomic distances.

In addition, the interatomic distances R_1(Ge-Ge) and R_2(Ge-Si) obtained from the EXAFS data are in agreement with the interatomic distances derived from the spatial distribution of elastic deformation calculated within the valence force field (VFF) model. The calculation was performed by the procedure based on the use of the Green function of the "atomistic" elastic problem, which was further developed in [15]. The calculation was performed for the QD model with sharp Ge/Si boundaries.

In conclusion, a few words about the first attempt at extracting from X-ray absorption spectra the spectral information about the energy structure of the free states of the quantum dot itself. XANES spectra of two samples were compared. One sample was synthesized using the conventional conditions. The other sample was doped with boron (B) by a special procedure so that Ge in nanoclusters had a substantial deficit of electron charge.

A detailed comparative analysis of the obtained spectra showed the presence of a maximum A in the spectrum of the doped sample at a distance of -1.8 eV from the point of inflection of the main GeK absorption edge (the position of the GeK absorption edge) (Fig. 3). Such maximum is absent in the sample with no dopant. The maximum has the intensity of the order of 0.2% of the absorption jump in the main GeK absorption edge. It was suggested that the maximum is due to the appearance of free levels in the quantum dot at a depth of the order of 1.1 eV from the bottom of the

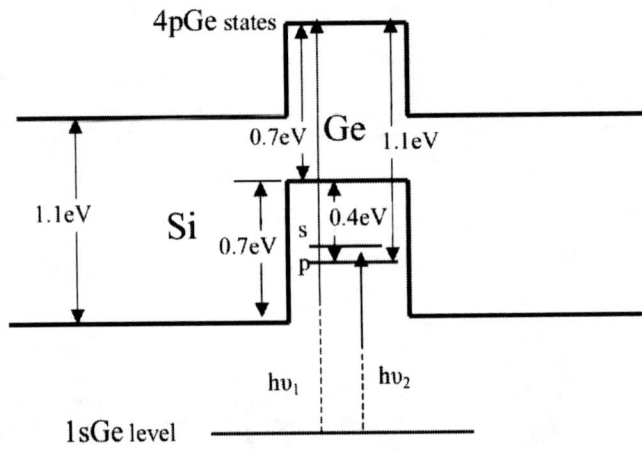

FIGURE 4. Scheme of hole states and X-ray absorption electron transitions in the Ge quantum dots with boron dopants

Ge conduction band (Fig. 4) in accordance with previous experimental and calculation results [16]. The figure shows the relative difference intensity (δ) for the spectra of the doped and pure samples as a function of energy (E).

CONCLUSION

The microstructural parameters of Ge/Si heterosystems largely influenced by the elastic deformation at the boundaries arising from a mismatch of the lattice parameters of the nanocluster and substrate was detected by the direct method showing that EXAFS spectroscopy is very perspective tool to study materials containing nanostructures.

ACKNOWLEDGMENTS

Financial support from the Russian State Scientific and Engineering Program on "Basic Researches in Physics" (theme 01.40.01.09.04, public contract 40.072.1.1.1176) is greatly appreciated.

REFERENCES

1. Sayers D.E., Stern E.A. & Lytle F.W., *Phys. Rev. Lett.*, **27**, 1204 (1971).
2. Rehr J.J. and Ankudinov A.L., *J.Synchrotron Radiation*, **8**, 61 (2001).
3. Shchukin V.A. and Bimberg D., *Appl.Phys. A.*, **67**, 687 (1998).
4. Barenco A., Deutsch D., Ekert A. and Jozsa R., *Phys. Rev. Lett.*, **74**, 4083 (1995).
5. Yakimov A.I., Dwurechenskii A.V., Nikiforov A.I. and Pchelyakov O.P., *JETP Lett.*, **68**, 125 (1998).

6. Yakimov A.I., Dwurechenskii A.V., Nikiforov A.I. and Pchelyakov O.P., *Physics of Low- Dim. Struct.*, **3/4**, 99 (1999).
7. Yakimov A.I., Adkins C.J., Boucher R., Dvurechenskii A.V., Nikiforov A.I., Pchelyakov O.P. and Biskupskii G., *Phys. Rev. B*, **59** 12598 (1999).
8. Yakimov A.I., Dvurechenskii A.V., Stepina N.P. and Nikiforov A.I., *Phys.Rev.B*, **62** 9939 (2000).
9. Rieger M.M. and Vogl P., *Phys. Rev., B* **48** 14276 (1993).
10. Koninsberger D.C. and Prins R., *X-ray Absorption: Principles, Applications, Techniques of EXAFS, SEXAFS and XANES*, New York, Wiley 1988, pp.710.
11. Kakar S., van Buuren T., Treusch R., Heske C., Himpsel F.J., Chase L.L. and Terminello L.J., in Abstracts of MRS 1998 Spring Meeting, V2.5, San Francisco, California 1998, p.350.
12. Oyanagi H., Sakamoto K. and Shioda R, *J. Phys. IV France*, **7**(C2) 669 (1997).
13. Wei S., Oyanagi H., Sakamoto K., Takeda Y. and Pearsall T.P., *J.Synchrotron Rad.*, **6** 790 (1999).
14. Binsted N., Campbell J.W., Gurman S.J. and Stephenson P.C., *SERC Daresbury Lab. Rep.*, (1991).
15. Nenashev A.V. and Dvurechenskii A.V., *JETP*, **117**(9) 570 (2000).
16. Dvurechenskii A.V., Nenashev A.V. and Yakimov A.I., *Nanotechnology*, **13** 75 (2002).

X-ray spectromicroscopy of clusters heated by fs laser radiation.

A.Ya. Faenov[1], A.I. Magunov[1], T.A. Pikuz[1], I.Yu. Skobelev[1], F. Blasco[2], F. Dorchies[2], C. Stenz[2], F. Salin[2], G.C. Junkel-Vives[3], J. Abdallah, Jr.[3], T. Auguste[4], S. Dobosz[4], P. D'Oliveira[4], S. Hulin[4], P. Monot[4], E. Biémont[5], P. Quinet[5], S. Hansen[6], A. Shlyaptseva[6], U.I. Safronova[7] and K.B. Fournier[8]

[1]Multicharged Ions Spectra Data Center of VNIIFTRI, Mendeleevo, Moscow region, 141570 Russia
[2]CELIA, Universite Bordeaux 1, 33405 Talence, France
[3]Los Alamos National Laboratory, P.O. Box 1663, Los Alamos, New Mexico 87545, USA
[4] Centre D'Etudes de Saclay, DSM/DRECAM, CEA, 91191 Cif-sur-Yvette, France
[5]IPNE, Université de Liége, Sart Tilman, B-4000 Liége 1and Astrophysique et Spectroscopie, Université de Mons-Hainaut, Rue de la Halle, 15, B-7000 Mons, Belgium
[6]University of Nevada, Reno, MS 220, Reno, Nevada 89557
[7]University of Notre Dame, Notre Dame, IN 46566
[8]Lawrence Livermore National Laboratory, P.O. Box 808, L-41, Livermore, California 94550

Abstract. The review of systematic investigations of X-ray radiation properties of different clusters heated by short-pulse (35-1100 fs) high-intensive (10^{16}- 10^{18} W/cm^2) Ti:Sa laser radiation is presented. The cluster targets were formed by the adiabatic expansion in vacuum of an Kr or Ar gas jets produced by a pulsed valve with Laval or conical nozzles. The gas pressure is varied from 15 up to 100 bar. High spectrally ($\lambda/\delta\lambda$=4000-5000) and spatially (40-80 µm) resolved X-Ray spectra near resonance lines (4-2 transitions) of Ne-like ions of Kr, H- and He-like ions of Ar have been obtained and detailed spectroscopic analysis was consistent with a theoretical two-temperature collisional-radiative model of irradiated atomic clusters incorporating with an effects of highly energetic electrons. The role of laser prepulse for X-ray intensity emission investigated in details. X-ray spectra radiation from plasma with electron density more than $2x10^{22}$ cm^{-3} was observed. Big effect of fast electrons influence on the X-ray emission of He-like Ar and Ne-like Kr spectra was demonstrated. Comparison of data obtained under various experimental conditions clearly showed that for increasing X-ray output from plasma the most essential to increase size of clusters and has reasonable value of ps prepulse.

INTRODUCTION

At present the cluster targets based on the supersonic gas jets, expanding at high pressure to the vacuum chamber, are widely used for plasma creation by femtosecond pulses, first of all, in connection with different applications (e.g. controlled fusion,

CP652, X-Ray and Inner-Shell Processes: 19th International Conference on X-Ray and Inner-Shell Processes
edited by A. Bianconi, A. Marcelli, and N. L. Saini
© 2003 American Institute of Physics 0-7354-0111-X/03/$20.00

plasma sources of monochromatic and broad-bend x-ray radiation and charged particles laser acceleration [1-10]. Besides these the similar experiments could be performed for the investigations in the fundamental spectroscopy and plasma diagnostics. In particular, the promising direction is the study of unusual (from the viewpoint of traditional conditions of the laser-created plasma) emission spectra of multicharged ions such as radiation transition lines in hollow ions [11-18] and transitions in inner shells, i.e. the radiation decay of autoionizing states (AIS). These lines are observed as a rule at high plasma density and their relative intensities are very sensitive to the plasma parameters. It is known that the high-density plasma is created during the absorption of a super short laser pulse in a cluster target [10,19-21], so radiation from AIS could be very large in such case.

The specific conditions in the femtosecond laser-produced plasma, namely, a comparably low ionization stage and an abundance of the evident fraction of hot electrons (see, e.g., [21,22]) have to lead to satellite lines in the emission spectra due to radiation decay of AIS in ions with a number of electrons from 2 to 9 (for spectra near the resonance lines of He-like ions) or with 11-12 for spectra near resonance lines of Ne-like ions. The existence of these lines gives an additional opportunity for the x-ray spectroscopy diagnostics. However, this demands preliminary investigations of satellite lines themselves, first of all, line's identification, an accurate wavelength measurements and line intensities modeling.

In this paper review of systematic investigations of X-ray radiation properties of different clusters heated by short-pulse high-intensive Ti:Sa laser radiation is presented. The cluster targets were formed by the adiabatic expansion in vacuum of Kr or Ar gas jets produced by a gas puff with a pulsed valve. High spectrally and spatially resolved X-Ray spectra near resonance lines of Ne-like ions of Kr, H- and He-like ions of Ar have been obtained and detailed spectroscopic analysis was consistent with a theoretical two-temperature collisional-radiative model of irradiated atomic clusters incorporating with an effects of highly energetic electrons.

EXPERIMENTAL SETUP

The experiments were conducted at the laser installation CELIA in Bordeaux, France and at UHI-10 in Saclay, France. The CELIA laser source [23] produces terawatt-level 35-1100 fs pulses with energy 15 mJ at high repetition rates. The contrast ratio between the main and secondary pulses could be varied from 10 to 10^6. A 6 micron laser spot radius is obtained in vacuum with an f/2 off-axis mirror with a 75mm focal length. The maximum laser intensity on target with a 35 fs pulse duration is $1-2 \times 10^{17} W/cm^2$.

The UHI-10 laser source [6] is a 10Hz Ti:Sapphire system with 10TW peak power and the pulse energy about 600 mJ after recompression for 60 fs laser pulses. The contrast is measured with a high-dynamic cross-correlator and to be about 10^6 for ns time scale and about 10^5 at 1 ps on the main beam. The corresponding Rayleigh length in vacuum and laser pulse intensity are 600 μm and $7 \times 10^{17} W/cm^2$, respectively. It is necessary to underline that the peak laser intensity of the order of $10^{17} W/cm^{-2}$ was enough to produce by the OFI ions of Ar only up to the F-like ones.

Cluster targets were produced in the vacuum chamber by a pulsed supersonic gas jet. The gas targets at Saclay were formed by a conical nozzle with big input and output holes (1 and 5mm, respectively), while at Bordeaux a Laval nozzle with medium-sized input an output holes (0.8 and 2.4 mm, respectively) and conical nozzle with sizes 0.62 and 3.8 mm were used. A backing pressure from 15 bar up to 100 bar was used in Ar experiments and from 15 up to 35 bar in Kr experiments. Mathematical

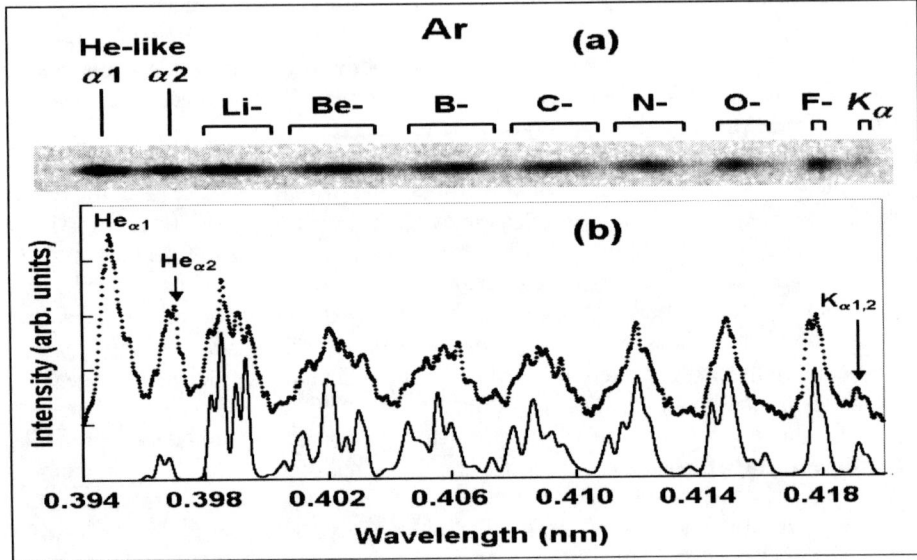

FIGURE 1. a) Typical image of spectra near resonance line of He-like Ar XVII, obtained in Bordeaux experiments. b) densitogram (above) and modelled (below) spectra .

modelling [21,22,24,25] of two-phase gas flow in such nozzles showed that the number of Kr atoms per cluster reached about 2.5×10^7 and 2×10^6 at Saclay and Bordeaux, respectively and about $10^5 - 10^7$ for Ar experiments.

The x-ray spectral measurements were conducted using an X-ray Focusing Spectrometer with Spatial Resolution (FSSR) [23,26,27]. A spherically bent mica crystal (R=150mm) and DEF X-ray film were used for the spectrometer in the Saclay experiments. The spectrometer was oriented in such way that spatial resolution in the direction of laser propagation was obtained. In the Bordeaux experiments, a spherically bent mica crystal (R=150 and 100 mm) and DEF X-ray film or X-ray CCD camera were used for the 2 spectrometers [23] and spatial resolution in the direction perpendicular and parallel to the laser propagation were obtained. The spectral resolution $\lambda/\delta \sim$ 5000 for the Saclay and $\sim$ 4000 for Bordeaux experiments have been reached. The spatial resolution was 30-40 μm at Saclay and 80 μm at Bordeaux experiments.

Typical spectrogram of Ar spectra, obtained on the film in Bordeaux experiments using 45 fs laser pulse is shown in Fig.1a. Strong emission of Li-...F-like dielectronic satellites is clearly seen. It is necessary to underline that at the same time intensity of

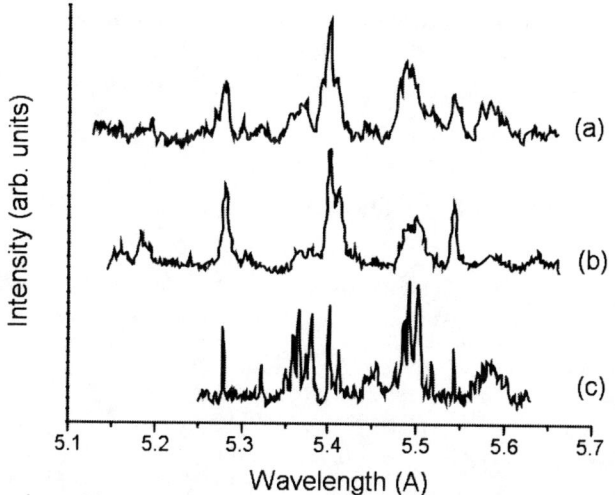

FIGURE 2. Experimental Kr laser plasma L-shell spectra: (a) Bordeaux fs laser plasma, (b) Saclay fs laser plasma, (c) nanosecond laser plasma [28].

K_α line is a very small. Examples of Kr spectra from Bordeaux and Saclay experiments are given in Fig. 2 (a) and (b), respectively. Figure 2 (c) shows a spectrum observed using the same type of FSSR spectrometer from an earlier experiment with Kr clusters irradiated with ns laser pulses [28]. Four distinct Ne-like lines and satellite structures from Na- and Mg-like Kr are evident in all of the experimental spectra in Fig. 2, while lines from F-like Kr are prominent only in the fs experimental spectrum from Saclay. The experimental spectra from the ns laser plasma has well-resolved lines and is dominated by Na-like emission, while both fs laser spectra have much broader lines and are dominated by Ne-like emission.

ATOMIC STRUCTURE CALCULATIONS

The calculation of energies, radiation and autoionization probabilities for the $1s^k2s^m2p^n(S'L')\ ^{2S+1}L_J$ levels in the Ar X–XVII ions were performed using the code based on the multiconfiguration Hartree-Fock method with the relativistic corrections (MHFR) [29,10,30]. The mixing of configurations shown above and those with the $3l$ electron in outer m-shell was included. For accounting the influence of other configurations the special procedure of optimization of electrostatic and spin-orbit interaction integrals was applied. This procedure consists in interpolation or extrapolation of the average energy of configuration along the isoelectronic sequences using experimental data for other ions (P VII-XIV, S VIII-XV, K XI-XVIII and Ca XII-XIX) available in literature. More details about the procedure one can find in [30].

Experimental wavelengths of Kr were determined using a relative calibration fixed by the Ne-like 4C line. Experimental relative intensities were determined by normalizing each spectrum to its most intense feature. The narrow Ne- and Na-like lines in the ns spectrum have wavelengths determined to within 1 mÅ and the broader

F-like lines in the Saclay fs spectrum have wavelengths determined to within 2.5 mÅ. Experimental wavelengths and relative intensities, theoretical wavelength values, and radiative decay rates for F-, Ne- and Na-like transitions were measured and calculated in [31].

KINETIC MODEL AND COMPARISON WITH EXPERIMENTS

The system of steady-state radiative-collisional rate equations was solved for uniform plasma with different values of plasma parameters. Multicharged argon ions

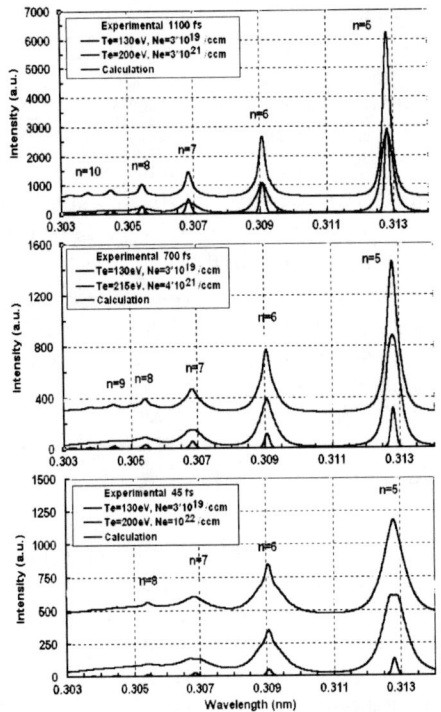

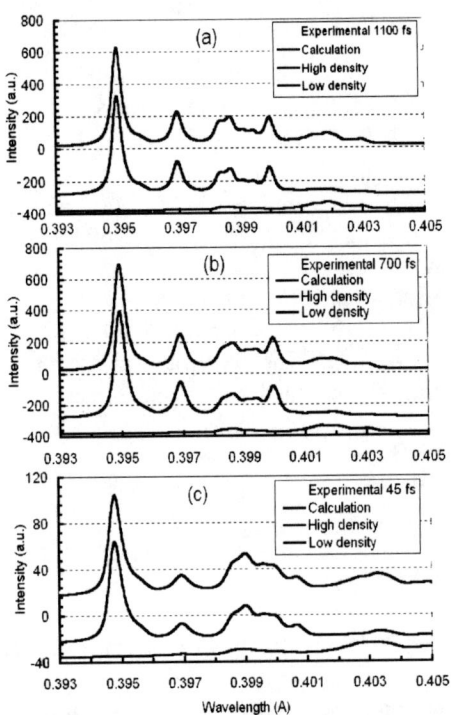

FIGURE 3. Comparison of the measured spectra of Ar plasma (top curve in each panel.) with the results of calculations (full modeling overlapped with experimental spectra; modeling with only high density plasma- middle spectra; modeling used only preplasma conditions- below spectra) for the Rydberg He-like lines of Ar XVII. a) 1.1 ps laser pulse, b) 700 fs pulse, c) 45 fs pulse. See for more details [19]

FIGURE 4. Comparison of the measured spectra of Ar plasma (above curve in each Fig.) with the results of model calculations (full modeling overlapped with experimental spectra; modeling with only high density plasma- middle spectra; modeling used only preplasma conditions- below spectra) for the spectra near He$_\alpha$ line of Ar XVII includes Li- and Be-like satellites. a) 1.1 ps laser pulse, b) 700 fs pulse, c) 45 fs pulse.

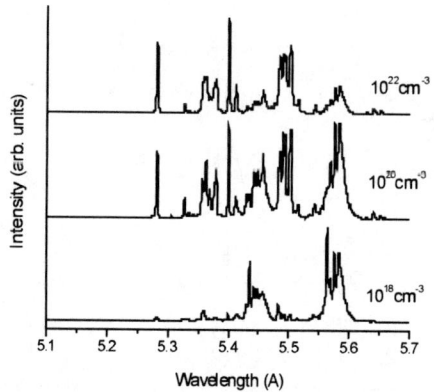

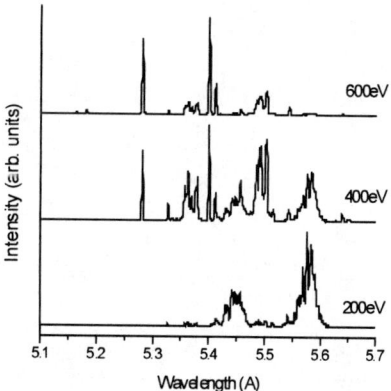

FIGURE 5. Modelled spectra with electron temperature T_e = 400eV, hot electron fraction f = 0, no opacity, and various electron densities.

FIGURE 6. Modelled spectra with electron density n_e = 10^{21}cm^{-3}, hot electron fraction f = 0, no opacity, and various electron temperatures.

with a total number of bound electrons m = 1,2,3 and 4 were taken into account [9,10,25]. Atomic configurations with principal quantum numbers n < 6 were considered, including auto-ionization states, for H-, He-, Li-, and Be-like ions (25 H-like levels, 59 He-like levels, 334 Li-like levels and 1188 Be-like levels). The rate coefficients for the electron collision processes were calculated using a model electron-energy distribution function, which includes a provision for hot electrons. The relatively long thermalization time for hot electrons make it possible to consider them as an electron beam with a Gaussian distribution centered around energy E_0= 5 keV. The emission plasma spectra were calculated for spectral region 3.93 - 4.05 Å, which was the region observed experimentally. Examples of results are presented in Fig.4. Calculations with this model [10,32] have shown that even a small amount of hot electrons with these parameters (around 10^{-7} – 10^{-5} in the case of Figs.3,4) makes a very big contribution not only to the satellites but also to the resonance lines of the He-like ions with atomic numbers Z_n = 10-20. Since E_{hot} is on the order of ionization potential of these ions, these electrons also efficiently excite the Rydberg states n^1P_1 with n ≥ 4 as well. Figure 3 shows the measured plasma emission spectra for various pulse durations in the range of the n^1P_1 - 1^1S_0 (n ≥ 5) transitions of the He-like argon ion. The line shapes of Ar XVII were determined with inclusion of the Stark shift in an ionic microfield, the impact broadening due to elastic electron-ion collisions, and the Doppler broadening [19]. The latter was taken into account in combination with the spectral resolution, which corresponded to the effective ion temperature T_i=2 keV. The distribution function for the ionic microfield was taken with regard to the ion correlations and the Debye screening. The fraction of hot electrons substantially affects the relative populations of Rydberg levels. One can see from Figs.3 and 4 that the results of independent calculations with the chosen plasma parameters reproduce well the experimental data – the smaller duration, the bigger plasma density, which makes

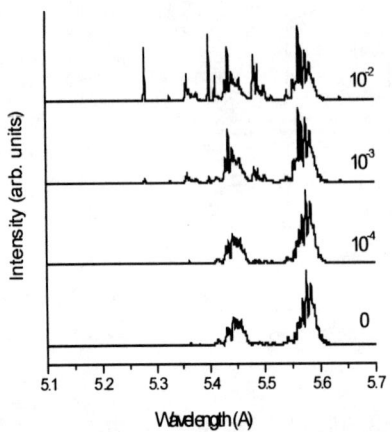

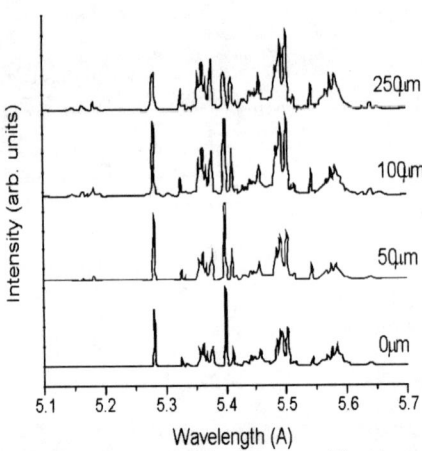

FIGURE 7. Modeled spectra with electron temperature T_e = 200eV, electron density n_e = 10^{20}cm^{-3}, no opacity, and various fractions of hot electrons.

FIGURE 8. Modeled spectra with electron temperature T_e = 500eV, electron density n_e = 10^{20}cm^{-3}, hot electron fraction f = 0, and various plasma diameters.

the main contribution to the observed spectra. Good coincidence between theoretical modeling and experimental results could be seen also from Fig.1b. The plasma parameters N_e=2·10^{22} cm-3, T_e=200 eV, T_h=5 keV and f=10^{-4} were taken from the results the kinetic calculations for the argon plasma, created under the same conditions [20,22].

For modeling of Kr clusters spectra the time-dependent collisional-radiative kinetics model calculates the populations of all the energy levels relevant to the construction of L-shell spectra. The model includes ground states of all ionization stages, from the bare ion to the neutral Kr atom, and fine-structure levels for 249 F-, 157 Ne-, 995 Na- and 914 Mg-like ions levels. Details of the configurations considered are given in [31].

The electron distribution function, taken into account in modelling, is composed of a Maxwellian portion at the bulk electron temperature T_e and a Gaussian tail with a width of 200eV centred about the energy T_{hot}, which we take to be 5keV, as in previous analyses of K-shell Ar [10] The dependence of modelled spectra on electron temperature is given in Fig.5. The ionization balance increases with the electron temperature so that higher ionization stages dominate the spectra, but there is a little change in the relative intensities of lines within a single ionization stage. The electron density dependence of modelled spectra is shown in Fig.6. Increasing density increases the ionization balance significantly and also changes the shape of the Na- and Mg-like satellite structures and the relative intensities of Ne-like lines. In particular, increasing density tends to amplify the long-wavelength satellites within a group. The dependence of modeled spectra on the fraction of hot electrons f is given in Fig.7. Increasing the fraction of hot electrons changes both the ionization balance and the shape of the satellite structures and tends to amplify short-wavelength satellite lines. In contrast to studies of K-shell Ar emission, hot electron fractions less than 10^{-3} do not have significant effects on the Kr L-shell spectra. This is because at low temperatures (<200eV), the Kr plasma is ionised only up to the Si-like ionisation stage. Emission from ionisation stages beyond Mg are not modelled here, so the effects of hot electrons on the modelled emission spectra are not apparent until the hot electron fraction is sufficient to affect at least the Mg-like charge state. At $f \sim 10^{-2}$, the effects of hot electrons are pronounced and Ne-like lines dominate the spectra.

The relatively large oscillator strengths of the Ne-like 4C and 4D lines suggest that they may be more susceptible to absorption than the 4G and 4F lines. In our modelling opacity is included with a simple absorption model [31] that treats the plasma as a homogeneous slab. Since the line opacity in this model is dependent on the density of ions in the lower level of transitions and their oscillator strengths, opacity effects increase the 4G/4D and 4F/4D Ne-like line ratios dramatically. Including opacity also increases the relative intensity of lines from other ionisation stages, which tend to have both smaller oscillator strengths and less populated lower levels. These effects are illustrated in Fig. 10, which shows modelled spectra at 500 eV and 10^{20} cm^{-3} with various plasma diameters. The emitting regions of both the ns and fs laser plasmas have a measured size of about 200 μm. The clusters in the ns plasma have enough time to expand to form a relatively homogeneous plasma with an electron density near 10^{20} cm^{-3}. In contrast, the dense ($n_e > 10^{21}$ cm^{-3}) clusters in the fs laser plasma persist during emission. The clusters have diameters around 0.06-0.07 μm and the distance between

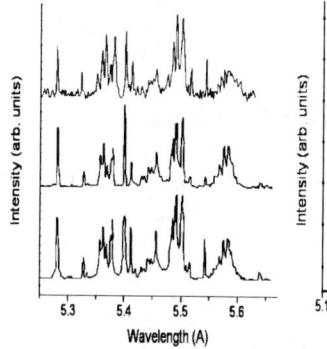 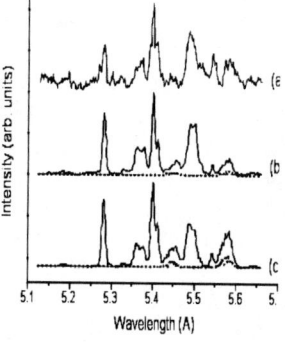 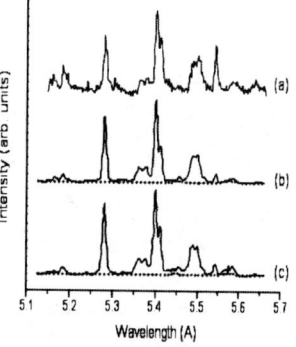

FIGURE 9. Comparison of experimental ns spectrum (a) and modelled spectra with and without opacity. The modelled spectrum in (b) has $T_e = 400$eV, $n_e = 1\times10^{21}$cm^{-3}, $f = 0$ and no opacity effects. The modelled spectrum in (c) has $T_e = 600$eV, $n_e = 8\times10^{19}$cm^{-3}, $f = 0$ and a plasma size of 200μm.

FIGURE 10. Bordeaux experimental fs spectrum (a), modelled spectrum without hot electrons($T_e = 525$eV, $n_e = 2\times10^{21}$ cm^{-3}, $f = 0$) in (b), modelled spectrum with hot electrons($T_e = 300$eV, $n_e = 2\times10^{21}$ cm^{-3}, and $f = 2\times10^{-2}$) (c). The dashed lines show the contribution of the bulk and prepulse plasma regions, which are included in the modelled spectra (solid lines) in (b) and (c).

FIGURE 11. Saclay experimental fs spectrum (a), modelled spectrum without hot electrons($T_e = 650$eV, $n_e = 2\times10^{21}$ cm^{-3}, $f = 0$) (b), modelled spectrum with hot electrons($T_e = 400$eV, $n_e = 2\times10^{21}$ cm^{-3}, and $f = 4\times10^{-2}$) (c). The dashed lines show the contri- bution of the bulk and prepulse plasma regions, which are included in the modelled spectra (solid lines)

clusters is around 1.5 μm [22], so along a line of sight the effective absorbing size of the fs plasma is around 8 μm.

The best-fitting modelled spectra to the ns laser plasma spectrum are given in Fig.9 (b) and (c). We could clear see that the modelled spectrum without opacity, shown in Fig.9(b), has several problems with the fit. The Ne-like lines are too intense relative to the Na- and Mg-like satellites. Also, the Ne-like 4G and 4F lines are much smaller than in the experiment and again, this cannot be resolved by changing the electron temperature, density, or the fraction of hot electrons. All these difficulties are resolved by including opacity, as shown in Fig.9(c). The large oscillator strengths of the 4C and 4D lines make them more susceptible to self-absorption than most other lines in the spectrum, so that the final fit with opacity has good agreement with the experimental spectra. No better fit to the ns spectrum can be obtained by including hot electrons.

The best fit for the Bordeaux spectrum without hot is shown in Fig.10(b). The modelled ratios of Na-like to Ne-like lines are in good agreement with experiment. Emission from cool regions of the inhomogeneous plasma contributes narrow Mg-like lines that bring the modelled spectra into better agreement with the experimental spectrum. This emission has been included in both modelled spectra and its relative magnitude is indicated by the dotted lines in Fig.10(b). However, even with this contribution the spectrum without hot electrons has too little Mg-like emission. We have shown that hot electrons are useful for spreading out the ionization balance so that larger fractions of Mg-like ions can coexist with the same emission ratios of Na- to Ne-like Kr. This is shown in Fig.10(c),where the synthetic spectrum included hot electrons is a better fit to experiment. If to use the single-temperature modelling in theoretical spectrum the Na-like satellite structure at 5.49 Å slopes is reversed compare with the experimental spectrum. This is precisely the effect of hot electrons, especially large fractions at small electron temperatures (see Fig. 7). The best fit with hot electrons gives good agreement with experiment for both the shape of the Na-like satellite structure and the relative intensities of the Mg- and Na-like features.

The experimental Saclay fs Kr spectrum is shown in Fig.11(a). The evidence of hot electrons in this experimental spectrum is the presence of both Mg- and F-like lines with significant intensity. This indicates a spreading out of the ionization balance characteristic of hot electrons. The best fit obtained without including hot electrons, given in Fig.11(b), has ratios of Na-like to Ne-like lines that are in good agreement with experiment, but both Mg- and F-like features are too weak in the high-temperature spectrum. The high temperature implies a large absolute intensity so that the contribution of Mg-like emission from the preplasma and the bulk plasma does not significantly improve the fit. The best fit with hot electrons, given in Fig.11(c), has an electron temperature of 400eV and 4% hot electrons. It has a approximately the same ratios of F- and Na- like line intensities as the spectrum without hot electrons and a better fit to the experimental spectrum of the Mg-like satellites. Note that higher temperatures and larger fractions of hot electrons are required for reasonable fits of the F- and Mg-like structures for the conical nozzle than for the Laval nozzle. This is in agreement with the recent analyses of Ar plasma from different nozzles [25,32].

CONCLUSIONS

Systematic experimental and theoretical investigations [5,6,9,10,19-25,31,32] of high-resolved Ar and Kr X-ray spectra carried out during past 4 years allow to do the following summary:

i) The interaction of unltrashort laser pulses with clusters strongly depends on the parameters of laser prepulse and duration of main laser pulse.

ii) For typically used nowadays Ti:Sa laser systems with contrast 10^{-6} plasma ionization state of obtained plasma is managed mainly by laser prepulse.

iii) High contrast 35-80 fs laser pulses interaction with big size clusters causes creation cold "overcritically" dense plasma and appearance in spectra relatively intensive complicated satellite structures of multielectron multicharged ions.

iv) For some practical applications laser prepuslse could play positive role – increasing X-ray emission of plasma and producing ions with greater charges.

The spectral resolution achieved in these measurements is limited by the Doppler shift in moving ions that does not allow direct measurement of the dominating autoionizing width in some lines from the observed profiles. However, estimations show that already with the double increasing in spectral resolution and autoionizing width such a measurement can be made possible. These conditions may be met in plasma, produced from clusters of heavier atoms. The verification of this possibility is of doubtless interest.

ACKNOWLEDGMENTS

The work was partly supported by Fond Europeen de Developpement Economique Regional and Conseil Regional d'Aquitaine (France), the NATO grant #PST.CLG.977637, Award No. RP1-2328-ME-02 of the U.S. Civilian Research & Development Foundation, the auspices of the U.S. Department of Energy by University of California Lawrence Livermore National Laboratory under contract No. W-7405-Eng-48. Work of S. H. and A. S. was supported in part by DOE, SNL, and UNR.

REFERENCES

1. McPherson, A., Luk, T.S.,Thompson, B.D. et al., *Phys.Rev.Letters* **72**, 1810 (1994).
2. Ditmire, T., Donnelly,T., Rubenchik, A.M. et al., *Phys. Rev. A* **53**, 3379 (1996).
3. Lezius, M., Dobosz, S., Normand, D., Schmidt, M., *Phys. Rev. Letters* **80**, 261 (1998).
4. Ditmire, T., Zwelback, J.,Yanovsky, V.P. et al., *Nature* **398**, 489 (1999).
5. Dobosz, S., Schmidt, M., Perdrix, M. et al., *JETP* **88**, 1122 (1999).
6. Auguste, T., D'Oliveira, P., Hulin, S. et al., *JETP Letters* **72**, 38 (2000).
7. Parra, E., Alexeev, T., Fan, J. et al., *Phys. Rev. E* **62**, 35931 (2000)
8. Milchberg, H.M., McNaught, S.J., Parra,E., *Phys. Rev.E.* **64**, 056402 (2001).
9. Junkel-Vives, G.C., Abdallah, J., Jr, Blasco, F. et al., *Phys. Rev. A* **63**, 021201R (2001).
10. Abdallah, J. Jr., Faenov, A.Ya., Skobelev,I.Yu. et al., *Phys. Rev. A* **63**, 032706 (2001).
11. Armour, I.W., Fawcett, B.C., Silver, J.D., Trabert, E., *J. Phys. B.* **13**, 2701 (1980).
12. Briand, J.P., Billy, J.P.L., Charles, P. et al., *Phys. Rev. Letters* **65**, 159 (1990); *Phys. Rev A* **43**, 565 (1990).

13. Winter, H., Aumay, F., *J. Phys. B.* **32**, R39 (1999).
14. Faenov, A.Ya., Abdallah, J.,Jr.,Clark, R.E.H. et al., *Proceedings of SPIE* **3157**, 10 (1997).
15. Urnov, A.M., Dudau, J., Faenov, A.Ya. et al., JETP Letters **67**, 489 (1998).
16. Faenov, A.Ya., Magunov, A.I., Pikuz, T.A. et al., Physica Scripta **T80**, 536 (1999).
17. Rosmej, F., Faenov, A.Ya., Pikuz, T.A. et al., *J. Phys. B* **32**, L107 (1999).
18. Abdallah,J., Jr., Skobelev, I.Yu., Faenov, A.Ya. et al., *Quantum Electronics* **30**, 694 (2000).
19. Magunov, A.I., Pikuz, T.A., Skobelev, I.Yu. et al., *JETP Letters* **74**, 375 (2001)
20. Junkel-Vives, G.C., Abdallah, J., Jr., Blasco, F. et al., *Phys. Rev. A* (2002) submitted
21. Skobelev, I.Yu., Faenov, A.Ya., Magunov, A.I. et al., *JETP* **121**, 73 (2002).
22. Skobelev, I.Yu., Faenov, A.Ya., Magunov, A.I. et al., *JETP* **121**, 966 (2002).
23. Blasco, F., Stenz, C., Salin, F. et al., *Rev. Sci. Instrum.* **72**, 1956 (2001)
24. Boldarev, A.S., Gasilov, V.A., Blasco, F. et al., *JETP Letters*, **73**, 514 (2001)
25. Junkel-Vives, G.C., Abdallah, J., Jr., Auguste,T. et al., *Phys. Rev. E* **65**, 036410 (2002)
26. Faenov, A.Ya., Pikuz, S.A., Erko, A.I.et al., *Physica Scripta* **50**, 333 (1994)
27 Skobelev, I.Yu., Faenov, A.Ya., Bryunetkin, B.A. et al., *JETP* **81**, 692 (1995).
28. Dyakin, V.M., Skobelev, I.Yu., Faenov, A.Ya. et al., *Quantum Electronics* **27**, 691 (1997)
29. Cowan, R.D., *The theory of atomic structure and spectra*, University of California Press, Berkeley, 1981.
30. E. Biemont, E., P. Quinet, P., Faenov, A.Ya. et al., *Physica Scripta* **61**, 555 (2000).
31. S. Hansen, S., A. Shlyaptseva, A., Faenov, A.Ya. et al., *Phys. Rev. E.* (2002) submitted
32. Junkel-Vives, G.C., Abdallah, J., Jr., Blasco, F. et al., J. Quant. Spectrosc. Radiat.Transf. **71**, 417 (2001)

X-ray Absorption Spectroscopy in Mineralogy: A Review

Annibale Mottana

Università degli Studi Roma Tre, Dipartimento di Scienze Geologiche,
Largo S. Leonardo Murialdo 1, 00146 Roma
&
Istituto Nazionale di Fisica Nucleare, Laboratori Nazionali di Frascati,
Via Enrico Fermi 40, 00044 Frascati RM, Italy

Abstract. The number of mineral species known to date rapidly approaches 4000, and yet they represent but a small fraction of all the known inorganic and organic compounds. Nevertheless, minerals represent an ideal field of activity for X-ray absorption spectroscopy (XAS), because the investigation of their crystal-chemical peculiarities takes an enormous advantage of the property of this method of being atom-selective, even in the presence of a wide range of competing atoms located in similar structural environments. As a matter of fact, XAS on minerals proved to be a useful probing method as early as for W. Kossel's pioneer studies of in the 1930's, just after the fine structures occurring at and near the absorption edge had been first detected. However, XAS did not really become consolidated in mineral studies until the 1980's, when synchrotron sources became available to users. A concise, but complete review of the historical and recent applications of XAS to minerals and to their analogues synthesized for geological/geophysical purposes i.e., to better understand the mechanisms by which the Earth evolves, is here given. Special reference will be made to transition metals (Ca, Ti, Cr, Mn, Fe, Ni) which absorb in the hard X-ray spectral region (> 4 KeV) and to the geologically-significant elements (O, Na, Mg, Al, Si, S and K) which absorb in the soft X-ray region (500-4000 eV).

INTRODUCTION

X-ray Absorption Spectroscopy (XAS) has recently become such a widespread method of mineral characterization as to make a whole generation of modern scientists believe that it is as traditional as X-ray fluorescence (XRF) spectroscopy and the X-ray diffraction (XRD) methods.

Indeed, this is not so. On one hand, after pioneer studies by C.G. Barkla and H.J.G. Moseley in the years between 1905 and 1915, XRF spectroscopy developed as a quantitative method after 1945 [1] and is now widespread only for rock analysis, having been replaced by the electron microprobe (EMPA, [2]) in mineral chemical studies. On the other hand, after the pioneer studies of M. von Laue and W.L. Bragg between 1912 and 1919, XRD methods developed impetuously from 1920 to 1960 [3]

CP652, *X-Ray and Inner-Shell Processes: 19th International Conference on X-Ray and Inner-Shell Processes*
edited by A. Bianconi, A. Marcelli, and N. L. Saini
© 2003 American Institute of Physics 0-7354-0111-X/03/$20.00

and kept growing steadily later on, so that they are now the reference tool for the structural characterization of all crystalline substances.

By contrast, although the absorption of X-rays was first investigated by Barkla as early as 1906, the first detection of a fine structure at the absorption edge was made only as late as 1916 [4] and the related absorption lines at the various edges were investigated even later [5,6]. Furthermore, after having been preliminarily interpreted for the XANES part [7] and, ten years afterwards, for the EXAFS one [8], this fine structure proved to be of such a difficult experimentally recording as to distract from further attempts, so that XAS as a method remained stagnant for nearly half the century that followed. Indeed, XAS was revitalized in the early seventies, but first on theoretical grounds only, to became operational after a major breakthrough in the type of X-ray source took place i.e., when conventional tubes were superseded and synchrotron radiation became the standard source of continuous X-rays, near the end of 1970's [9]. XRD and XRF appear to be mature methods, now, having reached their theoretical upper plateau and depending for their future developments upon external rather than internal forces (improvements in detectors, monochromators and sources, as against in the theory). By contrast, XAS is not yet completely theoretically understood and has not achieved its full potential exploitation even at the present time, for the reasons that follow (cf. [10-12]).

(i) Over the past 20 years, many scientists (cf. [13-18]) have struggled to reach a unifying theory of X-ray absorption i.e., a theory that would unify the one-electron single scattering (EXAFS) and multiple scattering (XANES) theoretical treatments of the phenomenon initially worked out by Sayers, Lytle and Stern [19-21] and by Dehmer and Dill [22-24], respectively, thus creating a consistent set of algorithms that would translate into a computer code. Note, in particular, the sequence of theoretical contributions by C.R. Natoli and his co-workers [13-15, 25-28]. They show how XAS theory progressively improved to cope with experimental evidence, but it always was and still is unable to cope with certain fine details occurring in the spectra. The EXAFS treatment is by now well founded, but, although significant improvements in the interpretation of the edge-region structure (i.e., the XANES features) are evident (see below), a general consensus have been not reached yet [29]. Consequently, the theoretical calculation codes too, that are necessary for the full reproduction of the whole absorption spectrum, appear to be still incomplete.

(ii) The intensity of the X-ray source (practically, after 1974: a synchrotron) is improving at a steady rate over the years (Moore rule) and is the major responsible for such a state of the art. Not only such an increase in intensity reflects onto the spectra with an enhanced signal-to-noise ratio, thus making the detection of even weakest effects feasible and increasingly accurate, but it also promotes resolution, given in addition to the singular properties of synchrotron radiation. All this is beneficial in that it makes the spectroscopic information very accurate, but it complicates greatly its interpretation so as to defy understanding.

(iii) Finally, the greatly increased number of study cases worked out so far has created the conditions for pointing out a number of exceptions that confirm the rule, on one hand, but, on the other hand, also a substantial number of others which defy understanding with the present theories. This is forewarning for a possible change of paradigm [30] i.e., of the very basic theories taken for granted in all present-day XAS investigations.

Summarizing: XAS historical development over the past 20-25 years occurred via a continuous equilibrium between experimental and theoretical improvements, to be best described as a sort of systematic running of one after the other, and resulting into a substantial upgrading of the whole art.

The overall state of such an art was reviewed frequently over the past, often quite extensively (e.g., [31]), but predominantly under the physicist or chemist points of view. I am now going to focus in detail on XAS application to minerals, the crystalline compounds that Nature offers to investigation. This subject too has been reviewed several times, although never extensively nor completely: e.g., [10,11] mostly centered their reviews on research carried out in the U.S.A, [32] in France, [33,34] in the U.K., [35,36] in Italy. Indeed, so far XAS studies of minerals and their synthetic analogues have been mainly carried out by groups from those four countries (i.e., those where a synchrotron radiation facility was readily available), with additional contribution by minor research groups in Canada, Japan, Germany and Portugal. Elsewhere XAS spread out either rather recently or at a non-regular time sequence. None has felt the need of reviewing the works of these minor groups for the international community and because of this I feel my duty to go through the entire evidence and review it *in toto*.

Nevertheless, I will start with a warning note: no one could ever review XAS in Mineralogy while ignoring completely the progresses of the XAS method in Physics and Chemistry. Indeed, although physicists and chemists mostly indulge in operating with simple compounds, usually pure in composition and artificial, they often are actually dealing with the man-made analogues of minerals. Therefore, without realizing, they contribute to the progress of Mineralogy too, and mineralogists benefit from them. In addition, physicists contribute to mineral-devoted XAS also by helping in the set up and in the maintenance of the apparatus: this is not at all a cheap contribution at a modern synchrotron radiation facility!

SYSTEMATICS OF MINERALS

The number of mineral species acknowledged by the International Mineralogical Association (I.M.A.) in the year 2000 was ca. 3850, and is now approaching 4000, the yearly growth rate being 50-60 species. Therefore, minerals are just a very small fraction of all solid compounds (ca. 100,000 inorganic substances and 400,000 organic ones, most of them man-made). However, they are very significant for Science, as for a long time they have been the only crystalline materials available and, even now, are among the most challenging solid substances for both their chemical and structural peculiarities. As a matter of fact, minerals often display new or unusual structure types to crystallographers (who mostly operate by XRD, on single crystals as well as on powders), thus constituting prototypes for future artificial products. Moreover, many minerals are different from their ideal end members, being often quite impure in their chemical compositions as they selectively concentrate into their structural sites all chemical impurities (minor elements) occurring in the natural environment they grow in.

Consequently, minerals offer certain interesting cases for XAS studies that artificial products cannot offer:

(i)   When pure in composition (i.e., close to an end member), they allow studying the local geometry of the sites where each individual major atom is located under the condition of (near) internal equilibrium, so that the structural asset of the compound can be probed by XAS locally, on a short range i.e., for volumes three order of magnitude smaller than those probed by XRD methods. In addition, XAS provides information on the electronic properties of each atom, which no XRD method does, such as its density of local and partially empty states and effective charge density. This kind of information may appear to be secondary for a mineralogist oriented to study the regional distribution of minerals in the Earth, but it is not for those who study minerals in the environment; it will certainly become more and more important in the next future for investigations devoted to the sustainable use of natural raw materials.

(ii)   When impure, minerals make it possible to identify the structural location of each different chemical impurity (minor element) and independently study it owing to the well-known XAS property of being chemical selective. In other words, XAS permits studying each atom independently on all others and this, in turn, allows gathering information on such important data as partitioning among sites, deformation at and electronic properties of each site, again on a short to very short range, albeit with lesser accuracy than for major elements. Nevertheless, this short range information has important consequences on the long-range geometry determined by the XRD methods, and leads to constraints upon the inferred crystal-chemistry and geochemistry that would not be conceived otherwise; e.g., it provides clues on the geographical variation of a given mineral species, an information that is of primary importance for mineral exploitation and dressing.

(iii)   Finally, there are certain important clues on crystal-chemistry that can be best retrieved using XAS: e.g., the oxidation state (valence) and oxidation ratio of transition atoms, one at a time even when two or more occur together in the examined structure.

Experimentally speaking, there never was, or is, any difference between minerals and other solid samples (salt, glass, etc.), provided they were prepared as powders (as it is usually the case). Only occasionally problems arise which depend on sample orientation (critical for single crystals). In such a case, the mineralogist well-known experience on optics becomes an important moment in sample preparation for spectra recording.

Most minerals are aluminosilicates (the most abundant atoms occurring in Nature being O, Si, and Al, in order of decreasing percentage) of alkaline and chalk-alkaline earths (Ca, Na, Mg, K, in the same order) i.e., light atoms that need either special monochromators [37] or special crystals [38] because their absorption lines lie in the soft X-ray range (< 4 KeV). The most significant heavy atoms occurring in the Earth's crust are the transition elements Fe, Ti, Mn, Cr and Zr. All of them but Fe are to be classed among "minor" or "trace" atoms. Yet, heavy atoms (as well as the rare earth elements and the radioactive ones) were studied first and best by mineralogists working with XAS, both for practical instrumental reasons (no need of special apparatus, nor of vacuum or of inert gas filled experimental chambers) and for their significance in geochemistry, being indicators in the "discriminant" analysis of igneous rocks [39].

Consequently, when reviewing the impact of XAS on Mineralogy, it is opportune to draw a separation not only between studies performed using conventional X-ray sources and synchrotron radiation, but also between those carried out on heavy atoms (mostly concerning the transition elements but also U and Th) and those on light atoms (i.e., the so-called "rock-forming" elements).

The first separation is both a time boundary (the "Brehmsstrahlung" from conventional X-ray tubes was used in XAS from 1920 till 1980 and occasionally still is - especially for EXAFS, where the rotating anode source, introduced in 1960, provides excellent spectra even now, but for heavy atoms only, as it was never strong enough to allow recording spectra for light atoms), and a quality boundary (white synchrotron radiation not only allows recording spectra all over the entire range of the electromagnetic spectrum with excellent signal-to-noise ratio and high resolution, but it does so over the very low energy range, down to few eV i.e., for very light atoms). Yet, the number of XAS studies on minerals involving light atoms (or the L-lines of heavy atoms, most of which absorbing at <1 KeV) is still rather small.

TIME SEQUENCE OF XAS STUDIES ON MINERALS

The first absorption spectrum on record for any atom and any mineral is probably Fig. 2 in [7]. Alternatively, it was the "Spektrogram" taken by E. Wagner [40] that Kossel quoted in that paper ([7] p. 317) and Wagner himself (in [6] p. 632) stated to have been recorded on rock salt, but I could not retrieve. Actually, Kossel's spectrum had been experimentally recorded by "Herr Stenström", just as W. Friedrich and P. Knipping had recorded the first diffraction plate at the suggestion of M. von Laue [3].

This pioneer spectrum (Fig. 1) shows the fine structure at the uranium M-edge in a mix of U, Th nitrates deposited on sandpaper. It consists of three features in the range 2900-3100 cm^{-1} plus an additional single feature at ca. 3500 cm^{-1}. In energy terms (5548, 5182, 4303, 3728 eV) they correspond to the M_1 3s, M_2 3p$_{1/2}$, M_3 3p$_{3/2}$,

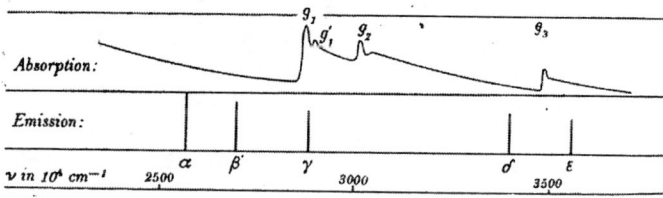

FIGURE 1. The absorption spectrum of uranium at the M edge recorded on a U-Th nitrate as reported by W. Kossel (1920 Fig. 2, cf. [7]). This is possibly the first published XAS spectrum on any mineral and was taken using a conventional discharge tube as source (see text).

M_4 3d$_{3/2}$ lines of U [41]. The M_5 3d$_{5/2}$ line at 3552 eV was not detected. It was on the basis of this meager experimental evidence that Kossel worked out its theory about X-ray absorption for what we now call the XANES region of the spectrum.

In the following years XAS became forgotten but for very simple solid systems, usually chemical compounds taken directly from a bottle in the store cabinet. Yet XAS spectra of minerals kept being recorded, albeit irregularly, and some of them are remarkable for resolution, despite of the conventional source used; e.g., those for REE in oxides [42], or for Si and Al in a variety of minerals including some rock-forming ones [43].

The first XAS experiment on a synchrotron source X-ray beam line was carried out at SSRL [44] and did not involve minerals, nor did those that rapidly followed (mostly in the same lab) as soon as information on the outstanding potentials of the new source spread out among physicists.

Consequently, it was only in 1978 when the first modern, synchrotron-derived XAS study on any mineral was performed, by Gordon Brown's group at SSRL [45]. As a matter of fact, the investigated material was not a true mineral, but synthetic NaFe^{3+}Si$_2$O$_6$, "acmite", a polysilicate end member the characteristics of which are such as to be safely extrapolated to its natural counterpart, aegirine, at least from the crystal-structure standpoint, although considerably deviating from it from the chemical one. This study (Fig. 2) is a good witness of the major reason of interest of XAS for the mineralogists of that time. The energy positions of the "white line" (i.e., the first and most intense feature occurring in the Fe K-edge region of the spectra) of acmite glass and of crystalline acmite synthesized from it are compared to determine the "chemical shift" (i.e., the shift in energy) that marks the change in coordination of Fe^{3+} from four-fold to six-fold.

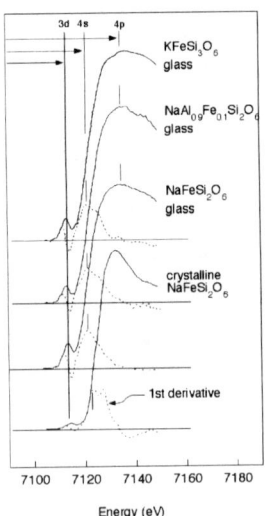

FIGURE 2. Fe K edges of three silicate glasses and the model compound "acmite", showing also the first derivative spectra (adapted from Brown et al., 1978, cf. [45]). This is the first XAS spectrum on any mineral taken using a synchrotron source (see text).

The eighties. Following Brown's model, several studies on glass and solid having the same composition followed, at the Fe K edge as well as at those of other transition atoms such as V, Cr, Ni, Co, etc. They were carried out, in particular, at LURE by the French group led by Georges Calas [46-48], but also at SSRL and NSLF [49-52]. These investigations produced twofold evidence:

(i)   features, particularly at the pre-edge (PE), increase in energy ca. 2-5 eV on going from the absorber low to high oxidation state;

(ii)   the intensity of the absorber "white line" decreases, and that of the pre-edge increases with coordination decreasing.

Furthermore, XAS showed to be able to detect the occurrence of multiple coordination for the same atom, as well as to pick up some unusual coordinations particularly for "network-forming" and "network-modifying" atoms in synthetic and natural glasses and in metamict minerals [53-55].

These studies were often combined with studies on "model" compounds (i.e., minerals and mineral analogues having the valence and coordination of their relevant transition atoms well determined by XRD methods), again to establish the energy dependence of the main features occurring in the XANES region of the absorber atom, when constrained in the structure of solids, upon oxidation state and coordination. This line of investigation established that:

(i)   the whole edge region, including the pre-edge, undergoes a negative "chemical shift" on going from high to low oxidation;

(ii)   the pre-edge feature energies are insensitive to bond length, but their intensities decrease strongly with increasing coordination;

(iii) the main-edge features are sensitive to both bond length and coordination, and shift to lower energies as bond length increases;

(iv)   the main-edge features may vary as a function of the symmetry of the site where the absorber is located: in particular, they increase with its distortion;

(v)   the number of features to be fitted in the pre-edge of a transition atom increases with the coordination of the site occupied by it;

(vi)   atoms located in centrosymmetrical configurations (e.g., octahedral sites) do not exhibit pre-edge, or they show a weak structure, but only when they are located in highly distorted sites.

Outstanding examples of such an investigation line are e.g., [50,56]. However, the substantial contribution due to the Italian school operating at the ADONE synchrotron (at that time one of the most advanced facilities in the world) should not be forgotten: the investigating scientists were not mineralogists, but physicists; as such, they always carried out their studies on simple synthetic analogues of minerals, and disregarded natural materials as being "dirty" [57,58], but contributed anyway to introducing XAS to mineralogists.

Indeed, in those very years we started at ADONE a project aiming at studying the spectral modifications related to changes in the M2 and M1 site-geometries of a series of clinopyroxenes, first in the binary join diopside-jadeite, then in the ternary system diopside-hedenbergite-johannsenite [59-61] (Fig. 3). This line of research on mineral groups and families aimed at investigate the influence of crystal structure

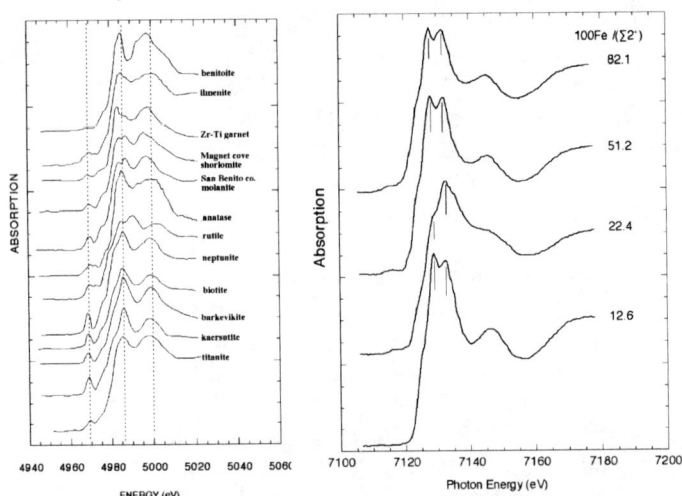

FIGURE 3. Two ways of making use of the XANES spectra. Left panel. Ti K edges of "model" compounds selected to detect oxidation, coordination and site distortion (adapted from Waychunas, 1987, cf. [50]). Right panel: Fe K edges of four clinopyroxenes in the system diopside-hedenbergite-johannsenite to detect variation in the geometry of the M1 octahedron as a function of Fe occupancy, the M2 polyhedron being always filled by Ca (adapted from Davoli et al., 1988 cf. [61]).

geometry on the spectral features for homo- or isostructural compounds, and for both one absorber atom or for two or more atoms displaying diadochic relationships to the first one. This line was independently followed to detect the accommodation of REE in epidotes [62] and of various metal atoms in sulfides [63-66]. By contrast, other researchers studied mineral sets belonging to the same family to the purpose of determining such properties as the electronic hopping in biotites during deprotonation [67] or to compare calculated vs. experimental spectra in feldspars [68].

The first studies on sheet silicates introduced the problem of the spectral discrepancy due to the anisotropic orientation in the sample, which was solved by using single crystals or films of oriented crystallites to record <u>polarized spectra</u> parallel or orthogonal to the electric vector of synchrotron radiation, which is highly linear polarized in the plane of the ring [69,70].

Another technical improvement occurred in 1987, when the first XANES spectra at the O K edge appeared [71]. Although performed on simple oxides and by a group of physicists, these studies were of interest to mineralogists because they could be compared usefully to parallel studies on the same materials at the Mg and Fe K edges [72]. Despite their great interest for geological and geochemical research, recording spectra at the O K edge remained seldom ever since because of the difficulties inherent in operating the Grasshopper apparatus [73].

The combined properties of XAS and synchrotron radiation allowed for the first time to study natural and synthetic <u>semi-amorphous to very fine-grained precipitates</u> occurring as crusts on the surface of many rocks. They produced information on their short-range structures and on the oxidation state of their constituting atoms, mostly 3d transition elements. Thus the location of Mn, Co, and Ni impurities in manganese oxide crusts from laterites could be identified [74,75] and the maturation of Fe^{3+}-bearing hydroxides deposited from gels was followed at different stages [76].

All these studies on semi-amorphous materials, as well as those at the PE and XANES regions of various "model" minerals mentioned above, were often performed in combination with, or supplemented by EXAFS studies having their main aim to confirm the "single scattering" origin of EXAFS modulations and deduce from it a structural information not to be easily gathered by XRD methods. They could receive computational support from the highly successful theoretical approaches that had developed during the middle years 1970's [77-79] and were undergoing upgrade in the early 1980's [80-82], thus creating the conditions for the further popularity of EXAFS. By contrast, there was a persistent underrate against XANES, that went on being considered useful only for the "fingerprinting" identification of closely related structures, rather than for accurate quantitative determinations, as it turns clearly out e.g., by Natoli's rule of bond length determination $(Er-E_h) \cdot d^2 = cost$ [25] defined in those very years.

The nineties. There was no break in the use of synchrotron-derived XAS among the few mineralogists operating at that time. Actually, there was a systematic spreading of the technique, particularly for the aspects of Mineralogy bordering on Geochemistry or overlapping with it, which include studies on glasses and amorphous substances. Moreover, at that time XAS had become fully acknowledged even by many XRD crystallographers as being another method that would provide useful, otherwise non-obtainable data on bond lengths and coordinations by means of the EXAFS part of the spectrum. By contrast, the XANES part was still underrated as a "fingerprinting" tool on the way of upgrading to give semi-quantitative data; in particular, the PE part of the XANES spectrum could, in the case of transition metals (especially Fe), provide reliable data not only on coordination, but also on the oxidation ratio, although with a fairly large error.

Another substantial change intervening during this decade is in the extending of XAS studies to lower and lower energies, both for the K lines of the involved atoms (light elements), and for the incoming use of the L lines (transition metals, heavy elements). Such a trend is partly due to the opportunity offered to mineralogists of having second generation synchrotron available for their research (the physicists being most concerned in developing their own research in the third generation ones), but it also reflects the widespread understanding that certain sources no longer on the far front of Physics could still provide novel, unexpected information to all Material Sciences - Mineralogy included - than conventional sources did.

Studies on <u>mineral groups and families</u> continued to unravel the changes undergone by XAS spectra as a result of changing the chemical composition with the structure being substantially the same. They were extended to K edges such as those of S, Co, Cu and Zn, or to L edges such as those of Ag, Cd, and Sb. The investigated materials were first miscellaneous minerals [83], then, in a systematic way, sulfur compounds such as thiospinels [84], chalcopyrites [85,86], tetrahedrites [87,88], pyrites [89,90], as well as the industrial mineral wolframite [91]. The above sequence of references reflects the application of the potentials of L-edge XAS best revealed by the British school operating at SRS, Daresbury [34,92]. By contrast, the French and Italian schools mainly devoted themselves to the K lines of low-energy elements that are significant in the Geosciences: Mg, Al and Si [93-95], and Na, Mg and Al [96-99], respectively. Off course, there were also studies about the K edges of transition elements [100-104], among which two [100,101] should be pointed out because they showed that Ti can indeed be present in four-fold coordination in the polysilicate chains, thus solving a long standing controversy on its proxy with Si. Another study [102] contributed to the problem of Fe^{3+} in four-fold coordination in the micas.

A scientific intermediate position, that was made possible by the use of grating monochromators operating at a rather weak source such as ALADDIN I (1 GeV), was taken by the Canadian school. They were able to investigate not only low-energy K edges such as those of Al, Si and P, but even very very low-energy ones such as that of B (188 eV), as well as at the corresponding L edges. In this way, the Canadian group could approach and solve mineralogical problems absolutely impossible to deal with by other methods (the only alternative one being ELNES), such as the structural characterization of SiO_2 polymorphs from stishovite to opal, in particular confirming the special characteristics of $^{[6]}Si$ in stishovite and the close similarity between opal and cristobalite [105-108], as well as the four-fold coordination of B in danburite. They also could point out the surface damage created by the rapid conversion of $^{[4]}B$ to $^{[3]}B$ taking place in danburite due to its reaction with moisture or by its grinding or polishing [109,110].

Among the mineral groups, the one that benefited most from XAS studies is the pyroxenes. Omphacites were studied for the order-disorder distribution of Ca-Na and Mg-Al at the M2 and M1 sites, respectively [97,99] (Fig. 4), while orthopyroxenes for their Fe partitioning and for the bonding configuration acquired by the octahedral site at both room and high temperature [111-113]. The purpose was always to establish a geothermometer and to define the intrinsic crystal chemistry, using the information also to model the structure from the theoretical point of view [94]. Another well-studied mineral group is the garnets, which were studied at many edges, and particularly at the Ti and Fe K edges to solve the problem of Ti valence in "schorlomites" [83,114,115]. This black variety of garnet was found to belong mostly to the andradite species, but occasionally also to grossular [114]. Its coloring agent, Ti, is entirely tetravalent and in octahedral coordination, whereas the little V present, also octahedral in coordination, is trivalent [115]. Staurolites were studied at the Fe, Mn, Zn, and Ti K edges [116] for both the PE, XANES and EXAFS regions in order to assign all major and minor atoms in the different sites of such complex structures. Fe, Zn and Mn are mostly divalent in the tetrahedral site T2, but some substantial octahedral Zn is also present; by contrast Ti is tetravalent and in octahedral coordination.

Metamict minerals were also studied, but fairly rarely: the local environments of Zr, Th and U in zircon and thorite were found to undergo changes with increasing metamictization [117] i.e., much the same result as it had been obtained in an early study on zircon at the Zr and Nb *K* edges [118].

Obviously, most if not all these studies involved the systematic investigation of "model" compounds suitable to be used as reference for unknown minerals. In addition, spectra for such "model" compounds were recorded for rare atoms during studies at the margin of Mineralogy *sensu stricto*, and yet of high importance in Geochemistry, such as those related to contaminating mine tailings or nuclear waste: U [119], As [120], Cd [121] and Pb [122]. As a matter of fact, this kind of application of XAS was not new, having already had some 10-15 years of history in Chemistry and Environmental studies. However, only recently did mineralogists became involved, and this occurred after a preliminary stage during which various soils, glasses, gels and other semi-amorphous materials reacting with aqueous solutions had been studied only for their geochemical implications (e.g., [123-126]).

A large number of such studies were carried out on both parts of the spectrum i.e., for XANES as well as for EXAFS. However, most of them still relied only on the latter one for quantitative structure data (bond lengths, Debye-Weller factors, etc.) and used the former part for the electronic and chemical properties (valence, oxidation ratio, etc.). Nevertheless, these combined studies, when carried out on mineral series, were particularly useful in that they could show when or where a substitutional solid solution was continuous or involved partitioning and clustering. Furthermore, they were able to show the amount of distortion at each individual site involved in the substitution that was related to the electronic properties of the absorbing atom.

A special kind of approach to the study of mineral series was taken by our Italian school that chose to investigate pyroxenes, garnets, olivines and other "rock-forming"

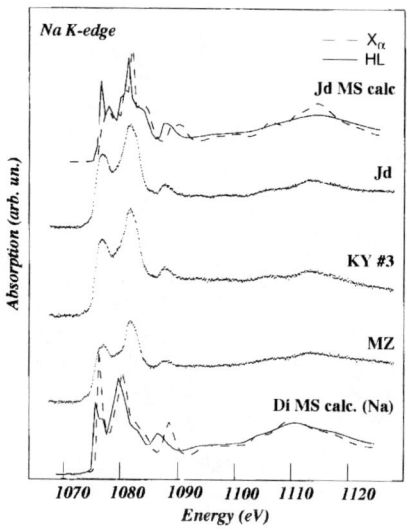

FIGURE 4. Na *K* edges of jadeites with C2/c symmetry, compared with the calculated spectra of reference jadeite and diopside, respectively (adapted from Mottana et al., 1997, cf. [97]).

minerals. We combined our XANES experimental spectra with calculated spectra computed from the structural data measured for the same (or very similar) minerals, and interpreted the observed discrepancies by differentiating what is possibly due to inadequacy of the theory (we constantly used Natoli's one-electron multiple-scattering approach and computer codes [14,27]) and what is undoubtedly due to the geometrical distortion undergone by the studied sites because of the atomic substitution involved in the solid solution [96-99,127-131]. In this way, having had the substantial support of theoretical calculations at several edges on many minerals, in particular garnets [129], we could demonstrate (Fig. 5):

(i)   the volume probed by the photoelectron around the photoabsorber extends to ca. 5-7 coordination shells (ca. 0.8-15 nm) depending upon the kind of structure involved, its type of order, and the channeling effects that may occur, but it is never so narrow or limited as to confirm the lasting prejudice against XANES as being only a local probe;

(ii)   the XANES region of the absorption spectrum consists of a first sector close to the edge that is sensitive to fairly long-range interactions (Full Multiple Scattering, FMS, region [26]) and extends in energy to ca. 25-30 eV above the threshold, and of a second sector (Intermediate Multiple Scattering, IMS, region [26]) that extends in energy to at least 50-60 eV above threshold. The IMS region merges into the Single Scattering, SS, regime of the EXAFS region. This IMS region is sensitive to order on the intermediate scale and may provide quantitative information on the atom distribution around the photoabsorber;

(iii)   where two sites containing the same atom contribute to absorption, the experimental spectrum is the sum of the weighted contributions of two independent

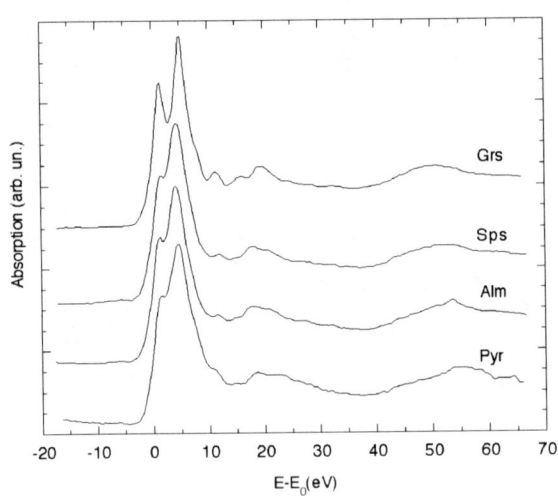

FIGURE 5. Variation of the Al *K* edges of garnets showing their dependence upon the type of eight-fold coordinated atoms (adapted from Wu et al., 1996, cf. [129]).

spectra and can be deconvoluted by computing these two spectra separately starting from small clusters having their geometries, then adding them up in different amounts until the experimental spectrum is reproduced. This allows determining the relative significance of the two sites and the partitioning of the absorbing atom between the two sites;

(iv) the pre-edge region varies in intensity depending upon coordination: a semi-quantitative evaluation of the amount of mixed coordination can be made, at least for Ti and Fe.

Many researchers tackled the problem of quantitative evaluation of the PE region in high Z elements, particularly Fe, the most important petrogenetic transition metal, and for a variety of silicates and oxides, among which amphiboles and olivines [132-136]. Their attempts took advantage either of the full synchrotron beam (this being the favorite method for glasses) or of micro-XANES (also called SmX, synchrotron micro X-ray spectroscopy) i.e., an intensively focussed beam scaled down in size to 10 x 10 μm, that allows studying mineral grains in thin section. The results were mostly semi-quantitative, as they depended greatly upon the kind of minerals used for calibration, but they were encouraging and worth pursuing, the more so as preliminary attempts on Mn and Cr [137,138] could confirm the validity of the PE method.

Another research line that benefited greatly from the combined use of the XANES and EXAFS treatments of the whole experimental spectrum was the study of minerals either very fine-grained or semi-amorphous i.e., those impossible to study by single-crystal XRD diffraction. These studies started by evaluating the involvement of minor or trace atoms in crusts or fine-grained ores for geochemical reasons e.g., Co, Cu and Ni in lithiophorite [139], but rapidly moved into the field of Mineralogy in that they could contribute to the understanding of the crystal chemistry of several major and minor atoms in a variety of hydrous or anhydrous Mn and Fe oxides [140-142]. The same type of investigation was carried out, repeatedly, on calcite [143-145], and the influence of the local strain induced by several impurity atoms (including U^{4+}, cf. [144]) substituting at random in the octahedral site onto the topology of the entire calcite structure could be precisely modeled.

The study of layer minerals via polarized XAS was one of the most significant research themes during this decade, as soon as the angular dependence of X-ray absorption spectra (that had been already known in the experiments for more than 20 years [146]) became definitively understood on a theoretical basis [147-149]. This technique, that required to have suitable single crystals available, was applied to mineral families such as the clay minerals and the micas [150,151], and produced results complementary, but much better resolved, than those obtained on disordered powder by the conventional method [152-157]. Most recently, a comprehensive treatment of polarized EXAFS applied to fine-grained minerals has been proposed [158,159] and the theoretical extent of the effect that polarization produces on the PE region of the XAS spectrum has also been studied [135].

The years nineties saw a substantial progress in the development of apparatus for high T-P XAS studies i.e., studies carried out under conditions close to those occurring in situ in the Earth, rather than on quenched materials. The first high-T furnaces that were developed were mainly used to investigate the glass-liquid transition in synthetic materials, but soon they were combined with studies on synthetic solids analogous to minerals, used as both reference standards and as actual study cases. Temperatures as high as 2100 K were reached [160-162]. More or less at the same time high-P presses were developed, but rapidly lost their interest because of the incoming of the diamond anvil cell (DAC). This small, but highly flexible apparatus was adapted in such a way as to let the X-ray beam impinge transversally the sample squeezed between the two diamonds rather than across them, so as to avoid the diffraction peaks (glitches) of the diamonds. Alternatively, the whole cell was rotated with respect to the beam in order to change the angle between the anvils and the beam and consequently the energy of the diffracted photons. The maximum pressure obtained was 6 GPa, and the most important results concerned not so much the observation of various phase transitions, but the amorphisation of the material above a certain P [163,164]. A simple addition to DAC [165] made then possible to carry out XAS measurements at high temperature i.e., as a function of combined P and T, although with large error margins.

Another substantial improvement in the apparatus that affected XAS study on minerals positively was the development of micro-XAS, which is preliminary to the event of a XAS microprobe (SmX). In modern Mineralogy this instrument is needed to complement effectively the nearly universal use of EMPA for chemical analysis on spots down to less than 1 μm in diameter, the ever growing use of TEM for the structural characterization at the nm level, and, finally, that of SIMS for trace element and isotopic analyses which make it possible to perform radiometric dating at the same very reduced spatial scale. Spectroscopic methods are hindered, under this point of view, by their intrinsic wave lengths, and also by the need of maintaining adequate signal-to-noise ratio and resolution when concentrating the impinging radiation over such small areas/volumes without spoiling the studied material or surface. In the case of XAS, such a technical problem was initially solved by making use of synchrotron radiation focussed beams (see above) combined with recording systems based on the total electron yield (TEY) or fluorescence yield (FY) modes [110,166]. They were effective for most applications when sources of relatively low energy were used, but were not when fit onto third generation sources, where the available high brilliance allows high flux densities. For these, a variety of focusing devices, all based on optics (zone plates, multilayer mirrors, Fresnel disks, etc.), have been developed. However, most of them fail of being useful to Mineralogy because they dramatically restrict the energy range of investigation, thus making EXAFS studies impossible. In most cases, the XANES spectra themselves are too short to be suitable of interpretation. Even when the mechanical problems involved in the adjustment of the sample and in its scanning are solved, these methods present problems that greatly complicate the acquisition of XAS spectra at a sub-μm resolution [167]. Research in this field has recently pointed towards tapered capillary optics as the workable tool to produce spectra over the whole energy range required by EXAFS (ca. 500 to 1000 eV) and with the required resolution for XANES in the near-edge region (10^4). Some results obtained on simple substances such as metals at third generation sources like APS,

Chicago, ESRF, Grenoble, and ANL, Berkeley, appear to be promising [168,169], but still out of the current interest of most mineralogists. This is not so much for the P-T range they can cover (up to 40 GPa and 1700 K), but because XAS can only be recorded on a high-energy range (> 7 KeV). Preliminary attempts at lower energies (down to ca. 5 KeV) have been limited or unsatisfactory (down to ca. 3 KeV) so far.

The new millennium. The turning of the century saw XAS becoming a more and more widespread tool among mineralogists; in particular for those exploiting two of its intrinsic abilities, on the line already investigated before:
1. determining the oxidation state and ratio of transition atoms;
2. determining the atom oxidation state, coordination and site location in very fine grained, semi-amorphous or amorphous natural phases.

Furthermore:
3. extracting quantitative information from polarized spectra has been investigated in the specific case of layer silicates.

The studies about oxidation are all based on the quantitative deconvolution of the experimentally-recorded pre-edge features into gaussian or lorentzian components and on the correlation of energies and intensities of these components to those of known reference materials (oxides and silicates). According to a by now well-tested procedure, they start from the simultaneous evaluation of glasses and simple minerals [132-134,170] and work out various ways of applying the general method to more and more complex minerals such as the rock-forming silicates [171-174]. Most studies deal with Fe, the commonest transition atom in Nature, but a few ones concern Cr, an atom that is most intriguing for environmental protection [175,176] because it occurs as $^{[4]}Cr^{6+}$ in aqueous solutions, and reduces to $^{[6]}Cr^{3+}$ when entering the structure of reacting Fe^{2+}-bearing silicates [177,178]. Micro-XANES (SmX) has recently been used to determine the Fe^{3+} and Fe^{2+} partitioning among several silicate phases in thin section (on a area of 10 x 15 μm), over a range from as high as 90% in micas to as low as 0-2% in garnets, and with an accuracy within ± 5-10%, despite the problem of surface orientation still needs to be better considered [174]. This method will be applied soon to rock-forming Cr-bearing minerals, as all the theoretical and practical bases of such a technique are already there [175].

The use of diamond anvil cell has spread out among spectroscopists despite the very many problems involved in the manipulation of the apparatus, in its set up at the synchrotron line and in the evaluation of the recorded spectra. DAC combines the potential of reaching rather high T, P with a very wide spectrum of operational conditions, so that its application is bound to become as common for minerals as it is now for geochemical fluids [179-182].

XAS has again and again proved to be increasingly useful to study fine-grained and semi-amorphous materials, especially metamict minerals as well as those which underwent amorphisation either because they were submitted to very high P, or because they were involved in a very high T contact metamorphism. Often, this way of making use of XAS is nothing more than a continuation of the initial study on the relationships between glassy and crystalline material by which Gordon Brown had started involving XAS into Mineralogy more than two decades ago [45]. A significant example of such a line of research concerns a new model of zircon metamictization proposed as a part of a study of site modifications around Zr in glasses under the combined influence of water and pressure [183]. This work also shows a number of Zr K edges of "model" compounds. The need for reliable spectra on "model" minerals is far from being exhausted: new data that account for both a large number of case studies and experimental improvements (spectral width, resolution, etc.) can be found in many papers: Fe [170, 184], B [185], Cu [186], Mn [187]. In this line of approach, the use of combined information from the PE, XANES and EXAFS regions usually leads to conclusions which compare favorably with those extracted by other methods. However, occasionally, it is EXAFS alone that is being used to characterize very fine grained minerals occurring as varnish, crusts, dendrites or, most generally, atoms dispersed in a selectively non-absorbing medium of any kind: Pb [188], Eu [189], Mn [187,190,191], Fe [192].

Most of these applications of XAS explore unconventional materials the mineralogical significance of which is dubious – so to say, but which are to taken into consideration anyway because they witness the most modern trend of Mineralogy, that is currently struggling to cover new fields such as biomineralogy, gemology, environmental mineralogy, etc. [193]. However, the "classical" aspects of Mineralogy have not been forgotten, and study on rare, as well as on common rock-forming minerals are constantly being pursued by mineralogists using XAS spectroscopy.

The temperature dependence of cation ordering in mineral solid solutions has been studied for synthetic orthopyroxenes [194,195] by both recording and computing the whole enstatite-ferrosilite series at the Mg and Fe K edges, the latter cation being constrained to be Fe^{2+} by the intrinsic oxidation-reduction conditions of the piston-cylinder apparatus. The linear dependence of the intensity ratio of the two major XANES peaks with composition (En mol%) has been documented together with the deviation from ideality of the M1 and M2 site occupancies; these two results open the way to the use of XANES results as geothermometer and geospeedometer, as suggested some time ago [111]. Composition and temperature dependence have been studied also for synthetic Ni-Mg olivines [196], showing that this mineral series does indeed show kinetics effects, but also that it is possible to define for it the intersite exchange energy and the blocking temperature on cooling.

The changes undergone by XAS spectra as a function of the angular dependence of the sample (polarized XAS or P-XAS) have been the subjects of many studies, particularly for clay minerals and the micas. The former minerals have been used in the form of turbostratic highly-oriented, self-supporting films treated as being a single crystal during quantitative analysis. As a typical example, the study was carried out on nontronite for both the PE features and the EXAFS part of the polarized spectra [197,198], however together with a variety of other methods (XRD, IR, ME, texture analysis) in order to reach definitive results. It is notable that not only the spatial distribution of the Fe, Al and Mg atoms in and out the octahedral plane could be determined, but also the preferred alignments of Fe-Fe and (Al,Mg)-(Al,Mg) pairs

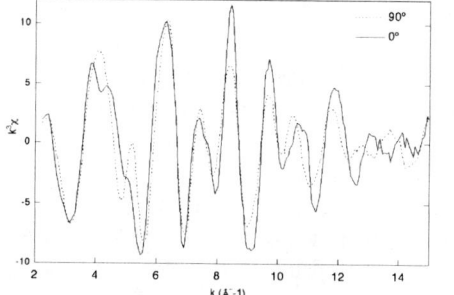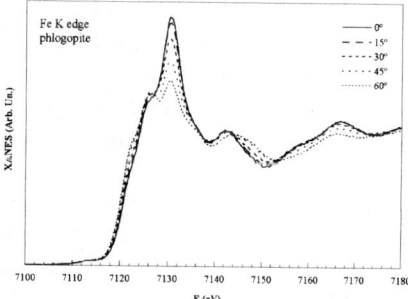

FIGURE 6. Left panel: polarized Fe *K*-edge EXAFS for nontronite (adapted from Manceau et al. 1998, cf. [151]). Right panel: polarized Fe *K*-edge XANES spectrum of phlogopite (adapted from Mottana et al., 2002, cf. [200])

along certain structural directions; such a result implies the segregation of Fe in small *trans*-octahedral domains separated by empty octahedra.

Micas have been the subject of many investigations, which produced not only "model" spectra for the majority of the mica species acknowledged by I.M.A. at most significant *K* edges (Mg, Al, Si, K, Cr, Fe), but also a systematic data bank that includes both powder samples and oriented samples taken at different angles. Single crystals (blades obtained by cleaving mica crystals as large as 5 x 5 mm) have been used. They were impinged by a synchrotron radiation strongly polarized in the horizontal plane first at right angle, than rotated along the vertical axis by as much as 75° i.e., so as to be almost parallel to the impinging beam. Not only a strong variation of the intensities in the PE and XANES features was observed, but also their energy displacements and even the appearance of new features (Fig. 6). These new features are interpreted as being activated by third (or higher) order multiple-scattering pathways progressively involved in the photoelectron scattering process. The examination and evaluation of one individual case (tetra-ferriphlogopite) has already appeared [199], while the overlook of all results can be found in Ref. 200.

CONCLUSIONS

After its rebirth in the early seventies and the initial random application to miscellaneous minerals and amorphous compounds, XAS has become a most powerful tool to unravel certain intimate properties of natural solid materials by exploiting to the best its property of being chemically selective (as all spectroscopic methods are). Through synchrotron-induced XAS, mineralogists now have a technique available that can selectively determine with a high level of accuracy the oxidation state and ratio of anyone of the many atoms which occur in a mineral of any grain size (or even apparently amorphous), either when they are in essential amounts or in minor viz. trace amounts. Furthermore, XAS can locate each atom in the structure by defining its coordination shell, determining its bond lengths to the atoms nearby, and evaluating the extent of the order that the given atom and its first coordination shell acquire over a fairly long distance (up to several coordination shells).

Information from XAS complements well the information obtained via methods based on XRD and, despite being less accurate in the determined values, it actually completes our knowledge of the structural asset of the examined mineral to a scale of magnitude three orders smaller than the XRD methods. Thus, information from XAS establishes itself in between the very local information that TEM offers to a mineralogist and the long-range one that he best determines when solving a structure using single-crystal XRD. Moreover, all this extensive XAS information can be determined on very small volumes, at a few micrometer size, and at high pressures and temperatures.

This is entirely on the line that modern Mineralogy is aiming at. Consequently, it is easy to foresee that XAS will continue to spread among mineralogists and will acquire further importance in all types of research on minerals and related compounds, including those related to the life sciences (Biomineralogy).

ACKNOWLEDGEMENTS

This review would have not been possible unless preceded by a long series of cooperative studies with many colleagues and students. Ivan Davoli, first, introduced XAS to me at ADONE, in the early days when this synchrotron was in the front line of second generation facilities. Then I could get advantage of the cooperation of Augusto Marcelli, Eleonora Paris and many others, till the former students and present young co-workers Giannantonio Cibin and Francesca Tombolini. They not only help me now in moving forward with the study of minerals, but were also essential in preparing and polishing this review. Our joint research activity at synchrotron facilities is supported by C.N.R., Rome (contract CNRC00C9DF001).

REFERENCES

1. Jenkins, R., de Vries, J.L. *Practical X-ray spectrometry*, Macmillan, London (1970).
2. Castaing, R. *Application des sondes électroniques à une méthode d'analyse ponctuelle chimique et cristallographique.* Thèse, Paris (1951).
3. Ewald, P.P., ed. *Fifty years of X-ray diffraction*, Oosthoek, Utrecht (1962).
4. Kossel, W. *Verhandlungen der Deutschen Physikalische Gesellschaft*, **18**: 339 (1916).
5. Fricke, H. *Phys. Rev.* **16**, 202 (1920).
6. Hertz, G. *Physikalische Zeitschrift*, **21**, 630 (1920).
7. Kossel, W. *Z. Phys.* **1**, 119 (1920).
8. Kronig, R.deL. *Z. Phys.* **70**, 317-323 (1931); II: **75**, 191-210 (1932); III: **75**, 468 (1932).
9. Lytle, F.W. *J. Synchrotron Rad.* **6**, 123 (1999).
10. Brown, G.E. Jr., Calas, G., Waychunas, G.A., Petiau, J. in *Spectroscopic methods in Mineralogy and Geology [Chapter 11]* edited F.C. Hawthorne. *Rev. Mineral.* **18**, 431 (1988)
11. Brown, G.E. Jr., Parks, G.A. *Rev. Geophys.* **27**, 519 (1989).
12. Davoli, I., Paris, E. in *Absorption spectroscopy in Mineralogy*, edited by A. Mottana, F. Burragato. Elsevier, Amsterdam, 1990, pp. 206.
13. Kutzler, F.W., Natoli, C.R., Misemer, D.K., Doniach, S., Hodgson, K.O. *J. Chem. Phys.* **73**, 3274 (1980).

14. Natoli, C.R., Benfatto, M., Brouder, C., Ruiz Lopez, M.Z., Foulis, D.L. *Phys. Rev. B* **42**, 1944 (1990).
15. Filipponi, A., Di Cicco, A., Tyson, T.A., Natoli, C.R. *Solid State Commun.* **78**, 265 (1991).
16. Zabinsky, S.I., Rehr, J.J., Ankudinov, A., Albers, R.C., Eller, M.J. *Phys. Rev. B* **52**, 2995 (1995).
17. Filipponi, A., Di Cicco, A. *Task Quarterly* **4**, 575 (2000).
18. Benfatto, M., Congiu Castellano, A., Daniele, A., Della Longa, S. *J. Synchrotron Rad.* **8**, 267 (2001).
19. Sayers, D.E., Lytle, F.W., Stern, E.A. *Adv. X-ray Anal.* **13**, 248 (1970).
20. Sayers, D.E., Stern, E.A., Lytle, F.W. *Phys. Rev. Lett.* **27**, 1204 (1971).
21. Sayers, D.E., Lytle, F.W., Stern, E.A. *J. Non-Cryst. Solids* **8-10**, 401 (1972).
22. Dehmer, J.L., Dill, D. *Phys. Rev. Lett.* **35**, 213 (1975)
23. Dill, D. Dehmer, J.L. *J. Chem. Phys.* **61**, 694 (1974)
24. Dehmer, J.L., Dill, D. *J. Chem. Phys.* **65**, 5327 (1976).
25. Bianconi, A., Dell'Ariccia, M., Gargano, A., Natoli, C.R. in *EXAFS and Near Edge Structure*, edited by A. Bianconi, L. Incoccia, S. Stipcich. Springer, Berlin, 1983, pp. 57.
26. Natoli, C.R., Benfatto, M. in *EXAFS and Near Edge Structure IV*, edited by P. Lagarde, D. Raoux, J. Petiau. *J. Phys. (Paris)* **47-C8**, 1986, pp. 11.
27. Tyson, T.A., Hodgson, K.O., Natoli, C.R., Benfatto, M. Phys. Rev. B **46**, 5997 (1992).
28. Natoli, C.R. Invited talk to 19th International Conference on X-ray and Inner-Shell Processes (Roma, 24-28 June 2002)
29. Rehr, J.J., Albers, R.C. *Rev. Mod. Phys.* **72**, 621 (2000).
30. Kuhn, T.S. *The structure of scientific revolutions.* University of Chicago Press, Chicago, 1962.
31. Crozier, E.D. *Nucl. Instrum. Meth. B* **133**, 134 (1997).
32. Calas, G., Manceau, A., Combes, J.M., Farges, F. in *Absorption spectroscopy in Mineralogy*, edited by A. Mottana, F. Burragato. Elsevier, Amsterdam, 1990, pp. 171.
33. Henderson, C.M.B., Cressey, G., Redfern, S.A.T. *Radiat. Phys.Chem.* **45**, 459 (1995).
34. Schofield, P.F., Henderson, C.M.B., Cressey, G., van der Laan,G. *J. Synchrotron Rad.* **2**, 93 (1995).
35. Mottana, A. *Atti Accademia Nazionale Lincei* s. IX, **2**, 103 (1991).
36. Artioli, G. *Notiziario Neutroni e Luce di Sincrotrone* **2**, 4 (1997).
37. Hussain Z., Umbach, E., Shirley, D.A., Stohr, J., Feldhaus, J. *Nucl. Instrum. Meth.* **195**, 115 (1982).
38. Wong, J., Shimkaveg, G., Goldstein, W., Eckart, M., Tanaka, T., Rek, Z.U., Tomkins, H. *Nucl. Instrum. Meth. A* **291**, 243 (1990) .
39. Pearce, J.A., Cann, J.R. *Earth Planet. Sc. Lett.* **12**, 339 (1971).
40. Wagner, E. *Physikalische Zeitschrift*, **18**: 432 (1917).
41. Williams G.P. in *X-ray data booklet* [Section 1.1], edited by A. Thompson & D. Vaughan, Lawrence Berkeley National Laboratory University of California, Berkeley, 2001, pp. 1.
42. Vainshtein, Z.E., Blokin, S.M., Bril', M.N., Staryi, I.B., Naderno, Yu.B. *Russ. J. Inorganic Chem.* **10**, 14-19 (1965).
43. Brytov, I.A., Konashenok, K.I., Romashchenko Yu. N., *Geochem. Int.* **16**, 142 (1979).
44. Eisenberg, P., Kincaid, B., Hunter, S., Dayers, D.E., Stern, E.A., Lytle, F. in *Proceedings of the 4th International Conference on Vacuum Ultraviolet Radiation Physics*, edited by E.E. Koch, R. Haensel & C. Kunz, Pergamon, Oxford, 1974, p. 806 .
45. Brown, G.E. Jr., Keefer, K.D., Fenn, P.M. in *Abstract Programs, Geological Society of America Annual Meeting* **10**, 373 (1978).
46. Calas, G., Levitz, P., Petiau, J., Bondot, P., Loupias, G. *Rev. Phys. Appl.* **15**, 1161 (1980).
47. Bonnin, D., Muller, S., Calas, G. *B. Minéral.* **106**, 467 (1982).
48. Calas, G., Petiau, J. *B. Minéral.* **106**, 33 (1983).
49. Waychunas, G.A., Apted, J.M., Brown, G.E. Jr., *Phys. Chem. Mineral.* **10**, 1 (1983).
50. Waychunas, G.A. *Am. Mineral.* **72**, 89 (1987).
51. Wong, J., Lytle, F.W., Messmer, R.F., Maylotte, D.H. *Phys. Rev. B* **30**, 5596 (1984).

52. Greegor, R.B., Lytle, F.W., Sanstrom, D.R., Wong, J., Schultz, P. *J. Non-Cryst. Solids* **55**, 27 (1983).
53. Bianconi, A., Fritsch, E., Calas, G., Petiau, J. *Phys. Rev. B* **32**, 4292 (1985).
54. Binsted, N., Greaves, G.N., Henderson, C.M.B. *Contrib. Mineral. Petr.* **89**, 103 (1985).
55. Greegor, R.B., Lytle, F.W., Ewing, R.C., Haaker, R.F. *Nucl. Instrum. Meth. B* **229**, 587 (1983).
56. Calas, G., Petiau, J. *Solid State Commun.* **48**, 625 (1983).
57. Balzarotti, A., Comin, F., Incoccia, I., Piacentini, M., Mobilio, S., Savoia, A. *Solid State Commun.* **35**, 145 (1980).
58. Belli, M., Scafati, A., Bianconi, A., Mobilio, S., Palladino, L., Reale, A., Burattini, E. *Solid State Commun.* **35**, 355 (1980).
59. Davoli, I., Stizza, S., Durazzo, A., Mottana, A. *Periodico di Mineralogia* **52**, 637 (1983).
60. Davoli, I., Paris, E., Mottana, A., Marcelli, A. *Phys. Chem. Mineral.*, **14**, 21 (1987).
61. Davoli, I., Paris, E., Mottana, A. in *Synchrotron Radiation Applications in Mineralogy and Petrology*, edited by S.S. Augustithis. Athens, Theophrastus, 1988, pp. 97.
62. Cressey, G., Steel, A.T. *Phys. Chem. Mineral.* **15**, 304 (1988).
63. Sainctavit, Ph., Calas, G., Petiau, J., Karnatak, R., Esteva, J.M., Brown, G.E.Jr. *J. Phys. C* **8**, 411 (1986).
64. Sainctavit, Ph., Petiau, J., Calas, G., Benfatto, M. Natoli, C.R. *J. Phys. C* **9**, 1109 (1987).
65. Charnock, J.M., Garner, C.D., Pattrick, R.A.D., Vaughan, D.J. *Phys. Chem. Mineral.* **15**, 296 (1988).
66. Charnock, J.M., Garner, C.D., Pattrick, R.A.D., Vaughan, D.J. *Mineral. Mag.* **53**, 193 (1989).
67. Guttler, B., Niemann, W., Redfern, S.A.T. *Mineral. Mag.* **53**, 591 (1989).
68. McKeown, D.A. *Phys. Chem. Mineral.* **16**, 678 (1989).
69. Manceau, A., Combes, J.M. *Phys. Chem. Mineral.* **15**, 283 (1988).
70. Kaiser, P., Bonnin, D., Frétigny, C., Cortes, R., Manceau, A. *J. Chim. Phys.* **86**, 1699 (1989).
71. Nakai, S., Mitsuishi, T., Sugawara, H., Maezawa, H., Matsukawa, T., Mitani, S., Yamasaki, K., Fujikawa, T. *Phys. Rev. B* **36**, 9241 (1987).
72. Jackson, W.E., Knittle, E., Brown, G.E.Jr., Jeanloz, R. *Geophys. Res. Lett.* **14**, 224 (1987).
73. Davoli, I., Paris, E., Mottana, A. *Atti Accademia Nazionale Lincei.* s. 8, **82**, 527 (1988).
74. Manceau, A., Bonnin, D., Kaiser, P., Frétigny, C. *Phys. Chem. Mineral.* **16**, 180 (1988).
75. Manceau, A. *Am. Mineral.* **74**, 1386 (1989).
76. Combes, J.M., Manceau, A., Calas, G., Bottero, J.Y. *Geochim. Cosmochim. Ac.* **53**, 583 (1988).
77. Ashley, C.A., Doniach, S. *Phys. Rev. B* **11**, 1279 (1975).
78. Lee, P.A., Pendry, J.B. *Phys. Rev. B* **11**, 2795 (1975).
79. Sayers, D.E., Stern, E.A., Lytle, F.W. *Phys. Rev. B* **11**, 4835 (1975).
80. Lee, P.A., Citrin, P.H., Eisenberger, P., Kincaid, B.M. *Rev. Mod. Phys.* **53**, 769 (1981).
81. Teo, B.K. in *EXAFS spectroscopy, techniques and applications*, edited by B.K. Teo and D.C. Joy, New York, Plenum Press, 1981, pp. 13.
82. Boland, J.J., Halaka, F.G., Baldeschwieler, J.D. *Phys. Rev. B* **28**, 2921 (1983).
83. Groot, F.M.F. de, Figueiredo, M.O., Basto, M.J., Abbate, M., Petersen, H., Fuggle, J.C. *Phys. Chem. Mineral.* **19**, 140 (1992).
84. Charnock, J., Garner, C.D., Pattrick, R.A.D., Vaughan, D.J. *Am. Mineral.* **75**, 247 (1990).
85. Sainctavit, Ph., Petiau, J., Flank, A.-M., Ringeisen, J., Lewonczuk, S. in *Conference Proceedings 2nd European Conference on Progress in X-ray Synchrotron Radiation Research*, edited by A. Balerna, E. Bernieri, S. Mobilio, Bologna, SIF, **25**, 1990, pp. 829.
86. McKeown, D.A. in *X-ray Absorption Fine Structure, Proceedings of the VI XAFS Conference*, edited by S.S. Hasnain, Chichester, Ellis-Horwood, 1991, pp. 346.
87. Laan G., van der, Pattrick, R.A.D., Henderson, C.M.B., Vaughan, D.J. *J. Phys. Chem. Solids* **53**, 1185 (1992).
88. Pattrick, R.A.D., van der Laan, G., Vaughan, D.J., Henderson, C.M.B. *Phys. Chem. Mineral.* **20**, 395 (1993).
89. Mosselmans, J.F.W., Pattrick, R.A.D., van der Laan, G., Charnock, J.M., Vaughan, D.J., Henderson, C.M.B., Garner, C.D. *Phys. Chem. Mineral.* **22**, 311 (1995).

90. Charnock, J.M., Henderson, C.M.B., Mosselmans, J.F.W., Pattrick, R.A.D. *Phys. Chem. Mineral.* **23**, 403 (1996).
91. Schofield, P.F., Henderson, C.M.B., Redfern, S.A.T., van der Laan, G. *Phys. Chem. Mineral.* **20**, 375 (1993).
92. Cressey, G., Henderson, C.M.B., van der Laan, G. *Phys. Chem. Mineral.* **20**, 111 (1993).
93. Ildefonse, Ph., Calas, G., Flank, A.M., Lagarde, P. *Nucl. Instrum. Meth. B* **97**, 172 (1995).
94. Cabaret, D., Sainctavit, Ph., Ildefonse, Ph., Flank, A.-M. *J. Phys.-Condens. Matt.* **8**, 3691 (1996).
95. Cabaret, D., Sainctavit, Ph., Ildefonse, Ph., Flank, A.-M. *Am. Mineral.* **83**, 300 (1998).
96. Mottana, A., Murata, T., Wu, Z., Marcelli, A., Paris, E. *J. Electron Spectrosc.* **79**, 79 (1996).
97. Mottana, A., Murata, T. Wu, Z.Y., Marcelli, A., Paris, E. *Phys. Chem. Mineral.* **24**, 500 (1997).
98. Mottana, A., Robert, J.-L., Marcelli, A., Giuli, G., Della Ventura, G., Paris, E., Wu, Z.Y. *Am. Mineral.* **82**, 497 (1997).
99. Mottana, A., Murata, T., Marcelli, A., Wu, Z.Y., Cibin, G., Paris, E., Giuli, G. *Phys. Chem. Mineral.* **27**, 20 (1999).
100. Paris, E., Mottana, A., Della Ventura, G., Robert, J.L. *Eur. J. Mineral.* **5**, 455 (1993).
101. Quartieri, S., Antonioli, G., Artioli, G., Lottici, P.P. *Eur. J. Mineral.* **5**, 1101 (1993).
102. Cruciani, G., Zanazzi, P.F., Quartieri, S. *Eur. J. Mineral.* **7**, 255 (1995).
103. Artioli, G., Pavese, A., Bellotto, M., Collins, S.M., Lucchetti, G. *Am. Mineral.* **81**, 603 (1996).
104. Wu, Z.Y., Marcelli, A., Mottana, A., Giuli, G., Paris, E., Seifert, F. *Phys. Rev. B* **54**, 2976 (1997).
105. Li, D., Bancroft, G.M., Kasrai, M., Fleet, M.E., Feng, X.H., Tan, K.H., Yang, B.X. *Solid State Commun.* **87**, 613 (1993).
106. Li, D., Bancroft, G.M., Kasrai, M., Fleet, M.E., Secco, R.A., Feng, X.H., Tan, K.H., Yang, B.X. *Am. Mineral.* **79**, 622 (1994).
107. Li, D., Bancroft, G.M., Kasrai, M., Fleet, M.E., Feng, X.H., Tan, K.H. *Am. Mineral.* **79**, 785 (1994).
108. Li, D., Bancroft, G.M., Fleet, M.E., Feng, X. H., Pan, Y. *Am. Mineral.* **80**, 432 (1995).
109. Li, D., Bancroft, G.M., Fleet, M.E., Hess, P.C., Yin, Z.F. *Am. Mineral.* **80**, 873 (1995).
110. Kasrai, M., Fleet, M.E., Muthupari, S., Li, D., Bancroft, G.M. *Phys. Chem. Mineral.* **25**, 268 (1998).
111. Mottana, A., Paris, E., Davoli, I., Anovitz, L.M. *Rendiconti Fisici Accademia dei Lincei*, s. 9, **2**, 379 (1991).
112. Farges, F., Guyot, F., Andrault, D., Wang, Y. *Eur. J. Mineral.* **6**, 303 (1994)
113. Closmann, C., Knittle, E., Bridges, F. *Am. Mineral.* **81**, 1321 (1996).
114. Mottana, A., Marcelli, A., Giuli, G., Paris, E., Scordari, F., Schingaro, E. *Rendiconti Fisici Accademia dei Lincei*, s. 9, **7**, 251 (1996).
115. Locock, A., Luth, R.W., Cavell R.G., Smith D.G.W., Duke M.J.M. *Am. Mineral.* **80**, 27 (1995).
116. Henderson, C.M.B., Charnock, J.M., Smith, J.V., Greaves, G.N. *Am. Mineral.* **78**, 477 (1993).
117. Farges, F., Calas, G. *Am. Mineral.* **76**, 60 (1991).
118. Nakai, I., Akimoto J, Imafuku M, Miyawaki R, Sugitani Y. *Phys. Chem. Miner.* **15**, 113 (1987).
119. Thompson, H.A., Brown, G.E.Jr., Parks, G.A. *Am. Mineral.* **82**, 483 (1997).
120. Foster, A.L., Brown, G.E.Jr., Tingle, T.N., Parks, G.A. *Am. Mineral.* **83**, 553 (1998).
121. Parkman, R.H.Ju., Charnock, J.M., Bryan, N.D., Livens, F.R., Vaughan, D.J. *Am. Mineral.* **84**, 407 (1999).
122. Morin, G., Ostergren, J.D., Juillot, F., Ildefonse, Ph., Calas, G., Brown, G.E.Jr. *Am. Mineral.* **84**, 421 (1999).
123. Knapp, G.S., Veal, B.W., Lam, D.J., Paulikas, A.P., Pan, H.K. *Mater. Lett.* **2**, 253 (1984).
124. Farges, F., Ponader C.W., Calas G., Brown G.E.Jr. *Geochim Cosmochim Ac.* **56**, 4205 (1992).
125. Cotter-Howells, J.D., Champness, P.E., Charnock, J.M., Pattrick, R.A.D. *Eur. J. Soil Sci.* **54**, 393 (1994).
126. Farquhar, M.L., Vaughan, D.J., Hughes, C.R., Charnock, J.M., England, K.E.R. *Geochim. Cosmochim. Ac.* **61**, 3051 (1997).
127. Paris, E., Wu, Z.Y., Mottana, A., Marcelli, A. *Eur. J. Mineral.* **7**, 1065 (1995).
128. Wu, Z.Y., Mottana A., Marcelli A. Natoli C.R., Paris E. *Phys. Chem. Mineral.* **23**, 193 (1996).

129. Wu, Z.Y., Marcelli A., Mottana A., Giuli G., Paris E., Seifert F. *Phys Rev. B* **54**,2976 (1996).
130. Mottana, A., Paris E., Marcelli A., Wu Z.Y., Giuli G. *Mitt. Öster. Mineral. Ges.* **141**,35 (1996).
131. Wu, Z.Y., Marcelli, A., Mottana, A., Giuli, G., Paris, E. *Europhys.Lett.* **38**, 465 (1997).
132. Bajt, S., Sutton, S.R., Delaney, J.S. *Geochim. Cosmochim. Ac.* **58**, 5209 (1994).
133. Delaney, J.S., Bajt, S., Sutton, S.R., Dyar, M.D. *Geochemical Society Special Publication* **5**, 165 (1996).
134. Delaney, J.S., Dyar, M.D., Sutton, S.R., Bajt, S. *Geology* **26**, 139 (1998).
135. Cabaret, D., Joly, Y., Renevier, H., Natoli, C.R. (1999) *J. Synchrotron Rad.* **6**, 258 (1999).
136. Dyar, M.D., Delaney, J.S., Sutton, S.R., Schaefer, M. *Am. Mineral.* **83**, 1361 (1998).
137. Sutton, S.R., Delaney, J.S., Bajt, S., Rivers, M.L., Smith, J.V. *Lunar and Planetary Science* **24**, 1385 (1993).
138. Sutton, S.R., Jones, K.W., Gordon, B., Rivers, M.L., Bajt, S., Smith, J.V. *Geochim. Cosmochim. Ac.* **57**, 461 (1993).
139. Manceau, A., Llorca, S., Calas, G. *Geochim. Cosmochim. Ac.* **51**, 105 (1987).
140. Manceau, A., Gorshkov, A.I., Drits, V.A. *Am. Mineral.* **77**, 1133 (1992).
141. Manceau, A., Gorshkov, A.I., Drits, V.A. *Am. Mineral.* **77**, 1144 (1992).
142. Silvester, E., Maceau, A., Drits, V.A. *Am. Mineral.* **82**, 962 (1997).
143. Pingitore, N.E., Lytle, F., Davies, B.M., Eastman, M.P., Eller, P.G., Larson, E.M. *Geochim. Cosmochim. Ac.* **56**, 1531 (1992).
144. Sturchio, N.C., Antonio, M.R., Soderholm, L., Sutton, S.R., Brandon, J.C. *Science* **281**, 971 (1998).
145. Reeder, R.J., Lamble, G.M., Northrup, P.A. *Am. Mineral.* **84**, 1049 (1999).
146. Heald, S.M., Stern, E.A. *Phys. Rev. B* **16**, 5549 (1977).
147. Pettifer, R.F. in *2nd European Conference Program Synchrotron Radiation Research*, edited by A. Balerna, E. Bernieri, S. Mobilio, Bologna, SIF, **25**, 1990, pp. 383.
148. Pettifer, R.F., Brouder, C., Benfatto, M., Natoli, C.R., Hermes, C., Ruiz López, M.F. *Phys. Rev. B* **42**, 37 (1990).
149. Brouder, C. *J. Phys-Condens. Matt.* **2**, 701 (1990).
150. Manceau, A., Bonnin, D., Stone, W.E.E., Sanz, J. *Phys. Chem. Mineral.* **17**, 363 (1990).
151. Manceau, A., Chateigner, D., Gates, W.P. *Phys. Chem. Mineral.* **25**, 347 (1998).
152. Manceau, A. *Can. Mineral.* **28**, 321 (1990).
153. Ildefonse, Ph., Kirkpatrick, R.J., Montez, B., Calas, G., Flank, A.-M., Lagarde, P. *Clay. Clay Miner.* **42**, 276 (1994).
154. Cruciani, G., Zanazzi, P.F., Quartieri, S. *Eur. J. Mineral.* **7**, 255 (1995).
155. Drits, V.A., Dainyak, L.G., Muller, F., Besson, G., Manceau, A. *Clay Miner.* **32**, 153 (1997).
156. Ildefonse, Ph., Sainctavit, P., Calas, G., Flank, A.M., Lagarde, P. *Phys. Chem. Miner.* **25**,112 (1998).
157. Doyle, C.S., Traina, S.J., Ruppert, H., Kandelewicz, T., Rehr, J.J., Brown, G.E.Jr. *J. Synchrotron Rad.* **6**, 621 (1999).
158. Manceau, A., Chateigner, D., Gates, W.P. *Phys. Chem. Miner.* **25**, 347 (1998).
159. Manceau, A., Schlegel, M., Chateigner, D., Lanson, B., Bartoli, C., Gates, W.P. in *Synchrotron X-ray Methods in Clay Science*, edited by D. Schulze, P. Bertsch, J. Stucky. Clay Mineral Society of America, **9**, 1999, pp. 69.
160. Seifert, F., Paris, E., Dingwell, D.B., Davoli, I., Mottana, A. *Condens. Matt. Mat. Commun.* **1**, 115 (1993).
161. Farges, F., Fiquet, G., Andrault, D., Itié, J.-P. *Physica B*, **208&209**, 263 (1995).
162. Andrault, D., Itié, J.-P., Farges, F. *Am. Mineral.* **81**, 822 (1996).
163. Itié, J.-P. *Phase Transit.* **39**, 81 (1992).
164. Itié, J.-P., Polian, A., Martinez, D., Briois, V., Di Cicco, A., Filipponi, A., San Miguel, A. *J. Phys. IV France* **7**, Colloque C2, 31 (1997).
165. Andrault, D., Peyronneau, J., Petit, P.E., Itié, J.-P. *Terra Nova* **5**, 361 (1993).
166. Sutton, S.R., Bajt S., Delaney J., Schulze D., Tokunaga T. *Rev. Sci. Instrum.* **66**,1464 (1995).

167. Heald, S.M., Brewe, D.L., Barg, B., Kim, K.H., Brown, F.C., Stern, E.A. *J. Phys. IV France* **7**, Colloque C2, 297 (1997).
168. Hausermann, D., Hanfland, M. *High Pressure Res.* **14**, 223 (1996).
169. Heifets, E., Hemley, R.J., Bernasconi, M., Ulivi, L., Chiarotti, G. in *Proceedings of the International School of Physics Enrico Fermi*, Course **147**. Bologna, SIF (in press).
170. Galoisy, L., Calas, G., Arrio, M.A. *Chem. Geol.* **174**, 307 (2001).
171. Petit, P.-E., Farges, F., Wilke, M., Solé, V.A. *J. Synchrotron Rad.* **8**, 952 (2001).
172. Wilke, M., Farges, F., Petit, P.-E., Brown, G.E.Jr., Martin, F. *Am. Mineral.* **86**, 714 (2001).
173. Dyar, M.D., Delaney, J.S., Sutton, S.R. *Eur. J. Mineral.* **13**, 1079 (2001).
174. Dyar, M.D., Lowe, E.W., Guidotti, C.V., Delaney, J.S. *Am. Mineral.* **87**, 514 (2002).
175. Bajt, S., Clark, S.B., Sutton, S.R., Rivers, M.L., Smith, J.V. *Anal. Chem.* **65**, 1800 (1993).
176. Peterson, M.L., Brown G.E.Jr., Parks G.A., Stein C.L. *Geochim. Cosm. Ac.* **61**, 3399 (1997).
177. Brigatti, M.F., Lugli, C., Cibin, G., Marcelli, A., Giuli, G., Paris, E., Mottana, A., Wu, Z.Y. *Clay. Clay Miner.* **48**, 272 (2000).
178. Brigatti, M.F., Galli, E., Medici, L., Poppi, L., Cibin, G., Marcelli, A., Mottana, A. *Eur. J. Mineral.* **13**, 377 (2001).
179. Bassett, W.A., Anderson, A.J., Mayanovic, R.A., Chou, I-M. *Chem. Geol.* **167**, 3 (2000).
180. Bassett, W.A., Anderson, A.J., Mayanovic, R.A., Chou, I.-M. *Z. Kristallogr.* **215**, 711 (2000).
181. Hoffmann, M.M.,Darah J.G.,Heald S.M.,Yonker C.R., Fulton J.L. *Chem.Geol.* **167**,89 (2000).
182. Anderson, A.J., Jaynetti, S., Mayanovic, R.A., Bassett, W.A., Chou, I-M. *Am. Mineral.* **87**, 262 (2002).
183. Farges, F., Rossano, S. *Eur. J. Mineral.* **12**, 1093 (2000).
184. Gualtieri, A.F., Moen, A., Nicholson, D.G. *Eur. J. Mineral.* **12**, 17 (2000).
185. Fleet, M.E., Muthupari, S. *Am. Mineral.* **85**, 1009 (2000).
186. Cheah, S.-F., Brown, G.E.Jr., Parks, G.A. *Am. Mineral.* **85**, 118 (2000).
187. Scheinost, A.C., Stanjek, H., Schulze, D.E., Gasser, U., Sparks, D.L. *Am. Mineral.* **86**, 139 (2001).
188. Morin, G., Juillot, F., Ildefonse, Ph., Cals, G., Samama, J.-C., Chevallier, P., Brown, G.E.Jr. *Am. Mineral.* **86**, 92 (2001).
189. Rakovan, J., Newville, M., Sutton, S. *Am. Mineral.* **86**, 697 (2001).
190. McKeown, D.A., Post, J.E. *Am. Mineral.* **86**, 701 (2001).
191. Reiche, I., Vignaud, C., Champagnon, B., Panczer, G., Brouder, C., Morin, G., Solé, V.A., Charlet, L., Menu, M . *Am. Mineral.* **86**, 1519 (2001).
192. Refait, P., Abdelmoula, M., Trolard, F., Génin, J.-M-.R., Ehrhardt, J.J., Bourrié, G. *Am. Mineral.* **86**, 731 (2001).
193. Hemley, R.J. *Science* **285**, 1026 (2000).
194. Wu, Z., Paris, E., Giuli, G., Mottana, A., Seifert, F. *J. Synchrotron Rad.* **8**, 966 (2001).
195. Giuli, G., Paris, E., Wu, Z., Mottana, A., Seifert, F. *Eur. J. Mineral.* **14**, 429 (2002).
196. Henderson, C.M.B., Redfern, S.A.T., Smith, R.I., Knight, K.S., Charnock, J.M. *Am. Mineral.* **86**, 1170 (2001).
197. Manceau, A., Lanson, B., Drits, V.A., Chateigner, D., Gates, W.P., Wu, J., Huo, D., Stucky, J.W. *Am. Mineral.* **85**, 133 (2000).
198. Manceau, A., Drits, V.A., Lanson, B., Chateigner, D., Wu, J., Huo, D., Gates, W.P., Stucky, J.W. *Am. Mineral.* **85**, 153 (2000).
199. Giuli, G., Paris, E., Wu, Z., Brigatti, M.F., Cibin, G., Mottana, A., Marcelli, A. *Eur. J. Mineral.* **13**, 1099 (2001).
200. Mottana, A., Marcelli, A., Cibin, G., Dyar, D.M. in *Micas: crystal chemistry and metamorphic petrology*, edited by A. Mottana, F.P. Sassi, J.B. Thompson Jr., S. Guggenheim. *Rev. Mineral. Geochem.* **46**, 2002, pp. 371.

Non-equilibrium-state x-ray absorption spectroscopy: a local structure study of photo-induced phase transition

H. Oyanagi[*], T. Tayagaki[+] and K. Tanaka[+]

[*]National Institute for Advanced Industrial Science and Technology, 1-1-1 Umezono, Tsukuba, Ibaraki 305-8568, Japan
[+]Department of Physics, Graduate School of Science, Kyoto University, Kyoto 606-8502, Japan

Abstract. We describe non-equilibrium-state x-ray absorption spectroscopy focusing on local structure of photo-excited states trapped at low temperature. For this purpose, a novel Ge 100 pixel array detector with a packing density of 88% was developed. The local structure of photo-induced phase of Fe(II) spin crossover complex, [Fe(2-pic)$_3$]Cl$_2$EtOH (2-pic=2-aminomethyl pyridine), was investigated at low temperature (T <150 K). The use of pixel array detector and high-flux synchrotron x-ray source (multipole wiggler) successfully provided x-ray absorption spectra with high quality, *in-situ*, during the photo-excitation. It was found that *the photo-induced phase under optical pumping at low temperature (T < 50 K) has an octahedral geometry with the elongated Fe-N distance (2.16 ± 0.01 Å)*, stabilizing the high spin state (S=2) configuration. No indication of symmetry breaking of FeN$_6$ clusters upon LS↔HS spin-state switching was observed. It was demonstrated that the technique is a promising means to probe the local structure of non-equilibrium state such as trapped excited states or metastable states.

INTRODUCTION

X-ray absorption spectroscopy (XAS) is a powerful local probe currently recognized as a standard technique complementary with crystallography. Recent interests in local structures of non-equilibrium states such as photo-excited states have motivated in-situ x-ray studies using high brilliance/flux x-ray photon sources, *e.g.*, undulators and wigglers and lasers.[1,2] The effect of photo-excitation on the local structure and hence functions in some biological systems has been studied since early days of synchrotron radiation research.[3,4] The fluorescence detection XAS is useful in probing "x-ray thin" photo-excited specimen in order to maximize efficiency of photo-excitation in a visible photon region. Use of a grazing-incidence geometry has solved an intrinsic problem, *i.e., thickness mismatch.*[5] Using a high brilliance x-ray beam, a grazing-incidence fluorescence XAS experiments on photo-excited state became feasible.[6] For fluorescence XAS applications, a high energy resolution detector is generally required in order to filter out the signal from elastic peaks and/or other fluorescence lines. A standard x-ray detector is a solid state detector (SSD). On the other hand, diffraction

CP652, *X-Ray and Inner-Shell Processes: 19th International Conference on X-Ray and Inner-Shell Processes*
edited by A. Bianconi, A. Marcelli, and N. L. Saini
© 2003 American Institute of Physics 0-7354-0111-X/03/$20.00

techniques, with less strict demands for energy resolution, have been successfully applied to ultra-fast timing experiments with a 50 psec time resolution, allowing the real-time observation of photo-induced structural changes.[1,2] For XAS applications, however, a high efficiency Ge detector over a wide energy range (4-60 keV) is needed. Since the throughput of a single element SSD is limited by a linear amplifier dead time, a multi-element SSD with a high packing fraction is a practical approach.[7,8] Although some progress has been made, a monolithic pixel array detector (PAD) is an ultimate strategy for improving the throughput and packing ratio. We have recently developed Ge PAD with 100 segments with a packing fraction of 88%[9].

Photo-induced phase transitions form rich class of materials where photo-stimulation eventually results in a global phase transition with pronounced changes in physical properties. Multistability of a ground sate and photo-excitation induce macroscopic transitions from a true ground state to metastable states. [Fe II (2-pic)$_3$]Cl$_2$EtOH (2-pic=2-aminomethyl-pyridine) (hereafter abbreviated as Fe-pic) is a typical spin-crossover complex[10,11] which shows a thermally induced first-order phase transition from a low-spin (LS, S=0) to a high spin (HS, S=2) state. Fe-pic also shows a photo-induced phase transition at much below the critical temperature of thermally induced HS to LS transition. Light-induced excited spin state trapping (LIESST) has been observed indicating that the excited HS state can be trapped at low temperatures[12,13]. Ogawa et al. demonstrated non-linear characteristics such as threshold intensity, incubation period and phase separation[14].

Whether the structure of photo-induced phase is the same with that of thermally induced one or not is becoming a matter of hot discussion.[15] In this context, a three-step model to describe ground state multistability by three adiabatic states, taking into account the system with broken symmetry, has been proposed.[16] Tayagaki & Tanaka have recently shown that Raman spectra of photo-induced phase have additional lines indicating lowering of symmetry, i.e., symmetry breaking in photo-converted HS phase.[15] Much attention has been paid to the local structure (symmetry) of iron atom, which can be studied by XAS, i.e., x-ray near-edge absorption structure (XANES) and extended x-ray absorption fine structure (EXAFS). Previous XAS study found that photo-excited Fe-pic shows intermediate XANES and EXAFS between the LS and HS phases, indicating that a multi-domain structure is formed by photo-excitation.[17] In this paper, we report the local structure of completely photo-converted HS phase Fe-pic probed by the Fe K-edge XANES and EXAFS.

EXPERIMENTAL

In-situ XAS under laser irradiation

The experimental setup is described elsewhere[18]. Essential feature of the present setup is the detector (Ge PAD). Synchrotron radiation from 27-pole wiggler magnet (BL13) at the Photon Factory was used. In Fig. 1, a schematic of principle of XAS for optically pumped systems and experimental arrangement are illustrated. Photo-excited state

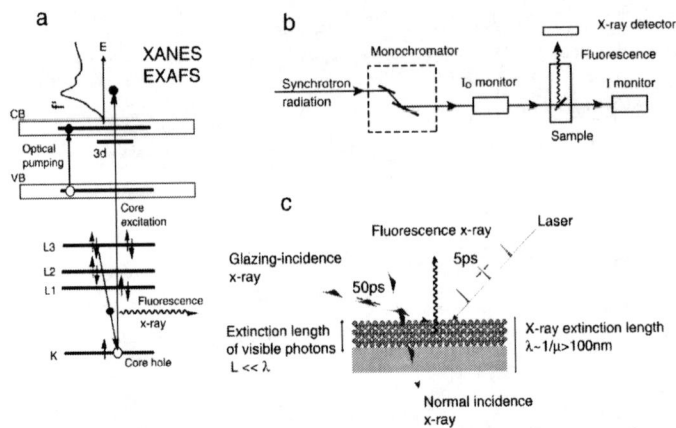

FIGURE 1. Principle of fluorescence-detected XAS probing the local structure of excited state (a), schematic experimental setup for XAS in a fluorescence mode (b), and surface-sensitive fluorescence geometry for matching the x-ray extinction with that of visible photons used for photo-excitation (c).

which can be probed if the lifetime is much longer than the photoelectron lifetime (10^{-15} sec). Photo-induced metastable state of materials with mutistable ground state has a long lifetime at low temperature where thermal excitation is negligible. A conventional XAS experimental setup is a transmission mode but an optimum thickness in a hard x-ray region (>4 keV) is orders of magnitude greater than that of visible photons. Thus sample preparation to optimize optical pumping and x-ray probing efficiency is difficult. Using a fluorescence detection and a grazing incidence geometry, one can reduce x-ray probing depth to several nanometer if the total reflection regime is chosen. In the present case, fine powder specimen was used to ensure that fluorescence yield can be used for monitoring absorption. Figure 2 shows the dimension of Ge PAD. A 10x10 Ge array was fabricated from pixels (4.7x4.7 mm) with 5 mm thickness. The output of each Ge pixel was pulse-height analyzed and counted by a "hybrid" electronics for which the maximum throughput after a dead time correction was 80 kcps per channel using a 0.5 μsec shaping time providing a total maximum count rate of 8 MHz.

Fe-pic powder sample was synthesized in an inert atmosphere as described elsewhere[10,11]. A fine powder sample was mounted on an aluminum holder of a closed-cycle helium cryostat which can rotate on a precision goniometer (Huber 420) so that the incidence angle is controlled with a minimum step of 10^{-4} degree. The wiggler x-radiation from MPW #13 of Photon Factory was used (Fig. 2). The energy and maximum positron current were 3 GeV and 400-500 mA, respectively. A directly water-cooled silicon (111) double crystal monochromator was used. The energy resolution of XAS measurements was better than 2 eV at 9 keV, calibrated from the near-edge features of copper metal at 9.8 keV. As a photo-excitation source, 100 mW diode pumped solid state (DPSS) laser (532 nm) was used. A light-guide was used to introduce the green light onto the specimen (Fe-pic fine powder). During the photo-

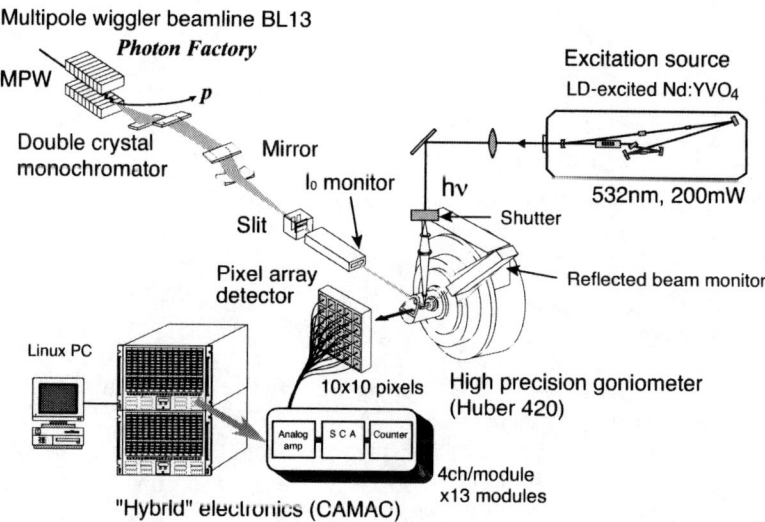

FIGURE 2. XAS measurement system under light irradiation at Photon Factory (BL13B). A high precision two-axis goniometer (Huber 420) controls a grazing incidence angle. Fiber optics are used for photo-excitation by solid-state CW laser.

excitation, the sample temperature was kept at 5.1 K within ±0.1 K. The power density of 532 nm slightly defocused laser light on the sample surface was 6.7 mW/cm^2.

Ge pixel array detector

A largest experimental error in XAS is a systematic noise. In a fluorescence mode, it usually comes from the elastic/inelastic scattering. The fluorescence signal therefore must be separated from scattered photons and other fluorescence lines. The former can be minimized by collecting a signal over a cone-like solid angle away from the scattering plane. Since the best detector position (solid angle) is geometrically limited, an ideal detector should have a high packing fraction and small pixel size. As discussed below, however, the criterion on energy resolution behaves against the geometrical requirement and the pixel dimensions and arrangement should be compromised.

The pixel array detector is equipped with 100 resistive charge sensitive preamplifiers PSC 954. The energy output is fed into a "hybrid" CAMAC module which integrates analog circuits (linear amplifiers and single-channel analyzer, SCA) and digital circuits (logic and memory). Parameters of both analog and digital circuits (amplifier gain, shaping time, SCA windows *etc*) are computer-controlled and data-based. The energy resolution depends on the two factors, *i.e.*, the electric resolution as a function of capacitance of detector and FET and the Fano resolution, Poisson statistics from the generation of *electron-hole* pairs in a detector. The capacitance of detector dependent on a pixel arrangement rises sharply as the interpixel length of pixel array decreases. In

Ge pixel array detector (PAD)

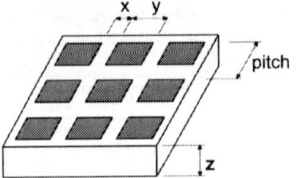

Dimensional parameters

pitch: p=5mm
thickness: z=5mm
interpixel length: x=300 mm
pixel length: y=4.7mm

FIGURE 3. Schematic of Ge pixel array detector and dimensional parameters used as a high efficiency fluorescence x-ray detection.

the present design shown in Fig. 3, the interpixel length was 300 μm. From full-width at half maximum (FWHM) values for peaks ^{55}Fe (5.9 keV) and ^{241}Am (59 keV), the energy resolution was evaluated. The FWHM for 5.9 keV radiation was 215 eV.

The expected energy resolution was 210 eV (5.9 keV) in good agreement with the experimental value (215 eV). It was found that the Ge PAD covers a wide energy range (4-60 keV) being an ideal x-ray detector for undulator sources at third generation synchrotron radiation facilities. The maximum throughput (80 kcps) was achieved after a dead time correction. We note that the use of digital signal processor (DSP) would improve the performance in throughput upto 200 kcps[19]. The total throughput of Ge PAD potentially exceeds 20 MHz. Moreover, the systematic noise arising from scattering/diffraction can be reduced by filtering out the channels those suffer from the non-statistical noise.

RESULTS & DISCUSSION

XANES and symmetry

The normalized Fe K-edge XANES spectra for the low temperature, high temperature and photo-induced phases are compared in Fig. 4. Data for low-temperature LS and photo-converted HS phases were taken at 5.1 K, while high-temperature HS phase was measured at 150 K. The critical temperature for thermally induced spin state transformation Tc2 is 121 K.[10,11] Characteristic XANES features commonly observed for the three phases are labeled as A, B, C and P. The arrow (Co) indicates the feature (7.1670 keV) used to measure a HS fraction as will be discussed. A weak pre-edge feature P is a dipole-forbidden transition (1s-3d). The intensity of pre-edge feature is symmetry-sensitive as distorted coordination geometry would result in mixing of p states, which enhances a transition. Note that the feature B is observed as a weak shoulder for high temperature and photo-induced phases, while in the low temperature phase, it appears as a separate peak shifted to a higher energy side. Peaks A and C also shift to higher energy in the low temperature LS phase. Upon the first-order phase

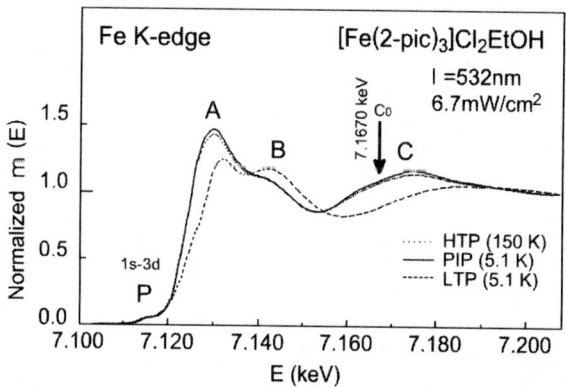

FIGURE 4. Fe K-edge XANES spectra for [Fe(2-pic)$_3$]Cl$_2$EtOH. The dotted line, dashed line and solid lines indicate the specra for high-temperature phase (HTP), low-temperature phase (LTP) and photo-induced phase (PIP). LTP and PIP were measured at 5.1 K while HTP was recorded at 150 K.

transition from the HS to LS state, the variation of the iron-nitrogen distance sharply decreases. The energy shift of feature C in the high temperature and photo-induced HS states toward a low k direction is due to the expansion of FeN$_6$ octahedron by *ca.* 0.2 Å according to the k•R=constant rule.[20] Surprisingly, the photo-induced phase has XANES features very similar to those of high temperature phase, inspite of a large temperature difference (5.1 K *vs.* 150 K).

A single *d* electron placed in an octahedral field where five-fold degenerate 3*d* orbitals split into the *eg* orbitals and *t2g* orbitals. Because of the electrostatic interactions between these orbitals and negative ligands (nitrogen atoms), the *t2g* orbital becomes more stable. The energy difference between the *eg* orbital and *t2g* orbital is determined by ligand field which is related to the iron-nitrogen distance R$_{Fe-N}$. In Fe-pic, the unit cell contains four octahedral FeN$_6$ clusters which are connected each other *via* hydrogen bonds. In a weak ligand field, the true ground state takes a HS state (S=2), while in a strong ligand field, it takes a LS state (S=0). In Fe-pic, the cooperative change of the spin state occurs at Tc2. The nature of cooperativity is believed to be the spin-lattice interaction mediated by vibrational modes and lattice strain. The variation of R$_{Fe-N}$ is consistent with the magnetic susceptibility measurement. Because of strong multiscattering (MS) of photoelecrons, XANES features are symmetry-sensitive. Absence of energy shift for characteristic features upon the photo-conversion of spin state indicates that there is no appreciable symmetry change in FeN$_6$ clusters. We note that the 1*s*-3*d* transition peak (P) was not dependent on the HS-LS transition, consistent with the absence of symmetry change upon the phase transformation.

In summary, the XANES spectra indicate that *the geometry of nitrogen atoms of HS state, a nearly octahedral symmetry (Oh) of high temperature phase, is preserved in the photo-induced phase.* Using the relative intensity of normalized XANES feature C, dynamical process of photo-conversion from the LS to HS states at 5.1 K was investigated. As demonstrated in Fig. 5, the efficiency of the photo-conversion strongly depends on the excitation photon flux[16]. The LS-to-HS conversion is slow when a 1/10

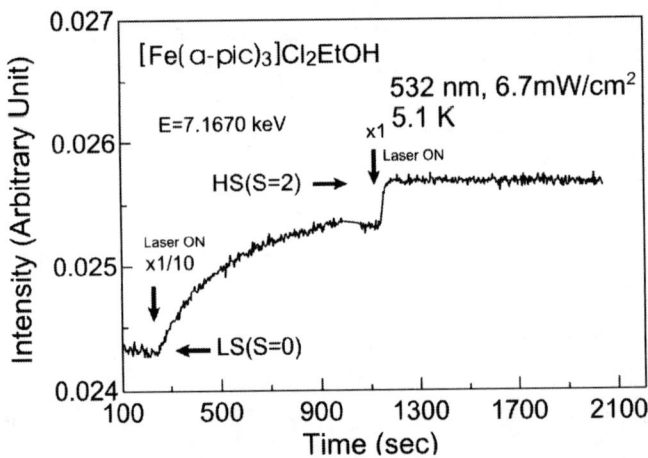

FIGURE 5. The normalized intensity of XANES features (C_0) of $[Fe(2\text{-pic})_3]Cl_2EtOH$ as a function of elapsed time after the illumination by laser (532 nm, 6.7 mW/cm^2). Positions indicated by x1/10 and x1 are the opening of a shutter with a ND filter (1/10) and without a ND filter. Nonlinear characteristic of photo-induced phase transition is demonstrated.

ND filter was used for 532 nm, 6.7 mW/cm^2 laser irradiation but quick without a ND filter. The converted HS fraction measured by XANES intensity at feature C_0 is strongly influenced by the excitation photon flux. The non-linear nature of photo-conversion is explained by a HS-LS two-level system where the potential barrier for HS-to-LS decay process becomes higher as the HS fraction increases, stabilizing the HS state[21].

EXAFS and local structure of FeN$_6$ cluster

Figure 6 shows the Fourier transform (FT) magnitude of the first-shell EXAFS oscillations $\chi(k)$ times k^2 for low and high temperature phases measured at 5.1 K and 150 K, respectively. After a smooth background (atomic absorption) was subtracted from the normalized fluorescence yield spectra, EXAFS oscillations were extracted by fitting a cubic spline function. A standard Fourier transform was performed over the k-range of 4-16 Å^{-1} using a Hanning window. Figure 6 shows that the nearest neighbor peak shifts to larger R by *ca.* 0.2 Å on going from the LS to HS state. The average R_{Fe-N} value was determined by a least-squares curve fit using theoretical back-scattering amplitude and phase shift functions calculated by FEFF 7 code[22]. The FT magnitude of the first shell EXAFS for the high temperature HS phase is significantly smaller than that of low temperature LS phase because of a thermal disorder contribution to the mean-square relative displacement (σ^2).

The least-squares curve fit analysis showed that $R_{Fe-N} = 2.00 \pm 0.01$ Å in the low temperature phase while that of high temperature phase is 2.17 ± 0.01 Å. The results are in good agreement with the average crystallography data, *i.e.*, 2.00 Å and 2.17 Å, respectively.[23] The large change in R_{Fe-N} is consistent with the spin state of the two

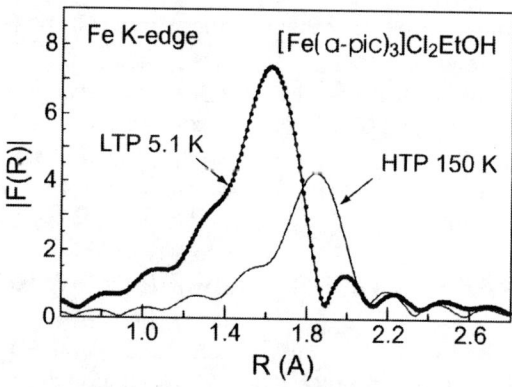

FIGURE 6. Magnitude of the Fourier transform of the EXAFS oscillations $\chi(k)$ times k^2 for [Fe(2-pic)$_3$]Cl$_2$EtOH. The first shell contribution (Fe-N) is shown. LTP and HTP data were recorded at 5.1 K and 150 K, respectively.

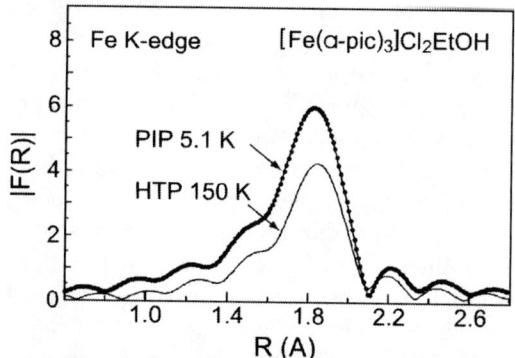

FIGURE 7. Magnitude of the Fourier transform of the EXAFS oscillations $\chi(k)$ times k^2 for [Fe(2-pic)$_3$]Cl$_2$EtOH under photo-excitation (PIP). For comparison, the results for HTP is shown.

phases reflecting the ligand field strengths. However, we found no indication of a local lattice distortion such as Jahn-Teller (JT) distortion which would appear as a decrease of FT peak magnitude as a result of interference between the two distances, *i.e.*, elongated and shortened bonds[24]. It was concluded that the original nitrogen coordination around iron atom is conserved in three phases. A structural difference between the LS and HS phases is not found in symmetry but in the metal-ligand distance. It should be noted that the ethanol molecules are ordered in the low temperature LS phase but disordered in the high temperature HS phase.[23]

Figure 7 shows the magnitude of Fourier transform for the photo-induced HS phase in comparison with that of initial low temperature LS phase. Both data were measured at 5.1 K. Compared to the initial LS phase, the Fe-N peak shifts to a larger R direction upon photo-excitation. The Fe-N peak positions for the two phases in Fig. 7 are quite similar. The curve fit analysis gave $R_{Fe-N} = 2.16 \pm 0.01$ Å in photo-converted HS phase which is close to that of high temperature HS phase (Fig. 6). A slightly larger value of

R_{Fe-N} in high temperature HS phase is due to a thermal expansion (T=150 K *vs* 5.1 K). Determined disorder parameters (σ^2) for the two phases in Fig. 7 are essentially the same. This indicates that the structural disorder due to a thermal motion or vibrational amplitude is not modified by photo-induced spin conversion.

To summarize, *the local structure of photo-induced HS phase is very similar to that of thermally induced HS phase.* The strong ligand field in photo-induced phase is thus due to a symmetrical compression of FeN_6 cluster keeping the original (*Oh*) symmetry.

Local structure of photo-induced phase

In Fig. 8, the local structure of Fe-pic derived from the EXAFS results are schematically illustrated. First, the local structures of thermally induced HS and LS phases were consistent with the crystallographic data.[23] The iron-nitrogen distance increases by 0.17 Å upon the LS$\leftrightarrow$HS magnetic transition at 121 K. From temperature dependence of XANES spectra, it was found that the laser irradiation induced a similar structural variation below much lower temperature, T_{C1}=50 K. The local structure of photo-induced HS phase was essentially the same with that of high temperature HS phase. Thus we conclude that *the averaged local structure of photo-converted HS phase is an expanded but undistorted FeN_6 cluster. Any indication of symmetry breaking or lattice distortions of FeN_6 clusters was not observed.* Raman spectra of photo-induced phase exhibited peaks attributed to the vibrations of ligand molecules and those of FeN_6 cluster where the new lines observed upon photo-switching are attributed to the vibrations of ligand molecules.[15] The absence of symmetry breaking in XANES is not surprising since XANES is dominated by the strong multiple scattering within the FeN_6 cluster and the Raman spectra of photo-converted HS phase keep the features ascribed to the undistorted FeN_6 cluster.

However, the elongated R_{Fe-N} in the photo-induced HS phase would give rise to a bond-stretching stress which should propagate to outer ligand molecules (picolylamine). Moreover, as the relaxation process at low temperature is slow, the distortion could be frozen with a large amount of local lattice strain which would be different from that of the thermally induced HS phase. If the Raman spectra detected such a molecular distortion, rich additional lines in the Raman spectra in photo-converted HS phase can be explained. Indeed, all the additional Raman lines disappear at higher temperature than 40 K $\sim$ T_{C1}[15]. The ligand molecular distortions may affect the longer-range interaction via hydrogen bonds such as O-H····Cl and N-H····Cl. In case of high temperature phase, disordered arrangements of ethanol molecules contribute to delocalize the lattice strain caused by the expansion of the R_{FeN} as expected from the contribution to the entropy variation ΔS (*ca.* 20%) upon the phase transformation. In fact, the rotational ground state of of EtOH was found to be strongly correlated to the spin state transformation[10].

Moreover, the spin phase transition takes place when the disordered ethanol site reaches a critical population[23]. Although the extra Raman lines observed in photo-converted HS phase were ascribed to a metastable phase stabilized by a cooperative Jahn-Teller distortion[15], the strained ligand molecules due to significant stretching of

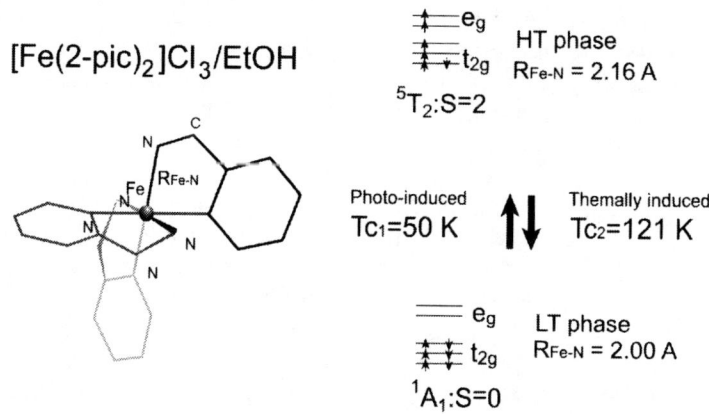

[Fe(2-pic)$_2$]Cl$_3$/EtOH

e_g
t_{2g} HT phase
R_{Fe-N} = 2.16 A
$^5T_2$:S=2

Photo-induced
T_{C1}=50 K

Themally induced
T_{C2}=121 K

e_g LT phase
t_{2g} R_{Fe-N} = 2.00 A
$^1A_1$:S=0

FIGURE 8. Local structure and spin state of [Fe(2-pic)$_3$]Cl$_2$EtOH. At Tc2~121 K, the thermally induced phase transition is observed while at Tc1~50 K, the photo-induced transition is observed. The excited HS state is trapped at lower temperature.

R_{FeN} may cause infra-active symmetry lowering. Infrared absorption data indeed support symmetry breaking in picolylamine molecules[15]. The infra-active modes of the picolylamine molecules may strongly couple with the electronic state in photo-induced HS phase *via* hydrogen bonds. It is likely that cooperativity in the LS-HS macroscopic phase transition is traced to the intermolecular interaction mediated by local lattice distortions in ligand and solvate molecules beyond FeN$_6$ clusters.

SUMMARY AND CONCLUSION

A novel Ge pixel array detector with 100 segments was developed for XAS experiments in a fluorescence mode. Each pixel is a 5 mm thick square (4.7x4.7 mm in dimension) separated by 0.3 mm. Almost all pixels (99%) were in operation with a high energy resolution (240 eV at 5.9 keV). The maximum throughput using a hybrid electronics with a 0.5 μsec shaping time was 80 kcps after a dead time correction, giving a total count rate of 8 MHz. The Ge pixel array detector can cover a wide energy range (4-60 keV) matching with the energy range of gap-tuned undulators at third generation storage rings. Apart from its superior characteristics in high efficiency, it can provide spectra with minimum systematic noise due to scattering and/or diffraction.

We have studied the local structure of photo-induced phase transition of spin crossover complex [Fe(2-pic)$_3$]Cl$_2$EtOH *in-situ* under laser excitation. It was found that the XANES features for photo-induced and thermally induced HS phases are quite similar. The EXAFS results indicate that the iron-nitrogen distance R_{Fe-N} in photo-converted HS phase is the same with that of high temperature phase. Combining the two experimental results, it was concluded that the *photo-induced phase is a trapped high-spin state which is stabilized at temperatures below 50 K.* The local lattice distortion within FeN$_6$ in photo-converted HS phase is negligibly small. The results may look apparently against the Raman scattering experiment which observed symmetry lowering in the photo-induced phase[15]. However, since the additional Raman lines observed in photo-converted HS phase are attributed to the softened vibrational modes of the picolylamine molecules, the preserved *local symmetry* may not disagree with the Raman results. The coupling of the spin states of the iron atom leading to a macroscopic spin conversion is mediated by lattice vibration mode and lattice strain. Cooperativity of the phase transition could arise from the interaction between the FeN_6 clusters *via* distorted ligand molecules linked by hydrogen bonds. Symmetry breaking of ligand molecules associated with the lattice strain and disordering of ethanol molecules may not be surprising taking the slow relaxation process into account. Whether the two-domain nature found in our previous work is related to an intermediate phase with local pairing of LS and HS states[25] is an interesting topics for future studies. The present XAS experiments revealed the *symmetric and local* distortion but *asymmetric and more distant* counterpart should be investigated to understand the nature of photo-induced spin conversion phenomena.

ACKNOWLEDGMENTS

The development of Ge pixel array detector is a collaboration with C. Fonne, D. Gutknecht, P. Dressler, R. Henck, M-O. Lampert, S. Ogawa and K. Kasai to whom the authors express their greatest thanks. The authors also would like to thank Y. Shimoi for helpful discussions and S. Mohamed for supporting experiments. A part of this work was financially supported by the Budget for Nuclear Research of the Ministry of Education, Culture, Sports, Science and Technology, based on the screening and counseling by the Atomic Energy Commission. The authors express their thanks to S. Koshihara, M. Kamada, K. Yonemitsu, T. Kawamoto and K. Nasu for useful discussion and encouragement.

REFERENCES

1. S. Techert, F. Schotte and M. Wulff, Phys. Rev. Lett. **86**, 2030 (2001).
2. A.M. Lindenberg, I. Kang, S.L. Johnson, T. Missalla, P.A. Heimann, Z. Chang, J. Larsson, P.H. Bucksbaum, H.C. Kapteyn, H.A. Padmore, R.W. Lee, J.S. Wark and R.W. Falcone, Phys. Rev. Lett. **84**, 111 (2000).
3. L. Powers, B. Chance, M. Chance, B. Campbell, J. Friedman, S. Khalid, C. Kumar, A. Naqui, K.S. Reddy and Y. Zhou, Biochemistry **26**, 4785 (1987).
4. B. Chance, R. Fischetti and L. Powers, ibid **22**, 3820 (1983).
5. H. Oyanagi, A. Kolobov and K. Tanaka, J. Synchrotron Rad. **5**, 1001 (1998).
6. A. Kolobov, H. Oyanagi and K. Tanaka, Phys. Rev. Lett. **87**, 145502-1 (2001).
7. S.P. Cramer, O. Tench, M. Yocum and G.N. George, Nucl. Inst. Meth. **A266**, 586 (1988).
8. H. Oyanagi, M. Martini and M, Saito, Nucl. Inst. Meth. **A403**, 58 (1998).
9. H. Oyanagi, C. Fonne, D. Gutknecht, P. Dressler, R. Henck, M-O. Lampert, S. Ogawa and K. Kasai, in preparation.
10. P. Gutlich; J. Phys. (Paris) Colloq. C2 (1979) 378.; P. Gutlich, A. Hauser and H. Spiering, Angew. Chem. **33**, 2024(1994).
11. G.A. Renovitch, W.A. Baker Jr., J. Am. Chem. Soc. **89**, 6377 (1967).
12. S.A. Keneman, Appl. Phys. Lett. **19**, 205 (1971).

13. J.S. Berkes, S.W. Ing and W.J. Hillegas, J. Appl. Phys. **42**, 4908 (1971).
14. Y. Ogawa, S. Koshihara, K. Koshino, T. Ogawa, C. Urano and H. Takagi, Phys. Rev. Lett. **84**, 3181 (2000).
15. T. Tayagaki and K. Tanaka, Phys. Rev. Lett. **86**, 2886 (2001).
16. T. Luty, *Proc. of Int. Conf. on Photoinduced Phase Transitions, their Dynamics and Precursor Phenomena*, 2001, Tsukuba.
17. H. Oyanagi, T. Tayagaki and K. Tanaka, J. of Nanoscience and Nanotechnology, in press.
18. H. Oyanagi, K. Haga and Y. Kuwahara, Rev. Sci. Instrum. **67**, 350 (1996).
19. B. Warburton, private communication.
20. C. Natoli, *EXAFS and Near Edge Structure* ed. by A. Bianconi, L. Incoccia and S. Stipcich, Springer-Verlag (1982), p.43.
21. S. Koshihara, Y. Takahashi, H. Sakai, Y. Tokura and T. Luty, J. Phys. Chem. B **103**, 2592 (1999).
22. J.J. Rehr, S.I. Zabrinsky and R.C. Albers, Phys. Rev. Lett. **69**, 3397 (1992).
23. M. Mikami, M. Konno, and Y. Saito, Acta Cryst. **36**, 275 (1980).
24. A. Bianconi, N.L. Saini, A. Lanzara, M. Missori and T. Rossetti, Phys. Rev. Lett. **76**, 3412 (1996).
25. N. Sasaki and T. Kambara, Phys. Rev. B **40**, 2442 (1989).

Local structural distortion and electronic modifications in PrNiO$_3$ across the metal-insulator transition

C. Piamonteze*[†], H. C. N. Tolentino*, A. Y. Ramos***, N. E. Massa[‡], J. A. Alonso[§], M. J. Martinez-Lope[§] and M. T. Casais[§]

*Laboratorio Nacional Luz Sincrotron, Caixa Postal 6192, 13084-971 Campinas, SP, Brazil
[†]Instituto de Fisica Gleb Wataghin, UNICAMP, Caixa Postal 6165, 13083-970 Campinas, SP, Brazil
**LMCP Laboratoire de Mineralogie-Cristallographie de Paris, UMR 7590 CNRS, Paris,France
[‡]Laboratorio Nacional de Investigacion y Servicios en Espectroscopia Optica, Centro CEQUINOR, Departamento de Quimica y Departamento de Fisica, Universidad Nacional de La Plata, Casilla de Correo 962, 1900 La Plata, Argentina
[§]Instituto de Ciencia de Materiales de Madrid, Consejo Superior de Investigaciones Cientificas, Cantoblanco, E-28049 Madrid, Spain

Abstract.
Local electronic and structural properties of PrNiO$_3$ perovskite were studied by means of X-ray Absorption Spectroscopy at Ni K and L edges. The EXAFS results at Ni K edge show a structural transition from three different Ni-O bond-lengths at the insulating phase to two Ni-O bond-lengths above T$_{MI}$. These results were interpreted as being due to a transition from a structure with two different Ni sites at the insulating phase to one distorted Ni site at the metallic phase. The Ni L edge spectra show a remarkable difference between the spectra measured at the insulating and metallic phases that indicates a decreasing degree of hybridization between Ni3d and O2p bands from the metallic to the insulating phase.

INTRODUCTION

Transition metal oxides with perovskite structure have attracted considerable attention due to their interesting physical properties, such as high Tc superconductivity in cuprates and colossal magnetoresistance in manganites. Rare earth nickelates RNiO$_3$ (R= rare earth) undergo a very sharp metal-to-insulator (MIT) phase transition as temperature decreases [1] (only LaNiO$_3$ is a purely metallic member of the serie) making them potentially suitable for applications in optical switches and actuators. RNiO$_3$ systems have a distorted perovskite structure, where the NiO$_6$ octahedra tilt and rotate to fill the empty space left around the rare earth. Both the degree of distortion, measured by the Ni-O-Ni angle, and the MI transition temperature, T$_{MI}$, increases as the R ion becomes smaller. Therefore, the electronic localization has been associated with an increasing structural distortion. Looking at RNiO$_3$ electronic structure, most of the experimental results point to a charge transfer gap [2, 3], which is in accordance with the Zaanen-Sawatzki-Allen (ZSA) scheme [4]. For charge transfer insulators, the gap is governed by Δ, the energy necessary to move one electron from O 2p to Ni 3d band. These

CP652, X-Ray and Inner-Shell Processes: 19th International Conference on X-Ray and Inner-Shell Processes
edited by A. Bianconi, A. Marcelli, and N. L. Saini
© 2003 American Institute of Physics 0-7354-0111-X/03/$20.00

bands are hybridized but their overlap may depend on the structural distortion. Thence, the degree of hybridization between Ni 3d and O 2p bands should play an important role at the gap opening. The interplay among charge transfer energy, hybridization and bandwidths leads to a strict correlation between electronic and structural properties. $RNiO_3$ systems present an antiferromagnetic transition that, for R=Pr and Nd, takes place almost simultaneously with the electronic localization [5], whereas for smaller R ions T_{MI} and T_N are very far apart [6]. The magnetic structure found for these systems consists of both ferro and antiferromagnetic couplings between Ni^{3+} ions. Such arrangement requires the existence of two different Ni sites. However, neutron diffraction results, for R=Pr, Nd and Sm, show a structure where the unique Ni site presents its Ni-O bond-lengths very close to each other. Across the transition, the NiO_6 octahedra remains almost the same and a slight change at Ni-O-Ni angles was observed [7]. For systems with smaller R ions, Alonso et al. [8, 9] observed a monoclinic distortion at the insulating phase that leads to the interpretation in terms of the existence of two unequivalent Ni sites. Moreover, when crossing the electronic transition, the Pbnm symmetry is established at the metallic phase [10]. Nevertheless, in $NdNiO_3$ film samples, resonant X-ray scattering showed the existence of a long range ordered ground state with two different Ni sites [11] at the insulating phase and by raman spectroscopy a symmetry modification across the MI transition is seen [12].

In this work we have measured EXAFS at Ni K edge and XANES at Ni L edge for a $PrNiO_3$ sample across its MIT. EXAFS measurements were done aiming to study the changing on NiO_6 octahedra across T_{MI}. X-Ray Absorption at Ni L edge probes the Ni 3d density of unoccupied states. Therefore, it provides information on the electronic structure of these systems, more specifically on the degree of hybridization between Ni 3d and O 2p bands.

EXPERIMENT

Polycrystalline powder sample of $PrNiO_3$ (T_{MI}= 130K) was prepared by a wet-chemistry technique as described elsewhere [13]. EXAFS measurements were carried out at Ni K edge (8333eV) at the D04B-XAS1 beam line of LNLS, Campinas, Brazil [14]. The measurements were done in transmission mode using a Si111 channel-cut monochromator. Including beam divergence for a vertical beam profile at sample holder of 0.8mm, we have a total energy resolution of 2.4eV at Ni K edge energy. EXAFS spectra were measured up to 9430eV corresponding to a maximum k value of 16 $Å^{-1}$. Sample temperature was lowered using a closed cycle He cryostat. XAS measurements at Ni L_{III} (853eV) and L_{II} (870eV) edges were performed at D08A-SGM beam line at LNLS. The energy resolution, was around 0.3 eV. Measurements were taken in total electron yield mode by collecting the current flow to the sample. The sample was cooled by means of a cold finger connected with the sample holder filled with liquid nitrogen. In both experimental arrangements the sample temperature was measured with a thermocouple located as close as possible to the sample.

TABLE 1. Parameters obtained from the EXAFS data fit. E_0 shift was maintained at -1.8eV for all temperature values.

T(K)	N_1	R_1 (Å)	N_2	R_2 (Å)	N_3*	R_3 (Å)[†]	σ^2	S_0^2
8	1.0(2)	1.82 (1)	3.8(1)	1.935(4)	1.4	2.07	0.00115	0.9
150	0	-	4.1(3)	1.917(4)	1.6	2.00	0.00155	0.7
300	0	-	4.1(5)	1.91(1)	1.9	2.01	0.00159	0.7

* fixed by the constraint $N_1+N_2+N_3= 6$

[†] fixed by the constraint $\langle R \rangle = 1.94$Å

RESULTS

Structural Properties - EXAFS on Ni K edge

Figure 1 presents the EXAFS signal ($\chi(k)$) and its Fourier Transform (FT), performed between 3.5 and 16Å^{-1}, for $PrNiO_3$ sample at three different temperatures. The more remarkable feature observed on $\chi(k)$ signal appears between 15Å^{-1} and 16Å^{-1}, where the 8K spectra presents a sudden increase of the EXAFS oscillations. From the FT spectra (figure 1b) the differences between the metallic and insulating phases can be more clearly visualized. >From 300K to 150K the FT amplitude increases, which is expected since the damping due to thermal disorder has decreased. However, this behavior is not maintained for the 8K FT amplitude, which presents a decrease of the first peak. It should be noted that this amplitude decrease is present only in the first peak of 8K FT spectrum, which corresponds to the O neighbor shell. Moreover, there is an amplitude increasing at approximately 1.3Å for the spectra measured at 8K.

The fitting process of EXAFS data was carried out on real space between 1.23 and 1.84Å using the program code FEFFIT [15]. The total coordination number was maintained fixed at 6 and the average Ni-O bond-length given by $\langle R \rangle = (N_1 * R_1 + N_2 * R_2 + N_3 * R_3)/(N_1 + N_2 + N_3)$ was fixed at 1.94Å, which is the average distance found from neutron diffraction experiment [7]. Table 1 shows the parameters obtained from the fitting of $PrNiO_3$ data. At the insulating phase $PrNiO_3$ show three Ni-O bond-lengths which differ about 0.14Å to 0.25Å among them. These results are very different from that observed with neutron diffraction [7]. By EXAFS it is not possible to distinguish between the configuration of one Ni site with three different distances from that of two different Ni sites. Nevertheless, the hypothesis of one single site seems unlikely, since it would lead to a extremely distorted NiO_6 octahedra. The simplest model we can figure out with the results presented by table 1 is a configuration with two Ni sites in the same proportion, one being composed of 4x(R_2) and 2x(R_1); and the other site with 4x(R_2) and 2x(R_3). Above the electronic transition, at the metallic phase, the shorter Ni-O bond-length disappears, remaining a single distorted site. A point that is quite surprising from the analysis is the big modification of the S_0^2 factor across the transition. This could, in principle, be explained by a strong modification on the electronic structure, which changes the electron correlation and the many body excitations. However, we are not totally confident on this explanation and the differences can be accommodated by the error bars of 20%.

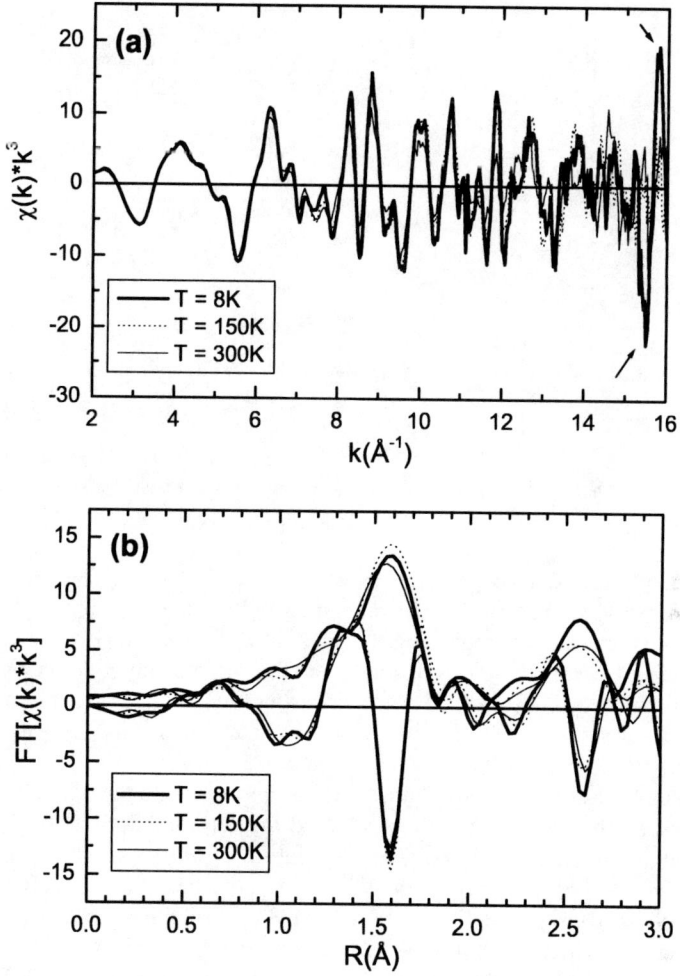

FIGURE 1. (a) EXAFS signal ($\chi(k)$) weighted in k^3 for $PrNiO_3$ (T_{MI}=130K) at all measured temperatures. (b) The amplitude and imaginary parts of Fourier Transform EXAFS signal.

Electronic Properties - XANES at Ni L_{II}, L_{III} edges

The Ni L edge spectra measured at both electronic phases of $PrNiO_3$ are showed on figure 2. Comparing the spectrum at room temperature with reported results [2] it is seen a small amount of Ni^{2+} shown by a small increase of the feature located around 853eV. This is due to a sample aging that we detected comparing these measurements with other taken with the same sample one year before [16]. The same aging was detected

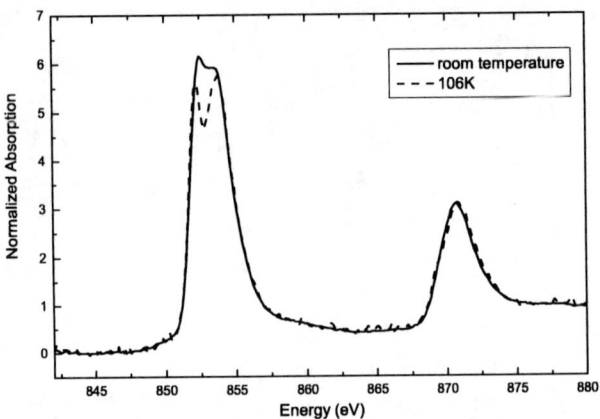

FIGURE 2. X-Ray Absorption Spectra on Ni L edge for sample $PrNiO_3$(T_{MI}= 130K).

by measurements on Ni K edge. The results presented on the previous section come from measurements carried out at the same period with that of reference [16], i. e., with fresh samples. Some aspects regarding the spectra intensity discussed on [16] were not observed here and we attribute this difference to the sample aging. However, we believe that the qualitative effects observed when lowering the sample temperature across T_{MI} are not affected by this contamination. Therefore, in this section only the qualitative differences observed across the *MI* transition are discussed.

>From figure 2 it is seen a remarkable difference on the sub-structure at Ni L_{III} edge (853eV) between the spectra at the metallic and insulating phases. The structure shape of Ni L_{III} edge has been attributed as being due to multiplet contributions [3]. From the metallic to the insulating spectra it is observed a big increase on the multiplet splitting of Ni L_{III} edge. Qualitatively, a broadening of the spectra structure can be associated with an increasing degree of covalence [17, 2]. Since $RNiO_3$ perovskites have a charge transfer gap, a decrease in the degree of hybridization at the insulating phase when compared to the metallic one, is expected. Another information important to be noticed is that Ni L edge absorption spectroscopy probes the very localized Ni 3d bands. Therefore, the modifications observed on figure 2 can only be explained by a modification on Ni local structure, expected on the light of the EXAFS results.

CONCLUSIONS

The EXAFS results indicate a structural transition from a unique distorted site at the metallic phase, to two different distorted sites below T_{MI}. The reason why such structure was not observed by neutron diffraction may be attributed to a non-cooperative character of the octahedra distortion along the crystal. Nevertheless, Raman [12] and infra-red spectroscopy [18] have already indicated differences on the octahedra symmetry across the phase transition. The Ni L edge results indicate an increasing multiplet splitting when going from metallic to insulating phase that can be associated to a decrease on the degree of covalence . In accordance with the K edge results, the modifications observed by X-ray absorption spectra at Ni L edge can only be explained by a modification at Ni local structure due to the spacial localization of Ni 3d bands. Therefore, our results show that on top of the inter-octahedra modification seen by neutron diffraction, there is also an intra-octahedra structural transition that leads to a modification on the degree of hybridization across the electronic transition.

ACKNOWLEDGMENTS

This work has been partially supported by LNLS/ABTLuS/MCT. The funding agency from the State of São Paulo, Brazil (FAPESP) is acknowledge for giving a PhD grant to CP (PROC. 00/00789-3).

REFERENCES

1. Lacorre, P., Torrance, J. B., Pannetier, J., Nazzal, A. I., Wang, P. W., and Huang, T. C., *J. Solid State Chem.*, **91**, 225 (1991).
2. Medarde, M., Fontaine, A., Muñoz, J. L. G., Rodríguez-Carvajal, J., de Santis, M., Sacchi, M., and Rossi, G., *Phys. Rev. B*, **46**, 14975 (1992).
3. Mizokawa, T., Fujimori, A., Arima, T., Tolura, Y., Mōri, N., and Akimitsu, J., *Phys. Rev. B*, **52**, 13865 (1995).
4. Zaanen, J., Sawatzky, G. A., and Allen, J. W., *Phys. Rev. Lett.*, **55**, 418 (1985).
5. García-Muñoz, J. L., Rodríguez-Carvajal, J., and Lacorre, P., *Phys. Rev. B*, **50**, 978 (1994).
6. Rodríguez-Carvajal, J., Rosenkranz, S., Medarde, M., Lacorre, P., Fernandez-Díaz, M. T., Fauth, F., and Trounov, V., *Phys. Rev. B*, **57**, 456 (1998).
7. García-Muñoz, J. L., Rodríguez-Carvajal, J., Lacorre, P., and Torrance, J. B., *Phys. Rev. B*, **46**, 4414 (1992).
8. Alonso, J. A., García-Muñoz, J. L., Fernández-Díaz, M. T., Aranda, M. A. G., Martínez-Lope, M. J., and Casais, M. T., *Phys. Rev. Lett.*, **82**, 3871 (1999).
9. Alonso, J. A., Martínez-Lope, M. J., Casais, M. T., García-Muñoz, J. L., and Fernández-Díaz, M. T., *Phys. Rev. B*, **61**, 1756 (2000).
10. Alonso, J. A., Martínez-Lope, M. J., Casais, M. T., García-Muñoz, J. L., Fernández-Díaz, M. T., and Aranda, M. A. G., *Phys. Rev. B*, **64**, 94102 (2001).
11. Staub, U., Meijer, G. I., Fauth, F., Allenspach, R., Bednorz, J. G., Karpinski, J., and Kazadov, S. M., *Phys. Rev. Lett.*, **88**, 126402 (2002).
12. Zaghiroui, M., Bulou, A., Lacorre, P., and Laffez, P., *Phys. Rev. B*, **64**, 81102 (2001).
13. Alonso, J. A., Martínez Lope, M. J., and Hidalgo, M. A., *J. Solid State Chem.*, **116**, 146 (1995).
14. Tolentino, H. C. N., Ramos, A. Y., Alves, M. C. M., Barrea, R. A., Tamura, E., Cezar, J. C., and Watanabe, N., *J. Synchrotron Rad.*, **8**, 1040–1046 (2001).
15. Newville, M., Ravel, B., Hakel, D., Rehr, J. J., Stern, E. A., and Yacoby, Y., *Physica B*, **208&209**, 154 (1995).
16. Piamonteze, C., Tolentino, H. C., Vicentin, F. C., Ramos, A. Y., Massa, N. E., Alonso, J. A., Martinez-Lope, M. J., and Casais, M. T., *Surf. Rev. Lett.*, **9**, in press (2002).
17. de Groot, F. M. F., Fuggle, J. C., Thole, B. T., and Sawatzky, G. A., *Phys. Rev. B*, **42**, 5459 (1990).
18. Massa, N. E., Alonso, J. A., Martínez-Lope, M. J., and Rasines, I., *Phys. Rev. B*, **56**, 986 (1997).

Strain Effects in $La_{0.7}Sr_{0.3}MnO_3$ Films by X-ray Absorption Spectroscopy

A.Y. Ramos [a,b], N.M. Souza Neto [a,c], C. Giacomelli [a], H.C.N. Tolentino [a], L. Ranno [d], E. Favre-Nicolin [d]

[a]LNLS Laboratório Nacional de Luz Síncrotron, CP6192, 13084-971 Campinas, SP, Brazil
[b]LMCP Laboratoire de Minéralogie –Cristallographie, UMR 7590 CNRS, Paris, France
[c]IFGW,Universidade Estadual de Campinas CP6165, 13083-970 Campinas SP, Brazil
[d]Laboratoire Louis Néel, UPR5051 CNRS, Grenoble France

Abstract. We report on Mn K-edge X-ray absorption study, in plane and out of plane, of $La_{0.7}Sr_{0.3}MnO_3$ films, epitaxially grown on a tensile substrate $SrTiO_3$ by laser ablation. From Extended X-ray Absorption Fine Structure in the film plane we observe a small increase of Mn-Mn distances with respect to relaxed film. In addition, a small distortion of the MnO_6 octahedron is evidenced from Extended and Near Edge Absorption measurements. The respective amplitudes found for these two effects are on the same order, so that no modification of the Mn-O-Mn angle is evidenced.

INTRODUCTION

The discovery of colossal magnetoresistance (CMR) in doped manganites [1] has renewed the interest in these manganese perovskites. From a fundamental point of view they show a variety of interesting phenomena due to the interplay of several, opposite magnetic interactions highly correlated with the crystal structure and the hole doping (metal-insulator transition, charge ordering, orbital ordering, phase segregation...). High quality thin films of manganites can be grown using deposition techniques similar to the ones developed for high-T_C superconductors. This open up the large possibilities to the design of tunable magnetic devices [2]

As low cost magnetic sensor should be operated at temperature close to room temperature, special interest is paid to perovskites showing a ferromagnetic transition close to 300K. This is the case of Sr-doped manganites of composition close to $La_{0.7}Sr_{0.3}MnO_3$ (LSMO, Tc =380K). In LSMO, there is no s electron and the band of mainly oxygen 2p character lies just 2eV below the Fermi energy, which result in a fully polarized conduction band of 3d character. This feature is very interesting for spin-tunnel junctions and some first prototypes have been realized with quite high yields at low temperature [2]

The synthesis of thin films on slightly mismatched substrates has shown the significant sensitivity of manganite properties to distortions. It is well known that the versatility of these properties in the bulk manganites is correlated to variations of the

CP652, X-Ray and Inner-Shell Processes: 19th International Conference on X-Ray and Inner-Shell Processes
edited by A. Bianconi, A. Marcelli, and N. L. Saini

© 2003 American Institute of Physics 0-7354-0111-X/03/$20.00

local structure of Mn ions, - variations of the local Jahn-Teller distortion within the MnO_6 octahedral and/or in the Mn-O-Mn angle (octahedral tilt) [3]. For bulk rare-earth manganese oxides the local structure such as Mn-O-Mn angle and Mn-O bond length can be varied by changing the doping concentration or by applying hydrostatic pressure. The substrate induced crystallographic distortions are anisotropic unlike the distortions induced by hydrostatic pressure or cation substitution. However there is no consensus on how these strains are related to the local atomic organization around the manganese atoms, in spite of the importance of this link in understanding the magnetoresistance in perovskites [4]. At the local scale, unusual and complex effects may be expected, as, for example a static anisotropic distortion of the MnO_6 octahedron, which could lead to an increase of the Jahn-Teller splitting of the e_g levels and thus tends to localize the electrons, contrary to the effect of hydrostatic pressure. On the other hand it can also be suggested that the modifications of the magnetic properties are more related to modifications in the Mn-O-Mn angle.

The present work deals with the structural characterization at the local scale, of the strain-induced modifications around the manganese ions in an epitaxially grown LSMO thin film. The linear polarization of the synchrotron light has been used to gain selective information about these modifications in the plane and out of plane of the film.

EXPERIMENTAL

The manganite films have been grown on single crystalline substrates using the pulsed laser deposition method. The growth was performed using a deposition temperature of 750°C under a 300 mTorr oxygen atmosphere. The two substrates used in the present study were $SrTiO_3$ (STO) and MgO (MO). LSMO has a pseudo cubic structure (space group is R-3c) with a=0.387nm [5]. MO and STO have cubic structures with a=0.421nm and a=0.3905nm respectively. On the MO substrate, the high lattice mismatch is (9%) and the growth is non pseudomorphic. The low lattice mismatch between LSMO and STO allows a pseudomorphic growth for film thicknesses below a relaxation critical thickness of about 100nm. For the fully constrained films, the in plane tensile stress is ε_{xx}= + 0.89%. The two films compared in this study have 60nm in thickness. The film grown on MgO (LSMO/MO) is fully relaxed (ε_{xx}= 0%) whereas the film grown on STO (LSMO/STO) is fully constrained with an in plane tensile stress. Further description of the growth conditions and structural characterization by X-ray diffraction can be found elsewhere, together with a characterization of their transport and magnetization properties [6,7]

The X-ray absorption experiments were performed at the XAS1 beamline of the LNLS (Laboratório Nacional de Luz Síncrotron) in Campinas, Brazil [8]. The data were collected at the Mn K edge (6539eV) using a Si (111) monochromator. The incident beam was monitored by an ion chamber and the data were collected in the fluorescence mode using a Ge 15-elements solid state detector from Camberra. Ideally the angle between electric field vector **E** and film surface equal to 0 for in plane measurements and 90 degrees for out of plane measurements. These limit values – and especially θ =90 – are hardly accessible experimentally. Actually the experiments are performed with the angle between electric field vector **E** and film surface equal to 5 and 80 degrees [9]

The EXAFS spectra were recorded in the range 6440 to 7300eV with energy step of 1 eV in the range 6540 to 6560 close to the edge and 2eV out of this range. The XANES were collected in the range 6440 – 6700eV with energy steps of 0.3eV. The total energy resolution for XANES was 1.60eV (including instrumental resolution and core hole width). The energy calibration was checked after each spectrum using a Mn metal foil. The XANES were normalized over an interval of 200 to 250 eV above the edge. In these conditions the edge structure in all spectra can be compared in position and intensity.

RESULTS AND DISCUSSION

XANES Measurements

In Figure 1 we show the Mn absorption K edge of the constrained sample for the direction in plane and out of the film plane. The inflection points in the two spectra are separated from about 0.5eV, whereas they are coincident in the fully relaxed LSMO/MO sample. In addition the main line at the edge is higher and broader in the film plane. Both effects are related to change in local structural and atomic structure [10-12]. The negative shift of the edge position in plane can be associated to an increase of the average distance Mn-O. A positive edge shift has been reported in $Nd_{0.5}Sr_{0.5}MnO_3$ films grown on a compressible substrate [13]. The amplitude of the shift is lower in our study, proportional to the difference in the lattice mismatch.

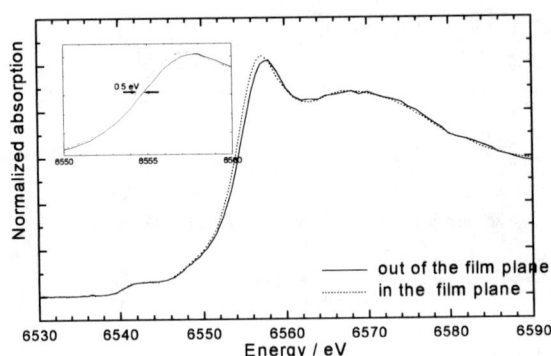

FIGURE 1. Near Edge X-ray absorption spectra at the Mn K edge for the fully constrained LSMO film in plane (solid line) and out of plane (dotted line) measurements. Insert : zoom

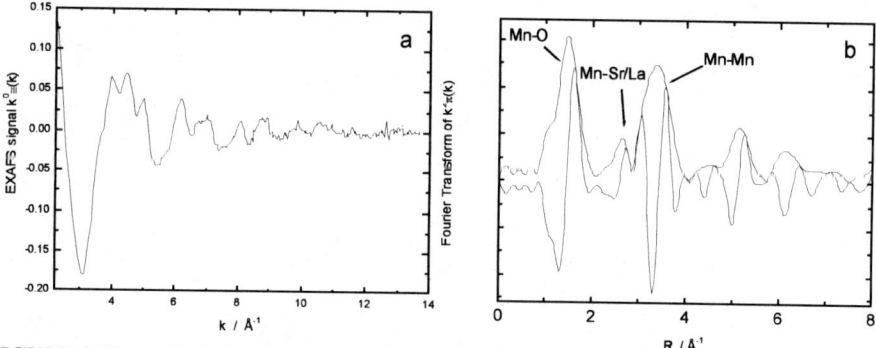

FIGURE 2. X-ray absorption spectra (a) and Fourier transform (b) for the fully constrained LSMO film (in plane measurements).

EXAFS Measurements

Figure 2 shows the in-plane EXAFS signal ($\chi(k)$) and the associated k^2-weighted Fourier transform (range 2.5-13.5Å) for the constrained samples in the plane of the measurements. Due to the symmetry of the atomic distribution around Mn (point group R-3c) the peak of the FT mainly ascribed to the manganese shell is sharp and well separated from those due to the (La, Sr) single scattering contribution, differently from the case of $La_xCa_{1-x}MnO_3$ [14-16].

We have checked by ab-initio simulations on the basis of the crystallographic structure of (La,Sr)MnO$_3$, that the contribution of multiple scattering to that peak are low. This allows in a first approximation to describe this peak as accounting only for the Mn neighbours. The EXAFS signal coming from the contributions above this peak have been subtracted from the total signal to isolate the information of the farther shells (figure 3). We observe a progressive small shift of the main oscillations. This shift can be associated to an increase of about 2% in the distance Mn-Mn in the constrained film [9].

The signal of this Mn peak has been analyzed by back-Fourier Transform in the R-range 2.95 to 3.95 Å and fit using backscattering amplitude and phases obtained from a bulk $La_{0.7}Sr_{0.3}MnO_3$ sample. The low range part ($<5Å^{-1}$) where the contribution of multiple scattering may be relevant has been discarded. The agreement between experimental and fitted curves is excellent leading to the determination of Mn-Mn distance with small relative uncertainties ($0.003Å^{-1}$). An increase of 0.025Å in the Mn-Mn distance is found in the constrained film (Table 1).

TABLE 1. Mn-O and Mn-Mn in-plane distances in LSMO films

Substrate	Mn-O (Å)	Mn-Mn (Å)
MO (ε_{xx}=0.0%)	1.948 ±0.005	3.898±0.003
STO (ε_{xx}=+0.89%)	1.94 ±0.01	3.873±0.003
	2.05 ± 0.01	

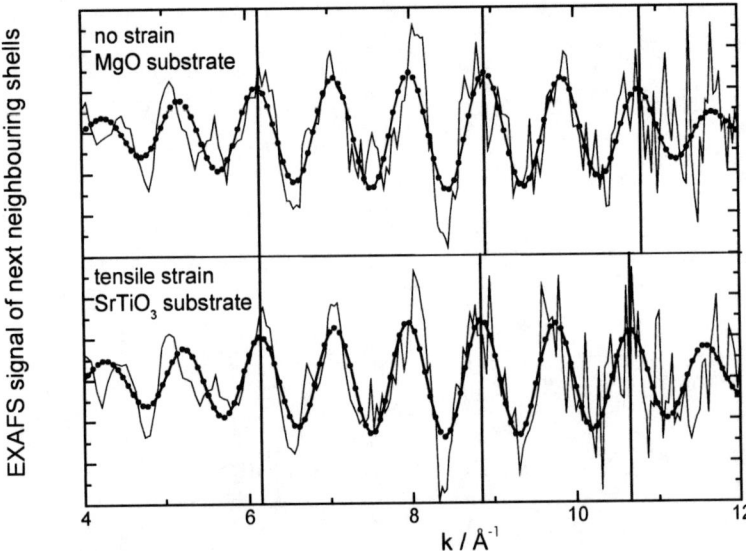

FIGURE 3. In plane EXAFS signal from the upper shells (R>3Å) in the constrained and relaxed films (full line). Dotted line: Fourier filtered signal. The small progressive shift is associated to an increase of the distance in the constrained film

The coordination shell was also back- Fourier transformed in the R-range 0.9 to 1.9 Å^{-1} and fitted using Mn-O amplitude and phases obtained from polycrystalline La$_{0.7}$Sr$_{0.3}$MnO$_3$ [Table 1]. For the relaxed films the coordination shell can be described as a single shell of 6 O at around 1.95 Å, as in the bulk compound [17], whereas in the constrained film an additional contribution of about one long Mn-O distance Mn (2.05Å) has to be introduced. The average Mn-O distance would then increase to about 1%. However the quantitative analysis of the first shell in mixed manganites is known to be a complex task, because of possible complex mixture of collective or non collective Jahn Teller distortion of the octahedra [14-18]. In addition the sensitivity of the analysis to long Mn-O bond is lower than to short Mn-O bonds because the XAS amplitude falls off as 1/r^2[19]. The extend of the increase of the distance is then only approximated, but in perfect agreement with the amplitude and the direction of the modification observed in the XANES spectra.

We should note that the observation of a longer Mn-O bond length implies an increase of the Mn-Mn distance, even when the Mn-O-Mn angle is maintained. As the amplitude of both effects observed here are of the same order, in a first approximation the angle Mn-O-Mn is found unaltered. These conclusions are not in agreement with the results recently reported in constrained La$_x$Ca$_{1-x}$MnO$_3$ films by [20]. Further measurements are needed to provide more accurate determination of the amplitude of the Mn-O and Mn-Mn increase and a definitive conclusion about any modification of Mn-O-Mn angle.

SUMMARY

In this work we present the results of a polarized X-ray Absorption Study of the local characterization of structural distortion around the manganese atoms in $La_{0.7}Sr_{0.3}MnO_3$ films fully constrained on a tensile substrate $SrTiO_3$. From EXAFS in the film plane a small increase of Mn-Mn distances with respect to relaxed film is reported. In addition, a small distortion of the MnO_6 octahedron is evidenced in EXAFS and XANES. The respective amplitudes found for these two effects are on the same order, so that there is no evidence of any modification of the Mn-O-Mn angle.

ACKNOWLEDGMENTS

This work is partially supported by LNLS/ABTLuS/MCT. AYR acknowledges the grant from CNPq. The solid state detector was granted by multi-users proposal from FAPESP under the contract 1999/12330-6.

REFERENCES

1. Jin, S., Tiefel, T. H, McCormack, M., Fastnact, R.A., Ramesh, R. and Chen, L., *Science* **264** (1994) 413.
2. Viret, M., Drouet, M., Contour, J.P., Fermon, C., Fert, A. *Europhysics. Lett.* **39** (1996) 3266
3. Fontcuberta, J., Martinez, B., Seffar, A., Piñol, s., Garica-Muñoz, J.L. and Obradors, X., *Phys. Rev. Lett.* **76** (1996) 1122.
4. Millis, A.J., Darling T. and Migliori, A., *J. Appl. Phys.* **83** (1998) 1588
5. Urushibara, A. Morimoto, Y. Arima, T. Asamitsu, A., Kido, G. and Tokura, Y., *Phys. Rev. B* **51** (1995) 14103.
6. Ranno, L., Llobet, A., Hunt, M.B,. and Pierre, *J. Appl. Surf. Sci.* **138-139** (1999) 228.
7. Ranno, L., Llobet, A, Tiron, R. and Favre-Nicolin E, *Appl. Surf. Sc* **188** (2002) 170.
8. Tolentino, H.C.N, Ramos, A.Y., Alves, M.C.M., Barrea, R.A., Tamura, E. Cezar, J.C. and Watanabe, N., *J. Synchrotron Rad.* **8** (2001)1040.
9. Ramos, A.Y., Giacomelli, C., Favre-Nicolin, E. and Ranno, L. *Physica B*, in press
10. Tyson, T.A., Mustre de Leon, J., Conradson, S.D., Bishop, A.R., Neumeier, J.J., Röder, H. and Zang Jun, *Phys. Rev. B*. **53** (1996) 13 985.
11. Ignatov, A.Y., Ali, N. and Khalid, S., *Phys. Rev. B* **64**, 014413(16) 2001.
12. Bridges, F., Booth, C.H.. Kwei, G.H., Neumeier, J.J., Snyder, J. Mitchell, J. Gardner, J.S. and Brosha, E. *Phys. Rev. B* **63**, 214405(14) (2001).
13. Qian, Q., Tyson, T.A., Kao, C.-C., Prellier, W., Bai, J., Biswas, A. and Green, R.L. *Phys. Rev. B* **63**, 244424(4) (2001).
14. Subias, G., Garcia, J., Blasco, J. and Proietti, M.G., *Phys. Rev. B* **57** (1998) 748
15. Woo, H. Tyson, T.A. Croft, M., Cheong, S-W., and Woicik, J.C., *Phys. Rev. B.* **63** (2001)134412
16. Lanzara, A., Saini, N.L., Brunelli, M., Natali, F. and Bianconi, A., *Phys. Rev. Lett.* **81** (1998) 878
17. Mastelaro, V.R., de Souza, D.P.F. and Mesquita R.A., *X-ray Spectrometry* **31** (2002)154.
18. Massa, N.E., Tolentino, H.C.N., Salva, H., Alonso, H., Martinez-Lope.M.J., Casais, M.T., J. Magn. Magn Mat. **233** (2001) 91
19. Bridges, F. Booth, C.H., Kwei, G.H., Neumeier, J.J., Snyder, J., Mitchell, J., Gardner, J.S. and Brosha, E., *Phys. Rev. B* **63** (2001) 2144-5
20. Miniotas, A., Vailionis, A., Svedberg, E.B. and Karlsson, U.O., *J.Appl. Phys.* **89**, 2134-2137 (2001).

Probing physics in local lattice displacements: the case of inhomogeneous state and superconductivity in the copper oxides

N. L. Saini[1], H. Oyanagi[2] and A. Bianconi[1]

[1]INFM Unit, Dipartimento di Fisica, Universita' di Roma, "La Sapienza", Roma, Italy
[2]National Institute of Advanced Industrial Science and Technology, Tsukuba Central 2, 1-1-4 Umezono, Tsukuba, Ibaraki 305, Japan

Abstract. Local lattice displacements in the copper oxide superconductors are determined by polarized Cu K-edge extended x-ray absorption fine structure (EXAFS) measurements. Temperature dependent local atomic displacements show anomalies, at the T_c and a temperature T_s where the charge inhomogeneous state appears, as revealed by a change in the correlated Debye-Waller factor (DWF) of the Cu-O bonds. While the DWF shows a clear drop at the T_c, an order parameter like up-turn appears at the T_s. The anomalies shows-up with different amplitude, depending on the superconducting transition temperature of the system and the micro-strain in the electronically active CuO_2 plane. The measured Cu-O displacements appear to be closely related to the kink structure in the angle resolved photoemission experiments. The results are discussed to find some correlation between electron-lattice coupling, inhomogeneous charge state and superconductivity in the copper oxides.

INTRODUCTION

Structurally the case of copper oxides showing high T_c superconductivity is non-trivial and knowledge of long-range crystallographic structure is not enough to explain their basic properties unlike the simple solids such as normal metals. Indeed, the basic characteristics of these doped oxides depend strongly on the local atomic structure, as revealed by a series of experiments. In fact, the importance of the electronically active CuO_2 plane in these oxides has created major interest to study the electronic versus structural behavior of this structural unit. Even if the fundamental character of the superconducting order parameter with a charge 2e remains intact, understanding of the mechanism is stagnated by interplaying low temperature orders, related with the charge, spin and lattice degrees of freedom, that can compete or coexist with the superconductivity. Indeed, these oxides manifest self-organization of various degrees of freedom (stripes) at a mesoscopic length-scale, the phenomena which has been a point of recent debate in the field [1].

The pending problems in the copper oxides are the nature of the coupling mechanism responsible for creating the pairs in the CuO_2 plane, driving force for the inhomogeneous charge-state and any correlation between the superconductivity and the charge inhomogeneity. Recent experiments support a key role of local electron-lattice interactions [2-8], and therefore it becomes important to distinguish and quantify the lattice

CP652, *X-Ray and Inner-Shell Processes: 19th International Conference on X-Ray and Inner-Shell Processes*
edited by A. Bianconi, A. Marcelli, and N. L. Saini
© 2003 American Institute of Physics 0-7354-0111-X/03/$20.00

displacements that could be associated with the superconductivity and the charge inhomogeneity. The quantitative value of the displacements should also help us to distinguish proper model based on elactron-lattice interaction [9].

The main experimental probes used to determine the local displacements in these complex oxides are the pair distribution function (PDF) analysis of neutron and x-ray diffraction, extended x-ray absorption fine structure (EXAFS) and ion channeling [2-7]. All these techniques have their own limitations to provide information on the quantitative atomic displacements, however, the results on the local lattice displacements determined by these techniques agree quite well even if there are ambiguities in the measured magnitude due to small disorder. However, recent advances in the materials growth and development of new experimental techniques at higher experimental facilities (synchrotron radiation sources and neutron sources) have brought closer the outcome of the mentioned experimental techniques.

The EXAFS spectroscopy, a fast ($\sim 10^{-15}$ sec) and local (~ 5-6Å) tool [10], has been widely exploited to study the copper oxides. Availability of the high brilliance and polarized x-ray synchrotron radiation sources has been an added advantage for the technique allowing quantitative determination of the directional atomic displacements around a selective site in the copper oxides [6, 7, 11-13]. In fact, with recent technical advances, the EXAFS spectroscopy offers unique approach to pin point short-range atomic displacements and their dynamics.

Here we have used Cu K-edge EXAFS to determine local atomic displacements in the superconducting copper oxides with the aim to explore possible implication of these atomic displacements in their basic properties such as the intrinsic local charge and structural inhomogenieties and superconductivity. We have exploited the EXAFS with high k-resolution to determine temperature dependent distribution of the local lattice distortions (dynamic and static) in the electronically active Cu-O networks. The correlated Debye-Waller factor (DWF) of the Cu-O bonds (σ^2) has been taken as an order parameter of the local displacements, revealing anomalous change across the charge stripe ordering temperature appearing as an up-turn with different amplitude, depending on the superconducting system. In addition, there is a drop of local displacements at the T_c with variable amplitude that depends on the T_c of the material. The local displacements depend on the elastic fields due to local strain and are found to be closely related to the kink structure in the band dispersion as observed by angle resolved photoemission [8]. The results provide a clear indication that the local electron-lattice coupling is one of the main ingredients for the inhomogeneous state of the copper oxides and there is an intimate relationship between superconductivity and the charge inhomogeneities.

EXPERIMENTAL

Polarized Cu K-edge x-ray absorption measurements on single crystals samples were performed at the beamlines BM29 and BM32 at the European Synchrotron Radiation Facility (ESRF), Grenoble and BL13B of Photon Factory, Tsukuba. At the BM29 the synchrotron radiation emitted by a Bending magnet source at the 6 GeV ESRF storage ring was monochromatized by a double crystal Si(311) monochromator. For temperature dependent measurements the samples were mounted in a closed cycle two stage He cryostat. Fluorescence yield (FY) off the samples was collected using 13 Ge element solid state detector to measure the absorption signal at the BM29. A Si(111) crystal was used as monochromator and a 30-element Ge x-ray detector array was used to measure the absorption spectra at the BM32. At the BL13B the synchrotron radiation emitted by a 27-pole wiggler source at the 2.5 GeV Photon Factory storage ring was monochromatized by a double crystal Si(111) and sagittally focused on the sample. The spectra were recorded by collecting the fluorescence photons using a 19-element Ge x-ray detector array. The sample temperatures were controlled and monitored within an accuracy of ±1 K. As our standard experimental approach, several absorption scans were collected to limit the noise level to the order of 10^{-4}. Standard procedure was used to extract the EXAFS signal from the absorption spectrum [10] and corrected for the x-ray fluorescence self-absorption before the analysis. Further details on the experiments and data analysis could be found in our earlier publications [11-14].

RESULTS AND DISCUSSION

Copper oxide superconductors are heterogeneously structured materials having alternated layers of body centered cubic (bcc) CuO_2 layers and rock-salt face centered cubic (fcc) M-O (M= Ba, Sr, La) layers [15, 16]. The mismatch between the two sub-lattices is conventionally estimated by $1-t=[r(A-O)]/\sqrt{2}[r(Cu-O)]$ where r(A-O) (i.e., r(La-O), r(Sr-O) and r(Ba-O)) and r(Cu-O) are the respective bond lengths and t is the Goldschmidt tolerance factor [16]. Due to the lattice mismatch the CuO_2 sheets are under compression and (M-O) layers under tension. The elastic fields due to the mismatch, determined by the micro-strain, play important role in the physics of these copper oxides as demonstrated experimentally [17]. Here we focus on the local atomic displacements as a function of the micro-strain in the CuO_2 plane due to the lattice mismatch to see the influence of it, in the light of the charge inhomogeneities and the superconductivity. We have used $La_2CuO_{4.1}$ (LCO) ($T_c\sim 40$ K), $Bi_2Sr_2CaCu_2O_{8+\delta}$ (Bi2212) ($T_c\sim 87$ K) and $HgBa_2CuO_{4+\delta}$ (Hg1201) ($T_c\sim 94$ K) as representatives for the La-based, Bi-based and Hg-based families. These systems contain respectively the La-O, Sr-O and Ba-O as rock-salt layers, sustaining different chemical pressure on the CuO_2 planes, and the doping is through interstitial oxygen ions in the block layers.

Fig. 1 shows Fourier transforms (FT) of the representative polarized Cu K-edge EXAFS measured on single crystals with E vector of the plane polarized x-rays falling parallel to the CuO_2 square plane. The FT provide a global atomic distribution around the absorbing Cu atom in the measured systems and peaks appear due to scattering of the photoelectron, ejected at the Cu site, with the near neighbor atoms. The main peaks in the FT are denoted by Cu-O, Cu-M (M=La, Sr(Ca), Ba(Ca) for the LCO, Bi2212 and Hg1201 systems) and Cu-O-Cu, appearing due to scattering of the ejected photoelectron at the Cu site with the nearest in-plane oxygen atoms (at ~1.9 Å), M atoms (sitting at ~3.2 Å and 45° from the direction of the photoelectron) and the next Cu atom (at ~ 3.8 Å), respectively. There are evident differences in the FT of the EXAFS spectra measured on different systems. The major differences appear around the Cu-M (M=La, Sr(Ca), Ba(Ca)) peak due to different block-layers. The absolute differences may be difficult to extract because of complex interference effects due to different origins of the backscattering in the three systems at the Cu-M position and interference with the Cu-O-Cu multiple scattering.

Here we focus our attention on the atomic displacements in the electronically active CuO_2 plane (i.e. the in-plane Cu-O bond). In the in-plane polarized Cu K-edge EXAFS the signal due to the Cu-O bond distances is well separated from the longer bond contributions and can be easily extracted and analyzed separately. In this work we have used the 'standard procedure' for the analysis of EXAFS data considering a single distance for the coordination shell, where the effective DWF includes all distortion effects, taking into account both static and dynamic distortions. This standard approach is adopted to make a direct comparison of the temperature dependent distortions in different systems, where the correlated DWF of the Cu-O pairs, σ^2 is a suitable order parameter of the local CuO_2 distortions. Quantitative value of the σ^2 depends on technical aspects (experimental geometry and analysis), however, this is irrelevant for the temperature dependence. Within the reported uncertainties, we have ensured the quantitative values for the DWF in different systems by measuring the three systems in same experimental conditions and applying the same data analysis procedure.

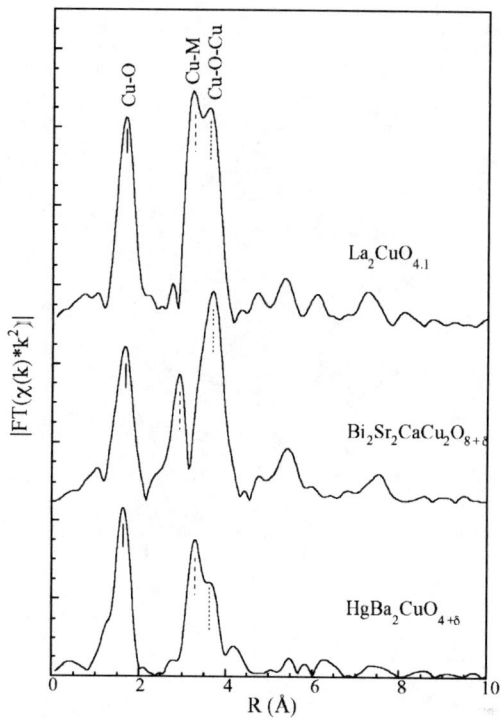

FIGURE 1. Fourier transforms of the Cu K-edge EXAFS measured on the LCO (upper), Bi2212 (middle) and Hg1201 (lower) with varying micro-strain.

We have determined the σ^2 by modeling the Cu-O EXAFS considering a single Cu-O bond, as revealed by diffraction measurements. The Cu-O EXAFS was simulated in the same k (k=3-17Å^{-1}) range for all the systems. The number of parameters which may be determined by EXAFS is limited by the number of independent data points: N_{ind} ~ $(2\Delta k \Delta R)/\pi$, where Δk and ΔR are respectively the ranges in k and R space over which the data are analyzed. In the present case Δk=14 Å^{-1} and ΔR=1 Å give N_{ind}~9 for the first shell EXAFS. Except the radial distance R and the σ^2, all other parameters were kept constant in the conventional least squares paradigm following the standard approach and our experience on the similar systems [6, 7, 11-14]. The average distances were independent of temperature and found to be similar to the one determined by the diffraction experiments on the three systems.

The Cu-O DWF (σ^2), has been used to make a systematic comparison between the systems with variable micro-strain. The temperature dependence is shown in Fig. 2 for the LCO ($T_c\sim40$ K), Bi2212 ($T_c\sim87$K) and Hg1201 ($T_c\sim94$K) systems representing the three different families of the superconducting copper oxides.

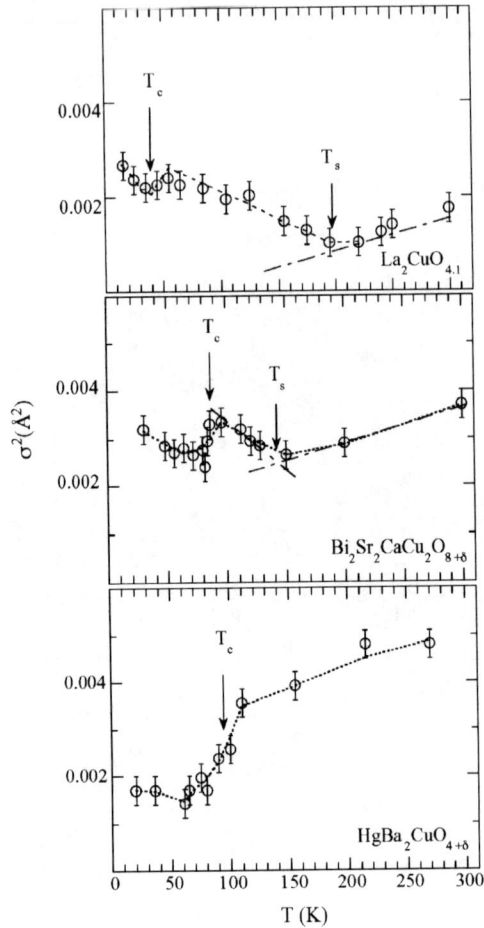

FIGURE 2. Temperature dependence of the Cu-O σ^2 determined by EXAFS; LCO (upper), Bi2212 (middle) and Hg1201(lower). The dashed line is a guide to the eyes. The resulting σ^2 shows abnormal temperature dependence with an increase below a temperature T_s followed by a decrease around the superconducting transition temperature T_c. The error bars represent the average estimated noise level.

From the temperature dependence of σ^2 we can easily define at least two anomalous temperatures. There is an anomalous increase at a temperature T_s followed by a decrease around the superconducting transition temperature T_c. The increase at T_s appears in the LCO and Bi2212 systems, however, the Hg1201 system does not show any evident up turn. On the other hand, the drop in σ^2 at the superconducting transition temperature T_c appears common to all the systems (however, less evident in the LCO system). The Hg1201 system manifests a large decrease in the σ^2 around the superconducting

466

transition temperature. Here we should mention that, apart from the static and dynamic distortions of the CuO$_2$ lattice, σ^2 contains contribution from the thermal vibrations. However, in the present case the thermal contribution to σ^2 should be similar for all the systems and hardly affects the present discussion.

It is known that at the appearance of any charge density wave like instability the DWF shows an anomalous change as found in several density wave systems [18]. Indeed the temperature dependence of the σ^2 shows an anomalous up-turn at a temperature T$_s$, due to the instability (driven by a particular local lattice distortion in the CuO$_2$ plane [7, 11]). Here we provide a ready reference of the La$_{1.48}$Sr$_{0.12}$Nd$_{0.4}$CuO$_4$ (LNS) system [19-22] in which static charge stripe order has been observed. Fig. 3 shows temperature dependence of the σ^2 determined by Cu K-edge EXAFS in this model compound [23].

The σ^2 of the model LNS system shows an anomalous up-turn at ~60K where the charge stripe ordering is known to occur as shown by several experimental techniques in this system [19-22]. Considering evidences of charge stripe ordering in the model LNS system, we assign the anomalous up-turn in the σ^2 to a charge instability giving charge stripe ordering. The results are also consistent with the charge stripe ordering in the LCO system below ~190 K, revealed by x-ray diffraction [24].

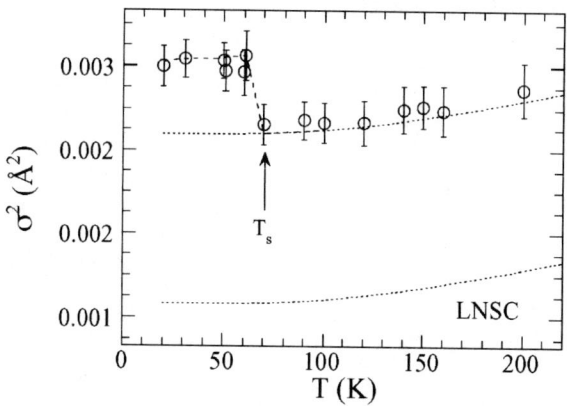

FIGURE 3. Temperature dependence of the Cu-O pairs σ^2 in the LNS system (symbols). Expected temperature dependence of the σ^2 for a fully correlated motion of Cu and O, calculated by Einstein model, is shown by lower dotted line, A constant value of 0.00145 is added to guide the temperature dependence of the experimental σ^2 (upper dotted line). The dashed line across the charge stripe order temperature (T$_s$) is guide to the eyes [23].

Therefore, the anomalous increase in σ^2 is due to stripe ordering in the CuO$_2$ plane of the LCO (T$_s$~190 K) and Bi2212 (T$_s$~140 K) systems. Recently Sharma et al [3] have further confirmed the results and found a clear up-turn in the temperature dependence of the excess displacements (a parameter similar to the DWF measuring dynamic and static distortions) measured by ion-channeling on the YBCO system at the stripe ordering temperature. In fact, below this temperature the pair distribution function becomes larger than that due to thermal fluctuations and the formation of striped phase should give an asymmetric bond length distribution due to splitting of the Cu-O bonds as demonstrated earlier [6, 7].

The lower temperature anomaly in σ^2 appears around the superconducting transition temperature T_c. The correlated DWF σ^2 shows an anomalous decrease around the T_c. This is a clear indication that the appearance of the superconducting state is accompanied by a decrease of the instantaneous local atomic displacements pointing towards a key role of local electron lattice interactions in the superconducting pairing. The drop is found to be maximum for the Hg-based compound where the block-layers are Ba-O with smaller micro-strain in the CuO_2 plane than the case of Bi2212 (Sr-O) and LCO (La-O).

It is interesting to note that the two anomalies appear with different amplitudes, depending on the system and there appears proportionality like correlation between the two amplitudes and the micro-strain in the CuO_2 plane. Fig. 4 shows amplitudes of the two anomalies, given by the drop in the DWF at T_c and the up-turn at the T_s as a function of the micro-strain in the studied systems. Here the micro-strain ε, has been estimated by measuring the average <Cu-O> bond-lengths by the EXAFS analysis. The micro-strain $\varepsilon = 2(d_0 - <R_{CuO}>)/d_0$ determines the relative compression of the average in-plane bonds, $<R_{Cu-O}>$, and is directly related to the chemical pressure. The d_0 is the Cu-O equilibrium distance, i.e., Cu-O bond length for an unstrained CuO_2 plane. The d_0 is measured to be ~1.985 (± 0.005) Å on an undoped model system $Sr_2CuO_2Cl_2$, which is consistent with others [25]. The d_0 is taken to be 1.97 Å throughout this paper considering the correction due to effect of hole doping on the Cu-O bonds (~ 0.16 doped holes per Cu site) [15].

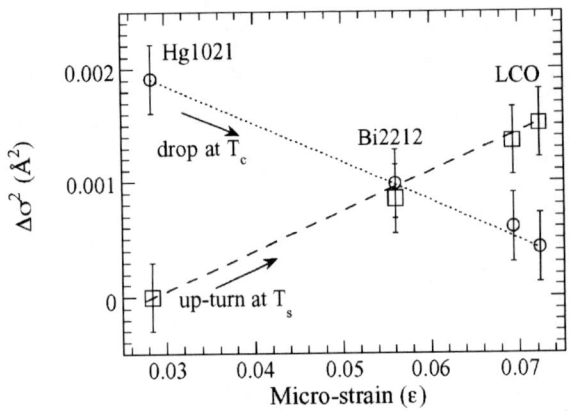

FIGURE 4. The drop in the DWF ($\Delta\sigma^2$) at the T_c (open circles) and the up-turn at the T_s (open squares) are plotted as a function of the micro-strain, determined for different systems.

It is clear from the plot that, while the up-turn in temperature dependent σ^2 increases at the T_s, the drop at the T_c shows a clear decrease with increasing micro-strain. This behavior has direct implication on the relation between the charge inhomogeneous state and the superconductivity in the copper oxides. It has been shown that up-turn at T_s is due to asymmetric pair distribution function derived by splitting of the Cu-O bonds [6, 7]. Therefore the amplitude of the up-turn of σ^2 is directly proportional to the barrier height in the multi-well potential. The present results show that the height of the barrier increases, and hence the charge stripe ordering, which is strongly tied to the electron-lattice interaction, gets stronger with increasing the micro-strain. Lets analyze this aspect and recall the recent angle resolved photoemission spectroscopy (ARPES) results on the high T_c superconductors. High-resolution ARPES results on high T_c superconductors reveal a kink in the dispersion, defined by an abrupt change of electron velocity at 50-80 meV [8]. This kink has been interpreted to be due to phonons associated with the movement of the oxygen atoms. To further enlighten, we have plotted the ratio of the two velocities, i.e., the dressed velocity and the bare velocity as a function of micro-strain in Fig. 5, indicating increasing trend of the electron-lattice coupling as a function of the micro-strain. This behavior of the electron-lattice interaction, analogous to the amplitude of the up-turn in the σ^2 (Fig. 4) suggests that the kink structure in the electron dispersion is closely related to the Cu-O displacements, with asymmetric bond distribution, driving the system in an inhomogeneous charge state with self-organization in stripes. This observation further suggests that the energy scale of ~50-80 meV, revealed by the ARPES measurements, seems to be related to the charge inhomogeneous state. However, it is still to be explored the relationship between the energy scale and the superconductivity.

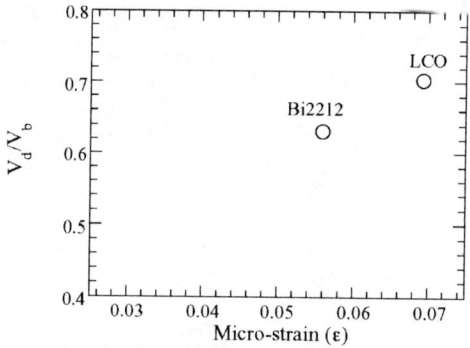

FIGURE 5. Ratio of the dressed electron velocity V_d and bare electron velocity V_b is shown as a function of the micro-strain. The two velocities are taken from the ref. [8], determined by high resolution ARPES data.

Let us now make a comment on the drop in the σ^2 at the superconducting transition temperature T_c, which depends on the micro-strain. Incidentally this anomalous decrease in the local Cu-O displacements is also found by ion channeling experiments, measuring excess displacements on different systems, showing a drop of variable amplitude, that depends on the superconducting transition temperature [3]. In fact, higher the micro-strain, smaller the drop at T_c and smaller the transition temperature. This eventually contradict the fact that higher micro-strain means stronger electron-lattice coupling and hence to expect higher T_c. Therefore the Cu-O local displacements should have two components. At the superconducting transition the drop, revealing decreased Cu-O displacements at T_c could be due to decrease of the incoherent part of the displacements to transfer electron lattice interaction energy in the pairing mechanism entering into a coherent state. As a matter of fact, there seems to be an anomalous drop of local Ge-Nb displacements at the T_c in the Nb_3Ge intermetallic system, showing short coherence length superconductivity [26]. This further suggests an intimate relationship between the local lattice fluctuations and superconductivity in the short coherence superconductors having high transition temperature.

SUMMARY

In summary, we have measured local lattice displacements in copper oxide superconductors by high resolution polarized Cu K-edge EXAFS measurements to address the problem of inhomogeneous state, superconductivity and local distortions. The correlated DWF has been taken as an order parameter to determine the temperature dependent local and instantaneous Cu-O displacements. The choice is due to the fact that the DWF is related with the Debye frequency and has direct implication on the superconducting transition temperature. We have studied the Cu-O displacements as a function of chemical pressure on the CuO_2 plane, defined by the micro-strain. We find two temperatures, the superconducting transition temperature T_c and the charge stripe ordering temperature T_s, where the local Cu-O displacements show anomalous change with variable amplitude. While the amplitude of the drop of local displacements at the superconducting transition temperature decreases, the amplitude of the up-turn across the charge stripe ordering temperature gets increased with increasing micro-strain in the CuO_2 plane. The results are compared with the kink structure in the dispersion, seen by angle resolved photoemission experiments. We conclude that the kink structure and the related energy scale are due to Cu-O displacements and tied to the charge inhomogeneous state of the copper oxides. Furthermore, we find that the appearance of the superconducting state is accompanied by a decrease of the instantaneous local lattice distortions also in the Nb_3Ge intermetallic compound, indicating that the local displacements are the key to the superconductivity of the materials with small coherence length. Nevertheless, present experiments have direct implication on the correlating between electron-lattice interaction, inhomogeneous state and high T_c superconductivity in the complex copper oxides. At this stage it is speculative to predict precise role of the local lattice displacements, however, it appears that the local displacements control the fundamental electronic band structure near the singularity point (M point in the copper oxides and Γ point in the A15 intermetallics).

ACKNOWLEDGEMENTS

We would like to thank the ESRF staff for the help extended during the beamtimes. This research has been supported by the *Istituto Nazionale di Fisica della Materia* (INFM), by the co-financed project *"Leghe e composti intermetallici: stabilità termodinamica, proprietà fisiche e reattività"* of MURST and by the project *"5% Superconduttività"* del *Consiglio Nazionale delle Ricerche* (CNR).

REFERENCES

1. *Stripes and Related Phenomena*, eds. A. Bianconi, and N.L. Saini, (Kluwer Academics/ Plenum Publishers, New York, 2000); also *see e.g.* the speciale issues of *J. Superconductivity* vol. **10**, No.4 (1997) and *Int. J. Mod. Phys. B* vol. **14**, No.29-31 (2000).
2. A. Lanzara, G.-m. Zhao, N.L. Saini, A. Bianconi, K. Conder, H. Keller and K.A. Müller, *J. Phys.:Condens. Matter* **11**, L541 (1999).
3. R. P. Sharma, S.B. Ogale, Z.H. Zhang, J.R. Liu, W.K. Wu, B. Veal, A. Paulikas, H. Zhang and T. Venkatesan, *Nature* **404**, 736 (2000) and references therein; R. P. Sharma et al, unpublished (2002).
4. E. S. Bozin, G. H. Kwei, H. Takagi, and S. J. L. Billinge, *Phys. Rev. Lett.* **84**, 5856 (2000) and references therein.
5. R.J. McQueeney, Y. Petrov, T. Egami, M. Yethiraj, G. Shirane and Y. Endoh, *Phys. Rev. Lett.* **82**, 628 (1999).

6. N. L. Saini, A. Bianconi, and H. Oyanagi, *J. Phys. Soc. Jpn.* **70**, 2092 (2001).
7. A. Bianconi, N.L. Saini, A. Lanzara, M.Missori, T. Rossetti, H. Oyanagi, H. Yamaguchi, K. Oka and T. Ito, *Phys. Rev. Lett.* **76**, 3412 (1996).
8. A. Lanzara, P. V. Bogdanov, X. J. Zhou, S. A. Kellar, D. L. Feng, E. D. Lu, T. Yoshida, H. Eisaki, A. Fujimori, K. Kishio, J.-I. Shimoyama, T. Nodak, S. Uchida, Z. Hussain, and Z.-X. Shen *Nature* **412**, 510 (2001).
9. See e.g. S. R. Shenoy, V. Subrahmanyam, and A. R. Bishop, *Phys Rev. Lett.* **79**, 4657 (1997) and references therein.
10. *X Ray Absorption: Principle, Applications Techniques of EXAFS, SEXAFS and XANES* edited by R. Prinz and D. Koningsberger, J. Wiley and Sons, New York 1988.
11. A. Bianconi, N.L. Saini, T. Rossetti, A. Lanzara, A. Perali, M. Missori, H. Oyanagi, H. Yamaguchi, and Y. Nishihara, D.H. Ha, *Phys. Rev. B* **54**, 12018 (1996).
12. A. Lanzara, N.L. Saini, A. Bianconi, J.L. Hazemann, Y. Soldo, F.C. Chou and D.C. Johnston, *Phys. Rev.B* **55**, 9120 (1997).
13. N.L. Saini, A. Lanzara, H. Oyanagi, H. Yamaguchi, K. Oka and T. Ito and A. Bianconi, *Phys. Rev. B* **55** 12759 (1997).
14. A. Lanzara, N.L. Saini, M. Brunelli, F. Natali, A. Bianconi, P.G. Radaelli, S-W. Cheong, *Phys. Rev. Lett.* **81**, 878 (1998).
15. P. P. Edwards, G. B. Peakok, J. P. Hodges, A. Asab, and I. Gameson in, *High T_c Superconductivity: Ten years after the Discovery* (Nato ASI, Vol. **343**) ed E. Kaldis, E. Liarokapis, and K. A. Müller, (Dordrecht, Kluwer) (1996) p.135.
16. C.N.R. Rao and A. K. Ganguli *Chem. Soc. Rev.* **24**, 1 (1995); J. B. Goodenough, *Supercond. Science and Technology* **3**, 26 (1990); J. B. Goodenough and A. Marthiram *J. Solid State Chemistry* **88**, 115 (1990).
17. A. Bianconi, G. Bianconi, S. Caprara, D. Di Castro, H Oyanagi, N. L. Saini, *J. Phys. Condens. Matter,* **12** 10655 (2000) and references therein.
18. G. Grüner, '*Density Waves in Solids*' Frontiers in Physics Vol. **89**, (Addison-Wesley, USA, 1994).
19. N. Ichikawa, S. Uchida, J. M. Tranquada, T. Niemöller, P. M. Gehring, S.-H. Lee, and J. R. Schneider, *Phys. Rev. Lett.* **85** 1738 (2000) and references therin.
20. X.J. Zhou, P. Bogdanov, S.A. Kellar, T. Noda, H. Eisaki, S. Uchida, Z. Hussain, and Z.-X. Shen, *Science* **286**, 268 (1999).
21. T. Noda, H. Eisaki and S. Uchida, *Science* **286**, 265 (1999).
22. S. Tajima, T. Noda, H. Eisaki, and S. Uchida, *Phys. Rev. Lett.* **86**, 500 (2001).
23. N.L. Saini, H. Oyanagi A. Lanzara, D. Di Castro, S. Agrestini and A. Bianconi, F. Nakamura and T. Fujita, *Phys. Rev. B* **64**, 132510 (2001).
24. A. Bianconi, D. Di Castro, G. Bianconi, A. Pifferi, N. L. Saini, F. C. Chou, D. C. Johnston and M. Colapietro, *Physica C* **341-348** 1719 (2000).
25. L.L. Miller, X.L. Wang, S.X. Wang, C. Stassis, D.C. Johnston, J. Faber, Jr and C.-K. Loong, *Phys. Rev. B* **41**, 1921 (1990).
26. M. Filippi, N.L. Saini, H. Oyanagi and A. Bianconi, Int. J. Mod. Phys. B **16**, 1713 (2002).

Detection and characterization of trace element contamination on silicon wafers

Andy Singh, Katharina Baur, Sean Brennan, Takayuki Homma[1], Nobuhiro Kubo[1], and Piero Pianetta

Stanford Synchrotron Radiation Laboratory, 2575 Sand Hill Rd, Stanford, CA 94309, USA
[1]Waseda University, Dept. of Applied Chemistry, Shinjuku, Tokyo 169-8555, Japan

Abstract. Increasing the speed and complexity of semiconductor integrated circuits requires advanced processes that put extreme constraints on the level of metal contamination allowed on the surfaces of silicon wafers. Such contamination degrades the performance of the ultrathin SiO_2 gate dielectrics that form the heart of the individual transistors. Ultimately, reliability and yield are reduced to levels that must be improved before new processes can be put into production. It should be noted that much of this metal contamination occurs during the wet chemical etching and rinsing steps required for the manufacture of integrated circuits and industry is actively developing new processes that have already brought the metal contamination to levels beyond the measurement capabilities of conventional analytical techniques. The measurement of these extremely low contamination levels has required the use of synchrotron radiation total reflection x-ray fluorescence (SR-TXRF) where sensitivities 100 times better than conventional techniques have been achieved. This has resulted in minimum detection limits for transition metals of 8×10^7 atoms/cm^2. SR-TXRF studies of the amount of metal contamination deposited on a silicon surface as a function of pH and oxygen content of the etching solutions have provided insights into the mechanisms of metal deposition from solutions containing trace amounts of metals ranging from parts per trillion to parts per billion. Furthermore, by using XANES to understand the chemical state of the metal atoms after deposition, it has been possible to develop chemical models for the deposition processes. Examples will be provided for copper deposition from ultra pure water and acidic solutions.

INTRODUCTION

Total reflection x-ray fluorescence in conjunction with synchrotron radiation (SR-TXRF) has demonstrated sensitivities for transition metals that are 50 times better when compared to conventional x-ray sources [1]. With the high flux, low divergence, and linear polarization of a synchrotron x-ray source, fluorescence signals are enhanced while background contributions from elastic and inelastic scattering are reduced. With the synchrotron TXRF facility at the Stanford Synchrotron Radiation Laboratory (SSRL), detection limits of 8E7 atoms/cm^2 for transition metals on silicon

CP652, X-Ray and Inner-Shell Processes: 19th International Conference on X-Ray and Inner-Shell Processes
edited by A. Bianconi, A. Marcelli, and N. L. Saini
© 2003 American Institute of Physics 0-7354-0111-X/03/$20.00

surfaces have been achieved [1], allowing the semiconductor industry to conduct high sensitivity surface analysis for process development.

While TXRF can determine the amount of contamination on a silicon wafer surface, it is possible to exploit the broadband nature of synchrotron radiation to tune the excitation energy through an absorption edge of interest. From the x-ray absorption near edge structure (XANES), information on the unoccupied density of states can be acquired and the oxidation state of the contaminant of interest can be determined. By measuring the fluorescence yield using the grazing incidence SR-TXRF geometry, trace levels of impurities can be analyzed, allowing for trace contamination studies where knowledge of chemical information and impurity amount are needed.

One such relevant study involves trace copper metal deposition onto silicon wafer surfaces from ultra pure water (UPW) solutions that are commonly used by semiconductor manufacturers in post-clean rinse steps. Ever since the implementation of copper interconnect technology in chip processing, understanding how copper ions in solution interact with silicon surfaces has been of the utmost importance to the semiconductor industry, particularly since copper is a fast diffuser in silicon. For metal contamination in solution, there are two alternative reaction pathways that depend on the pH value of the solution. In lower pH solutions, it is expected that metal ions are electrochemically reduced and deposited on the surface as metallic particles [2]. For example, copper metal is reductively deposited on the silicon surface via the following chemical reaction:

$$Cu^{2+} + e_{Si}^- \Rightarrow Cu^{1+} + e_{Si}^- \Rightarrow Cu^0$$

Conversely, in higher pH solutions, an oxide layer readily forms on the silicon surface, and metal ions are precipitated and included into the oxide layer as a metal oxide/hydroxide [3]. In ultra pure water solutions where the pH is neutral, it is proposed that both reaction pathways can occur simultaneously, and that the deposition mechanism is sensitive to other factors such as the dissolved oxygen content. By utilizing both SR-TXRF and XANES, it is possible to investigate copper deposition mechanisms in ultra pure water solutions.

EXPERIMENTAL

Samples were prepared by cleaving silicon wafers into 20 cm x 10 cm pieces, which were subsequently cleaned using a 4:1 sulphuric acid/hydrogen peroxide mixture (H_2SO_4 = 96 vol %; H_2O_2 = 30 vol %) for 10 minutes followed by a rinsing in ultra pure water (UPW). These samples were then dipped into a solution of 0.5% HF for 1 minute to prepare a clean, hydrogen terminated surface. In order to study the influence of oxygen, de-oxygenated ultra-pure water solutions were prepared by Argon sparging, reducing the dissolved O_2 content to 0.3 ppm, while air saturated ultra-pure water solutions contained 3.4 ppm of dissolved oxygen. Copper in a 2% nitric acid matrix was then introduced at concentration levels of 10 and 100 ppb and surface contamination was then accomplished by immersing the hydrogen terminated silicon samples into these solutions.

To determine the surface concentration of copper, these samples were first analyzed with SR-TXRF at beamline 6-2 at the Stanford Synchrotron Radiation Laboratory (SSRL), using the setup shown in Fig. 1. The synchrotron radiation was monochromatized using a high flux, double multilayer monochromator set at an energy of 11.0 keV, giving a high excitation cross section for transition metals. The angle of incidence was 0.1°, which is below the critical angle for total external reflection resulting in a high surface sensitivity. A semiconductor (Si(Li)) detector was mounted so that the face of the detector was aligned perpendicular to the polarization vector of the linearly polarized synchrotron radiation. This geometry reduces the background contribution due to elastically scattered x-rays. Samples were measured for a standard counting time of 1000 seconds.

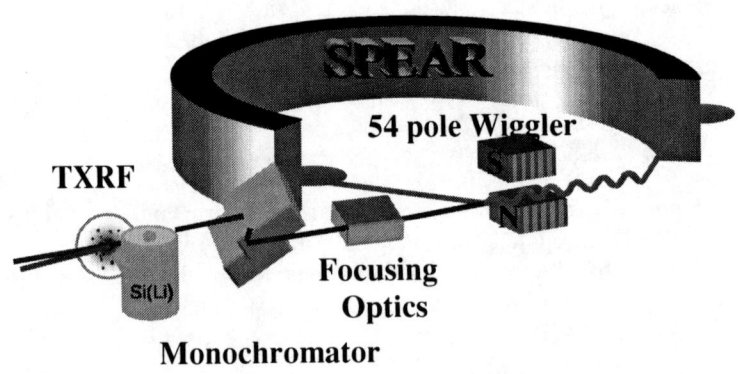

Figure 1. Schematic of the SSRL TXRF facility showing the SPEAR2 storage ring, BL6-2 54 pole wiggler, torroidal focusing mirror, double multilayer monochromator, Si(Li) detector oriented along the polarization vector of the incident radiation and the vertically mounted wafer.

Before absorption edge measurements were conducted on the silicon samples, reference XANES spectra of CuO and Cu_2O, that were prepared as thin powder samples in cells with Kapton tape windows, as well as a 5 μm thick copper foil were measured in transmission. This gave representative spectra for copper in its 3 oxidation states. The silicon samples that were contaminated in de-oxygenated solutions were then analyzed with XANES. These measurements were carried out at the same beamline, using a high-resolution double-crystal (Si 111) monochromator with an energy resolution of 0.89 eV at 8.9 keV. The excitation energy was tuned through the copper 1s absorption threshold (8.979 keV) in the same grazing incidence TXRF geometry, monitoring the fluorescence yield originating from the decay of the excited core state vacancy. The spectra were obtained between 8.940 keV and 9.2 keV at increments of 0.5 eV between 8.97 keV and 8.99 keV near the copper K edge, 1 eV above the edge between 8.99 to 9.060 keV, and 5 eV between 9.06 and 9.2 keV. The signal was integrated for 10 seconds per point for each scan and up to 30 scans were accumulated on the low concentration samples to achieve a high signal to noise ratio. Finally, a control sample with copper metal on silicon that was prepared by immersing a pre-cleaned silicon sample in 2% hydrofluoric acid spiked with a 1000 ppb of copper was measured in order to evaluate possible systematic errors in the sample preparation as well as the effectiveness of using XANES for trace metal contamination on silicon surfaces. This sample was also used to determine the oxidation, if any, of the metallic copper during the deposition in the solution and subsequent transfer to the measurement chamber.

RESULTS

Synchrotron Radiation Total Reflection x-ray Fluorescence

A synchrotron radiation TXRF spectrum of a silicon wafer with very low levels of iron and nickel contamination is shown in Fig. 2. In this spectrum, a high-energy peak at 11.0 keV, which corresponds to the elastic scattering of the primary synchrotron radiation, can be seen. The asymmetric tail on the low energy side of this peak corresponds to the Compton scattering of the incident synchrotron radiation, forming the background in the high-energy region. In the low energy region, the background is dominated by the emission of bremsstrahlung radiation, created by photoelectrons in the wafer. The spectrum shows the iron, nickel, and chlorine K_α fluorescence lines. The chlorine comes from the cleaning solutions used on this wafer, while the iron and nickel are trace contaminants deposited at levels 2.2E8 and 1.5E8 atoms/cm^2, as quantified by measuring peak intensities relative to that of a standard wafer with a known concentration. A minimum detection limit (MDL) of 1.2E7 atoms/cm^2 was achieved for a counting time of 20.4 hours, which corresponds to a MDL of 8.9E7 atoms/cm^2 if a standard counting time of 1000 seconds were to be used.

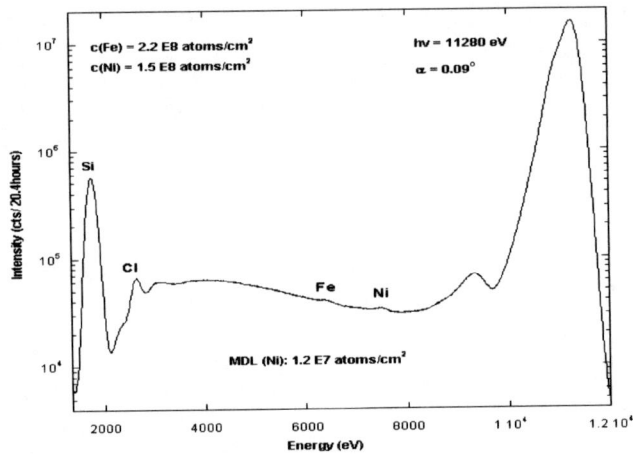

Figure 2. TXRF spectrum of an unintentionally contaminated wafer taken for 20.4 hours of counting time at an excitation energy of 11.28 keV. The MDL for this spectrum is 1.2×10^7 atoms/cm^2.

Synchrotron radiation TXRF measurements on the silicon samples that had been submerged in 10 and 100 ppb copper contaminated ultra pure water solutions are shown below in Fig. 3. The spectrum shows the copper K_α and K_β fluorescence lines from the intentional contamination as well as the Fe K_α fluorescence line due to unintentional iron contamination at levels below 1E9 atoms/cm^2. It was determined that the copper concentration for the 10 ppb deoxygenated and oxygenated samples shown on the left of Fig. 3 were 4E10 atoms/cm^2 and 9.3E11 atoms/cm^2, respectively. For the 100 ppb deoxygenated and oxygenated samples shown on the right of Fig. 3, the copper concentration was found to be 8.4E11 atoms/cm^2 and 6.3E12 atoms/cm^2, respectively.

From the SR-TXRF measurements, it can be seen that that there is a two order of magnitude increase in the amount of copper deposited on silicon in deoxygenated solutions when the copper concentration in the ultra pure water solution is increased from 10 to 100 ppb. Conversely, with the same increase of copper in the UPW solution, the amount of copper deposited from oxygenated solutions remains nearly the same. This drastic increase in the amount of deposited copper in deoxygenated solution can be attributed to the deposition mechanisms occurring at the silicon surface. In deoxygenated solutions, it would be expected that the primary mode of copper deposition would be a reductive one, resulting in copper metal cluster formation while in oxygenated solutions, copper precipitates into the growing oxide as copper oxide/hydroxide. At low concentrations, such as 10 ppb, the concentration in the UPW is not high enough for metal cluster growth in deoxygenated solutions, and any deposited copper is re-dissolved by the UPW solution. Therefore, the only copper that is deposited at 10 ppb occurs by precipitation into the oxide. This results in higher copper deposition amounts for the air saturated solution, since an oxide grows more readily when compared to the deoxygenated solutions. Alternatively, at higher concentrations, such as 100 ppb, copper cluster growth via the reductive mechanism can occur and therefore the amount of copper deposited is higher for the deoxygenated solutions since copper is deposited both reductively and incorporated into the oxide.

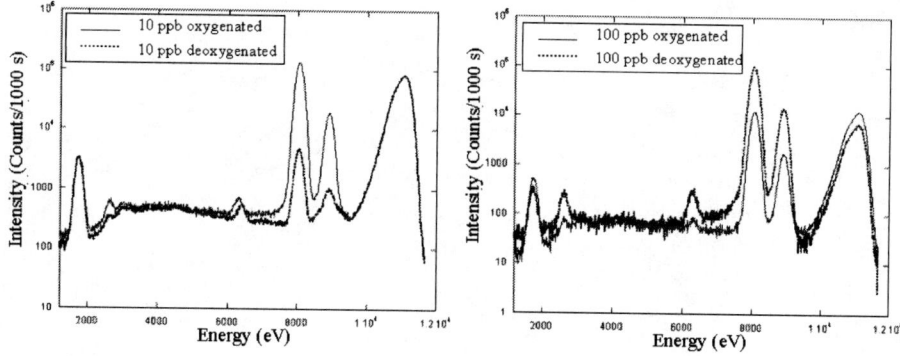

Figure 3. TXRF spectra of silicon samples dipped in oxygenated and deoxygenated UPW spiked with 10 and 100 ppb copper.

X-Ray Near Edge Absorption Spectroscopy

Fig. 4 shows the copper 1s absorption spectra of the copper reference samples (Cu, Cu_2O and CuO) having the oxidation state 0, I, and II, respectively. The copper 1s absorption edge was determined by the first inflection point, corresponding to 8.979 keV [4], and each spectrum was energy calibrated with respect to a copper reference metal spectrum that was measured periodically. A linear background fit to the pre-edge region was subtracted, and each spectrum was normalized with a spline fit done from the post-edge region starting at 9 keV up to 9.2 keV. It can be seen that the copper K absorption edge shifts to higher binding energies with increasing oxidation state of the copper sample. In addition to this energy shift, the detailed shape of the structure in the near edge region is strongly dependent on the atomic structure around the absorbing atoms.

Also plotted in Fig. 4 is the copper metal on silicon sample that was created by contaminating a clean silicon sample in 2% HF spiked with copper at a concentration of 1000 ppb. Since reductive deposition dominates in low pH solutions, it would be expected that any copper deposited in 2% HF would metallic in nature [5]. This is confirmed by the XANES spectrum which clearly shows that the metal on silicon sample resembles the copper metal foil spectrum. However, the energy position of the absorption edge is slightly shifted by 0.35 eV towards higher binding energies as compared to the copper metal foil reference. Except for this shift, the spectrum corresponds well in fine structure below as well as above the ionization potential, demonstrating the utility of using reference samples to identify the oxidation state of copper contamination on silicon samples. Moreover, the data from this sample shows that the sample preparation is effective and that copper metal on silicon can be measured with minimal environmental interference using the cleanroom setup at SSRL.

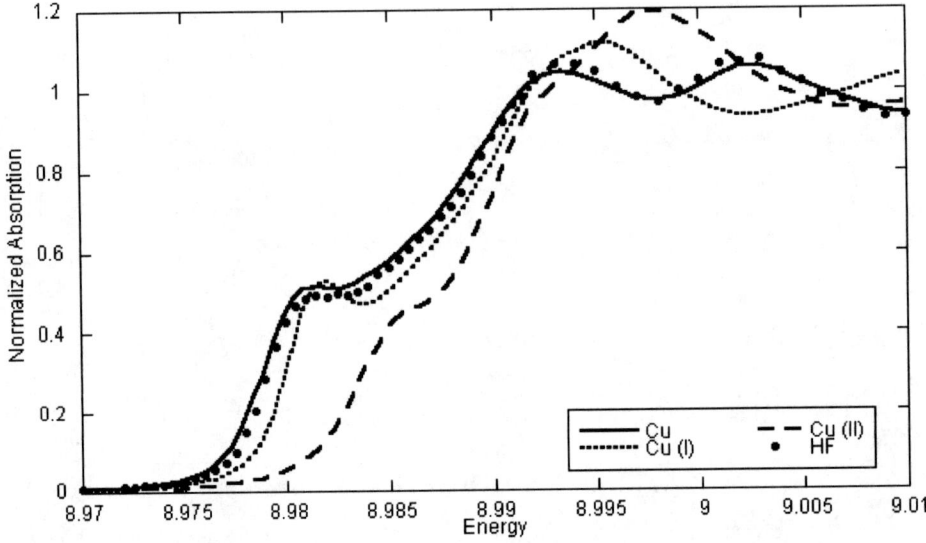

Figure 4. Copper K XANES spectra taken in transmission for Cu, Cu$_2$O, and CuO are shown. Each of these reference samples represents copper in different oxidation states. Also shown in dots is the XANES fluorescence yield data for the silicon sample dipped in 2% HF, spiked with 100 ppb copper.

Fig. 5 shows the copper 1s absorption spectra of silicon samples that were contaminated in deoxygenated UPW solutions with copper concentration of 10 and 100 ppb. A linear background subtraction and normalization using a spline fit to the post edge region was again conducted. Linear combination fits using the Cu, Cu$_2$O and CuO reference samples are also plotted in Fig. 5, with the percent contribution of each component plotted as well. These fits were conducted from 8.975 keV to 9.040 keV by performing a chi-squared minimization procedure with the WinXAS program [6] between the XANES data and a linear combination of the reference samples.

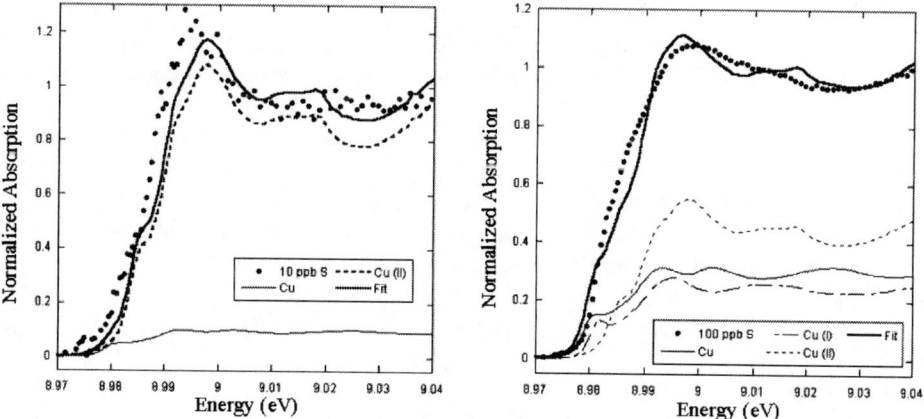

Figure 5. Copper K XANES spectra are shown for silicon samples dipped in 10 and 100 ppb copper contaminated UPW that was deoxygenated. A linear combination fit showing the percent contribution from the three copper reference spectra are also displayed with the overall fit depicted as the bold solid line.

From these linear combination fits of the x-ray absorption spectra it can be seen that the copper deposited for deoxygenated solutions with 10 ppb of copper are almost completely oxide in character (90% CuO and 10 % Cu metal), while the sample contaminated in 100 ppb deoxygenated ultra pure water is mixed between the copper metal, Cu(I) and Cu(II) oxidation states with a contribution of 30 %, 24%, and 46 %, respectively. This supports the TXRF data in that copper is primarily deposited as an oxide from deoxygenated UPW solutions at 10 ppb, since the amount of copper in the UPW is not enough for copper metal cluster growth. However, at higher concentrations, if cluster growth is energetically favorable, both oxidative and reductive mechanisms occur simultaneously giving rise to mixed oxidation character, which is seen for the sample contaminated in 100 ppb deoxygenated UPW solution.

CONCLUSION

It has been demonstrated that the amount and chemical state of copper deposited from ultra pure water solutions depend strongly on the dissolved oxygen content and copper concentration in the solution. With SR-TXRF, it was found that the amount of copper deposited on silicon from deoxygenated UPW solutions greatly increased when the amount of copper in solution was augmented from 10 to 100 ppb. With XANES spectroscopy in a grazing incidence geometry, it was possible to distinguish the chemical state of trace contaminants deposited onto silicon surfaces from solution by using the chemical shifts of the absorption edges along with other spectral features. For the deoxygenated solutions, it was found that copper deposited from a 10 ppb copper contaminated solution was 90% oxide in character while the copper deposited from a 100 ppb copper contaminated solution had copper in all three oxidation states.

Future studies will analyze air-saturated solutions in order to develop a complete representation of the surface deposition chemistries occurring in UPW solutions. Moreover, other systems involving trace metal deposition from solution could be examined as well. For example, co-deposition mechanisms, where other contaminants in solution have affected copper deposition, have been studied [7]. With XANES, the chemical state of trace contaminants deposited from such solutions can be directly determined, providing more insights into trace metal deposition phenomena.

ACKNOWLEDGEMENTS

We would like to thank the staff at SSRL for their expert technical assistance. This work was performed at SSRL, which is supported by the Department of Energy, Office of Basic Energy Sciences. The support of SIWEDS is also acknowledged.

REFERENCES

1. P. Pianetta, K. Baur, A.Singh, S. Brennan, J. Kerner, D. Werho, and J. Wang. Pianetta, Application of Synchrotron Radiation to TXRF Analysis of Metal Contamination on Silicon Wafer Surfaces, *Thin Solid Films*, 1999.
2. H. Morinaga, M. Suyama and T. Ohmi, , *J. Electrochem. Soc.*, **141**, 2834 (1994).
3. H. Morinaga, M. Aoki, T. Maeda, M. Fujisue, H. Tanaka and M. Toyoda, *Mat. Res. Soc. Symp. Proc.*, **477**, 57 (1997).
4. E. Gullickson in X-ray Data Booklet edited by A. Thompson and D. Vaughan (Lawrence Berkeley Labs, Berkeley, 2001), p. 1-38.
5. T. Homma and W. Chidsey, Nucleation of Trace Copper on the H-Si(111) Surface in Aqueous Fluoride Solutions", *J. Phys. Chem. B*, **102**, 7919-7923 (1998).
6. T. Ressler, *J. Synchr. Rad.* **5**,118-122 (1998).
7. T. Homma, J. Tsukano and T. Osaka, "Induced Codeposition of Trace Metals on H-Si(111) Surface of Buffered Fluoride Solutions", *Third Electrochemical Technology Applications in Electronics Proceedings*, v. 99-34, 95-101 (1999).

A polarized XANES investigation
of Mg-rich trioctahedral micas

F. Tombolini[1], G. Cibin[1], A. Marcelli[1], A. Mottana[1,2],
M.F. Brigatti[3] and G. Giuli[4]

[1] Laboratori Nazionali di Frascati, Istituto Nazionale di Fisica Nucleare, Via Enrico Fermi 40,
I-00044 Frascati RM (Italy)
[2] Dipartimento di Scienze Geologiche, Università Roma Tre, Largo S. Leonardo Murialdo 1,
I-00146 Roma RM (Italy)
[3] Dipartimento di Scienze della Terra, Università di Modena e Reggio Emilia, Largo S. Eufemia 19,
I-41100 Modena MO (Italy)
[4] Dipartimento di Scienze della Terra, Università di Camerino, Via Gentile III da Varano,
I-62032 Camerino MC (Italy)

Abstract. In this work we analyse the polarized Mg and Fe K-edge XANES spectra obtained on single crystals of phlogopite, (sample from Franklin, New Jersey), and of tetra-ferriphlogopite (sample from Tapira, Alto Paranaíba, Brazil). These crystals show composition close to the end-members, thus they present similar chemical composition in octahedral and interlayer position, but they differ for tetrahedral substitutions. Our aim is thus to discover the influence of the different chemical composition of the T sheet on the M sheet topology. To reach this goal, as the differences observed in the XANES spectra when changing the beam incidence angle are associated with the different contributions of the photoelectron pathways involving the absorber atom and its near and next-near-shell neighbours in the direction of the electric field vector, we investigated the angular dependence of the spectral XANES features at the Mg and Fe K edges.

INTRODUCTION

Mica 2:1 layer is built up by the packing of an octahedral sheet (M) between two tetrahedral T sheets. An interlayer cation (A), in our case potassium, separates two adjacent 2:1 layers. Si and Al usually occupy the T-sites, however a small Fe^{3+} amount can substitute Al. Only in tetra-ferriphlogopite Fe^{3+} together with Si occupies entirely tetrahedral positions together with Si. The most common cations, which occupy the M-sites, are Fe^{2+}, Fe^{3+}, Mg, Al and Ti. Any deformation, in either the T or M sheet, induced by an atomic substitution inevitably reflects on both layers, as they are forced to match by the deformation or tilting of the oxygen cage coordinating the substituting atom.

The mutual influence of tetrahedral and octahedral sheets was recognized by many authors. Lee and Guggenheim [1] demonstrated that Δz (which is a measure of the flattening of tetrahedral basal oxygen plane) is correlated to the different dimension

CP652, X-Ray and Inner-Shell Processes: 19th International Conference on X-Ray and Inner-Shell Processes
edited by A. Bianconi, A. Marcelli, and N. L. Saini
© 2003 American Institute of Physics 0-7354-0111-X/03/$20.00

of M1 and M2 octahedral sites. Moreover, tetrahedral and octahedral sheet lateral dimensions are linked one to the other by the following relationship:

$$\alpha = \cos^{-1}\left(\frac{\sqrt{3}}{2} \cdot \frac{{}^{[6]}\langle O - O \rangle_{unshared}}{{}^{[4]}\langle O - O \rangle_{basal}}\right)$$

where ${}^{[6]}\langle O - O \rangle_{unshared}$ is the mean value of octahedral triads, ${}^{[4]}\langle O - O \rangle_{basal}$ is the mean value of basal tetrahedral edges and α is the angle which measures the variation from 120° of the internal angles of the tetrahedral ring.

Octahedral chemical substitutions were found to affect the relative dimensions of M1 and M2 octahedral sites. The mean lateral octahedral dimensions were found to present a more complex chemical dependence involving the whole layer chemical composition. These observations are well confirmed by samples under examination. The difference between M1 and M2 octahedral site mean dimension is close to zero for both samples (Table 1). On the contrary mean octahedral dimension is greater in tetra-ferriphlogopite. Octahedral mean dimension were found to be affected by tetrahedral, octahedral and interlayer chemical composition [2].

Owing to the vastly different properties of Mg and Fe atoms in term of number of electrons, binding energies as well as electronegativity, the XANES spectra should undergo modifications that reflect the topological differences revealed by crystal structure refinements. These differences are of two completely different types: indeed, those deriving from the substitution of Al (or Fe^{3+}) for Si in the tetrahedral sheet are expected to be completely different from those deriving from the substitution of Fe^{2+} for Mg in the octahedral sheet. Furthermore, these differences should turn out to be emphasized when rotating the mica single crystal with respect to the impinging synchrotron radiation beam (polarization effect), as the ligand to absorber geometries involved are different in the T and M sheets.

In a recent manuscript [3] the Al and Si K-edge XANES spectra of a set of mica samples have been compared and the sizes of the T sites occupied by either Al or Si could be determined. In those systems Fe was in both the octahedral (${}^{[6]}Fe$) and

TABLE 1. Samples characterization

Chemical composition							
Phl-Fr[a]	$(Si_{3.12}Al_{0.88})(Al_{0.08}Fe^{2+}_{0.05}Ti_{0.01}Mg_{2.85})\ (Na_{0.01}K_{1.00})\ O_{10.22}(OH)_{0.46}F_{1.31}Cl_{0.01})$						
Tas 22-1[b]	$(Si_{3.05}Fe_{0.95})(Fe^{2+}_{0.17}Fe^{3+}_{0.08}Ti_{0.01}Mg_{2.73})\ (Na_{0.01}K_{0.99})\ O_{10.17}(OH)_{1.79}F_{0.08}).$						
Structural data obtained by XRD							
	<M1-O> (Å)	<M2-O>(Å)	a(Å)	b(Å)	octahedral sheet thickness (Å)	<T-O> (Å)	α(°)
Phl-Fr[a]	2.061	2.062	5.309	9.189	2.111	1.650	6.7
Tas 22-1[b]	2.086	2.086	5.362	9.288	2.151	1.680	11.5

[a] the complete unpublished crystal structure refinement is avalible on request
[b] see ref.[6]

tetrahedral ($^{[4]}$Fe) coordinations and in both oxidation states (Fe^{2+} or Fe^{3+}). XANES spectroscopy demonstrated that an increase of $^{[6]}$Fe induces an enlargement of the octahedral M sheet along the *a* and *b* axes and changes the matching conditions between the T and M mica layers. The major observable effect was a different geometrical arrangement of the tetrahedral network, with changes of structural parameters such as α. In addition, even the T size may change when increasing the Fe vs. Mg substitution in the octahedral sheet, although the tetrahedral topology is mostly affected by local chemical composition. Indeed, in the case of mica samples with the maximum amount of $^{[6]}$Fe, the Si-O tetrahedral distance appeared to be unaffected, whereas the Al-O tetrahedral distance was larger. This effect could be attributed to the lower bond strength of $^{[4]}$Al on surrounding anions.

Starting from the above local geometrical description of the mica tetrahedral sheet, the main goal of this work is to distinguish the corresponding site deformations of the M octahedral sheet associated with the different order, arrangements and chemical composition of the T sheet. To reach this goal we will analyse the polarized XANES spectra of two single-crystal mica samples that represent the end members of the join phlogopite-tetra-ferriphlogopite, where Al^{3+} and Fe^{3+} substitute Si, respectively. The chemical compositions of phlogopite (sample from Franklin, New Jersey; label: Phl-Fr) and tetra-ferriphlogopite (sample from Tapira, Alto Paranaíba, Brazil; label: Tas 22-1) are reported in Table 1.

EXPERIMENTAL METHODS

X-ray absorption experiments were performed at the Stanford Synchrotron Radiation Laboratory (SSRL) with the electron-storage ring operating at 3 GeV and electron current between 100 and 60 mA. Mg *K*-edge spectra were recorded at beamline 3-3, which was equipped with a pair of YB$_{66}$ crystals having a resolution of 0.35 eV at the Mg K edge. The spectra were recorded in the total electron yield mode

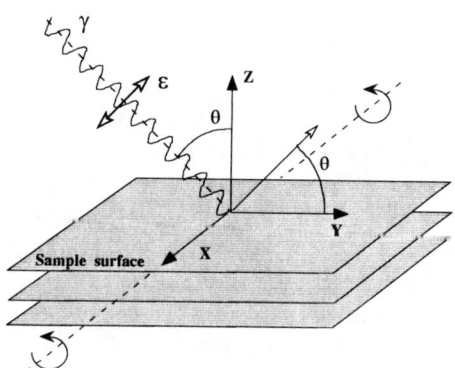

FIGURE 1. Coordinate system applicable to experiments with angular dependence on micas. XY is the plane where the sample lies and θ is the angle between the electric field vector ε and Z, normal to the surface.

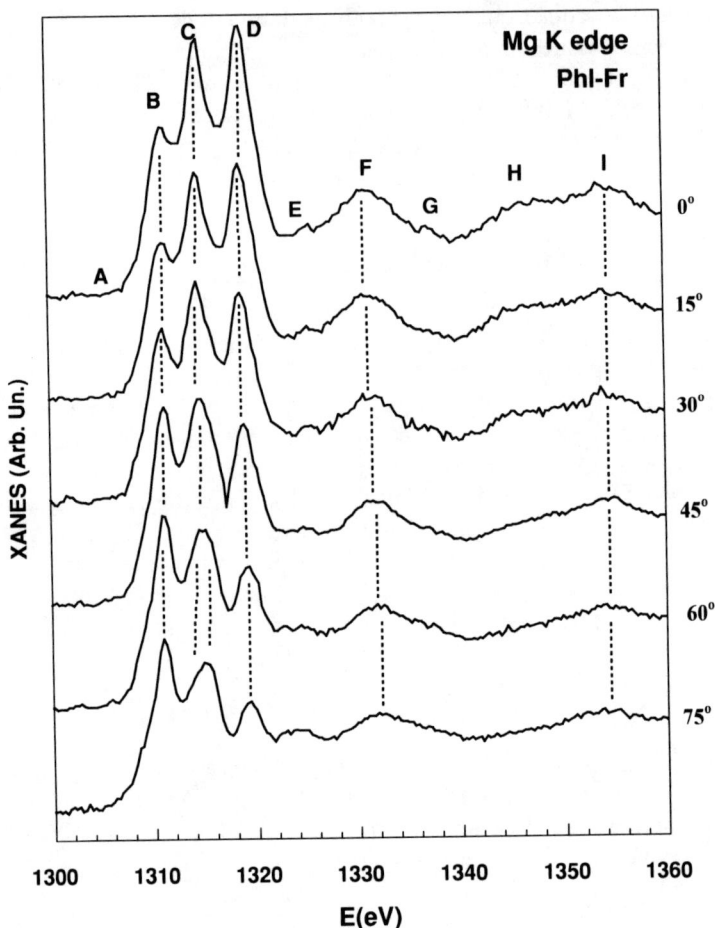

FIGURE 2. Mg K edge polarized XANES spectra of Phl-Fr from 0°, to investigate the in-M plane structure, up to 75° to investigate the out -M plane structure.

and energy-corrected as a function of the ring current, fitted and normalised to 1 at high energy (~50 eV above threshold [4]).

Fluorescence-yield and transmission *K*-edge spectra were recorded for Fe at beam line 4-1, which was equipped with a double-crystal Si(111) monochromator having a resolution of ~1.5 eV. Spectra were recorded at steps of 0.35 eV for 1 s using a Lytle detector with Soller slits, and with the sample compartment filled by N_2 [5]. The recorded spectra were fitted with a Victoreen polynomial function to account for the base line, and normalised to 1 at high energy.

For both edges polarized spectra were recorded at room temperature, at different angles of incidence of the linearly polarized X-ray beam, in the range from 0° (i.e., with the sample *a-b* plane perpendicular to the incoming beam) to 75° (see Fig. 1).

RESULTS AND DISCUSSION

XANES spectra at the Mg K edge

Mg mostly occupies the octahedral sheet for both samples, with limited Fe substitution. The investigation of the XANES spectra at the Mg and Fe K edges may help resolving the local distortions of the octahedral sites. The polarized XANES spectra of the phlogopite Franklin sample at the Mg K edge recorded as a function of the θ angle are shown in fig. 2. The orientation effects are very significant and affect all the spectral features present in the measured energy range. Fig. 2 points out that:
- the relative intensities of spectral features C and D with respect to B decrease monotonically as a function of the incidence angle. The broadening of features C and D changes too, and for C a clear double structure appears at the 45°, 60° and 75° angles of incidence;
- feature E, very weak at normal incidence (0°), increases its intensity and reaches its maximum going out of plane, at 75°;
- the intensity of feature F slightly decreases, while the energy of this peak shifts monotonically toward higher energies.
- the double XANES structures characterized by H and I (respectively at 1346 eV and 1354 eV) tends to weaken when going to higher incidence angles.

We shall now try to translate these observed spectral differences into information on local geometrical data. The differences observed in the XANES spectra when changing the incidence angle are associated with the different contributions of the photoelectron pathways (that are functions of the scalar products of the dipole field and the electric field of the incoming radiation) involving the absorber atom (Mg) and its near and next-near-shell neighbours in the direction of the electric field vector (perpendicular to the incident synchrotron radiation beam). At 0° incidence angle, the photoelectron only probes the in-plane structure i.e., all single and multiple scattering (MS) pathways that involve the Mg atoms in the M plane; by contrast, at 75° the geometrical distribution out of the M-plane is involved, so that it may be possible to extract information connected with the tetrahedral network.

Based on these consideration, we compared first the two end member spectra of phlogopite (Phl-Fr) and tetra-ferriphlogopite (Tas 22-1). These two micas have similar octahedral composition, the differences being located in the T sheet where the Al atoms substitutes Si in phlogopite whereas Fe substitutes Si in tetra-ferriphlogopite. In fig. 3 (top panel) we present the two normal incidence spectra (0°). The XANES spectra are very similar and only a slight contraction of the whole Tas 22-1 spectrum with respect to Phl-Fr can be observed.

This effect can be attributed to the expansion of the whole octahedral sheet containing Mg along the *a-b* axes in tetra-ferriphlogopite sample, as confirmed by XRD diffraction data (Table 1). XRD data [6] allow us to estimate the average structure of the octahedral sites. In both our samples Mg atoms occupy almost all the M sites (Phl-Fr 2.85, and Tas 22-1 2.73, both over 3.00 apfu), so that the octahedral site measured by XRD mainly reflects the Mg environment. Therefore, the spectral contraction of the Mg K edge polarized spectra (0°) reflects the *a-b* plane average octahedral expansion in agreement with the XRD data.

The comparison between the out-of plane (75°) XANES spectra of the two samples shows significant differences both at the edge and in the XANES spectral region. The tetra-ferriphlogopite (Tas 22-1) spectrum exhibits an overall broader structure. In addition, it shows a larger contraction with respect to the 0° spectrum. In both spectra the H feature is very weak and only one maximum (I) appears in the XANES region. The entire edge region is modified, but with a general correlation between the two samples except for peak E, present only in the phlogopite system. Moreover in tetra-ferriphlogopite the C peak can be clearly split in two components. The contraction of the Tas 22-1 spectrum at 75° is in agreement with the expansion of the Mg-octahedral site. However, the observed expansion is now in the direction of the c* axis (i.e., along [001]).

To be really useful, this detailed but qualitative interpretation of the behaviour of polarized XANES spectra has to be correlated by a quantitative analysis. To this purpose, the well-known Natoli's rule, that correlates interatomic distances with energy separation [7], can be used to relate the spectral energy contraction with the length of the recognized MS processes. In our case, we focus our attention on the atomic structure B in Fig. 3, the first peak at the edge, and the feature F, that lies in the Intermediate Multiple Scattering (IMS) region of these spectra. Both these structures are well defined in all spectra, so that measuring their energy value is easy: in particular, that of F can be determined with high accuracy (± 0.04 eV) via a gaussian fit in the energy range between 1325 and 1335 eV. The ratio between the difference in the energy peak position (ΔE) of structures F and B (± 0.02 eV) in the two compounds is therefore useful to estimate the expansion of the Mg atomic site (ΔR) probed by the photoelectron:

$$\Delta R_{Phl-Fr} / \Delta R_{Tas\ 22-1} \sim (\Delta E_{Tas\ 22-1} / \Delta E_{Phl-Fr})^{1/2}$$

The observed energy differences are summarized in table 2. If we now consider the polarized XANES spectra, the energy differences of the two samples in the two polarized configurations show a relative contraction in the *a-b* plane that is slightly

TABLE 2. Structural parameters of Phl-Fr and Tas22-1 samples obtained by Mg K-edge XANES.

Incidence angle	ΔE_{Phl-Fr}	$\Delta E_{TAS22-1}$	$(\Delta E_{Phl-Fr} / \Delta E_{TAS22-1})$	$\Delta R_{Phl-Fr} / \Delta R_{TAS22-1}$
0 °	19.90±0.06	19.60±0.06	1.015±0.006	0.992±0.003
75 °	21.00±0.06	20.30+0.06	1.035±0.006	0.983±0.003

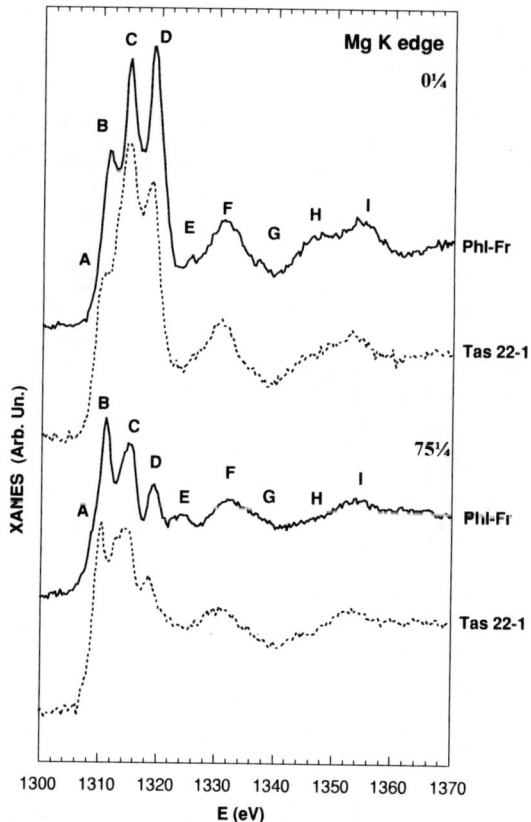

FIGURE 3. Comparison between Mg K edge polarized XANES spectra of Phl-Fr (continuous line) and Tas 22-1 (dots) at 0° (top) and 75° (bottom).

larger than that in the $c*$ direction. The $\Delta R_{Phl\text{-}Fr}/\Delta R_{Tas\ 22\text{-}1}$ ratio achieved at 0° is 0.992, in good agreement with the mean value 0.9897 of the contraction of a and b unit cell axes obtained by XRD (Table 1). Moreover, the contraction ratio 0.983 obtained at 75° is in agreement with the value 0.9814 obtained by XRD for the octahedral sheet thickness (2.111 Å and 2.151 Å for Phl-Fr and Tas 22-1, respectively). Thus, the values obtained by XANES (tab. 3) match quantitatively the XRD data (tab. 2) and support our model that Mg in tetra-ferriphlogopite occupies octahedra that are larger than those occurring in phlogopite. In particular, these octahedral sites have different shapes, being their expansion in the $c*$ direction larger than that on the a-b plane.

The origin of this difference in the octahedral sites can be attributed to the Fe occupancy in the tetrahedral network of the Tas 22-1. Indeed, a recent EXAFS investigation at the Fe K-edge [8] of these samples determined for tetra-ferriphlogopite a [4]Fe-O average distance 1.860 ± 0.015 Å. This value has to be compared with the average XRD T-O distance 1.680(2) Å, which averages [4]Si-O and [4]Fe^{3+}-O distances.

The larger tetrahedral sheet thus results in an increase of α and of the lateral octahedral dimensions in tetra-ferriphlogopite (Tas 22-1: α = 11.5°; Phl-Fr: α = 6.7°).

The change of the octahedral sheet thickness measured by XRD, and confirmed by the XANES polarized data at 75°, does not appear to be a direct consequence of the expansion of the octahedral site in the *a-b*-plane. Different mechanisms have to be taken into account to explain this observed feature, such as the composition of the octahedral anionic site, which is mostly occupied by F in phlogopite and by OH in tetra-ferriphlogopite. Work is in progress to clarify this mechanism of interaction between layers which are very important also for other layered systems.

COMPARISON BETWEEN XANES SPECTRA OF Phl-Fr
AT Mg AND Fe K EDGES

In fig. 4 we compare the Mg and Fe K-edge XANES spectra of phlogopite sample when oriented at 0° (in-plane M structure) and 75° (out-of-plane M structure).

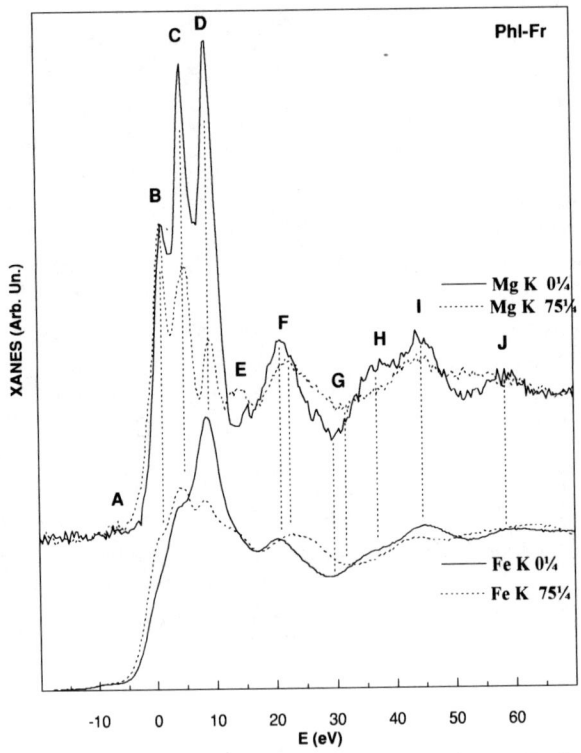

FIGURE 4. Comparison between XANES spectra of Phl-Fr at the Mg K edge (top) and at the Fe K edge (bottom).

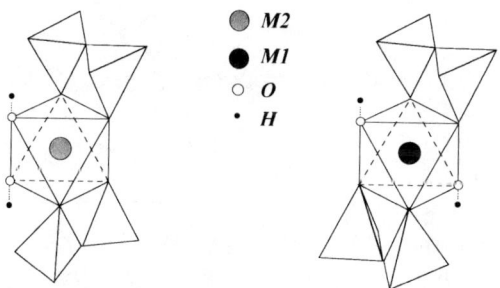

FIGURE 5. Different arrangement of the tetrahedral network related to the octahedral M1 site (right) and M2 site (left).

This comparison will eventually enhance the different local structures of the octahedron occupied by either Mg or Fe. According to the XAS theory, all features observed at the Mg K edge are present at the Fe K edge because the local geometry is essentially the same. The only significant differences occur in the intensities of the spectral features B, C, and D, all being close to the edge (full multiple scattering region of the spectrum, cf. [9]). In this energy region the variation of certain peak intensities can be ascribed to the different atomic potentials seen by the photoelectron. Moreover, the greater broadening of the features at the Fe K edge is known to be due to the different core-hole lifetimes, with additional input by the higher experimental resolution that characterises this high energy region with respect to the low energy Mg one.

Additional differences occur in the energy peak positions of features E, F, H, and I (Fig. 4). The energy shift of the F maximum at the Fe K edge on going from 0° to 75° is ca. 5 eV i.e., much more than the same shift when measured at the Mg K edge (~1.1 eV). By contrast, there is a change in the I maximum at the Fe K edge that is not appreciable at the Mg K edge. In other words, the I peak position remains constant (ca 45 eV above the edge threshold) in the Mg K edge spectrum, while it shifts ca. -4 eV at the Fe K edge: according to Natoli's rule [7] this is related to an expansion of the Fe octahedral site along the c axis. Such a behaviour can easily be understood in phlogopite Franklin, where Fe is present in a very low concentration (0.05 apfu): Fe atoms are located essentially in the large M1 site of the octahedral sheet, as inferred by XRD and confirmed by XANES [10]. As a consequence, the different angular behaviour of the Mg and Fe K edge XANES spectra can be related to the different local arrangements of the oxygen atoms surrounding either Fe or Mg (Fig.5). The larger anisotropy exhibited by the Fe K edge spectrum suggests also that the M1 site occupied by Fe, in contrast with many M1 and M2 sites occupied by Mg in the unperturbed octahedral sheet, is distorted by displaying a significant difference between the in-plane lateral dimension and the out-of-plane overall dimension. This latter value is essentially equal to the octahedral thickness in the direction of the $c*$ axis.

CONCLUSIONS

The use of polarizd XANES spectroscopy enhances the ability of this technique for integrating results obtained by XRD structural analysis and related to the long-range order characters of complex structures with additional data referring to the local structures (short to intermediate range order), thus adding important data to the full understanding of structural compounds. This information is almost impossible to be obtained by X-ray diffraction methods, in particular when low-Z elements are involved. In this case the in-plane and out-of-plane variations in two end member micas have been studied, and new information obtained by using an experimentally time-consuming set up that involves retrieval of many polarized XANES spectra at regular rotation intervals.

The main results obtained are:
- the possibility of independently determining the size of certain structural sites in both the T and M sheet of micas (e.g., the true rather than average size of tetra-ferriphlogopite);
- the possibility of determining the flattening direction of the octahedron in the M sheet that is involved in the Fe vs. Mg substitution;
- the mechanism of the interaction between the adjacent layers as a function of atomic concentrations has been also clarified.

Actually, these results are very important to characterize complex structures such as micas and more generally other layered systems. Thus, not only XANES spectroscopy confirms its importance as a local structural investigation method, but it contributes with quantitative data to the understanding of the mica compounds.

REFERENCES

1. Lee H.-L. and Guggenheim S. *Am. Mineral.* **66**, 350-357 (1981).
2. Brigatti M.F., Guggenheim S., Poppi, M. (2002). *submitted to Am. Mineral.*
3. Tombolini F., Marcelli A., Mottana A., Cibin G., Brigatti M.F. & Giuli G., *Int. J. Mod. Phys. B* **16**, 1673-1679 (2002).
4. Bianconi A."XANES spectroscopy" in *X-ray Absorption: Principles, Applications, Techniques of EXAFS, SEXAFS and XANES*, edited by Koningsberger & Prins, Wiley, New York, pp. 573-662.
5. Lytle F.W., Greegor R.B., Sandstrom D.R., Marques E.C., Wong J., Spiro C.L., YHuffmann G.P., Huggins F.E., *Nucl. Instr. Meth. in Phys. Res.* **A226**, 542-548 (1984).
6. Brigatti, M.F., Medici, L., Poppi, L., *Clays and Clay Minerals* **44**, 540-545 (1996).
7. Natoli C.R. "Near edge absorption structure in the framework of the multiple scattering model. Potential resonance on barrier effects" in *EXAFS and near edge structure* (Springer Series Chem Phys, Vol. 27), edited by A. Bianconi, L. Incoccia, S. Stipcich, Springer-Verlag, 1983, Berlin, pp. 43-47.
8. Giuli G., Paris E., Wu Z.Y., Brigatti M.F., Cibin G., Mottana A., Marcelli A., *Eur. J. Mineral.* **13** 1099-1108 (2001).
9. Benfatto M, Natoli CR, Bianconi A, Garcia J, Marcelli A, Fanfoni M, Davoli I., *Phys Rev B* **34**, 5774-5781 (1986).
10. Tombolini F., Brigatti M.F., Marcelli A., Cibin G., Mottana A. and Giuli G., *Eur. J. Mineral.* **14** (2002) *in press.*

Anisotropy and Correlation Effects in the Spectral Function of Graphite as measured by Electron Momentum Spectroscopy

M. Vos, A.S. Kheifets, V.A. Sashin and E. Weigold

Atomic and Molecular Physics Laboratories, Research School of Physical Sciences and Engineering, Australian National University, Canberra 0200 Australia

Abstract. Electron momentum spectroscopy measurements of graphite single-crystals are presented. Data were taken for incoming electrons with an energy of 50 keV, and both outgoing electrons with an energy near 25 keV. Spectra are presented for the major symmetry directions of graphite. To remove the effect of inelastic multiple scattering we use in all cases an identical deconvolution procedure, which consistently removed all intensity at high energy loss values. However the intensity becomes vanishing small only for binding energies of about twice the bandwidth. The shape of the observed spectra compare well with many-body calculations based on the cumulant expansion scheme but the intensity at high momentum is less than predicted by this theory.

INTRODUCTION

Electron momentum spectroscopy (EMS) provides very direct information about the electron wave function in atoms, molecules and solids [1]. Of these the solid state is the most challenging for a quantitative understanding. On a one-particle level the discrete electron states of atoms and molecules are replaced by dispersing states, described by the band structure. The electron-electron interaction is modified by the response (screening) of the material. The resulting final states that can be created by the annihilation of a target electron is therefore qualitatively different from those in atoms and molecules. EMS is, of course, well suited to study these spectral densities. Graphite single crystals are most interesting to study, as their electronic structure is very anisotropic, and their low atomic number ensures low levels of elastic multiple scattering. Moreover, the thin film preparation procedure is well established. Graphite has thus been an important testing ground for EMS spectroscopy over the years [2, 3]. In an earlier study we investigated the spectral function for a polycrystalline film [4]. In that case we compared the measurement with the angular average of the electronic structure for a given magnitude of electron momentum $|q|$. Here we want to study a single-crystal film, and expect to resolve details of the anisotropic electronic structure. For the energies used in the experiment (50 keV for the incoming electrons and 25 keV for both outgoing electrons) the plane-wave impulse approximation is well justified and the observed intensity for a given energy-momentum transfer to the target (ε, q) is proportional to the magnitude of the spectral function at that value.

CP652, *X-Ray and Inner-Shell Processes: 19th International Conference on X-Ray and Inner-Shell Processes*
edited by A. Bianconi, A. Marcelli, and N. L. Saini
© 2003 American Institute of Physics 0-7354-0111-X/03/$20.00

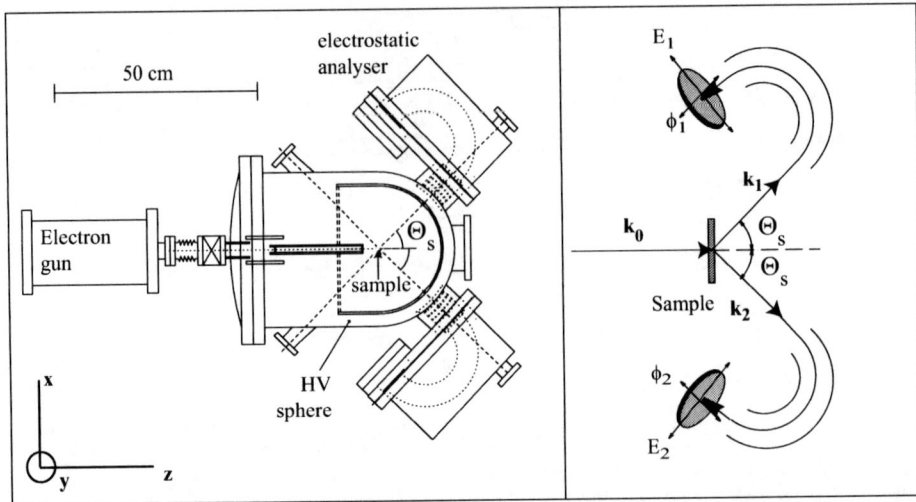

FIGURE 1. An outline of the experimental setup (left) and scattering geometry (right) An electron gun emits a well-collimated 25 keV electron beam. Upon entering the +25 kV high-voltage sphere (containing the target), the electrons are accelerated to 50 keV (and have a momentum k_0). Electrons emerging from the target after a near symmetric (e,2e) event (with momentum k_1, k_2) are analyzed by the electrostatic deflectors, near ground potential.

RESULTS

The EMS spectrometer was described in detail elsewhere [5] and it is outlined in fig. 1. It measures simultaneously electrons scattered over a range of azimuthal angles and energies. For the kinematics used a coincidence can only be detected if the recoil momentum is along the $y-$axis of the spectrometer. The magnitude of the recoil momentum q and separation energy ε can be calculated from the position of impact of both coincident electrons on the position sensitive detectors.

The sample is prepared in the usual way, by cleaving, followed by reactive ion etching in an Ar/O_2 mixture [3]. It could be rotated *in situ*. During the measurement the graphite diffraction pattern of the transmitted beam could be observed on a phosphor screen. This enables us to choose a thin part of the crystal and allows us to verify that the $y-$axis of the spectrometer is aligned with a major symmetry direction of the sample. In this way we collected data along the $\Gamma - M$ direction and the $\Gamma - K$ direction (see fig. 2). Along the $\Gamma - K$ direction the maximum of the $\sigma-$band (ie momentum value with minimal binding energy) is reached at a momentum value of $\simeq 1.3$ a.u. (1 atomic unit (a.u.) of momentum corresponds to $\simeq 1.89$ Å^{-1}.) This corresponds to the M point in the second Brillouin zone. Along the $\Gamma - M$ direction the maximum of the $\sigma-$band is reached at a momentum value of $\simeq 1.6$ a.u., which corresponds to the Γ point in the second Brillouin zone. The $\pi-$band has zero intensity under these conditions (transferred momentum perpendicular to surface normal, corresponding to a nodal plane of the $\pi-$band). The

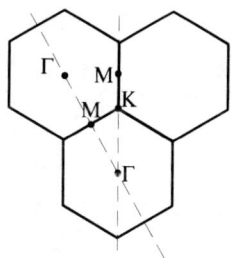

FIGURE 2. Projection of the Brillouin zones of graphite showing the various symmetry points measured in the experiment. Two measurements were done with the y−axis of the spectrometer aligned with either one of the two dashed lines.

alignment of the sample was checked, as described in [6] and some minor adjustments were made using electrostatic deflectors.

Some inelastic multiple scattering occurs. This can be corrected for by a deconvolution procedure as described in [7]. In this procedure one uses the measured energy loss spectrum of 25 keV electrons transmitted through the film to subtract the inelastic multiple scattering intensity from the measured spectra. The deconvolution has been applied to all data presented here. In fig. 3 we show the results of the measurements for both symmetry directions. In both directions we observe a well defined dispersing feature but extending over a larger range in the $\Gamma - M - \Gamma$ direction compared to the $\Gamma - K - M$ direction. A hint of a second dispersing feature is seen, shifted to lower binding energy compared to the main one. This is the π−band, which is not completely suppressed due to finite momentum resolution.

Within band structure theory each electron in a solid interacts with a periodic potential due to the average electron (and ion) density. This means that the solutions of the Hamiltonian are presented by the familiar Bloch waves. If electron-electron interaction is taken into account explicitly, the Bloch waves will decay and the peaks have an intrinsic width, i.e. they are 'life-time broadened'. We used the cumulant expansion scheme to take electron-electron interaction into account explicitly [4, 8]. This approach was very successful for the case of aluminum and noble metals [9, 10].

In fig. 3 we also display the result of the many-body calculation. The calculation is convoluted with a Gaussian of 1.5 eV. This is slightly more than the energy resolution of the experiment (better than 1.0 eV), but this value gave a better agreement with the experimental data. This apparent larger width of the spectra is probably due to the effects of finite momentum resolution in the directions perpendicular to the p_y−axis. Due to the extremely time consuming nature of these calculations we have not yet been able to investigate how finite momentum resolution influences the measurement.

Besides the position of the top of the valence band there is another difference between the two directions. The dispersing structure in the $\Gamma - M - \Gamma$ direction has a very smooth appearance and a gradual decrease in intensity. The small band gap ($\simeq 1$ eV) seen in band structure calculations at M ($\simeq 0.8$ a.u.) [11]) is not resolved, as it is smaller than the energy resolution and the life time broadening. Contrary, in the $\Gamma - K - M$ direction there is a slight kink in the dispersion, and a noticeable decrease of the intensity around

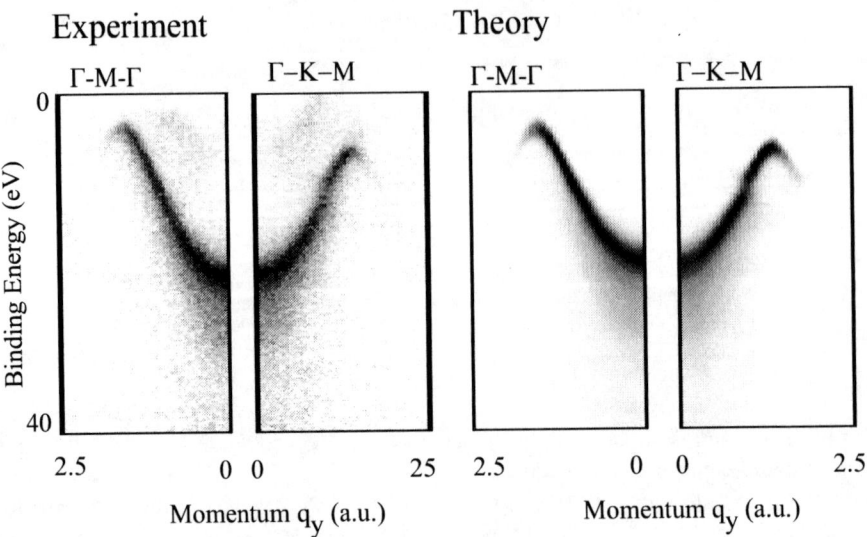

FIGURE 3. A comparison of the measured and predicted intensity distribution for two symmetry directions in graphite. Darker shades correspond to higher intensities.

0.9 a.u. These features are visible in both the experiment and theory. This is due to the band gap at the K point on the border of the first Brillouin zone (calculated gap $\simeq 2.5$ eV).

Note that the calculation predicts non-zero intensity below the main dispersing curve, at larger binding energies. This is indeed observed in the experiment. This intensity is due to electron-electron correlation, it is absent in the independent (mean field) electron theory which predicts a dispersing feature with no intrinsic width.

A more quantitative comparison of the measured and predicted peak shapes can be obtained from the spectra plotted in fig. 4. We present here the binding energy spectra at 0 a.u., 1.3 a.u. and 1.6 a.u. for both crystal orientations as shown in fig. 3. The anisotropy of the σ−band is clearly resolved. In fig. 4 we compare these data also with Linear-Muffin-Tin-Orbital (LMTO) calculations in the atomic sphere approximation [11]. The peak positions and peak heights are reasonably well described. The calculated bandwidth is slightly underestimated in the $\Gamma - K - M$ direction. The full potential LMTO predicts slightly larger ($\simeq 1.2$ eV) dispersion in both directions than calculations using the atomic sphere approximation [11]. Besides this self-energy effects, not included in these calculations, are known to increase the bandwidth of graphite (see, e.g. [12]).

Again we find a significant asymmetric broadening of the spectrum at zero momentum, both in the many-body theory and the experiment. The LMTO calculation predicts sharp features, and no intensity away from these features. The experiment peaks at similar (ε, q) values as the LMTO calculations. In the many-body theory life-time broadening decreases with decreasing binding energy and sharp features are only seen at small binding energies. This is indeed also seen in the experiment. In the many-body theory this decrease in life-time broadening causes the peak height at high momentum to increase

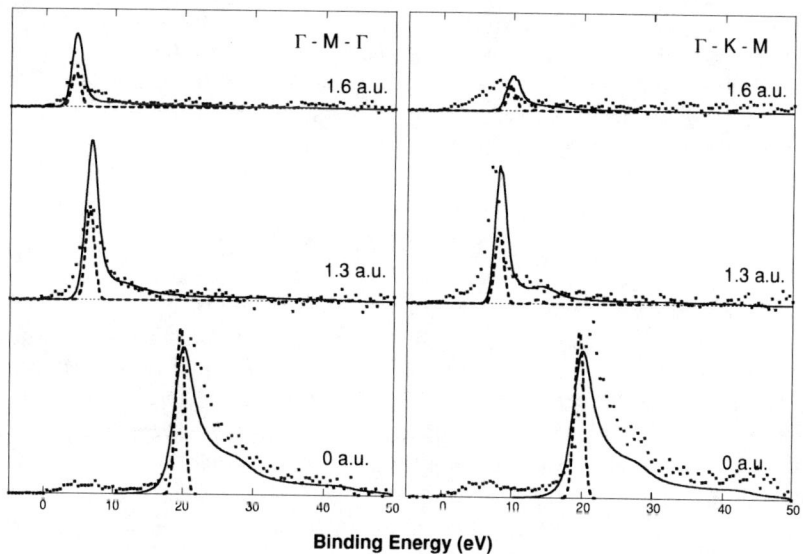

FIGURE 4. A comparison of spectra collected with the $y-$axis of the spectrometer aligned with the $\Gamma-M$ direction and $\Gamma-K$ direction. The experimental data (squares) are compared to LMTO calculations (dotted lines) and many-body perturbation theory based on the cumulant expansion scheme (solid lines). A single normalization factor was used to compare all 6 measurements with theory.

to values larger than that at zero momentum. This is never observed in the experiment. Thus, the agreement between the measurement and the many-body calculations is not perfect.

DISCUSSION

We present two different theories: a band structure theory and a many-body theory, but neither describes the experimental data completely. The LMTO theory predicts the dispersion, and, in the $\Gamma-M-\Gamma$ direction, the peak height quite well. It does not predict the observed intrinsic width and/or asymmetries in the spectral line shapes. The many-body theory describes the shape of the peaks well, but has some problems reproducing the intensities quantitatively, especially in the $\Gamma-M-\Gamma$ direction. The theory predicts maximum intensity near the top of the band, whereas in the experiment the maximum intensity is always found at zero momentum. This disagreement is an illustration of the power of this spectroscopy: a consistent description of a set of spectra is a demanding test.

The most likely flaw in the experiment is the alignment. We assume here that the thin film is completely flat, and that the $y-axis$ of the spectrometer is in the plane of the film. Any wrinkling of the film could cause substantial deviations, and such a distortion has the largest effect far from the origin (zero momentum). However the fact that the π electron intensity is almost absent is a clear indication that the measurement is such that the perpendicular component of the recoil momentum is small. Also we have done quite a number of experiments on single crystals. In all cases the energy-momentum combination with maximum intensity was at zero momentum. In contrast, for aluminum we do find that the maximum intensity is away from zero momentum, it being near the Fermi momentum k_F, for both the theory and experiment[9].

CONCLUSION

We presented here measurements of the electronic structure of graphite for electronic states with $q_{\Gamma-A} = 0$, i.e. states with vanishing momentum perpendicular to the graphitic plane. We resolve the anisotropy between the $\Gamma - M - \Gamma$ and $\Gamma - K - M$ directions. Substantial asymmetric broadening is seen in the spectra near zero momentum. Some disagreements were found between the predicted and measured intensity distributions.

We want to extend this work to include also states with $q_{\Gamma-A} \neq 0$. Here the π−band has non-vanishing intensity. By performing calculations on a finer mesh in k−space we can investigate the effect of finite momentum resolution on the data. Also we want to fully incorporate the self-energy corrections in the calculations. In this way we hope that a very detailed picture will emerge of the agreement between experiment and many-body theory for a very simple systems as graphite.

REFERENCES

1. Weigold, E., and McCarthy, I., *Electron Momentum Spectroscopy*, Kluwer Academic/Plenum, New York, 1999.
2. Gao, C., Ritter, A., Dennison, J., and Holzwarth, N., *Phys. Rev. B*, **37**, 3914–3923 (1988).
3. Vos, M., Fang, Z., Canney, S., Kheifets, A., McCarthy, I., and Weigold, E., *Phys. Rev. B*, **56**, 963 (1997).
4. Vos, M., Kheifets, A., and Weigold, E., *Phys. Rev. B*, **63**, 033108 (2001).
5. Vos, M., Cornish, G., and Weigold, E., *Rev. Sci. Instrum.*, **71**, 3831 (2000).
6. Vos, M., Canney, S., Lun, D., and Weigold, E., *Journal de Physique. IV*, **9**, Pr6–153 (1999).
7. Vos, M., Kheifets, A., and Weigold, E., "Electron Momentum Spectroscopy of Metals", in *Correlations, Polarization and Ionization in Atomic Systems, IAP Conference Proceedings 604*, edited by D. Madison and M. Schulz, American Institute of Physics, 2002, pp. 70–75.
8. Aryasetiawan, F., Hedin, L., and Karlsson, K., *Phys. Rev. Lett.*, **77**, 2268 (1996).
9. Vos, M., Kheifets, A., and Weigold, E., *J. Electron Spectrosc. Relat. Phenom.*, **114-116**, 1031–1036 (2001).
10. M.Vos, Kheifets, A., and Weigold, E., *AIP conference proceedings*, **604**, 70 (2002).
11. Kheifets, A., Lun, D., and Savrasov, S. Y., *J. Phys.: Condens. Matter*, **11**, 6779 (1999).
12. Heske, C., Treusch, R., Himpsel, F., Kakar, S., L.J.Terminello, Weyer, H., and E.L.Shirley, *Phys. Rev. B*, **59**, 4680 (1999).

Unified interpretation of pre-edge x-ray absorption fine structures in 3d transition metal compounds

Z.Y. Wu[1,2], C.R. Natoli[2], A. Marcelli[2], E. Paris[3], A. Bianconi[4] and N.L. Saini[4]

[1]Beijing Synchrotron Radiation Facility, Institute of High Energy Physics, Chinese Academy of Sciences, Beijing, 100039
[2]Laboratori Nazionali di Frascati, Istituto Nazionale di Fisica Nucleare, Via Enrico Fermi 40, I-00044 Frascati, Italy
[3]Dipartimento di Scienze della Terra and INFM, Universita' di Camerino, I-62032 Camerino, Italy
[4]Unitá INFM and Dipartimento di Fisica, Universitá di Roma "La Sapienza", P. le Aldo Moro 2, 00185 Roma, Italy

Abstract. Here we discuss origin of pre-edge features in the K-edge absorption spectra of transition metal atoms in octahedral coordination, in oxides and sulphides. We provide a unifying interpretation on the basis of multiple scattering simulations performed with different cluster models. We find that the pre-edge features arise due to hybridisation of the orbitals belonging to the central atom with the higher-shell metal orbitals. The results are obtained by performing multiple scattering simulations with cluster of size equal to the cation-cation plus the cation-anion bond lengths in order to ensure that the higher-shell metal atoms remain in the octahedral coordination. Within this framework, we are able to identify the electronic structure of the metal atoms and the ligand-field characters looking at position, shape and intensity of the different features observed in the XANES spectra.

INTRODUCTION

Knowledge of the relation between electronic structure and the atomic environment of 3d transition-metal oxides and sulphides has fundamental importance to understand transport and magnetic properties [1]. Recently, followed by the discovery of the high T_c superconductivity [2], colossal magnetoresistance effects [3] and other correlated phenomena, intense research has been devoted to the investigation of structural and electronic behaviour of these transition metal compounds to address a close interplay between electronic, magnetic and structural properties. Various high energy spectroscopy techniques, such as photoemission, inverse photoemission and x-ray absorption have been applied to investigate the electronic properties of these fascinating and still mysterious materials to find some relation between atomic scale structure related to the single particle excitation properties [4]. However, it is

CP652, X-Ray and Inner-Shell Processes: 19th International Conference on X-Ray and Inner-Shell Processes
edited by A. Bianconi, A. Marcelli, and N. L. Saini

© 2003 American Institute of Physics 0-7354-0111-X/03/$20.00

necessary to interpret these results with a suitable theory in order to clarify the structural environment and symmetry as well the electronic properties of these materials. Such analyses have been extensively carried out in case of photoemission techniques. Although x-ray absorption spectroscopy, as a tool of partial and local empty density of states (DOS), is a powerful method to investigate the electronic structure in these systems,[5] so far, just a few theoretical approaches have been published. In spite of these efforts, interpretation of the X-ray-absorption near edge structure (XANES), and in particular the pre-edge features, still remains an open and controversial problem [6-12]. For example, some weak but sharp pre-peaks appearing several eV below the main absorption edge at the metal K-edge XANES spectra of 3d transition metal oxides, that are dipole forbidden in the octahedral coordination, have been attributed to 1s to 3d or quadrupole transitions.

It is known that properties of transition-metal oxides are correlated to the geometric arrangement around the transition-metal site.[13,14] A quantitative interpretation of the features observed in these spectra is a prerequisite to a complete investigation of the geometrical environment around the photoabsorber as well as the electronic structure of metal-ligand atoms. In this paper we present a detailed theoretical analysis of the XANES spectra at the K edges of the transition metal. For the sake of clarity we summarize some experimental results of the metal K edges in Fig. 1. In left panel are compared the XANES spectra of the Ti K-edge of TiO_2 (anatase) and TiS_2, the Mn K-edge of MnO_2 and the Fe K-edge of Fe_2O_3. In the right panel, the Mn, Fe, Co, and Ni K-edge spectra for NaCl-type monoxides are shown.

Commercially available high purity powder samples were used for the measurements. X-ray absorption near-edge structure (XANES) spectra were recorded using the synchrotron radiation source of the BEPC storage rings of the Beijing Synchrotron Radiation Facility (BSRF), working at the typical energy of 2.2 GeV with an electron current of about 120 mA. The Ti, Mn, Fe, Co and Ni K edges were recorded in transmission mode using a Si(111) double crystal monochromator at the beam line 4W1B with 1.0 eV resolution.

The spectra reported in Fig. 1 display pre-edge features that were previously attributed to 1s to 3d or quadrupole transitions. However, in these compounds, the metal ions are coordinated with six ligand ions and the 1s to 3d transition is not allowed. We may than concentrate on the possible alternative interpretation of the near-edge structures in these systems using the multiple scattering (MS) theory, which is capable to interpret the modulations of the XAS spectra in a variety of systems.[19] In order to clarify the type of the transition involved and to identify the origin of the spectral features, we report several theoretical simulations, taking TiS_2 (d^0) and α-Fe_2O_3 (d^5) compounds as model examples, based on one-electron full multiple scattering theory.[20-26] We have used the Mattheiss prescription [27] to construct the cluster density and obtain the Coulomb part of the potential by superposition of neutral atomic charge densities using the Clementi and Roetti basis set tables.[28] For practical reasons we use the energy independent X_α exchange potential followed by a Lorentzian convolution to account for inelastic losses of the photoelectron in the final state and the core hole width. The total width of the Lorentzian is given by

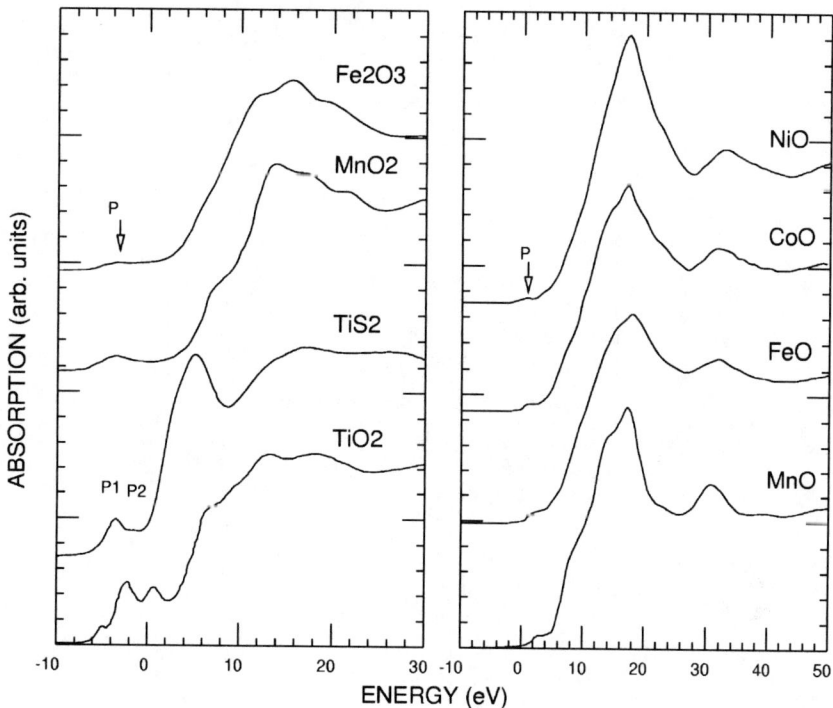

Figure 1. XANES spectra at the metal K edges. Left panel: Ti in rutile and anatase (TiO$_2$, and TiS$_2$); Mn in MnO$_2$ and Fe in α-Fe$_2$O$_3$. Right panel: K edge of monoxides.

$\Gamma_{tot}(E)=2I_m\Sigma(E)$ where $\Sigma(E)$ is a volume averaged value over the unit cell of the compound of self-energy $\Sigma(r,E)$ as suggested by Penn.[29] The use of the X$_\alpha$ exchange-correlation potential is not a limitation, since at low energies (around 30 eV above the onset of the absorption, the energy region we are interested to interpret) this potential and the real part of the Hedin-Lundqvist (H-L) one roughly coincide if the constant Σ is appropriately chosen. In order to simulate the charge relaxation around the core hole, in a photoabsorber of atomic number Z, we used the well screened Z+1 approximation (final state rule).[26] This method start considering the orbitals of the Z+1 atom and consequently truncating the charge density by using the excited electronic configuration of the photoabsorber with the core electron promoted to an empty orbital. We have chosen the muffin-tin radii according to the criterion of Norman [20] and allowed a 10% overlap between contiguous spheres to simulate the atomic bond.

TiS$_2$ is described in the 1T-CdI$_2$ type structure [31] and the sulphur atoms form a perfect octahedron around Ti. In the Fig. 2, we report the MS calculations at the Ti K edge of TiS$_2$ by using different atomic clusters containing increasing numbers of atoms, from 7 to 81, until convergence was achieved. The cluster has D$_{3d}$ symmetry with respect to the central titanium atom. For a minimal cluster composed of a Ti atom

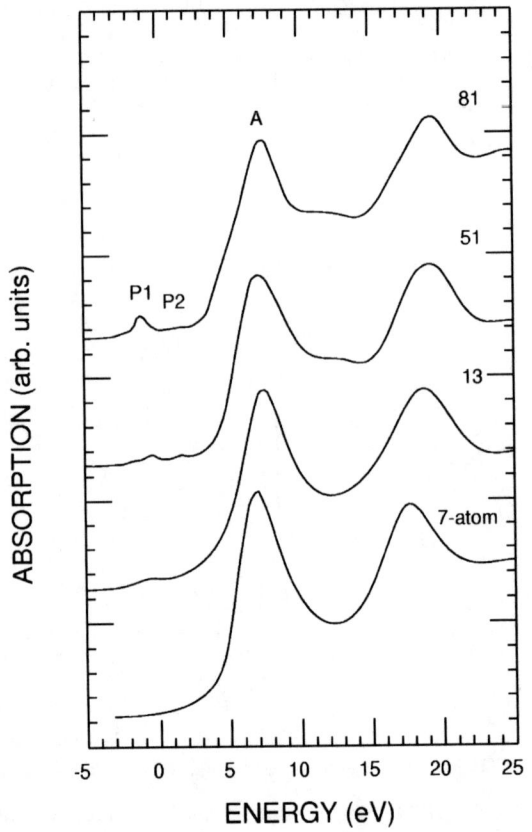

Figure 2. Theoretical XANES spectra as a function of the cluster size at the Ti K edge in TiS_2.

(emitter) surrounded by six sulphur (forming a perfect octahedron), the calculated absorption spectrum only gives rise to the peak A which corresponds, as expected, to transitions to Ti p-like states. Only after adding the second Ti shell, the pre-peak P appears. The presence of this feature has to be associated with the existence of unoccupied states due to the hybridisation of the central Ti p states with higher-neighbour Ti d orbitals. The increase from 13- to 51-atom cluster calculations is very clear: at the pre-edge two peaks arise. In fact, the 51-atom cluster includes an enough amount of outer-shell sulphur atoms (the cluster size is about Ti-Ti plus Ti-S bond lengths) to form an octahedral geometry around the Ti atoms surrounding the central atom in order to build t_{2g}- and e_g-like molecular orbitals. These d orbitals can re-combine symmetry-adapted, transforming them as the same irreducible representations of the photoabsorber,[32,33] and mixing with the p-states of the central atom (contribution P1 and P2). All the features observed in the experimental spectrum, in particular in the pre-edge region, are very well defined in the 81-atom cluster calculation indicating that the cluster size is sufficient to describe the bulk properties

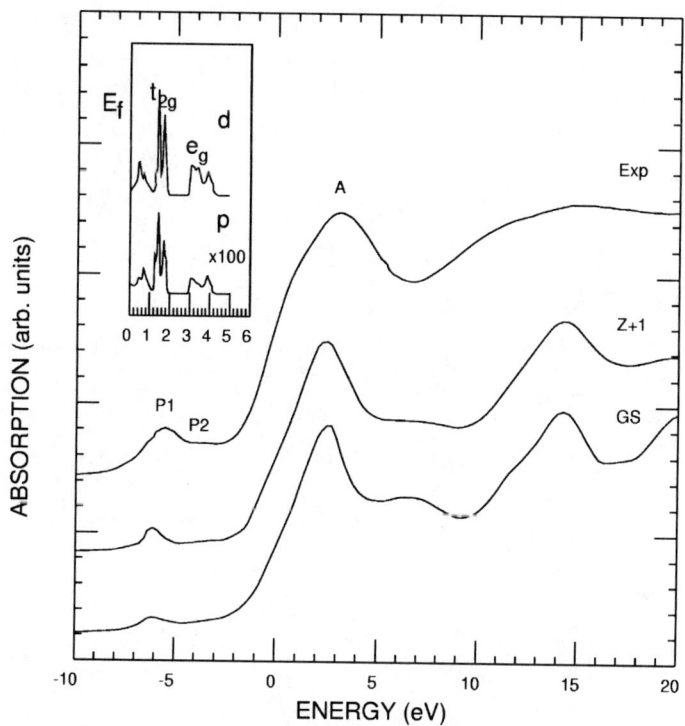

Figure 3. The last model used in Fig. 2 (81-atom cluster) with different potentials: full relaxed final state (Z+1) and ground state (GS) potentials. The inset gives the theoretical TB-LMTO Ti p-like density of states (DOS) (lower curve) multiplied by 100 and d-DOS (upper curve).

while the pre-peaks reflect the density of states associated with the medium-range order of solid.

It is appropriate to compare the near-edge data, such as the pre-edge structures with the unoccupied conduction-band density of states (DOS) of the compound [34]. In Fig. 3 we report experimental data and the MS XANES spectrum calculated using a 81-atom model compared with the calculated DOS just above the Fermi level E_f, as derived from tight-binding linear muffin-tin orbital (TB-LMTO) band-structure calculations which has been widely used for this type of systems, e.g., on 3d TM monoxides [35] and more recently extended to the investigation of high-T_c cuprates.[36].

There are two peaks in the Ti p-projected DOS, corresponding exactly to the P1 and P2 feature observed in the experimental data and in the MS calculation. From the corresponding d-projected DOS one can easily conclude that features P1 and P2 are due to dipole transitions to Ti 4p states hybridised with the crystal-field split of Ti 3d orbitals on neighbouring Ti atoms, i.e. t_{2g} and e_g band-like states. The energy separation between them is about 2.0(1) eV, nearly the same as the crystal-field d

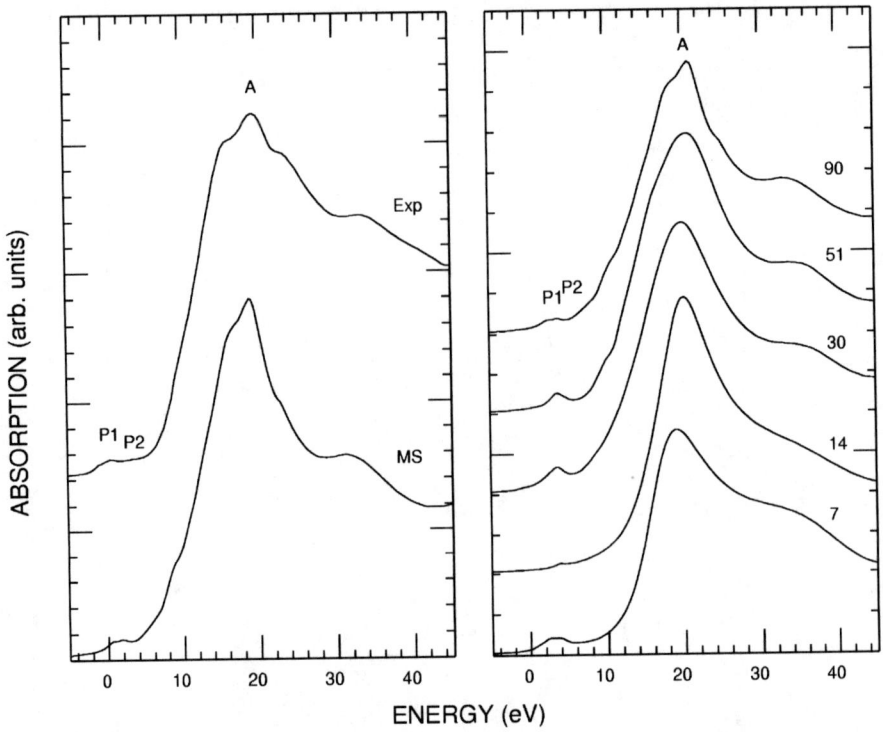

Figure 4. Left panel: the comparison of the experimental and full MS-calculation (using a 90-atom cluster). Fe K-edge XANES spectra for α-Fe$_2$O$_3$; right panel, MS calculation of the Fe K-edge XANES spectra for α-Fe$_2$O$_3$ as a function of the cluster size.

orbitals splitting and agrees well with Fischer [37] and other studies. [7,38,39,40] We present also ground state (GS) potential calculations, which ignores core hole relaxation effects (lower curve in Fig.3) and that, in the pre-edge region, is almost identical. This result is not surprising since the pre-edge features reflect the density of states associated with the medium-range order of the solid, as illustrated above. They should be rather insensitive to the details of the potential on the central Ti atom.

MS calculations of iron K-edge spectra of α-Fe$_2$O$_3$ (corundum) with different cluster models (7, 30, 51, and 90 atoms) are shown in the right panel of Fig. 4. In this compound, oxygen atoms are close packed and Fe^{3+} cations occupy nearly perfect octahedral sites. As in the case of TiS$_2$, the first model calculation does not give the pre-edge peaks. They appear only when cluster includes the next-nearest iron atoms (30-atom cluster calculation). All features observed in the experimental data are well reproduced in the 90-atom cluster calculation. As one may see in Fig. 4 when the cluster contains 51 and 90 atoms (i.e. the second Fe shell are in octahedral coordination with oxygen), the crystal-field splitting in the calculated spectra is not well defined like the behaviour at the O K-edge spectra because 13 iron atoms in the outer coordination shell, one at 2.88 Å, three at 2.95 Å and 3.35 Å, and six at 3.68 Å. Each of them is associated with its own t$_{2g}$-e$_g$ like splitting. The energy position will be

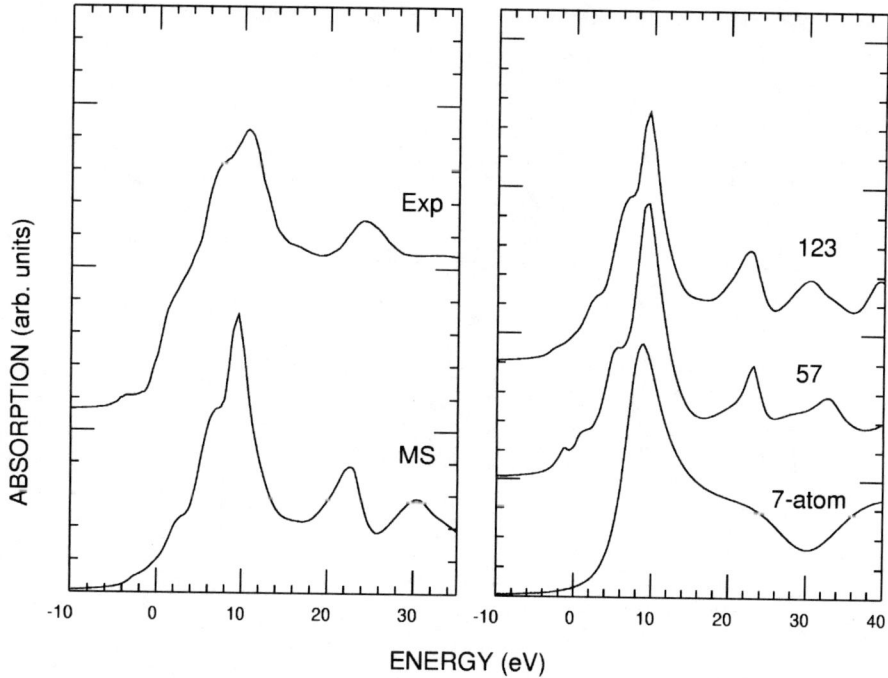

Figure 5. Left panel: comparison of the experimental and full MS-calculation (using 123-atom cluster) at the Mn K-edge XANES spectra for MnO; right panel, MS calculation of the Mn K-edge XANES spectra for MnO as a function of the cluster size.

modified by the overlap of these octahedral orbitals and by the interaction with the oxygen orbitals. The overall absorption consists of a mixing of several components leading to a loss of resolution.

The pre-edge features in other compounds have the same origin as illustrated by TiS_2 and α-Fe_2O_3. For example, in the monoxide MnO, the pre-edge peak is reproduced only using large atomic clusters (includind 57 and 123 atoms) as shown in Fig. 5. The intensity of these features decreases going from TiO_2 (rutile) to monoxides NiO and reflects the decreased number of unoccupied 3d states available for mixing with the central atom p-states and the increased cation-cation distance. This trend has been observed in the O and metal K edge in TM compounds by Colliex et al. [41], Kurata et al. [42], de Groot et al. [39] and Wu et al.[17].

Above results should not be confused with the results of metal in a tetrahedral site. The K-edge XANES of some transition atoms in tetrahedral coordination are characteristic roughly of a tetrahedral MO_4 (M=transition metal). The pre-edge peak in these systems with acentric tetrahedral symmetry is very intense and comparable to the transition to p states in the continuum above the absorption jump. It is now very well known that it is the result of a transition from 1s to the d part of final states T_2

(irreducible representation of the T_d point group) which includes not only the p but also the d base of the orbitals, is a dipolar transition partially allowed due to the mixing of the p and d orbitals. In Fig. 6 we report the calculated Ti K-edge spectrum (curve (b)) of a TiO_4 cluster (a perfect tetrahedron) compared with the experimental result (curve (a)) of Ti in glass from Ref. 12. A reasonable agreement has been obtained. Curve (c) is the calculation, suppressing the complete d base set in the final states of the photoabsorber, the preedge structure disappears as indicated above.

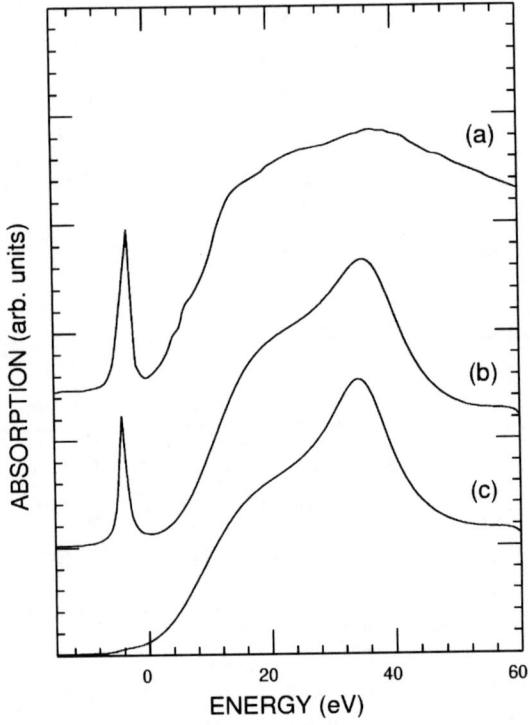

Figure 6. Comparison of experimental data (curve a) at the Ti K edge from Ref. 12. MS calculations in tetrahedral coordination (curve b). Curve (c) is the calculation without d base set in the final states (see text for more details).

In conclusion, by comparison the *ab initio* full MS computations and the experimental data at the metal K edge, all the transition features in the pre-edge region of TM compounds in octahedral coordination can be explained as due to the mixing of p orbitals of the absorbing atom with 3d orbitals of the higher-neighbouring metal octahedra. A good agreement between experimental data and theoretical ones has been achieved in the framework of the dipole approximation only. The core hole potential does not have a significant influence on the pre-edge peaks because they are dominated by higher-neighbouring metal orbitals.

We wish to emphasize that our comparison is precise and may be able to identify the higher-shell structural arrangement due to the position, shape and intensity of different features in the spectra and relates them with the electronic structures and ligand-field characters. The interaction of different higher-neighbouring octahedral configuration-dependent hybridisation mechanism can be involved to explain why the relative pre-edge intensity and energy separation are different in all 3d TM compounds. In fact, the amplitude for the p character of the central atom wave function varies for changing values of the M M distances and configuration of interactions.[43,44] We discovered a theoretical and simple formula that shows how the pre-edge features and their splitting depend on the medium-range order of solid and consequently the cluster size should be equal to metal-metal plus metal-anion bond lengths. Our results can be also extended with success to other transition-metal compounds and to more complex biomaterials.

ACKNOWLEGEMENTS

One of the authors (Z.Y. Wu) acknowledges the financial support of the *100-Talent Research Program* of the Chinese Academy of Sciences and of the *Outstanding Youth Fund* (10125523) of the National Natural Science Foundation of China.

REFERENCES

1. N.F. Mott in *Metal-insulator transitions*, Taylor and Francis 1990; N. Tsuda et al., *Electronic conduction in oxides*, Springer Series in Solid State Sciences Vol. 94 (Springer-Verlag, Berlin, 1991).
2. J.G. Bednorz and K. A. Muller, Z. Phys. **64**, 189 (1986).
3. R. von Helmont et al., Phys. Rev. Lett. **71**, 2331 (1993); R. Mahendiran et al., Appl. Phys. Lett. **66**, 233 (1995).
4. D. D. Sarma in: *Metal-Insulator Transitions Revisited* edited by P. P. Edwards and C. N. R. Rao, (Taylor and Francis, London (1995); D. D. Sarma and S. R. Barman in: *Spectroscopy of Mott Insulators and Correlated Metals*, Solid State Science **119**, edited by Y. Tokura and A. Fujimori, (Springer Verlag 1995), p126.
5. J. Garcia, J. Blasco, M.G. Proietti and M. Benfatto, Phys Rev **B 52**, 15823 (1995).
6. F. Antonangeli, M. Piacentini, R. Girlanda, G. Martino and E.S. Giuliano, Phys. Rev. **B 32**, 6644 (1985).
7. L.A. Grunes, Phys. Rev. **B 27**, 2111 (1983).
8. T. Uozumi, K. Okada, A. Kotani, O. Durmeyer, J.P. Kappler, E. Beaurepaire, and J.C. Parlebas, Europhys. Lett. **18**, 85 (1992).
9. D. Heumann, D. Hofmann, and G. Drager, Physica **B 208 & 209**, 305 (1995).
10. H. Maruyama, I. Harada, K. Kobayashi, and H. Yamazaki, Physica **B 208 & 209**, 760 (1995).
11. C. Brouder, J.P. Kappler and E. Beaurepaire, 2nd European Conference on Progress in X-ray Synchrotron Radiation Research, edited by A. Balerna, E. Bernieri and S. Mobilio, Vol. **25** (Bologna: SIF) 1990, p19.
12. F.W. Lytle, R. B. Greegor, and A.J. Panson, Phys. Rev. **B 37**, 1550 (1988).
13. D.D. Sarma, N. Shanthi, S.R. Barman, N. Hamada, H. Sawada, and K. Terakura, Phys. Rev. Lett. **75**, 1126 (1995).
14. S. Satpathy, Zoran S. Popovic, and Filip R. Vukajlovic, Phys. Rev. Lett. **76**, 960 (1996).
15. R. Brydson, H. Sauer, W. Engel, J.M. Thomas, E. Zeitler, N. Kosugi, and H. Kuroda, J. Phys.: Condens. Matter **1**, 797 (1989), and references therein.
16. M.J. Mckelvy and W.S. Glaunsinger, J. Solid State Chem. **66**, 181 (1987).

17. Z.Y. Wu, S. Gota, F. Jollet, M. Pollak, M. Gautier-Soyer, and C.R. Natoli, Phys. Rev. **B 55**, 2570 (1997).
18. K.M. Parida, S.B. Kanungo and B.R. Sant, Electrochimica Acta **26**, 435 (1981).
19. Z.Y. Wu, M. Benfatto, and C.R. Natoli, Phys. Rev. **B 45**, 531 (1992); Z.Y. Wu, M. Benfatto, and C.R. Natoli, Solid State Comm. **87**, 475 (1993); Z.Y. Wu, F. Seifert, B Poe, T. Sharp, J. Phys. Conden. Matter **8**, 3323 (1996); Z.Y. Wu, A. Mottana, A. Marcelli, C.R. Natoli and E. Paris, Phys. Chem. Minerals **23**, 193 (1996); Z.Y. Wu, A. Marcelli, A. Mottana, G. Giuli, E. Paris, and F. Seifert, Phys. Rev. **B 54**, 2976 (1996).
20. P.A. Lee and J.B. Pendry, Phys. Rev. **B 11**, 2795 (1975).
21. C.R. Natoli, D.K. Misemer, S. Doniach, and F.W. Kutzler, Phys. Rev. **A 22**, 1104 (1980); C.R. Natoli, and M. Benfatto, J. Phys. Collq. **47**, C8-11 (1986); C.R. Natoli, M. Benfatto, C. Brouder, M.Z. Ruiz Lopez, and D.L. Foulis, Phys. Rev. **B 42**, 1944 (1990); T.A. Tyson, K.O. Hodgson, C.R. Natoli, and M. Benfatto, Phys. Rev. **B 46**, 5997 (1992).
22. P.J. Durham, J.B. Pendry, and C.H. Hodges, Solid State Commun. **38**, 159 (1981); Comput. Phys. Comm. **25**, 193 (1982).
23. D.D. Vvedensky, D.K. Saldin, and J.B. Pendry, Comput. Phys. Commun. **40**, 421 (1986).
24. A. Bianconi, in X-ray Absorption: Principles, Applications, Techniques of EXAFS, SEXAFS, XANES, edited by R. Prinz and D. Koningsberger (Wiley, New York, 1988).
25. P.J. Durham, in X-ray Absorption: Principles, Applications, Techniques of EXAFS, SEXAFS, XANES, edited by R. Prinz and D. Koningsberger (Wiley, New York, 1988).
26. P.A. Lee and G. Beni, Phys. Rev. **B 15**, 2862 (1977).
27. L. Mattheiss, Phys. Rev. **A 134**, 970 (1964).
28. E. Clementi, and C. Roetti, Nuclear Data Tables, **Vol. 14**, 1974.
29. D.R. Penn, Phys. Rev. **B 35**, 482 (1987).
30. J.G. Norman, Mol. Phys. **81**, 1191 (1974).
31. F. Hulliger, Structure Chemistry of Layer-type Phases, ed. F. Levy, (D. Reidel Publishing Company, Dordrecht-Holland/Boston-USA, 1976); R.W.G. Wyckoff, *Crystal Structures* (John Wiley & Sons, Inc. 1963).
32. C.J. Ballhausen, in *An Introduction to Ligand Field Theory* (McGraw-Hill, New York, 1962).
33. S.F.A. Kettle, in *Symmetry and Structure* (John Wiley & Sons, New York, 1985).
34. Z.Y. Wu, G. Ouvrard, S. Lemaux, P. Moreau, P. Gressier,F. Lemoigno, and J. Rouxel, Phys. Rev. Lett. **77**, 2101 (1996); Z.Y. Wu, F. Lemoigno, P; Gressier, G. Ouvrard, P. Moreau, J. Rouxel, and C.R. Natoli, Phys. Rev. **B 54**, 11009 (1996).
35. O.K. Andersen, Phys. Rev. **B 12**, 3060 (1975); O.K. Andersen and O. Jepsen, Phys. Rev. Lett. **53**, 2571 (1984); O.K. Andersen, O. Jepsen, and M. Sob, in *Electronic Band Structure and Its Applications*, edited by M. Yussouff (Springer-Verlag, Berlin, 1986).
36. O.K. Andersen, O. Jepsen, A.I. Liechtenstein, and I.I. Mazin, Phys. Rev. **B 49**, 4145 (1994).
37. D.W. Fischer, J. Phys. Chem. Solids **32**, 2455 (1971), Phys. Rev. **B 5**, 4219 (1972); D.W. Fischer, Phys. Rev. **B 8**, 3576 (1973).
38. C. Sugiura, M. Kitamura, and S. Muramatsu, J. Chem. Phys. **84**, 4824 (1986).
39. F.M.F. de Groot, M. Grioni, J.C. Fuggle, J. Ghijsen, G.A. Sawatzky, and H. Petersen, Phys. Rev. **B 40**, 5715 (1989).
40. R. Brydson, H. Sauer, W. Engel, J.M. Thomas, E. Zeitler, N. Kosugi, and H. Kuroda, J. Phys.: Condens. Matter **1**, 797 (1989).
41. C. Colliex, T. Manoubi, and C. Ortiz, Phys. Rev. **B 44**, 11402 (1991).
42. H. Kurata, E. Lefevre, C. Colliex and R. Brydson, Phys. Rev. **B 47**, 13763 (1993).
43. Z.Y. Wu, G. Ouvrard, P. Gressier, and C.R. Natoli, Phys. Rev. **B 55**, 10382 (1997).
44. G.A. Waychunas, Am. Mineral. **72**, 89 (1987) .

VIII. BIOLOGICAL APPLICATIONS

A theoretical analysis of reflection of X-rays from water at energies relevant for diagnostics

Dusan Arsenović[*], Dragomir M. Davidović[+], Jovan Vukanić[+]

[*] Institute of Physics, Pregrevica 118, P.O. Box 57, Belgrade, Serbia
[+] The Vinča Institute of Nuclear Sciences, P.O Box 522, Belgrade, Serbia

Abstract. The reflection of X-rays from a semi-infinite water target, for energies used in X-ray diagnostics, is treated by the analog Monte Carlo simulation. In the developed procedure it was possible to calculate separately contributions of photons scattered, before reflection, fixed number of times with target electrons. It turned out that multiple collision type of reflection dominates at all energies investigated, whenever the absorption is small. The same process was also treated analytically as the classical albedo problem for isotropic scattering without energy loss. Very good agreement of results of the two approaches is obtained.

INTRODUCTION

Until recently, the problem of radiation protection and shielding was not specially theoretically treated and grounded, but the knowledge about it stemmed rather from immediate practice of radiologists and medical physicists and was usually represented in forms of tables and graphs [1]. With the development of nuclear technology, very general and powerful numerical codes for shield design and related problems were developed. However, very often, for concrete applications, the more adequate understanding of possible radiation damage may be achieved with simpler but physically more transparent approaches in which, for a given specific situation, only the most relevant processes are considered. In the present work, we treat in this spirit the process of reflection of X-rays for energies in the range used in radiological diagnostics. After explaining the theoretical background in the main lines, we give first Monte Carlo simulation results for backscattering parameters and than the same quantities obtained in the analytically solvable isotropic model. We compare both sets of results, and discuss their physical meaning.

CP652, *X-Ray and Inner-Shell Processes: 19th International Conference on X-Ray and Inner-Shell Processes*
edited by A. Bianconi, A. Marcelli, and N. L. Saini
© 2003 American Institute of Physics 0-7354-0111-X/03/$20.00

MONTE CARLO SIMULATION FOR BACKSCATTERING OF X-RAYS FROM WATER

In the analog Monte Carlo simulation, the complicated statistical process of penetration of photons through matter is considered as a series of finite number of the following elementary process: free passage of the photon on a given path, Compton scattering of the photon in a defined direction, and disappearance of photon due to photoabsorption. By knowing the probabilities of every of these processes separately which are related with the cross sections of the two mentioned processes, using the generator of random numbers, it is possible to follow step by step the history of the considered photon - until it disappears due to photoabsorption, or reaches the defined final states- in our case this is the photon reflection.

In details, the beam attenuation due to photoeffect and Compton scattering may be quantitatively described by the corresponding coefficients μ_a and μ_R, so that

$$dN_a = -\mu_a N dz, \quad dN_R = -\mu_R N dz. \tag{1}$$

Here, N is the number of photons and dN_a and dN_R are the corresponding decreases of photon numbers which are due to the two mentioned processes. One can see that from the initial number of photons N_0, $(\mu_0/\mu)N_0$ will be absorbed, while $(\mu_R/\mu)N_0$ will be scattered. At a penetration depth z, the numbers of survived photons from these two groups are

$$N_a = \frac{\mu_a}{\mu} N_0 e^{-\mu z}, \qquad N_R = \frac{\mu_R}{\mu} N_0 e^{-\mu z} \ . \tag{2}$$

Monte Carlo simulation of the considered process is such a procedure that from the generator of random numbers as a result gives a set of lengths $\{z_i\}$. These lengths are characterized by the fact that the ratio of number of those among them which are greater than z_0 and the total number of them is equal to $e^{-\mu z_0}$. Due to this, we can interpret z as freely traveled paths of various individual photons. The procedure of obtaining z is the following: if a series of homogeneously distributed random numbers a_i in the interval $(0,1)$ is given, we correspond to each of them the value of z, which is the root of the equation $\exp(-z_i) = a_i$. In this way, the traveled path of every individual photon is a random variable, which may be obtained if the set of random numbers is available. An analogous procedure may be applied for the angles φ and ϑ which are governed by the Klein Nishina Tamm formula [1]

$$d\sigma(\vartheta) = \frac{\pi e^4}{m_0 c^4} \left(\frac{1}{1+\alpha-\cos\vartheta} \right)^2 \left(1 + \cos^2\vartheta + \frac{\alpha^2(1-\cos\vartheta)}{1+\alpha(1-\cos\vartheta)} \right) \sin\vartheta d\vartheta \tag{3}$$

Here, α is the energy of incoming photon, given in the units of the rest energy of electron.

Since this expression is not explicitly dependent on φ, the azimuthal angles of scattering of every individual photons will be represented by the series of random numbers $\{\varphi_i\}$ which are homogeneously distributed in the interval $(0, 2\pi)$.

For the angle ϑ, in an analogous way as for the depth z, let us find the ratio of the number of particles σ_{ϑ_0} which are scattered through angles smaller than ϑ_0 and the total number of scattered particles (for all of them $\vartheta \leq \pi$). Firstly, we obtain analytic expression for σ_{ϑ_0} by integrating Eq. (3) from 0 to ϑ_0 so that we easily get $\sigma_{\vartheta_0}/\sigma_\pi$. Now again, as in the case of penetration lengths, one can correspond to the homogeneously distributed random numbers a_i in the interval $(0,1)$ the angles theta so that $a = \sigma_{\vartheta_0}/\sigma_\pi$. This means that for every random number a_i one has to solve the last equality in order to obtain ϑ_i.

The procedure is now simple. One follows the history of the considered photon. The first generated random number gives the length of penetration, the second and the third random numbers define the direction in which the photon is scattered, and so on.

We have followed the history of one million of photons incident normally on a semi-infinite water target, for different photon initial energies and obtained the number of photons backscattered in different directions with respect to the surface normal. The interval of directional cosines was 0.1. The results, for two different energies (10 keV and 60 keV) are given in the Table 1. These energies are in a sense limiting cases. Namely, for the energy of 10 keV, from one million photons 986131 photons were absorbed. Conversely, for initial energy of 60 keV almost one half of photons were reflected.

Analyzing Table 1 one can conclude that for E=60 keV, the angular distribution is close to the cosine distribution, which characterizes the multiple collisions. Conversely, when the absorption is dominant (E=10keV), the shape of the angular distribution approaches the one which exists in the single collision case.

There are great many powerful programs for Monte Carlo simulation, valid for very broad ranges of energies, which describe the corresponding processes in great details. However, they are extremely time consuming and their outputs are often not transparent. The important advantage of our program is that it is very handy and fast. The reason for this is the limitation of our considerations to two processes only, what is, for energies used in X-ray diagnostics, physically absolutely reliable and justified.

TABLE 1. Number of absorbed photons and the number of reflected photons in specified intervals of μ, obtained in Monte-Carlo simulation.

μ	0-0.1	0.1-0.2	0.2-0.3	0.3-0.4	0.4-0.5	0.5-0.6	0.6-0.7	0.7-0.8	0.8-0.9	0.9-1	Absorbed
10keV	134	416	641	887	1193	1424	1768	2109	2459	2838	986131
60keV	4248	13406	23236	32920	42765	52544	62782	72913	82910	92615	519661

COMPARISON OF X-RAY REFLECTION IN THE EXACTLY SOLVABLE MODEL - ISOTROPIC SCATTERING WITHOUT ENERGY LOSS - AND THE MONTE CARLO SIMULATION

Simple analysis of the Compton differential cross section shows that for small α, this cross section becomes nearly isotropic and the difference of energies of incident and scattered photons are negligible [1]. Due to this, at low photon energies, which are used in X-ray diagnostics, the following simplifying physical assumptions are justified:

(i) Photon scattering on atomic electrons is isotropic and without energy losses.

(ii) The absorption of photons in medium is characterized by the cross section, which may be treated as constant for a given incident energy.

The angular distribution of backscattered photons in the model where (i) and (ii) are exactly fulfilled is given by the Chandrasekhar exact solution [2]

$$R(\mu_0, \mu) = \frac{\omega}{2} \frac{\mu}{\mu + \mu_0} H(\omega, \mu_0) H(\omega, \mu). \qquad (4)$$

Here, μ_0 and μ are the directional cosines of the incident and reflected photon with respect to the target surface normal. $H(\omega, \mu)$ is the Chandasekhar H function whose tables exists for different values of the variable μ and the parameter ω. The parameter ω represents the ratio of the total cross section of Compton scattering to the total attenuation coefficient.

The total reflection coefficient follows by integrating the angular distribution from Eq. (4) over all exit directions [2]

$$R_N(\mu_0) = 1 - \sqrt{1 - \omega} H(\mu_0, \omega). \qquad (5)$$

We have compared the model exact results obtained from Eq. (4) with the results of our Monte Carlo simulation. We have chosen for discussion the results for photon energy of 60 keV when the parameter ω is great ($\omega = 0.936$) and almost one half of photons were reflected so that one could expect that our simplifying assumptions are fulfilled. As can be seen from Fig.1, where this comparison is represented, this is really the case, since the analytical and the results obtained by simulation agree within a few percent.

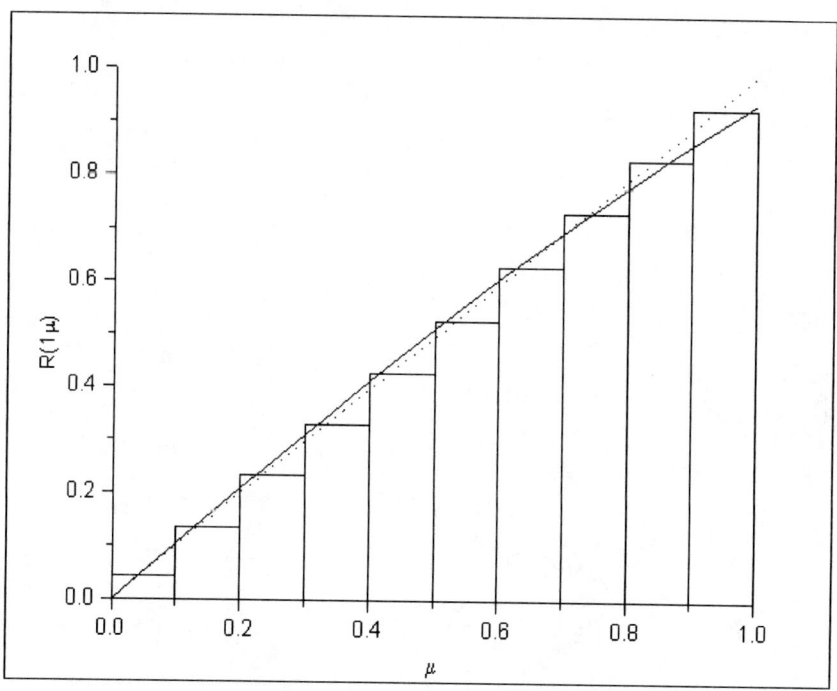

FIGURE 1. Angular distribution of backscattered photons for photon incident energy E=60keV.
_____ Results from isotropic model.
.... Results with the single backscattering anisotropy included.
Histogram from Monte-Carlo simulation.

For large values of the parameter, the complicated analytic expression (4) can be, using the features of Chandrasekhar function, with a very good accuracy represented as a following cosine distribution, also included in Fig. 1:

$$R(\mu_0, \mu) \cup 2R_N(\mu_0)\mu. \tag{6}$$

Fig. 2 compares total reflection coefficients for photon backscattering from water, obtained from Chandasekhar result (Eq. (5)) and by Monte Carlo simulation. It is visible that these analytical and simulation results agree again within a few percent. This is a surprisingly good agreement. Let us mention that for small values of the parameter ω (when high absorption is present) the reflection is dominated by single collisions, but the corresponding photon energies are then far below those used in X-ray diagnostics. The relative contribution of multiple backcattering events increases with increasing ω. So, in the limiting case $\omega = 1$ (scattering without absorption), the contribution of single collisions in photon reflection is only about 15%.

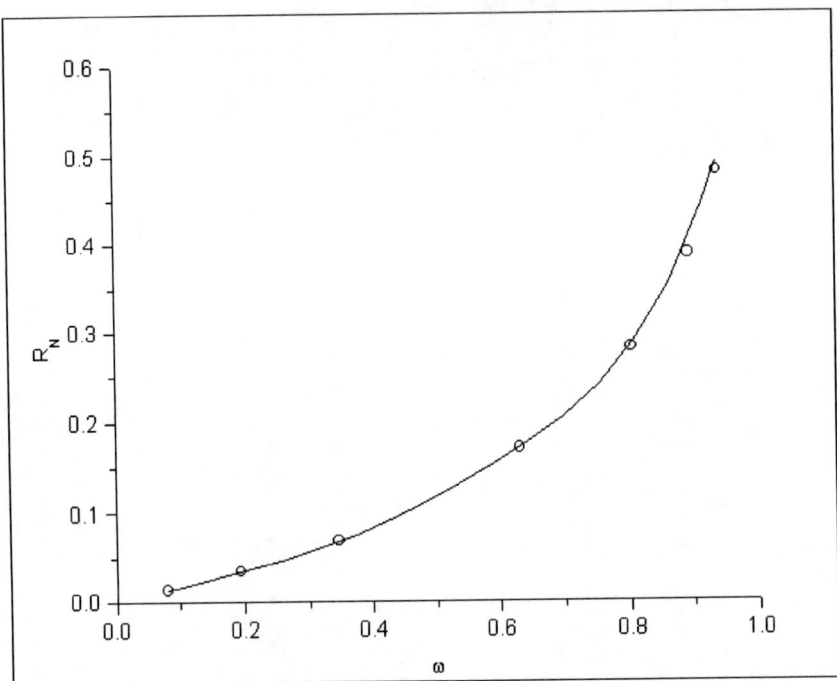

FIGURE 2. Total reflection coefficient as a function of the parameter ω. Comparison of Monte Carlo results and exact theory.

Altogether, our results show that, for energies used in X-ray diagnostics, results obtained from exactly solvable analytical model agree very well with our Monte Carlo simulation data and therefore the results of the model may be considered as reliable and accurate in this energy range.

REFERENCES

1. Shultis J. K. and Faw R.E., *Radiation Shielding*, Prentice Hall PTR, 1996.
2. Chandrasekhar S., *Radiative Transfer*, Dover, New York, 1960.

Synchrotron Radiation μ-X Ray Fluorescence on Multicellular Tumor Spheroids

E. Burattini[§○], G. Cinque[§], G. Bellisola*, G. Fracasso*, F. Monti[○] and M. Colombatti*

[§] Istituto Nazionale di Fisica Nucleare- Laboratori di Frascati, via E. Fermi 40, I-00044 Frascati
[○] Università degli Studi di Verona -Dipartimento di Informatica, Strada LeGrazie 15, I-37134 Verona
* Università degli Studi di Verona-Dip. Patologia, Sez. Immunologia, P.le A.L. Scuro, I-37134 Verona

Abstract. Synchrotron Radiation micro X-Ray Fluorescence (SR μ-XRF) was applied for the first time to map the trace element content on Multicellular Tumor Spheroids (MTS), i.e. human cell clusters used as an *in vitro* model for testing micrometastases responses to antitumoral drugs. In particular, immunotoxin molecules composed of a carrier protein (Transferrin) bound to a powerful cytotoxin (Ricin A), were here considered as representatives of a class of therapheutic macromolecules used in cancer theraphy. Spheroids included in polyacrylamide gel and placed inside quartz capillaries were studied at the ESRF ID22 beamline using a 15 keV monochromatic photon microbeam. Elemental maps (of Fe, Cu, Zn and Pb) on four groups of spheroids grown under different conditions were studied: untreated, treated only with the carrier molecule or with the toxin alone, and with the complete immunotoxin molecule (carrier+toxin). The results indicate that the distribution of Zn and, to some extent, Cu in the spheroid cells is homogeneous and independent of the treatment type. Total Reflection X-Ray Fluorescence (TR-XRF) was also applied to quantify the average trace element content in the spheroids. Future developments of the technique are finally outlined on the basis of these preliminary results.

INTRODUCTION

Immunotoxins (ITs) are powerful cytotoxic heteroconjugates capable of selectively killing tumor cells (1). ITs are obtained by linking chemically (or genetically) a carrier molecule to either a bacterial or a plant toxins. The vehicle molecule provides recognition and binding capacity to che cell membrane, while the associate toxin component catalytically affects cellular alterations leading to cell death. Unlike drugs used in chemotherapy, ITs are able to kill both proliferating and resting tumor cells, but their success in clinics depends largely on their diffusion within solid tissues.

The Multicellular Tumor Spheroids (MTSs) are currently applied in pre-clinical studies to estimate the optimal dose of radiation therapy and of new drugs, their penetration into solid tumors, and to assess the effectiveness of drug combinations inasmuch as this *in vitro* model maintains both the structural architecture and the differentiated functions typical of micrometastasis (2).

The detection of therapeutic macromolecules in MTS (and generally in tissues) by conventional biomedical imaging techniques is severely limited by the need for conjugation with radiotracers or fluorophores, and the poor spatial resolution available (3). We took advantage of the presence of some transition metals in Ricin toxin and in

CP652, *X-Ray and Inner-Shell Processes: 19th International Conference on X-Ray and Inner-Shell Processes*
edited by A. Bianconi, A. Marcelli, and N. L. Saini
© 2003 American Institute of Physics 0-7354-0111-X/03/$20.00

Transferrin carrier molecules to carry out this preliminary study on the multielemental spatial distribution in spheroids treated with such molecules, or with their IT coniugate, by applying XRF technique. Excitation obtained by a Synchrotron Radiation microbeam allowed the non-destructive analysis of MTS and the mapping of trace elements with micrometric resolution (3,4). A quantitative analysis was also accomplished on spheroids by conventional Total Reflection X-Ray Fluorescence to evaluate their average trace element content.

EXPERIMENTAL METHODS

Spheroids and ImmunoTransferrin

MTS were obtained from a monolayer culture of human breast carcinoma cell line MCF7 grown in a culture medium (RPMI 1640) with 10% fetal bovine serum (Bio Whitacker) and antibiotics according to Yuhas and co-workers (5). Briefly, spheroids were obtained from inoculating 10^6 MCF7 cells in 15 ml RPMI-FBS 10% in Petri dishes on a thin layer of agar (10 ml of a 0,75% solution of agar in RPMI-FBS 10%). Human Transferrin (hTfn) was selected as vehicle molecule of Ricin Toxin A-chain (RTA) because of the high number of receptors on MCF7 cell membrane. RTA is the toxic subunit of the heterodimeric toxin obtained from the seeds of *Ricinus communis* (2). Spheroids of about 200 μm diameter (~10^7 cells) were individually incubated during the last 24 h of culture with, respectively: IT (hTfn-RTA conjugate), RTA or hTfn alone at the highest molar concentration allowed before disgregation of the organic matrix occurred (10^{-9}, 10^{-9} and 10^{-11} M, respectively: see Fig. 1). Mock-treated samples were considered as controls. All MTS were separately washed twice in 5 ml of physiological solution (0.9 % NaCl) before analysis.

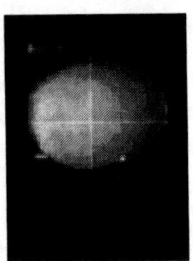 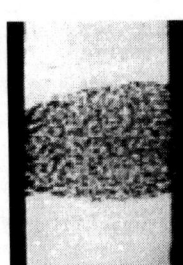 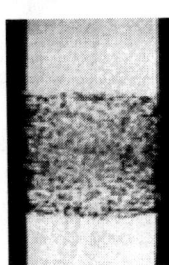 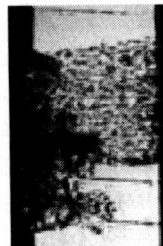

FIGURE 1. Images of spheroids inside capillaries (Ø ~200 μm) as seen by the CCD camera on ID22 table top (a), and from an inverted optical microscope in transmission (b, c and d): a) untreated, b) +Transferrin, c) +Ricin toxin, and d) +ImmunoToxin. Notice the effect of X-ray irradiation on gel in d).

SR micro X-Ray Fluorescence

Spheroids were fixed in 0.4% paraformaldehyde and rinsed in physiological solution. Ultrapure quartz capillaries (Hilgenberg), 10 μm thick and with ~200 μm bore diameter, were filled up with a 10% acrylamide solution, and single spheroids were placed within the capillaries with a micropipette (gel polymerization was achieved *in situ*). Sample drying was avoided by sealing the capillary extremities with paraffin followed by storage at 4 °C until irradiation.

Microspectroscopy experiments were carried out at the µ-FID (ID22) beamline of the European Synchrotron Radiation Facility in Grenoble (see e.g. 6). Capillaries containing the samples of interest (Fig. 1) were placed vertically on a Huber goniohead endowed with sub-micrometric resolution (remotely controlled) X-Y-Z movements. The signal-to-noise ratio for multielemental microanalysis was optimized using a 15 keV monochromatic photon beam. The microfocusing was achieved by a Fresnel Au/Si Zone-Plate ($M=10^{-2}$) coupled to a 10 µm pinhole placed immediately before the sample: the final beamspot dimensions were 1 µm x 10 µm (v x h) and the flux was over 10^9 ph s^{-1} at the sample. 2-D scanning was achieved on rastering by SR X-Y regions up to 400 x 400 µm^2 to fully cover the spheroid area (horizontal movement 45° respect to the beam). The X-ray fluorescence signal was collected for 50 s per point by a Si(Li) detector (Eurisys Mesures, $\Delta E=137$ eV fwhm at Mn K$_\alpha$ including 8 µm Be window) at 2 ÷ 5 cm from the sample in a 90° configuration. All multielemental maps were normalized to an average photon flux of $2 \cdot 10^9$ ph s^{-1} and to 50 s of acquisition time per pixel.

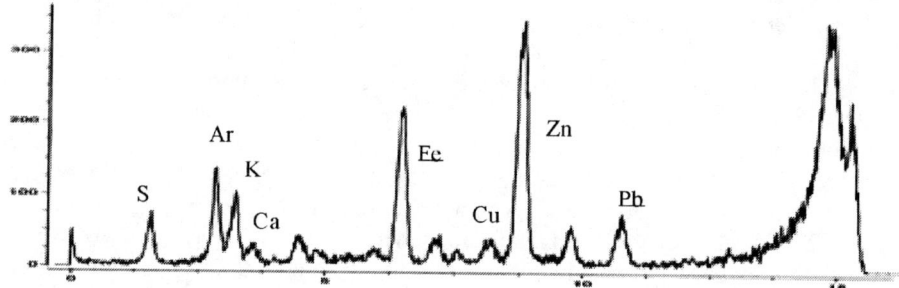

FIGURE 2. Typical SR µ-XRF spectrum of an untreated spheroid embedded in gel and inside a capillary: acquisition time 600 s. Underlined elements were also found (at lower concentration) outside the spheroid area.

Total Reflection XRF

Sets of 5 spheroids were washed twice in 5 ml isotonic solution (5% glucose), pipetted one by one within a volume of 1 µl, and then resuspended in 20 µl bidistilled H$_2$O directly onto sample holders: these were discs of amorphous Si (Seefelder Messtechnik) cleaned for trace element analysis and covered by 2 µl of an ultra pure silicon compound to increase surface wetting (Serva Feinbiochemica). Yttrium (Merck) was added as internal standard to all samples (5 µl of Y solution to a final concentration of 9.96 ppm). Blank samples consisted of 5 µl volumes of last isotonic solution used for spheroid washing. All samples were dried under vacuum at RT.

Quantitative analyses were performed by a TX2000 X-ray spectrometer from Ital Structures, equipped with a Philips X-ray tube and a Mo anode working at 40 kV and 30 mA (plus a W-Si multilayer mirror for selecting/redirecting the 17,4 keV K$_\alpha$ emission). Sample X-ray fluorescence was revealed by a standard Si(Li) detector of 20 mm^2 active area (Röntec) directly facing the sampleholder. Instrument calibration was obtained by processing standard solutions containing elements from P to Y, while analysis accuracy was checked with respect to certified concentrations of the Standard Reference Material 1577a Bovine Liver (NIST) (see e.g. 7). Spectra and quantitative analysis were elaborated by the EDXRF32 software (Ital Structures).

RESULTS AND DISCUSSION

SR μ-XRF Elemental Maps

The Immunotoxin here used was obtained by a chemical cross-link of human Transferrin, containing 2Fe^{++} per molecule, to Ricin A-chain as previously described (2). Preliminar mass spectrometry analysis of purified RTA revealed traces of Fe, Cu, Zn, Cd, Co and Pb. The experimental conditions of energy, flux and acquisition time here used for the SR beam were apt to a multielemental analysis of spheroids spanning a wide range of elements (K lines Z ≤ 40) and concentrations (from ng/g to μg/g). Following the XRF spectrum of the untreated spheroid (Fig. 2), the trace elements of interest selected for simultaneous mapping by SR microspectroscopy were K, Ca, Fe, Cu, Zn, and Pb. The presence of acrylamide and quartz, necessary to prevent any sample movement during SR rastering, gave around the spheroids a contamination of Fe and Pb quantified in circa half of the tissue content. All the following maps are contour plots giving in a gray scale the intensity of the fluorescence signal, which is directly proportional to the relative concentration of the element under investigation (3). Neglecting light element fluorescence because of their reabsorption by the organic matrix, only the most significant maps are shown.

In the case of the untreated spheroid (Fig. 3) Zinc and, to a lesser extent, Copper maps reveal both the spheroid shape and thickness as crossed by the SR beam: in fact, the 2-D round spheroidal shape in the plane perpendicular to the incident beam is accompanied by a progressive increase of the fluorescence counts from the outer to the inner layers of the spheroid in Fig. 3. A more careful analysis, consisting in the normalization of the X-Ray signals to both the ellipsoidal shape fitting the MTS and to the tubular shape fitting the capillary envelope, demonstrated that Zn and Cu are indeed homogeneously distributed within spheroid tissue. In other words, these metals were found to be useful tracers of the cells present within the spheroid, although some Cu is present in the quartz material as well. The distribution pattern of Zn, Cu, Fe and Pb in the fully immunotoxin treated MTS sample is shown in Fig. 4. The addition of 10^{-11}M hTfn-RTA to MCF7-MTS does not modify substantially the distribution of Zinc, which still appears homogeneously and exclusively accumulated within the spheroid independently from the type of treatment. Copper signal seems still correlated to the spheroid shape, even though partially hidden by the quartz capillary shielding. On the contrary, Fe map does not match the spheroid shape obtained after Zn analysis, as well as Pb distribution. Such a scenario is common to all other samples which underwent different treatment, and it reveals that neither the accumulation of transferrin, related to Fe signal, nor that of Ricin, possibly related to Pb, are detectable.

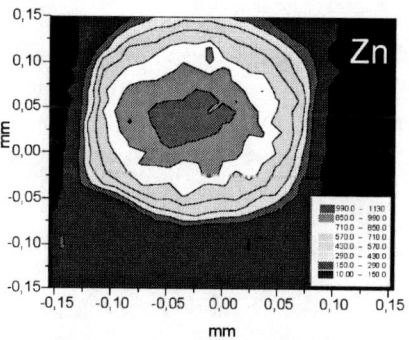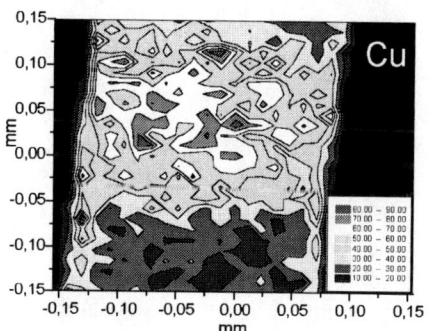

FIGURE 3. Untreated spheroid: micro X-ray fluorescence maps and trace element distribution.

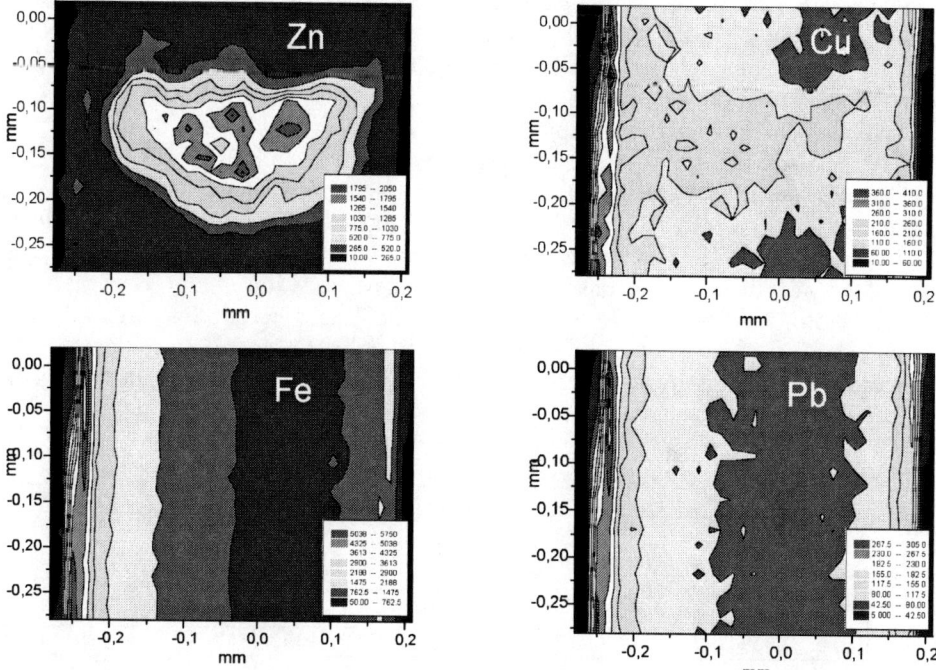

FIGURE 4. Immunotoxin treated spheroid: micro X-ray fluorescence maps of trace elements.

This seems due to both the low concentration of the antitumor drug - which is used in the present investigation at doses like those to be applied in the clinics - and to the contaminations of the capillary walls by traces of metals.

Assessment of MTS sensitivity to irradiation is possible by comparing microscope images of the samples before and after measurements, as illustrated in Fig. 1. The greatest effect is observed in the acrylamide gel whereas no evident damages are found in the untreated spheroid tissue. Minimal changes can be inferred in untreated

MTSs under SR irradiation as the main element profiles revealed by line scannings both inside and outside the spheroid region did not change before and after mapping. Indeed, a certain tissue softening is noticed after full immunotoxin treatment as evidenced by the irregular shape achieved by Zn and Cu maps of MTS in Fig. 4.

It is known that synchrotron microprobes applied to the elemental mapping of organic specimens allow spatial resolutions at least as high as other techniques employed in tumor imaging. To our knowledge, this is the first time that SR-μXRF has been applied to study the diffusion of molecules in a three dimensional model of solid micrometastasis. After this preliminary investigation, future developments of the technique can devised. Quantitative analysis by SR-μXRF could be done introducing an internal standard, such as an alkaline solution, in the sample. Future separate labelling of vector and toxin molecules with non naturally occuring elements (e.g. Ga, Y, I or Br) can provide the necessary tracers: use of appropriate bifunctional chelating agents will permit the introduction of many atoms per molecule and an easier detection by XRF apt to follow the immunotoxin transport within the three-dimensional tumor models. Finally, sample cooling with liquid nitrogen could allow further reduction of radiation damages and to perform 2-D scans hence avoiding the use of capillaries and gel for sample fixing.

As regards the quantitative analysis performed by TR-XRF on MCF7-MTS replicates, the results are shown in Table 1. These values refer to the wet weight of spheroids from which any contribution due to the buffer solution used for sample handling is carefully subtracted. A remarkable biological variability affects all elements identified in the tissue, and the highest variation among replicates was observed for Iron. Sulfur, Clorine, Potassium, and Zinc concentrations are in the ppm range, whereas Iron is at sub-ppm level in all cases. Compared to the control, the incubation of MTS with 10^{-11}M hTfn-RTA for 24 h does not significantly change the elemental concentration pattern of the tissue.

TABLE 1. Absolute trace element content (mean and error) of spheroids from MCF7 cell line.					
	Untreated I	Untreated II	IT-treated I	IT-treated II	IT-treated III	Weight av.
S (μg/g)	117,2 ± 23,0	98,2 ± 17,6	170,2 ± 24,7	111,0 ± 7,6	74,4 ± 10,4	102,9 ± 5,5
Cl (μg/g)	131,9 ± 43,5	72,8 ± 42,2	27,2 ± 47,9	20,4 ± 14,1	130,9 ± 22,3	56,9 ± 10,8
K (μg/g)	405,9 ± 9,6	406,7 ± 7,9	678,7 ± 11,8	307,2 ± 3,9	398,1 ± 5,6	370,4 ± 2,8
Fe (ng/g)	2610 ± 560	53 ± 500	549 ± 570	322 ± 170	190 ± 261	400 ± 130
Zn (μg/g)	3,02 ± 0,37	3,95 ± 0,31	4,72 ± 0,42	2,08 ± 0,12	2,19 ± 0,16	2,40 ± 0,09

It is worth observing that, in spite of the low background and the high detection limits characteristic of TR-XRF technique, no traces of metals other than Zn and Fe were detected in MCF7-MTS treated with hTfn-RTA conjugates. In particular Cu was successfully mapped in spheroids by the SR probe, but was not found neither in the control nor in the treated samples: since the Minimum Detection Limit of Copper is here about 50 ppb, it means that the μ-XRF analysis performed on spheroids had a better sensitivity of about one order of magnitude. We note that a comparison with existing studies on cancerous breast tissues composition is probably worthless because a cellular system grown *in vitro* strongly depend on the culture medium composition, and indeed our analysis on spheroids supplies rather different results with respect to other Authors (8). Given the overall congruence of our data, the weighted average of trace element concentration is given in Table 1 regardless to spheroid treatment: to our knowledge, this first assessment of tissue composition of human breast carcinoma cell line MCF7 is a necessary information for carrying out further studies on such a valuable tool for assaying anticancer theraphies as that represented by the MTS model.

ACKNOWLEDGMENTS

We acknowledge the staff of the beamline ID22 in Grenoble, as well as ESRF and EMBL Grenoble Outstation for support to the approved experiment LS995.

A special thank to Fabio Pasti, Alessandro Torboli and Manuel Fuentes of Ital Structures S.p.A. (Riva del Garda - TN), for the courtesy of TR-XRF measurements.

This work was partially supported by AIRC, Miur and Fondazione Cariverona.

REFERENCES

1. Colombatti, M., Dell'Arciprete, L., Chignola, R. and Tridente, G., *Cancer Res.* **50**, 1385-1391 (1990).
2. Fracasso, G., and Colombatti, M., *Crit. Rev. Oncol/Hemathol.* **36**, 159-178 (2000).
3. Sparks, C. J., in *Synchrotron Radiation Research,* edited by H. Winick and S. Doniak, New York: Plenum Press, 1980, p. 459.
4. Jones, K. W., "Synchrotron Radiation-Induced X-ray Emission" in *Handbook of X-Ray Spectrometry,* edited by R. E. Van Grieken and A. A. Markowicz, New York: Marcel Dekker, Inc., 2001, pp. 411-452.
5. Yuhas, J. M., Martinez, A. O., Ladman, A. J., *Cancer Res.* **37**, 3639-3643 (1977);
6. Bohic, S., Drakopoulos, M., Leitemberger, W., Rau, C., Simionovici, A., Snigereva, I., Snigerev, A., and Weitkamp, in *Proceedings SRI2000, Nucl. Instr. and Meth. A* **30-33** (2001).
7. Bellisola, G., Pasti, F., Valdes, M., and Torboli, A., *Spectrochim. Acta B* **54**, 1481-1485 (1999).
8. Majewska, U., Braziewicz, J., Banas, D., Kubala-Kunus, A., Gozdz, S., Pajek, M., Smok, J., and Urbaniak, A., *Biol. Trace Elem. Res.* **60**, 91-100 (1997).

Synchrotron Radiation and Energy Dispersive X-Ray Fluorescence Applications on Elemental Distribution in Human Hair and Bones

M.L. Carvalho, A.F. Marques and J. Brito

Centro de Física Atómica, Universidade de Lisboa, Av. Prof. Gama Pinto 2, 1649-003 Lisboa, Portugal
(e-mail: luisa@cii.fc.ul.pt)

Abstract. This work is an application of synchrotron microprobe X- Ray fluorescence in order to study elemental distribution along human hair samples of contemporary citizens. Furthermore, X-Ray fluorescence spectrometry is also used to analyse human bones of different historical periods: Neolithic and contemporary subjects. The elemental content in the bones allowed us to conclude about environmental contamination, dietary habits and health status influence in the corresponding citizens. All samples were collected *post-mortem.* Quantitative analysis was performed for Mn, Fe, Co, Ni, Cu, Zn, Br, Rb, Sr and Pb. Mn and Fe concentration were much higher in bones from pre-historic periods. On the contrary, Pb bone concentrations of contemporary subjects are much higher than in pre-historical ones, reaching 100 µg g-1, in some cases. Very low concentrations for Co, Ni, Br and Rb were found in all the analysed samples. Cu concentrations, allows to distinguish Chalcolithic bones from the Neolithic ones. The distribution of trace elements along human hair was studied for Pb and the obtained pattern was consistent with the theoretical model, based on the diffusion of this element from the root and along the hair. Therefore, the higher concentrations in hair for Pb of contemporary individuals were also observed in the bones of citizens of the same sampling sites. All samples were analysed directly without any chemical treatment

INTRODUCTION

The uptake of metals by the bones is a protective mechanism during chronic exposure because it limits their distribution to more sensitive tissues; on the other hand, their fixation to hair allows the use of this tissue as a recording filament of past exposure. Since the long residence time in bones, the concentration of heavy metals in tissue can be expected to reflect lifelong intake of some trace elements in the skeleton mineralised material. In the long-term, however, the bone reservoir of heavy metals constitutes a risk of endogenous exposure to toxins, as it returns to the blood stream, during the normal cycle of bone remodelling[1,2,3,4,5].

CP652, *X-Ray and Inner-Shell Processes: 19th International Conference on X-Ray and Inner-Shell Processes*
edited by A. Bianconi, A. Marcelli, and N. L. Saini
© 2003 American Institute of Physics 0-7354-0111-X/03/$20.00

The trace elements content in bones, from different historical periods, can give us information about environment and dietary habits by studying the elemental concentration in human remains of those periods[6, 7].

The correlation between elemental concentrations in different human tissues might be a powerful method in biological and clinical investigations. Study of the easily collectable tissues can allow determining the elemental contents in more difficulty accessible tissues.

The study of relationship between bone and hair elemental content is important because gives an approach to the balance of absorption, accumulation and excretion.

X-Ray fluorescence analysis with its capability to detect several elements simultaneously in very low concentrations, offers a very convenient tool for the study of trace elements in the human tissues.

Micro-analytical techniques were used with success to study the elemental distribution along the hair.

EXPERIMENTAL

Sample collection and preparation

All samples were collected *post mortem*. Hair samples were taken by tearing from the parietal area of the head. Prior to analysis samples were washed according to the IAEA protocol[8].

For each subject one sample of skull and hair was obtained. Age, sex, cause of death and specific diseases were registered. Bone samples were obtained by cutting a rectangle approximately 2 x 4 cm, at the temporal fossa of the skull. Pyrex beakers were used to hold the bones for 48 h in a lyophylizer and subsequently dried in an oven at $50°$ C for 48 h. The bones became easily friable in an agata mortar and a homogeneous powder was obtained.

For analysis, the powder was pressed into pellets 2.0 cm in diameter without any chemical treatment. Samples were of intermediate thickness and absorption corrections for incident beam and characteristic radiation were necessary.

Each pellet was glued on a mylar film, on a sample holder and placed directly on the X-ray beam for elemental determination.

A minimum of three replicates of each sample was made to reduce the error analysis.

Experimental set-up

Micro-SRXRF of the XRF station at LURE installed on the bending-magnet beam line D_1 of the DCI storage ring was used in the present measurements. The running conditions of the storage ring were 1.85 GeV positron energy and a 300 mA beam current with a 140h half-life.

The optical system of Bragg-Fresnel multilayer lenses (BFML)[9], combines a multilayer acting as a Bragg reflector to ensure the monochromatisation of the white incident beam, and a variable grating (Fresnel zone) to focus hard X-rays with submicron resolution. In order to reduce the size of the incident X-Ray beam, an entrance pinhole of 100 μm in diameter was added, 6.3 m upstream from the lens to reduce the X-ray source. After the beryllium window that isolates the X-ray beam line under vacuum from the rest of the experiment, a second pinhole (200 or 500 μm diameter) was installed to decrease the scattered radiation.

The sample is positioned in the image plane within an accuracy of 0.1 μm by 3 axis (x, y, z) remote controlled stage. This stage can be set perpendicularly to the beam for transmission tests or at 45° for fluorescence mapping. A small He/Ne red laser was adjusted to simulate the direction of X-rays in visible light to perform easy prealignment of the lens. A video colour microscope (magnification 700 ×) is used to precisely position the sample in the beam and to superimpose visible image and X-ray fluorescence maps[10].

The incident beam was of 14.3 keV in energy and the microprobe 100 μm in diameter. The fluorescence spectrum is recorded with a Si(Li) detector of resolution 160 eV FWHM with 13 mm^2 area. Pulses are processed through a Nucleus Multichannel Analyser card set in a PC computer. Quantitative calculations are obtained by fitting the spectra with WAPI3 and EPAIS codes[11], developed at LURE. For concentration calculation, sulphur used assumed as an internal standard. The sulphur average value was based on literature data, and on average concentrations determined from quantitative analysis carried out by X-Ray fluorescence technique on 20 different whole hairs. The values used in this work, expressed in mass percentage, are 5.0±0.6%. The analytical errors resulting from the entire procedure of concentration calculations are dominated by the sulphur biological variability that reflects in the uncertainty associated to the internal standard value, and was estimated to be approximately 15%. The detection limit (DL) was estimated to be 0.2 μg/g. The analysed points were approximately 1 mm from each other and from the bottom along 1 cm length.

The spectrometer used in this work for analysis of bones consists on an X-ray tube equipped with a changeable secondary target in molybdenum[12]. With this arrangement it is possible to obtain a monochromatic source.

The X-ray tube, the secondary target and the sample are in a triaxial geometry. This device allows a decrease in the background, taking the advantage of the effect of polarisation of the incident X-ray beam from the tube, and therefore improving the detection limits.

The characteristic radiation emitted by the elements present in the sample was detected by a Si(Li) detector, with a 30 mm^2 active area and 8 μm beryllium window. The energy resolution is 135 eV at 5.9 keV and the acquisition system is a Nucleus PCA card. Quantitative calculations are made through the fundamental parameters method[13]. The experimental parameters were obtained by calibration of the whole system by means of biological standard reference materials of known elemental concentrations. The absorption corrections are made using the information for coherent and incoherent scattering intensities and this influence is used to calculate a virtual matrix. The X-ray generator was operated at 50 kV and 20 mA and a typical acquisition time of 1000 s was used. A collimator of silver was placed in front of the

TABLE 1. Comparison of elemental concentration ($\mu g \ g^{-1}$) in NSB 1577a Bovine Liver measured in this work (± standard deviation) and the certified values

	K	Ca	Mn	Fe	Cu	Zn	As	Se	Br	Rb	Sr	Pb
Present Work	9946 ±168	120 ±9	10 ±1	192 ±13	159 ±5	125 ±5	≤0.6	≤0.6	8.8 ±0.4	12.5 ±0.5	≤0.6	≤0.8
Certified value	9960 ±70	120 ±7	9,9 ±0.8	194 ±20	158 ±7	123 ±8	0.047 ±0.006	0.71 ±0.07	9*	12.5 ±0.1	0.138 ±0.003	0.135 ±0.015

* non-certified value

detector in order to restrict the effective area of the detector by excluding regions close to the edges.

Accuracy tests

The accuracy was checked by analysis of a standard reference biological material. The elemental concentrations obtained for the standard reference material, NSB 1577a Bovine Liver are displayed in table 1. The results obtained in the present work are in very good agreement with the certified values.

RESULTS AND DISCUSSION

In table 2 the mean values and standard deviation (SD) for the elemental content of Mn, Fe, Co, Ni, Cu, Zn, Br, Rb, Sr and Pb in bone obtained in this work are listed, for citizens from contemporary and Neolithic periods. Values from literature are also listed for contemporary subjects[4,12] and from the Chalcolithic period[7], for comparison. From this table we can conclude that Pb in contemporary subjects is much higher than both in Neolithic and Chalcolithic periods, revealing the influence of environmental contamination on the present days. The enrichment factor for Mn in pre-historical bones can be attributed to a contamination, due a continuous uptake from the soil, where the human remains stayed for many years. The extremely high concentration of Fe in bones from the Neolithic period can be attributed also to contamination, from the

TABLE 2. Trace element concentrations ($\mu g \ g^{-1}$) in human bones from contemporary subjects and the Neolithic periods obtained in this work, together with values from the literature for contemporary citizens (ref. 12 and ref.4), and values for bones from the Chalcolithic period (ref.7)

	Mn	Fe	Co	Ni	Cu	Zn	Br	Rb	Sr	Pb
Contemporary (this work)	≤11	165±114	≤11	≤8	7±1	205±32	2.7±0.4	3.7±0.8	195±75	31±6
Contemporary (Ref. 12)	≤4	153±265			4.9±0.6	172±26		1.3±0.44	147±55	25±17
Contemporary (Ref. 4)	0.8	121	0.04			72	4.2			16
Chalcolithic (Ref. 7)	37	227			47	82	69		135	8.8
Neolithic	90±11	1430±20	≤11	≤7	35±7	320±10	50±10	8±1	95±10	≤3

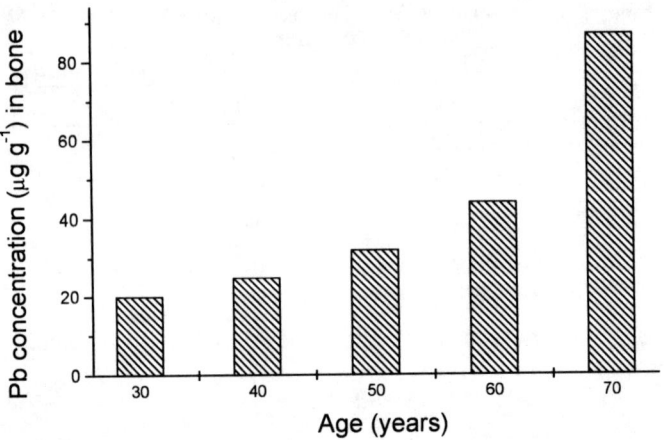

FIGURE 1. Pb concentrations in bone of contemporary citizens of different age, living in the same place near Lisbon

soil, considering that the bones remained buried for 8000 years. The highest concentration of Cu was found in bones from the Chalcolithic period. This can associated with some contamination from the Cu tools used for cooking during that period, also known by the copper age. The small amount of Sr in Neolithic bones may be related to the dietary habits very poor in fish and meat. The high Br levels both in Neolithic and Chalcolithic bones are associated to marine environment and confirm that these subjects lived close to the sea.

In Fig. 1 we show the Pb concentrations in bone for contemporary subjects living in the same place, but aged from 30 to 70 years. From this diagram we can conclude that Pb accumulates in bone and its concentration increases with the age of the individuals. The bone concentration of Pb is related to environmental pollution and subjects of the same age can contain different levels of Pb.

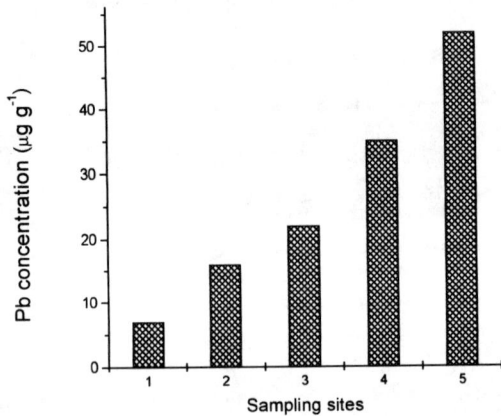

FIGURE 2. Pb concentrations in bone for individuals of the same age (40-50) years from different living places: (1- unpolluted place in the mountain, 300 km North Lisbon; 2- very low traffic small village, 60 km North Lisbon; 3- medium traffic small village, 40 km North Lisbon; 4- village, 20 km North Lisbon; 5- centre of Lisbon)

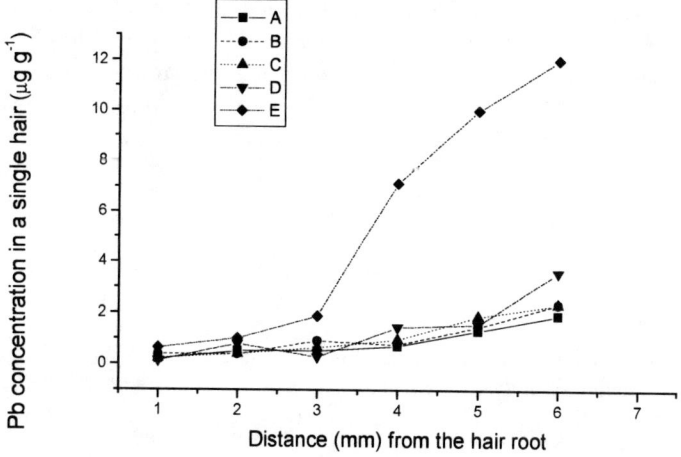

FIGURE 3. Pb distribution concentration in a single hair of different individuals, from the root along the hair

We display in Fig. 2 the Pb concentrations in bone of contemporary individuals of the same age, 40 to 50 years, submitted to increasing levels of pollution (1 to 5) corresponding: 1- unpolluted place in the mountain, 300 km North Lisbon; 2- very low traffic small village, 60 km North Lisbon; 3- medium traffic small village, 40 km North Lisbon; 4- village, 20 km North Lisbon; 5- centre of Lisbon. This diagram evidences the correlation between environmental pollution and Pb concentration levels in bone.

In Fig. 3 the distribution of Pb along the hair is presented. Single hairs of several contemporary individuals, A, B, C, D, E, were analysed by synchrotron microprobe from the root along the hair. Sample hairs were collected from individuals living respectively in sampling sites 1, 2, 3, 4 and 5 of Fig 2.

The obtained results show the diffusion of Pb along the hair. The distribution is not homogeneous and the elemental concentration of hair is influenced by external contamination. One of the major disadvantages in using hair as diagnostic for elemental concentrations is that no washing procedure removes completely the exogenous elements. So the hair analysis reflects not only the endogenous trace elements, but also the exogenous contamination.

CONCLUSIONS

This work demonstrates the usefulness of the diagnostic techniques applied to the study of trace elements concentrations in human tissues to the identification of accumulation patterns, as a result of a particular pollution or dietary impact. In particular, the results for bone show an increasing bone lead accumulation with both age and level of pollution, as demonstrated by in vivo studies. Furthermore, the pattern of lead accumulation in hair does agree with the reported increasing concentration with the distance to the scalp, as well as with the increasing level of pollution of the sampling sites.

We proved in this work that X-Ray Fluorescence is a suitable technique to analyse biological samples, and presents some advantages when compared to other techniques. The sample is not destroyed and there is no chemical preparation involved. Moreover, all the studied elements are detected simultaneously in a single analysis. Another advantage of the method is that a small amount of sample, a few grams, is needed for analysis. However this method is limited to elements heavier than aluminium.

REFERENCES

1. Bowen H.J.M., *Trace Elements in Biochemistry*. Academic Press, New York,1966.
2. Valkovic V., *Analysis of Biological Material for Trace Elements Using X- Ray Spectroscopy*. CRC Press, Boca Raton, FL, 1980.
3. Fraústo da Silva J.J. and R. J. P. Williams, *The Biological Chemistry of the Elements*, Lewis, Chelsea, MI, 1990.
4. *Report of the Task Group on Reference Man, International Communications on Radiological Protection*, Publ 23. Pergamon Press, Oxford, 1975
5. McNeill, F.E., Stokes, L., Brito, J.A.A., Chettle, D.R.. *Occup. Environ. Med.* **57**, 465-471 (2000)
6. Reiche I., Favre-Quattropani, L., Calligaro, T., Salomon, J., Bocherens, H., Charlet, L., Menu, M.. *Nucl Inst Meth B* **150**, 656-662,(1999)
7. Carvalho M.L., C. Casaca, T. Pinheiro, J. P. Marques, P. Chevallier and A. S. Cunha, *Nucl Inst Meth B* **168**, 559-565 (2000)
8. S. R. Yu, Paper IAEA/RL/50, IAEA, Vienna (1978)
9. Chevallier P., P. Dhez, F. Legrand, A. Erko, Yu. Agafonov, L. A. Panchenko and A.Yakshin, *J. Trace Microprobe Techn.* **14**, 517-539 (1996)
10. Chevallier P., K. Abbas and P. Sainfort, *X-Ray Spectrometry* **20**, 293-295 (1991)
11. Brissaud, J. X. Wang, P. Chevallier, *J. Radioanal Nucl. Chem.* **131**, 399-413 (1989)
12. Carvalho M. L., J. Brito and M. A. Barreiros, *X-Ray Spectrom.* **27**, 198-204 (1998)
13. Rindby A. *X- Ray Spectrom.* **18**, 113-120 (1989)

AUTHOR INDEX

A

Abdallah, Jr., J , 404
Abela, R., 112
Acerbi, E., 53
Ahmad, M., 301
Aleonard, M. M., 221
Alesini, D., 53
Alessandria, F., 53
Aliabadi, H., 195
Alonso, J. A., 450
Amusia, M. Y., 123
André, J.-M., 99
Armen, G. B., 0, 188
Arsenović, D., 509
Attallah, F., 221
Auguste, T., 404
Avaldi, L., 53

B

Baltzer, P., 71
Barni, D., 53
Bartolini, R., 53
Baur, K., 472
Bausk, N. V., 395
Bautista, M., 159
Beccara, S. a, 349
Beck, B., 131
Becker, J. A., 131
Beiersdorfer, P., 131
Bellisola, G., 515
Bellomo, G., 53
Bellucci, S., 355
Benfatto, M., 362
Berényi, D., 195
Bergmann, U., 250
Bertolucci, S., 53
Biagini, M. E., 53
Bianconi, A., 13, 462, 497
Biémont, E., 404
Birattari, C., 53
Biscari, C., 53
Bizau, J.-M., 141
Blancard, C., 141
Blasco, F., 404

Bobashev, S. V., 165
Boggaert, G., 221
Bonardi, M., 53
Boni, R., 53
Bonnelle, C., 99
Borchert, G. L., 112
Bosch, F., 206
Boscolo, I., 53
Boscolo, M., 53
Bosotti, A., 53
Botti, S., 355
Bourgeois, S., 267
Brennan, S., 472
Bridou, F., 99
Brigatti, M. F., 481
Brito, J., 522
Broggi, F., 53
Bruneau, J., 141
Burattini, E., 515

C

Calas, G., 378
Carbone, C., 53
Carreyre, J., 221
Carvalho, M. L., 522
Casais, M. T., 450
Castellano, M., 53
Catani, L., 53
Champeaux, J.-P., 141
Chandesris, D., 267
Chantler, C. T., 19, 370
Chemin, J. F., 221
Chen, M. H., 188
Chesnel, J.-Y., 195
Chiadroni, E., 53
Cialdi, S., 53
Cianchi, A., 53
Cibin, G., 481
Cinque, G., 515
Ciocci, F., 53
Clozza, A., 53
Colombatti, M., 515
Connerade, J.-P., 182
Cormier, L., 378
Cramer, S. P., 250